T0177706

Muon Spectroscopy - An Introduction

Muon Spectroscopy - An Introduction

Edited by

Stephen J. Blundell
University of Oxford

Roberto De Renzi
University of Parma

Tom Lancaster
University of Durham

Francis L. Pratt
ISIS Neutron and Muon Source

OXFORD
UNIVERSITY PRESS

Great Clarendon Street, Oxford, OX2 6DP,
United Kingdom

Oxford University Press is a department of the University of Oxford.
It furthers the University's objective of excellence in research, scholarship,
and education by publishing worldwide. Oxford is a registered trade mark of
Oxford University Press in the UK and in certain other countries

© Oxford University Press 2022

The moral rights of the authors have been asserted

Published in the United States of America by Oxford University Press
198 Madison Avenue, New York, NY 10016, United States of America

British Library Cataloguing in Publication Data
Data available

Library of Congress Control Number: 2021936416

ISBN 978–0–19–885895–9 (hbk.)
ISBN 978–0–19–885896–6 (pbk.)

DOI: 10.1093/oso/9780198858959.001.0001

Printed and bound by
CPI Group (UK) Ltd, Croydon, CR0 4YY

Links to third party websites are provided by Oxford in good faith and
for information only. Oxford disclaims any responsibility for the materials
contained in any third party website referenced in this work.

Preface

Muons, radioactive particles that can be produced by accelerators, have emerged as an important tool to study problems in condensed matter physics and chemistry. Beams of muons with all their spins polarized can be prepared and implanted in various types of material. The subsequent precession and relaxation of the spins of these muons can be used to investigate a variety of static and dynamic effects in a sample and hence to deduce properties concerning magnetism, superconductivity, molecular or chemical dynamics, and a bewildering array of other properties.

The technique has a slightly odd name: μSR. The μ stands, of course, for the muon and the S is the spin of the muon, because the spin angular momentum of the muon is what provides its magnetic moment, which is the property from which the muon's sensitivity to magnetic fields derives, and this is the key aspect of the interaction between the muon and the sample. But what about the R? The name was given to make it look a bit like NMR (nuclear magnetic resonance) and ESR (electron spin resonance), so you might guess that μSR is muon spin resonance. However, the muon is a multifaceted probe, and for reasons we will explain in detail in this book the technique is rarely performed resonantly, since the dominant effects are *rotation* (the Larmor precession of the muon, which we can observe directly) and *relaxation*. Thus, sometimes we will be dealing with muon spin rotation, sometimes with muon spin relaxation, and just occasionally with muon spin resonance. To avoid this ambiguity, we have called this book *Muon spectroscopy: an introduction*, but note that the technique will usually be referred to, as is conventional, by the name μSR.

The μSR technique was originally the preserve of a few specialists located in particle physics laboratories. Today it is used by scientists from a very wide range of science backgrounds and interests. Although there are some very good technical monographs and review articles describing the technique, we felt there was a need for a book designed for beginners of many different types: curious undergraduates, starting graduate students, or even experts in related fields who want to find out what all the fuss is about. We hope this is a book that many will be able to learn from, as well as learning to enjoy the beauty of the physics behind this wonderful probe.

For those wishing to explore more deeply into the μSR technique, each chapter is supplemented with further reading. More information is also available on the websites of the international muon facilities and also the International Society for Muon Spectroscopy (ISMS). The royalties of this book are being donated to the ISMS in order to contribute to the financial support for young scientists, particularly those from less wealthy countries, to attend the triennial international conference on muon spectroscopy.

This book grew out of a European workshop held during August 2019 at the Rutherford Appleton Laboratory in which scientists from across the world gathered for 2 weeks for an advanced school on muon spectroscopy. The school was enabled through the major support of ISIS, PSI and the EU SINE2020 project, along with additional support from ISMS, TRIUMF, and JPARC. The lecturers from this school consented to write material for the book based on their lectures. Inevitably, such an approach produces material using a range of styles and notations, some overlaps and some omissions. Therefore, the editors of this book resolved to take the submitted material, mix it up and revise it, reorder and reinvent parts, aiming to produce a final result which would look as if it came from a single author (the one exception being the historical appendix by Steve Cox; with a relatively new technique, a short history is inevitably a personal view). We are grateful for the forbearance and generosity of our team of contributing authors (listed on page vii) for allowing their material to be so roughly handled, but we hope that the final result is worth it. Our thanks are primarily to them, but also to Sönke Adlung and his team at OUP, our eagle-eyed copy editor Kate Blackham, as well as a number of kind friends and colleagues who read through the book and made numerous helpful comments, suggestions, and corrections: Pietro Bonfà, Stewart Clark, Jonathan Frassineti, Elena Garlatti, George Gill, Matjaž Gomilšek (who also produced Fig. 16.11), Alberto Hernandez Melian, Thomas Hicken, Ben Huddart, Gianrico Lamura, Muhammad Maikudi Isah, Ifeanyi John Onuorah, Giacomo Prando, Samuele Sanna, Toni Shiroka, Johnny Wilkinson, and Hank Wu, as well as our contributing authors. The book is better because of their efforts. Finally, we express our gratitude to the students who attended the school, pressed us all to think about how to explain the various aspects of μSR, or think again about what we thought we knew.

Stephen J. Blundell
Roberto De Renzi
Tom Lancaster
Francis L. Pratt
July 7, 2021

List of contributors

Peter J. Baker, STFC-ISIS Rutherford Appleton Laboratory, UK
Adam Berlie, STFC-ISIS Rutherford Appleton Laboratory, UK
Pabitra K. Biswas, STFC-ISIS Rutherford Appleton Laboratory, UK
Stephen J. Blundell, University of Oxford, UK
Pietro Carretta, University of Pavia, Italy
Stephen P. Cottrell, STFC-ISIS Rutherford Appleton Laboratory, UK
Stephen F. J. Cox, STFC-ISIS Rutherford Appleton Laboratory, UK
Pierre Dalmas de Réotier, CEA Grenoble, France
Roberto De Renzi, University of Parma, Italy
Adrian D. Hillier, STFC-ISIS Rutherford Appleton Laboratory, UK
Rustem Khasanov, Paul Scherrer Institute, Switzerland
Tom Lancaster, University of Durham, UK
Leandro Liborio, STFC-SCD Rutherford Appleton Laboratory, UK
James S. Lord, STFC-ISIS Rutherford Appleton Laboratory, UK
Martin Månsson, KTH Stockholm, Sweden
Iain McKenzie, TRIUMF, Vancouver, Canada
Francis L. Pratt, STFC-ISIS Rutherford Appleton Laboratory, UK
Thomas Prokscha, Paul Scherrer Institute, Switzerland
Zaher Salman, Paul Scherrer Institute, Switzerland
Simone Sturniolo, STFC-SCD Rutherford Appleton Laboratory, UK
Jun Sugiyama, CROSS-Tokai, Japan
Andreas Suter, Paul Scherrer Institute, Switzerland
Rui Vilão, University of Coimbra, Portugal
Koji Yokoyama, STFC-ISIS Rutherford Appleton Laboratory, UK

The book website is `https://musr.org/muon-spectroscopy-book`

Contents

The basics of μSR

This is a book about a set of closely related experiments that use muons to investigate matter. They are known as μSR, which stands for muon-spin relaxation, rotation, or resonance.

1.1 The key idea

The **muon** is a *radioactive* particle with a magnetic moment and it can be implanted into a sample.

We will now unpack the various parts of this sentence in more detail.

(a) The muon is a radioactive particle

The muon is radioactive and so it does not last very long. It has an average lifetime of $\tau_\mu = 2.2\ \mu$s, after which it decays into other particles. The special nature of the particle's decay allows us to read out the information that the muon probes.

[1] Real experiments work with either one muon at a time, or a pulse of 10^2–10^3 muons at a time. Therefore, accumulating 2×10^7 total counts might take ≈ 1 hour, though the time depends a lot on the beamline and the experiment, as we will discuss in due course.

Example 1.1

The radioactive decay of the muon means that the number of muons $n(t)$ in the sample a time t after implantation follows the equation

$$n(t) = n(0)\, e^{-t/\tau_\mu}. \qquad (1.1)$$

This means that if you were able to implant twenty million muons [$n(0) = 2 \times 10^7$] into the sample at $t = 0$ (a number of muons which is not uncommon[1] in a μSR experiment), the number remaining at $t = 20\ \mu$s, is $20 \times 10^6 \times e^{-20/2.2}$ which is over 22,000. So even though the muon decay time is 2.2 μs, you can measure muons that live a lot longer than that.

Note also that the number of muons $N(t)\, \mathrm{d}t$ that decay between t and $t + \mathrm{d}t$ is given by

$$N(t) = -\frac{\mathrm{d}n(t)}{\mathrm{d}t} = \frac{n(0)}{\tau}\, e^{-t/\tau_\mu} = N(0)\, e^{-t/\tau_\mu}, \qquad (1.2)$$

and $N(t)$ is the quantity that is recorded in a μSR experiment (see Fig. 1.1).

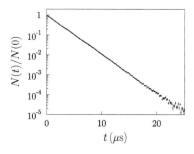

Fig. 1.1 The total counts as a function of time from a μSR experiment, illustrating the exponential decay predicted from eqn 1.2.

[2] This is because positive muons are often more effective for investigating the collective behaviour of matter, compared to negative muons which, owing to their resemblance to electrons, are more useful in investigating the chemical elements in a material. However, negatively charged muons are also used (see Chapter 22).

The muon is a subatomic particle and it can be positively or negatively charged. We usually employ positively charged muons in μSR measurements.[2]

(b) The muon has a magnetic moment

The muon is a spin-$\frac{1}{2}$ particle and has a small *magnetic moment*. Its magnetic moment allows it to interact magnetically with its environment and hence investigate the properties of the sample into which it has been implanted. A major part of this book will be to explore the nature and consequences of these magnetic interactions.

(c) The muon is implanted in the sample

Note the word *implant*. This is *not* a scattering technique, such as X-ray scattering or neutron scattering, in which an incoming particle is directed towards a sample along a particular direction and then a scattered particle detected.[3] This is an **implantation** technique in which the muon is directed towards the sample to be studied and it comes to a complete rest somewhere inside the sample.[4]

[3]Sometimes people mistakenly refer to this technique as 'muon scattering', which is incorrect.

[4]The site of the implanted muon is usually an interstitial position in a crystalline solid. This issue will be discussed in detail in Chapter 16.

1.2   The principles of the experiment

Larmor precession

The positive muons that we implant into materials in these experiments are always **spin polarized**, so all start with their spins aligned along a known direction. If there is a magnetic field at the point where a muon stops in a material, then the muon spin will undergo **Larmor precession** around the direction of the local field, at a frequency determined by the muon gyromagnetic ratio $\gamma_\mu = 2\pi \times 135.5$ MHz T^{-1}, as shown in Fig. 1.2. Specifically, the Larmor (angular) frequency ω is given by

$$\omega = \gamma_\mu B, \tag{1.3}$$

where B is the magnitude of the field at the muon site. Equation 1.3 is arguably the most important one in this book.

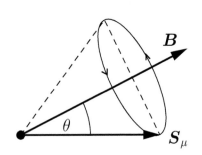

Fig. 1.2 In a magnetic field B the muon spin S_μ precesses around the field on a cone at an angular frequency of $\omega = \gamma_\mu B$.

[5]Many facilities use old-fashioned cgs units, and 2 mT translates to 20 G (i.e. 20 Gauss, where 10^4 G = 1 T). Such a calibration is often known as a 'T20', which means a transverse field of twenty Gauss. For more details on these units, see Appendix A.

[6]It is possible to do this though – see Chapter 20.

Example 1.2

Calculate the precession frequency for a muon in (a) the Earth's magnetic field ($\approx 50\,\mu$T), (b) 2 mT, the magnetic field commonly used to calibrate a μSR experiment,[5] and (c) 5 T, a relatively high magnetic field.

Solution: We use the formula for precession frequency ν

$$\nu = \frac{\omega}{2\pi} = \frac{\gamma_\mu B}{2\pi}, \tag{1.4}$$

and obtain (a) 6.8 kHz, (b) 0.27 MHz, and (c) 0.68 GHz. For experimental purposes, it is often more useful to write the expressions in terms of frequency (ν) units, rather than angular frequency (ω) units, although, as we will see later, it is often more convenient in theoretical work to use angular frequency.

As we have seen, a typical μSR experiment might measure times up to around $20\,\mu$s after muon implantation. Precession in the Earth's magnetic field [which would have a period of $1/(6.8$ kHz$) \approx 150\,\mu$s] is therefore too slow to measure directly and one would just observe the very beginning of a precession signal. On the other hand, a 5 T field produces very fast precession, and to measure this one would need very good timing resolution.[6]

Asymmetric emission of positrons

When a muon decays, a positron e^+ and two neutrinos (ν) are emitted via the reaction

$$\mu^+ \rightarrow e^+ + \bar{\nu}_\mu + \nu_e. \tag{1.5}$$

A property of this decay (which we will describe in more detail later) is that the positron is not emitted isotropically, with equal chance of it emerging in any direction. Instead it is emitted *preferentially along the direction of the muon spin at the moment of decay.* (The distribution is shown in Fig. 1.3.) Detecting the direction along which the positron is emitted allows us to work out in which direction the muon spin was pointing at the moment of its decay. Each muon lives for a different amount of time (according to the radioactive decay law in eqn 1.1) and so each muon decay gives rise to a positron whose detection contributes to a data point at a particular time t. If we make many such positron detections, taken over many million muon decays, we can build up a histogram. This histogram then tells us the average spin polarization of the large ensemble of muons as a function of time. The data set which we obtain from the experiment is simply the collection of measured average spin polarization values as a function of time after implantation.

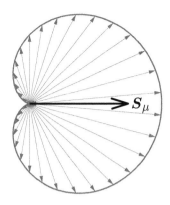

Fig. 1.3 The probability of positron emission in the direction of each grey arrow is indicated by its length. (for a particular value of emitted positron energy; see Appendix E.4 for more details). Positrons are preferentially emitted along the muon spin direction (shown in a black arrow). The figure is axially symmetric about the muon-spin direction (and so looks a bit like the surface of a cherry in three dimensions).

Example 1.3

A stopped spin-polarized ensemble of muons sits close to the origin of a set of Cartesian coordinates. Its spin initially lies along a unique direction (conventionally we call this z) and a B-field is applied along x. Assuming no other interactions, the muon spins will precess, undergoing circular motion in the y-z plane at an angular frequency $\omega = \gamma_\mu B$. We place positron detectors in front of (i.e. at some value of positive z) and behind (negative z) the origin. The average polarization $P(t)$ of the muon ensemble at a particular time t can be found by computing the normalized difference in the number $N_F(t)$ of positrons counted in the forward detector and the number $N_B(t)$ in the backward detector, both measured in a small interval centred on t, which is to say

$$P(t) \propto \frac{N_F(t) - N_B(t)}{N_F(t) + N_B(t)}. \tag{1.6}$$

The polarization in this case is proportional to a cosine function $P(t) \propto \cos \gamma_\mu Bt$, which is simply the circular precessing motion projected along the z direction. If we did not know the magnitude of the magnetic field we applied, we could measure it very accurately from the frequency of the oscillations in $P(t)$. An example of some data showing this effect can be seen in Fig. 1.4.

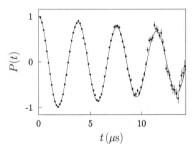

Fig. 1.4 Muon precession in a field of 2 mT.

The polarization $P(t)$ as a function of time t can be used to tell us about the magnetic field at the muon stopping site. If the magnetic field at the muon site fluctuates as a function of time, then we can also obtain information about the nature of those fluctuations. Much of this book describes how this and other electronic and chemical properties of matter can be extracted from the data.

1.3   Muon beams and spectrometers

In order to obtain enough muons to carry out one of these experiments we must create them in a high-intensity beam of particles. For this reason the experiments are carried out at facilities where such beams are made using particle accelerators. Muons are born 100% spin polarized because of the way they are produced from the decay of pions via a reaction, as described in the following example.

Example 1.4

In a muon facility, a beam of protons is fired into a target and this makes pions. The experiment is usually carried out using pions which are at rest in the surface of the target and which decay quickly into muons via the reaction

$$\pi^+ \rightarrow \mu^+ + \nu_\mu. \tag{1.7}$$

Since this reaction must conserve linear and angular momentum, and since the pion is initially at rest and has no spin or orbital angular momentum, we deduce that both the muon (μ^+) and the neutrino (ν_μ) must have opposite linear and angular momenta. Having opposite linear momenta just means the two particles head off in opposite direction. Because neutrinos always have a negative helicity (i.e. their spin is always observed to be antiparallel to their momentum) then the muon must also have negative helicity (so that *its* spin is also antiparallel to its momentum). For this reason, we are able to obtain a beam of fully spin-polarized muons.

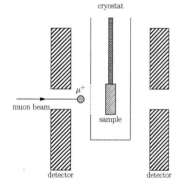

Fig. 1.5 A schematic of a μSR experiment.

As mentioned in Section 1.1, μSR measurements are *not* scattering experiments: the muons are always stopped in the material under investigation. The beam must therefore be of a low enough energy that the muon can be stopped in (typically) a few hundred milligrams of solid matter. The sample under investigation is placed in the beam, where the muons stop in the bulk of the material without any appreciable change in their spin orientation nor damage to the material. This is where the experiment begins, with a muon at rest in the sample in a known spin state. Muons are introduced into the sample one at a time or in a pulse. In either case, the number of muons in the sample is so small that there is effectively no possibility of muons interacting with each other: they are an ultra-dilute probe of matter.

The sample is usually mounted on a sample stick and placed in a **cryostat** so that its temperature can be varied. The cryostat has a window to allow the muon beam to enter. The positrons that emerge after muon decay are very energetic, travelling at a large fraction of the speed of light, and easily exit the sample and cryostat. They are detected in particle detectors placed around the cryostat. The apparatus comprising this array of detectors is known as a **spectrometer**. The output from these detectors is connected to the data acquisition system (which has to receive signals at a very high rate) and then to the analysis computers. Because the timing of the signals has to be measured to better than a nanosecond (and sometimes to a few picoseconds) it is worth remembering that the speed of light is around 0.3 m per nanosecond, so the lengths of cables and other paths for electrical signals have to be adjusted very carefully. A very simple schematic of the experiment, showing the main features of the spectrometer, is shown in Fig. 1.5 (we will discuss many more details of the experimental arrangement in Chapters 14 and 15).

Example 1.5

An example of a real muon spectrometer is the HiFi instrument at ISIS, as shown in Fig. 1.6 and Fig. 1.7. The sample is placed in a cryostat at the centre of the spectrometer, where it stops muons from the particle beam. The sample is surrounded by positron detectors, each of which consists of a piece of scintillator material linked optically through a plastic light guide to a photomultiplier.

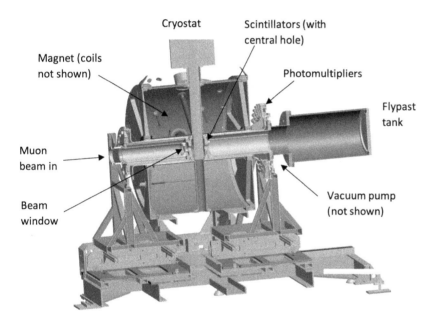

Fig. 1.6 An exterior view of the HiFi instrument at ISIS.

Fig. 1.7 A cutaway drawing of the HiFi spectrometer.

Muon spectrometers can be used at various user facilities around the world and the main ones are shown in Fig. 1.8. Two of these have pulsed muon beams (ISIS, based at the Rutherford Appleton Laboratory in Oxfordshire in the UK and J-PARC, located in Tokai, Japan) while the other two provide continuous muon beams (PSI, in Villigen, Switzerland and TRIUMF, in Vancouver, Canada). All four operate as user facilities and beamtime is available through competitive applications. There is also a continuous muon beam source MuSIC in Osaka, Japan, and further proposed muon facilities in Korea, China, and the US are at various stages of planning and/or building.

Fig. 1.8 User facilities where μSR experiments can be performed.

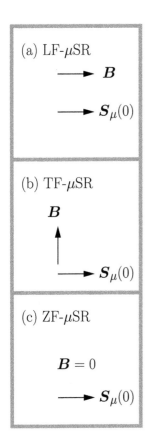

(a) LF-μSR

$\longrightarrow B$

$\longrightarrow S_\mu(0)$

(b) TF-μSR

B

$\longrightarrow S_\mu(0)$

(c) ZF-μSR

$B = 0$

$\longrightarrow S_\mu(0)$

Fig. 1.9 Experimental geometries for (a) longitudinal field (LF-μSR), (b) transverse field (TF-μSR), and (c) zero-field (ZF-μSR).

1.4   Experimental geometries

We will now introduce some important pieces of jargon concerning the type of experiment performed. When a magnetic field is applied to the sample, it is usually applied in one of two geometries:

- The magnetic field B is applied *parallel* to the direction of the initial muon spin polarization $S_\mu(0)$. This is known as the **longitudinal** geometry, and this is often referred to as **longitudinal-field** μSR, or LF-μSR for short.

- The magnetic field B is applied *perpendicular* to the direction of the initial muon spin polarization $S_\mu(0)$. This is known as the **transverse** geometry, and this is often referred to as **transverse-field** μSR, or TF-μSR for short.

These two geometries are shown in Fig. 1.9(a) and (b). Sometimes, however, experiments are performed with no applied magnetic field (and most spectrometers allow the user to make the magnetic field truly zero, nulling out the Earth's magnetic field). This configuration is known as **zero-field** μSR, or ZF-μSR for short [Fig. 1.9(c)].

1.5   What can we do with μSR?

We will spend a lot of this book looking at applications of the μSR technique which will answer this question. However, in this introductory chapter let us focus on a key principle: the muon is a **local probe**. This means that the muon tells us about its magnetic interactions at the particular point in the material where it stops. The muon often stops in a crystallographically unique position in a solid. However, because we measure the result of millions of muon implantations, there is likely to be a difference in the behaviour probed at the position of each muon. When we measure the polarization at each time t, we average over all of the muons that have been implanted in the sample and have interacted with the sample for that time t.

Example 1.6

Consider an ensemble of spin-polarized muons at rest at $t = 0$ sitting in slightly different transverse magnetic fields. They will precess at slightly different angular frequencies $\omega = \gamma_\mu B_i$, depending on the value of the field B_i at each muon site. The corresponding oscillations are progressively dephased as time elapses. The polarization, averaged over many muons, becomes washed out as time increases and consequently the precession signal decays. An example is shown in Fig. 1.4 for which a magnetic field of 2 mT was applied to a sample. The muons precess at an average frequency of 0.27 MHz (see Example 1.2), but because of these field variations the oscillations relax slightly as time increases. This relaxation can be used to estimate the size of the field variations inside the sample.

The decay of polarization described in the previous example is known as the **relaxation** of the polarization. There is lots of information hidden in the exact form that this relaxation takes, and accessing this information will allow us several insights into topics such as magnetism, superconductivity, electronic properties, and chemistry of materials.

Uses of μSR that we will describe in this book include the following:

- The detection of magnetic order.
- The observation of the presence of spin fluctuations.
- The determination of the penetration depth in a superconductor.
- The identification of the pairing symmetry in a superconductor.
- The measurement of the local electronic environment in a semiconductor, where the muon simulates a proton impurity.

In Part I of the book we will go through the basic principles of μSR in more detail. Part II describes specific science applications of the μSR technique. Part III is devoted to particular practical considerations that are important when performing μSR experiments. Part IV contains some more advanced topics in muon spectroscopy, while Part V describes the relationship between μSR and other, complementary techniques.

Chapter summary

- The muon is a radioactive particle with a magnetic moment and it can be implanted into a sample.
- The principle of the experiment is based on Larmor precession of the muon in a magnetic field and the asymmetric emission of positrons.
- A spectrometer is used to detect the positrons and the experiment can be performed in longitudinal field or transverse field geometries. Measurements can also be performed in zero field.
- There are many applications of μSR, as we will show in the rest of this book.

Part I

Elements of muon spectroscopy

In this introductory part of the book we set out the basic elements of the muon spectroscopy technique.

- In Chapter 2 we provide some background to the particle physics of the muon, as well as some perspectives of how muons can be used in science.

- When introducing the positive muon μ^+ into matter we find various types of implanted state: the muon can exist as a bare muon (positively charged), or can pick up an electronic charge to become the neutral muonium state, or even pick up two electronic charges and become a negatively charged species. These states, and the consequences of their formation, are explored in Chapter 3.

- The muon is a quantum particle and its quantum-mechanical interaction with its surroundings has important implications. In Chapter 4 we review the basic quantum-mechanical arguments underlying these interactions and discuss the consequences for muon spectroscopy.

- The experimental observable in a muon experiment is the relaxation function of the muon polarization. We give an introduction to these relaxation functions in Chapter 5.

2

Introduction

[1]I. I. Rabi (1898–1988). For details of the origin of Rabi's remark, see R. H. Bernstein and P. S. Cooper, Phys. Rep. **532**, 27 (2013).

The **muon** is a fundamental particle with some unique properties that have given it a special place in science. It has evolved from being a rather unexpected curiosity, a view taken shortly after its discovery in the 1930s, towards becoming a significant component of modern scientific research, where it has a firmly established role at the forefront of fundamental physics, as well as providing the basis for the powerful **muon spectroscopy** technique that is the topic of this book.

The slightly uncomfortable early days of the muon are aptly summarized by the comment "who ordered that?" that was made by the distinguished scientist Isidor Rabi[1] shortly after the muon's discovery. Rabi would go on to contribute to the founding of nuclear magnetic resonance, a technique that is closely related to muon spectroscopy, for which he was awarded a share of the 1944 Nobel Prize in Physics.

This introductory chapter looks at the muon from a number of different scientific viewpoints, introducing some basic ideas that will be explored in more detail in the chapters that follow.

2.1   Discovery of the muon

The story of the muon begins with its discovery in California in 1936 by Carl David Anderson and Seth Neddermeyer. These researchers were studying energetic cosmic rays via the tracks left behind in a cloud chamber. Such tracks show a curvature in magnetic field that can be used to determine the ratio of the momentum of the particles producing them to their charge, while the particle's velocity can be determined from the density of condensed water droplets in the tracks. From these two properties the mass-to-charge ratio can be deduced. Anderson had used the same technique a little earlier in the discovery of the positron, the first example of antimatter, that earned him a share of the 1936 Nobel Prize in Physics.

The newly discovered muon particle was found to have a mass that is around 200 times greater than that of an electron and 9 times less than that of a proton (Table 2.1). The muon can have positive or negative charge and is represented by the symbol μ. On the basis of this mass being intermediate between the light electron, classified as a **lepton**, and the heavy proton, classified as a **baryon**, the muon was initially classified as a **meson**. The meson had been predicted by Hideki Yukawa in 1935 as the particle responsible for the strong nuclear interaction. The mass

	charge	spin	mass	lifetime (μs)
e	$\mp e$	1/2	m_e	∞
μ	$\mp e$	1/2	$207\ m_e$	2.197
p	$\pm e$	1/2	$1836\ m_e$	∞

Table 2.1 The basic properties of the muon compared with those of the electron and proton. m_e is 0.511 MeV/c^2. The two charge states correspond to the particle and the antiparticle.

of the muon fitted the prediction for the Yukawa meson surprisingly well, but its relatively weak interaction with matter was inconsistent with expectations.[2] Yukawa's meson was later identified as the pion, represented by the symbol π, which is the short-lived parent of the muon, with an average lifetime of 26 ns. The muon itself is unstable with an average lifetime of 2.2 μs and positrons or electrons are produced in its decay. Thus the important sequence of events in the birth and death cycle of the muon was established:

$$\pi^{\pm} \rightarrow \mu^{\pm} \rightarrow e^{\pm}. \tag{2.1}$$

The properties of the muon are compared in Table 2.1 with the more familiar charged particles of matter, the electron and the proton.

2.2   The first muon application

The muon has the longest lifetime of any unstable elementary particle. This property led to the first experimental verification of the time dilation effect predicted by Einstein's special theory of relativity, thus providing an early demonstration of the use of the muon in resolving a scientific question.

This study was carried out by Bruno Rossi and David Hall, who were researchers based at the University of Chicago. In 1941 they went to Colorado to measure the count rates for energetic muons travelling vertically downwards at different altitudes. They compared the rates high up at Echo Lake (altitude 3240 m) with those down at Denver (altitude 1616 m) and found that the lifetime of the muons increased with the muon momentum exactly as predicted by Einstein. They deduced that the lifetime of the muon in its inertial reference frame was 2.4(3) μs, consistent with the more accurate value that we now have today (Table 2.1).

[2]As discussed in Section 2.3, the modern view taken in the **Standard Model** of particle physics, formulated in the latter half of the 20th century, classifies hadrons as a family of strongly interacting particles formed from bound states of quarks. The hadronic particles can be divided into baryons, which comprise an odd number of quarks (usually three), and mesons, which comprise an even number (usually a quark and an antiquark). Although Yukawa's intuition about the fundamental interactions being mediated by particles was correct, and the pion is well described by his predictions, the Standard Model identifies the gluon as the particle ultimately responsible for the strong force. An outline of the Standard Model is given in Appendix E.2.

Example 2.1

For a relativistic particle with velocity $v = \beta c$ the time dilation is determined by the Lorentz factor

$$\gamma = (1 - \beta^2)^{-1/2}, \tag{2.2}$$

and the lifetime of a high energy muon will be significantly increased by this Lorentz factor, going from τ_μ in the rest frame of the muon, to $\gamma\tau_\mu$ in the rest frame of a stationary observer. The muon energy is given by $E = \gamma m c^2$ and also by $E^2 = p^2 c^2 + m^2 c^4$. Combining these two equations, in units where $c = 1$, gives

$$\gamma = [1 + (p/m)^2]^{1/2}. \tag{2.3}$$

For an energetic cosmic ray muon with momentum $p = 800$ MeV/c and using the muon rest mass $m = 106$ MeV/c^2 the Lorentz factor can be evaluated as 7.6 and the measured muon lifetime therefore increases from 2.2 to 16.7 μs.

2.3 Muon perspectives

Particle physics

The particle physics view is based on the Standard Model classification of elementary particles. In this classification the muon is not a meson, but is itself an elementary particle. The muon and its associated neutrino are now recognized as two leptons within the six-membered lepton family (Fig. 2.1). The muon is subject to the electromagnetic interaction and the weak interaction.

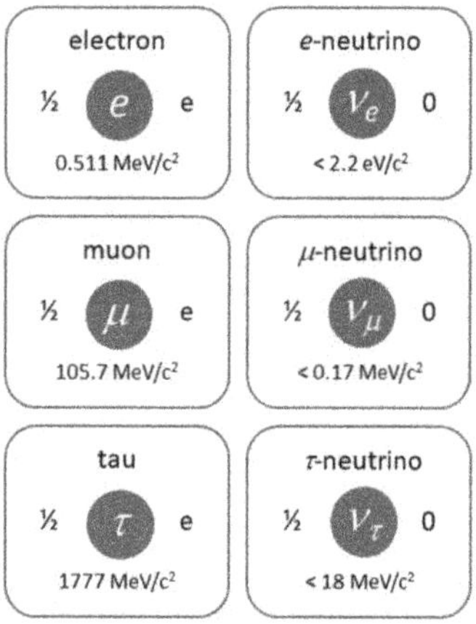

Its decay process is governed by the weak interaction and the decay produces two neutrinos in addition to the positron or electron, e.g. for the decay of a positive muon one has

$$\mu^+ \to e^+ + \nu_e + \bar{\nu}_\mu. \tag{2.4}$$

The weak force has an important characteristic, discovered in 1957 by C. S. Wu *et al.*, following the proposal of T. D. Lee and C. N. Yang. This characteristic is that **parity** (i.e. spatial inversion symmetry) is not conserved in weak decay processes. This parity non-conservation property leads to the negative helicity of neutrinos (see Example 1.4) and to the asymmetric positron emission shown in Fig. 1.3 that underpins μSR. Another important feature of eqn 2.4 is that virtually all muons follow this decay path. Thus muons form a very efficient source of the elusive neutrino particles. Such neutrinos can be directed towards special detectors at distant sites to enable mass determination. Purpose-designed high intensity muon beams can therefore be employed as 'neutrino factories' to help address key questions in cosmology such as the origin of Dark Matter.

The Large Hadron Collider (LHC) at CERN currently provides a major focus for high-energy particle physics research and notably provided the first experimental confirmation of the long-predicted Higgs boson in 2012, with subsequent studies continuing to establish further properties of the Higgs particle and its interactions. High energy muons are significant decay products of the Higgs particle and tracking these muons forms an important part of these particle physics experiments.

Particle physics is looking towards an extension of the energy frontier beyond LHC and in this context muon collider designs are actively being pursued. Muons have an advantage over hadrons in being elementary particles and they have less radiative energy loss compared to electrons, which is an important advantage in designing higher energy accelerators for future studies.

Another scientific frontier that muons are well suited to tackle is the departure of particle properties from the precise predictions of the Standard Model. The so-called muon $g - 2$ experiments have detected significant departures of the muon g factor from theory and the accuracy of the experiments is continuously being improved. Such research may reveal important new physics that is missing from the Standard Model.

Fig. 2.1 The muon is one of six leptons within the Standard Model classification of elementary particles. The spin, charge, and mass are shown for each particle.

A more detailed description of the muon particle physics that is relevant to implanted muon spectroscopy is given in Appendix E. The key features are the 100% spin polarization of muons produced by pions decaying at rest (see Appendix E.1 and E.3) and the asymmetric angular distribution of the emitted positrons with respect to the muon spin direction at the moment of decay (Appendix E.4). Both are essential for the μSR experiment since they give access to the initial and final directions of the muon spin, and hence its time evolution.

Implanted muons in matter

Moving on from high energy particle physics, we switch our attention to the study of low energy phenomena in materials with muons, using the technique known as muon spectroscopy. This technique is also given the label μSR, where μS stands for muon spin and R can be taken to stand for rotation, relaxation, or resonance, the three main types of muon spectroscopy measurement. Alternatively the R in μSR can simply be taken to stand broadly for 'research'. Making muons and transporting them to the sample under study are covered in Chapter 13 and a more detailed historical perspective of the μSR technique is given by Steve Cox in Appendix H.

In performing μSR studies it is necessary to slow the muons down to thermal energies in the process of implanting them in the sample of interest. This can be done by placing a sufficiently thick sample in the muon beam or by placing material in front of the sample to slow down or **degrade** the beam. Rapid thermalization and preservation of the muon polarization are key features of the implantation process. Positive muons are most often used here and the typical muons used in muon spectroscopy start off with an energy of 4 MeV, whereas the thermal energies associated with the sample are typically on the meV scale, so considerable energy needs to be given up by the muon. The muon energy loss starts with atomic ionization and excitation processes and proceeds to charge exchange, capture, and loss processes. The stopping process can be simulated using Monte Carlo methods, which are useful in designing experiments. This subject is covered in Chapter 17.

Muon states are often classified using the terms **diamagnetic** and **paramagnetic** which refer to the magnetic nature of the electrons to which the muon are coupled (because the electronic moment is much larger than the muon).[3] Thus we have that:

- a **diamagnetic muon state** occurs when the muon is uncoupled to any unpaired electron density; this could be because the muon is not strongly coupled to any electrons [i.e. the muon can be described as existing simply as μ^+, often written Mu^+ (see Chapter 3)] or because it is coupled to two electrons which are in a singlet state (as in Mu^-, to be discussed in Chapter 3).

- a **paramagnetic muon state** occurs when the muon is strongly coupled to a single unpaired electron [as in neutral muonium (written Mu^0, see below and also Chapter 3)].

[3]This terminology is consistent with the practice in NMR where the nature of the state one describes excludes the probe spin. In the same way, for μSR, our diamagnetic/paramagnetic distinction focuses just on the electrons to which the muon is coupled, and not on the magnetism of the muon. Thus a 'diamagnetic' state means an 'electronically diamagnetic' state and a 'paramagnetic' state means an 'electronically paramagnetic' state.

[4]It should be noted that states that
show this diamagnetic response may
actually be the result of a paramagnetic local environment with fast fluctuations.

The nature of the final muon states obtained in the sample will depend
on the type of sample being investigated (Fig. 2.2). In metals the final
muon state usually remains a positive muon, a diamagnetic muon state.
In insulators the diamagnetic positive muon state is one possibility, but
another state retaining an unpaired electron is also possible, a paramagnetic state. The simplest form of paramagnetic state is **muonium**,
a muon analogue of the neutral hydrogen atom. These different types
of state are identified by their response to a transverse field (TF), i.e.
a magnetic field applied perpendicular to the initial muon polarization.
Diamagnetic states produce a muon precession frequency that is very
close to that of a bare muon,[4] whereas paramagnetic states show significant shift with respect to the bare muon. The various muon charge
states and the transitions between them are discussed in more detail in
Chapter 3.

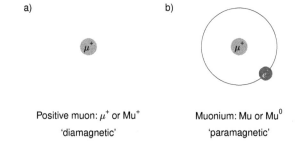

a)

b)

Positive muon: μ^+ or Mu^+

Muonium: Mu or Mu^0

'diamagnetic'

'paramagnetic'

Fig. 2.2 a) In muon spectroscopy terminology, a positive muon state in a local environment without unpaired electrons is termed diamagnetic. This state is identified
by a precession frequency in TF that is virtually indistinguishable from that of a
bare muon. b) The case where the TF precession frequency differs significantly from
that of a bare muon is called a paramagnetic muon state. The simplest example of
a paramagnetic state is muonium, where the positive muon captures an electron and
forms a bound state analogous to atomic hydrogen.

Condensed matter physics

A condensed matter physics perspective is at the heart of much of the
muon spectroscopy work carried out at the various muon user facilities
around the world. This perspective views the thermalized, implanted
muons as local probes of static and dynamic magnetism in the sample
under investigation. For studies of static magnetism the muon can often
be regarded as a local probe with a limited set of specific stopping sites.
The data can then be analysed in terms of these particular sites and
complementary site calculations can be made to aid the analysis. These
site calculation methods are discussed in Chapter 16.

Another situation that is often encountered is where the muon can
best be viewed as randomly sampling the internal magnetic environment within the material under study. A typical example here would
be superconducting vortex lattices having a long length scale compared
to any individual sites in the lattice (see Chapter 9). Magnetic sys-

tems with a large cell and many stopping sites could also fall into this category, as could systems with a high degree of disorder.

Besides observing static magnetism, the muon can also be used to study the fluctuating magnetic fields associated with various dynamical processes in the sample. This will be covered in detail in Chapters 5, 7, and 8. The range of fluctuation rates that can be observed using the muon depends on a number of factors, which are illustrated in Fig. 2.3. One factor is the coupling constant A (measured in frequency units) linking the muon to the internal magnetic fields that are fluctuating with rate ν. As we shall derive in Chapter 5, when the fluctuations are slow, so that $\nu < A$, the relaxation rate λ of the muon-spin polarization is proportional to ν. In the opposite limit of fast fluctuations where $\nu > A$ the muon-spin relaxation rate is proportional to A^2/ν. A second factor determining the limits of dynamical sensitivity is the range of muon relaxation rates that can be measured by a particular muon spectrometer, with a maximum relaxation rate λ_{max} determined by the time resolution of the source and detectors and a minimum relaxation rate λ_{min} determined by the background at a continuous source and by the muon flux and detector dead time at a pulsed source. This range of measurable relaxation rates defines two bands of sensitivity for each value of the coupling, as shown in Fig. 2.3, one for the slow-fluctuation limit and one for the fast-fluctuation limit. Notice that the gap between the two bands is called the **wipe-out** region, where the fluctuation rate leads to a relaxation that is too fast for the experimental apparatus and the signal is partially or totally missing.

It is useful to make a broad comparison of the range of fluctuation rates available via μSR with those accessible using other techniques. This is illustrated in Fig. 2.4 (on page 16) which shows that there can be considerable overlap between different methods. The μSR technique can be seen to provide a useful bridge between characteristically fast methods such as neutron scattering and characteristically slow methods such magnetic remanence and a.c. susceptibility.

When comparing μSR against bulk magnetic studies several advantages emerge. As a volume-averaged probe μSR is relatively immune to low concentration impurity phases that can dominate bulk susceptibility measurements at low temperature and phase segregated samples can easily be identified in muon data. μSR is also particularly well-suited to studying antiferromagnetism and allows measurements to be made in zero applied field. Another strong feature of μSR is its high sensitivity to very small magnetic fields and magnetic moments, typically being able to detect fields below 10 μT and moments below 0.01 μ_{B}. The technique has excellent compatibility with very low temperatures as a result of the incoming muons being able to penetrate the cryostat windows and the outgoing positrons being able to penetrate both the walls and windows of the cryostat. Further comparison of μSR and bulk measurement techniques is given in Chapter 23 and the complementary techniques of neutron and X-ray scattering are covered in Chapter 24.

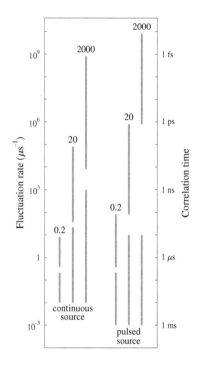

Fig. 2.3 Muon sensitivity to dynamics compared for continuous and pulsed sources and for three different values of coupling: 0.2, 20, and 2000 MHz. These reflect typical values for nuclear dipolar coupling, moderate electronic coupling, and very strong electronic coupling. The lower band is the slow fluctuation regime and the upper band is the fast fluctuation regime. Pulsed and continuous sources are discussed in Chapter 13.

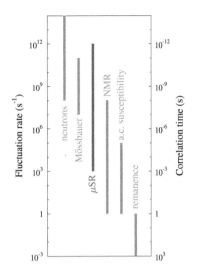

Fig. 2.4 Illustration of typical sensitivity to dynamics on different timescales for a range of methods in comparison with μSR.

	spin	magnetic moment	$\gamma/2\pi$ (MHz T^{-1})	method
e	$\frac{1}{2}$	$658.21\,\mu_p$	28025	ESR
μ	$\frac{1}{2}$	$3.1833\,\mu_p$	135.53	μSR
p	$\frac{1}{2}$	μ_p	42.576	NMR

Table 2.2 The magnetic properties and magnetic resonance method of the muon compared with those of the electron and proton. μ_p is 1.4106×10^{-26} J T^{-1}. NMR is nuclear magnetic resonance; ESR is electron spin resonance.

The many areas of condensed matter physics addressed by muon spectroscopy can be grouped into several major topics. Subsequent chapters cover magnetism (Chapters 6 and 7), dynamics (Chapter 8), superconductors (Chapter 9), semiconductors (Chapter 10), and ionic motion (Chapter 11).

Magnetic resonance

Another perspective of the muon views μSR as a member of the extended family of **magnetic resonance** techniques. Properties of the muon as a magnetic probe are summarized in Table 2.2 in comparison with those of the electron and the proton.

As a magnetic resonance technique, μSR most closely resembles NMR. In comparison with NMR there are many advantages of μSR, along with several disadvantages. Some of the advantages of μSR can be listed as:

- Having 100% spin polarization, in contrast to the very weak thermal polarizations obtained for NMR nuclei.
- A larger **gyromagnetic ratio** for the muon than for any nucleus.
- No need for specific nuclear isotopes, with muons being able to take measurements in any material.
- Being able to use an extended field range for measurements, spanning from zero field (ZF) up to 10 T (in the current generation of high field muon spectrometers).
- Compatibility with low temperature sample environments (e.g. helium dilution refrigerators).
- Potential access to a very wide range of **correlation times**, as discussed in the previous section.

On the other hand disadvantages of μSR compared to NMR can be listed as:

- Limited potential for spin manipulation by RF pulse sequences due to the short muon lifetime.
- The **muon stopping sites** need to be determined to fully interpret the data.
- There will be some perturbation of the system under study by the presence of the muon.

The relation between μSR and other forms of magnetic resonance is covered in more detail in Chapter 23.

Chemistry

The chemistry perspective is based on the realization that the chemical properties of Mu are analogous to those of the H atom, which is the simplest form of free radical and is thus strongly chemically reactive. Muonium therefore provides a highly reactive radioactively labelled light isotope of H. Studying Mu via muon spectroscopy allows the time dependence of chemical reactions of a hydrogenic species to be determined and transient initial states of reactions can also be followed. The existence of Mu significantly extends the hydrogen isotopic sequence of H, D, and T.

A further feature of muonium chemistry is that the low mass of Mu leads to large quantum zero-point energy so that Mu probes the upper reaches of the interatomic potential energy surface (PES) and can therefore provide a thorough test of chemical reaction theory. Vibrational excitation (e.g. with a laser) can be used to take the muon even higher up the PES (at least three times higher than the ground state). Muon spectroscopy using photoexcitation is covered in Chapter 19 and chemistry applications of muon spectroscopy are covered in Chapter 12.

Computational science

From a computational science perspective, the implanted muon problem is a variation on the hydrogen interstitial defect problem. Well-established computational methods such as Density Functional Theory (DFT) can be used to calculate the structure of the interstitial muon defect including the atomic relaxation around the muon site. The strong quantum character of the muon provides an extra challenge for computational studies, since standard computational methods such as DFT only give the classical site. Further computational procedures are required to properly model the quantum delocalization of the muon. Various computational approaches to determining muon sites and their properties are discussed in Chapter 16.

2.4   The μSR experiment

A view of the μSR experiment could perhaps start with understanding the basic features of a muon spectrometer, first introduced in the previous chapter. The muon spectrometer comprises a set of positron detectors placed around the sample to measure the time evolution of the muon spin polarization. In addition, magnetic coils are arranged around the sample position to supply external fields at the sample. Muon instrumentation is covered in more detail in Chapter 14, but here we just consider the simplest detector arrangement, as shown in Fig. 2.5a). The asymmetry of the positron emission is represented by the function $\Gamma(\theta) = 1 + a\cos\theta$, where a depends on the energy range of positrons seen by the detectors (see Appendix E.4). In this simple two-detector geometry a forward detector (F) is placed in the direction of the initial spin direction and a backward detector (B) is placed in the opposite orientation. The initial asymmetry a_0 seen by each detector is the average of $\Gamma(\theta) - 1$, computed over the detector surface facing the sample. In practice[5] we find $a_0 = 25\%$. The magnetic field direction is defined as the z-axis and in the longitudinal field (LF) orientation the initial spin direction is also along z. In this case the respective signals in the F and

[5]As discussed in Appendix E, this is lower than the theoretical maximum of 1/3.

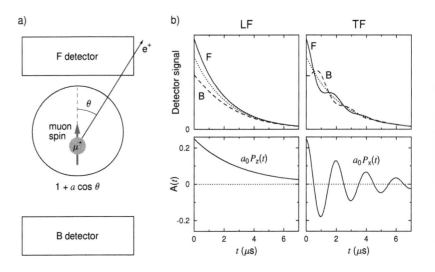

Fig. 2.5 a) Positron emission has an angular probability weighted towards the muon spin direction. A pair of detectors placed forward (F) and backward (B) with respect to the initial spin direction are used to measure the spin polarization. b) The time evolution of the F and B signals and the asymmetry function in the case of fields at the muon site that are either longitudinal or transverse with respect to the initial direction of the muon spin. The dotted lines in the upper plots represent the average signal.

B detectors, $N_F(t)$ and $N_B(t)$, are given by

$$N_F(t) = N_F^0 \exp(-t/\tau_\mu)[1 + a_0 P_z(t)],$$

$$N_B(t) = N_B^0 \exp(-t/\tau_\mu)[1 - a_0 P_z(t)], \tag{2.5}$$

where N_F^0 and N_B^0 are scaling parameters for the count rates of the F and B detectors, $P_z(t)$ is the longitudinal polarization function which is equivalent in this geometry to the longitudinal relaxation function $G_z(t)$ describing the relaxation of the muon spin in response to internal fields in the sample. The distinction between the polarization and the relaxation function will become clear when we discuss the transverse field experimental configuration. The polarization function is covered in detail in Chapter 5. In the detector signals the polarization function is superimposed on a large, exponentially decaying signal due to the muon lifetime [Fig. 2.5b)]. In order to extract the polarization or relaxation function, which is the key quantity of interest in muon spectroscopy, the forward-backward **asymmetry** $A(t)$ is evaluated as

$$A(t) = \frac{N_F(t) - N_B(t)}{N_F(t) + N_B(t)}$$

$$= \frac{(N_F^0 - N_B^0) + (N_F^0 + N_B^0)a_0 P_z(t)}{(N_F^0 + N_B^0) + (N_F^0 - N_B^0)a_0 P_z(t)}. \tag{2.6}$$

If the detectors are completely balanced in terms of geometry and sensitivity then $N_F^0 = N_B^0$ and a straightforward result is obtained

$$A(t) = a_0 P_z(t). \tag{2.7}$$

In practice, perfect balance is not possible and this results in different scaling factors for the two detectors. Compensation for this difference is then made by including a balancing factor $\alpha = N_F^0/N_B^0$ in the expression for $A(t)$, i.e.

$$A(t) = \frac{N_F(t) - \alpha N_B(t)}{N_F(t) + \alpha N_B(t)} = a_0 P_z(t). \tag{2.8}$$

The balancing factor is generally estimated for muons in a diamagnetic state by applying a small transverse field and adjusting α to give an $A(t)$ that oscillates symmetrically about zero. This F,B detector configuration is also used in the case where no external magnetic field is applied, known as zero field (ZF) mode. It should be noted that the F and B detectors here are logical detectors that must be mapped against the physical array of detectors in the instrument. For a muon instrument at a pulsed source these logical detectors typically represent a large group of physical detectors. In the case of a continuous source such grouping is not always required, but mapping from physical to logical detectors is still needed for dealing with different experimental configurations and allowing eqn 2.8 to be used consistently in each case.[6]

When in TF mode the initial polarization direction is defined as the x-axis, which is transverse to the magnetic field B_{TF} that is directed along the z-axis. In this case the polarization function describing the asymmetry becomes the transverse polarization function $P_x(t)$, which can be represented as the product of an oscillation term and a transverse relaxation function $G_x(t)$

$$P_x(t) = \cos(\gamma_\mu B_{TF} t) G_x(t). \tag{2.9}$$

For TF mode it is beneficial to have a number of detectors or detector groups distributed evenly around the field axis to maximize the signal, typically up to sixteen detector groups would be used to avoid significant signal loss through dephasing. The polarization function for each detector then involves a phase angle ϕ_i reflecting the phase offset with respect to the x-axis initial polarization direction (Fig. 2.6)

$$P_i(t) = \cos(\gamma_\mu B_{TF} t + \phi_i) G_x(t) \tag{2.10}$$

and the corresponding detector signals are given by

$$N_i(t) = N_i^0 \exp(-t/\tau_\mu)[1 + A_i(t)] \tag{2.11}$$

where N_i^0 is a scaling parameter for the detector count rate in the ith detector, and the detector group asymmetry functions are defined as $A_i(t) = a_i P_i(t)$, with the individual asymmetry parameters a_i specified to allow for variations in detector group geometry.

The muon polarization and relaxation functions describing the asymmetry are the core functions of the muon spectroscopy technique and the mathematical forms that they take in various experimental situations are described in Chapter 5. A quantum mechanical approach to understanding the evolution of the muon polarization is also presented in Chapter 4.

[6] For example, the physical F detector in a spectrometer is downstream and the physical B detector is upstream, whereas for surface muons (see Chapter 13) the initial muon polarization points upstream. The logical detector mapping in this case inverts the F and B detectors so that eqn 2.8 can be used to represent a positive initial asymmetry.

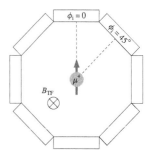

Fig. 2.6 Example arrangement of eight detector groups optimized for TF mode, indicating the detector phase ϕ_i.

Chapter summary

- The muon is a spin-1/2 lepton that interacts with matter and has a finite average lifetime of $\tau \approx 2.2$ μs.
- When stopped in a sample of matter, the muon can form diamagnetic and paramagnetic stopping states. Its spin interacts magnetically with its environment which, along with the nature of its decay, forms the basis of μSR spectroscopy.
- A μSR experiment can be carried out in LF, ZF, and TF modes. These measurements rely on the key concepts of muon polarization, relaxation, and asymmetry.

Further Reading

- A. Yaouanc and P. Dalmas de Réotier, *Muon Spin Rotation, Relaxation, and Resonance: Applications to Condensed Matter*, OUP (2011).
- S. J. Blundell, Contemp. Phys. **40**, 175 (1999).
- S. F. J. Cox, J. Phys. C **20**, 3187 (1987).
- K. Nagamine, *Introductory Muon Science*, CUP (2003).
- F. Halzen and A. D. Martin, *Quarks and Leptons*, John Wiley and Sons (1984).
- S. L. Lee, R. Cywinski, and S. H. Kilcoyne (editors) *Muon Science: Muons in Physics, Chemistry and Materials (Scottish Graduate Series)*, CRC Press (1999).

Exercises

(2.1) Calculate the average path length for the cosmic ray muon in Example 2.1. What would the path length be if there was no time dilation?

(2.2) A muon instrument is capable of measuring muon relaxation rates as fast as 20 μs^{-1} and as slow as 1 ms^{-1}. Calculate the fastest fluctuation rate that can be measured using this instrument when the coupling between the muon and the fluctuations is 2 MHz.

(2.3) A TF muon experiment is being carried out on a pulsed muon source instrument with a large number of detectors available in the muon spin rotation plane. A choice of detector grouping must be made: either four detectors at 90° intervals, eight detectors at 45° intervals, or sixteen detectors at 22.5° intervals. By considering the loss of asymmetry due to signal dephasing across each detector group, determine which of the groupings keep the fractional asymmetry loss below 3%.

Muon charge and spin states

The μSR experiment begins with a spin-polarized muon stopped in a material. In this chapter we discuss the states that the stopped muon can adopt, along with how these come to be realized during the implantation process. This is important as the stopped muon in its local enviroment is the fundamental probe that forms the basis of the measurement. A knowledge of the muon's stopping state (the **muon fate**) is the starting point that allows us to compute the effect of the interactions with the local environment on the muon-spin polarization, which is the subject of the next two chapters. More precisely, the energy scale that determines the muon fate is predominantly due to its charge, but the signature in the μSR experiment is determined by its spin state, diamagnetic or paramagnetic, according to Section 2.3. We shall cover mostly charge aspects in this chapter and consider spin in more detail in Chapter 4.

Let us start with the muon at rest in the material.[1] The stopped muon is found to exist in three different charge states. These are shown in Fig. 3.1 and summarized below.

[1]Later in the book we will consider cases in which the muon can hop from site to site, see Chapter 8.

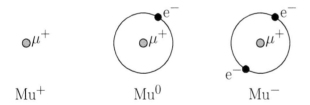

Fig. 3.1 The charge states of the muon: Mu^+, Mu^0, Mu^-.

[2]H^+ is sometimes called hydron, but it is simply a proton p^+.

[3]There is a difference between the physics and chemistry communities with notation. Physicists, schooled in particle physics, are more often likely to write the species as μ^+. Chemists, who are more aware of the analogy with hydrogen, tend to write Mu^+. In this book, we will use Mu^+ in the chapters that are concerned with semiconductors and chemistry (where the hydrogen analogy is more relevant) but revert to the physicist notation elsewhere, which is in keeping with the literature.

- **Mu^+**: The muon can exist on its own as a positively charged particle μ^+. This is analogous[2] to the hydrogen cation H^+, and for this reason this diamagnetic species is often written[3] with the symbol Mu^+.

- **Mu^0**: The muon can pick up an electron, and form an electrically neutral bound state μ^+e^- called muonium, given the symbol Mu. This is analogous to the hydrogen atom H. To avoid any ambiguity, we can write Mu as Mu^0 to emphasize that it is in the neutral, paramagnetic, state. Note that the chemically stable version of hydrogen is H_2, molecular hydrogen, but the muon flux is never large enough to produce Mu_2.

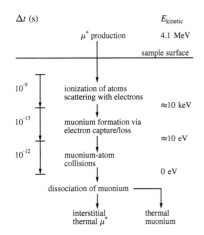

Fig. 3.2 The processes leading to a stopped muon following implantation of surface muons.

[4]This process can be described by the Bethe-Bloch equation, using ideas developed by Bethe and Bloch in the early 1930s.

[5]The electron affinity is ≈ 0.75 eV, and this is the (relatively small) energy needed to remove one of the electrons from Mu^- and turn it back into Mu^0. Note that Mu^0 is much more stable since ≈ 13.6 eV are needed to ionize it to Mu^+.

[6]This is not uncommon: any experimental detection has an initial blind window.

- **Mu^-:**   The muon can pick up two electrons and form the negatively charged bound state Mu^- which is analogous to the hydride ion H^-. This state is diamagnetic (an electron singlet).

These different charge states can be thought of as distinct chemical species, each determined by the nature of the electronic structure around the muon. However, the three states Mu^+, Mu^0, and Mu^- are idealizations. Even after complete muon thermalization, the final state may be weakly bound to neighbouring atoms or there may be screening effects in a metallic sample. This means that the actual spin and charge densities may depart from the simple pictures associated with these idealizations. But even in unsaturated molecules and magnetic solids, where the spin density becomes more and more delocalized, $\mathrm{Mu}^\pm$ and Mu^0 are useful proxies. We will therefore formulate our discussion in terms of the charge and spin state of the implanted muon, but it is important to keep in mind that the muon is a chemical species (see Chapter 12) as well as a magnetic entity.

3.1   State formation

Implantation process

Muons are usually implanted at energies of ≈ 4 MeV. A schematic of the implantation process that leads to a stopped muon is shown in Fig. 3.2. The dominant energy loss process by which the muons slow down is ionization of the sample atoms.[4] The muon may be able to capture one of these electrons, but it is likely to be stripped off again by a subsequent interaction. The charge cycling will stop once the muon's kinetic energy drops below the ionization energy, typically of order 10 eV, with the remaining energy lost as phonons. The charge cycling and slowing down is of order picoseconds and the muon's spin is essentially unaffected during this time because the Coulomb interaction does not couple to its magnetic moment, and the time spent as a neutral atom is far too short for the hyperfine interaction to have any effect. The muon then comes to rest slightly downstream of the track of ions, with a certain (material-dependent) probability of having an electron so forming neutral muonium (Mu^0). The relatively fragile negative ion[5] (Mu^-) is unlikely to be formed promptly in this way.

Prompt processes

Once it has stopped, the muon may undergo rapid charge-state changes or reactions on a timescale of order a few nanoseconds, i.e. before reliable data acquisition allows us to follow its spin relaxation.[6] The positive ion (Mu^+) will try to find a location with high electron density such as a lone pair on an atom (e.g. oxygen), a bond centred site, or close to an electronegative ion such as fluorine. It may attach to an ion or molecule, for example an oxide ion to form a singly negative hydroxide ion, or a water molecule to form the hydronium ion $\mathrm{H_2MuO}^+$ ($\mathrm{H_3O}^+$). In covalent solids Mu^+ may also promptly form a (spin singlet) bond, which implies a screening charge which is spin-unpolarized. This is typically the case in non-magnetic metals. In these cases our earlier classification by charge becomes less stringent, although, in insulators, there may still be a net positive charge overall. In molecular liquids and solids the neutral atom may react with a neighbouring molecule, perhaps attacking a double bond or lone pair site.[7] Alternatively it may find an interstitial space where it can remain as a free atom (this indeed happens also in instrinsic semiconductors[8]). In these two cases charge state changes may occur if the muon is able to capture an electron from the material, perhaps one of those liberated along the stopping track. An electron may also be returned to the material (e.g. to the conduction band) if appropriate.

If the muon is in the neutral (paramagnetic) state for a time comparable to the inverse hyperfine constant, before converting to diamagnetic Mu^+ or Mu^-, it will dephase in transverse field.[9] The result is a loss of apparent amplitude, accompanied by a phase shift. At low field when the Mu^0 fraction would give triplet preccesion, the phase shift and amplitude will be field-dependent, since the relevant timescale is the period of the triplet precession. There will also be loss of amplitude in zero field if the muon's electron is coupled to nearby nuclear spins.[10] These effects give the **missing fraction** where the total oscillating amplitude in a transverse field measurement does not add up to the full asymmetry expected of the instrument. The missing polarization is recovered in longitudinal field repolarization measurements.[11]

[7] See Chapter 12.

[8] See Chapter 10.

[9] We will examine the dynamics of neutral muonium in more detail in Chapters 4, 10, and 12.

[10] A similar dephasing occurs if Mu^+ is slow to convert to Mu^0 which is then observed in transverse field.

[11] Repolarization will be explored in Section 4.4.

Slower processes

Charge state changes may continue through the muon data-taking window (which is $\sim 10\mu s$). Reactions, site changes, and charge exchanges may be slowed down by energy barriers, and electrons or holes from the muon track will take time to diffuse to the muon. These changes generally result in relaxation of the muon polarization, usually exponential in form. In the case of Mu^0 converting to Mu^+ or Mu^-, in longitudinal field, there is no immediate change of polarization. The change can still be detected by techniques such as muon spin resonance with a delayed RF pulse. In high longitudinal field, the conversion of Mu^+ to Mu^0 causes only a small polarization change, but with the same time constant as in low longitudinal field or transverse field.

Charge cycling can occur at higher temperatures, for example in semiconductors where sufficient numbers of both holes and electrons are available for interaction. This will still be observed as an exponential relaxation. At high longitudinal field it will take many charge cycles to completely depolarize the muons, so a field scan will give relaxation to zero polarization at any field, with the relaxation rate falling off at high field. Analysis of this kind of data can give the charge cycling time and relative instantaneous occupation of the different states, keeping in mind that whenever charge exchange is effective also spin exchange may take place.

Metals

Among all materials, metals stand out because muonium (Mu^0) is never seen in any of them. The muon's electron orbitals overlap with the conduction electrons, with rapid spin exchange ensuring that any hyperfine interaction is averaged away and the muon appears to be in a diamagnetic state (like Mu^+). The conductivity should ensure that the muon's point charge is effectively screened on a longer scale, though there may be local distortion and charge redistribution around the muon site, and this has some experimental consequences.[12]

[12]Some of these will be discussed further in Chapter 23.

[13]See Exercise 3.1 for the derivation. The full quantum mechanical solution for the hydrogen atom (the proper way to treat the problem) yields identical results, but the derivation is much more involved.

[14]The reduced mass is conventionally given the symbol μ, which we use here and in the following example, but in the rest of the book we will retain the symbol μ to refer to either the muon or to magnetic moment.

[15]The implications for the basic properties are as follows: The Bohr radius a_H is *inversely proportional* to the reduced mass μ_H [because $a_H = 4\pi\epsilon_0\hbar^2/(\mu_H e^2)$]. The ionization energy is *directly proportional* to the reduced mass [because the ionization energy = $\hbar^2/(2\mu_H a_H^2)$ and $a_H \propto 1/\mu_H$]. The numerical values are listed in Table 3.1, and one can see that the variation is relatively small.

[16]A further property is the *electron affinity*, the energy released when an electron is added to a neutral species X to form the anion X^-. This quantity is more complicated to evaluate, but for similar reasons as we found for the ionization energy, this takes the same value (≈ 0.75 eV) for all the hydrogen analogues, including muonium.

3.2   Hydrogen analogues

The previous discussion highlighted the role of paramagnetic Mu^0. It will be helpful now to consider how it differs from hydrogen. The Bohr model for the hydrogen atom makes the following predictions.[13] The energy levels are given by $E = -(\hbar^2/2m_e a_0^2)(1/n^2)$ with $n = 1, 2, \ldots$ where the Bohr radius a_0 is given by $a_0 = 4\pi\epsilon_0\hbar^2/(m_e e^2)$. The ionization energy of hydrogen is then given by the energy to take an electron from the lowest level ($n = 1$) out to the continuum ($n = \infty$, where $E = 0$) and is equal to $\hbar^2/(2m_e a_0^2) = 13.6057$ eV. However, in all these formulae we have used the electron mass m_e and so we have been assuming the mass of the proton is infinite. Strictly, we should be using the *reduced mass*[14] μ_H. This has implications[15] for not only hydrogen, but all the hydrogen analogues, including muonium.

Example 3.1

The reduced mass μ_H for hydrogen is given by

$$\mu_H = \frac{m_e m_p}{m_e + m_p}, \tag{3.1}$$

and appears in the formula for the Bohr radius ($a_0 \propto \mu_H^{-1}$) and in the ionization energy ($\propto \mu_H^{-1}$). However, because $m_p \gg m_e$, $\mu_H \approx m_e(1 - m_e/m_p)$ and so the correction for the reduced mass is small. We can carry out the calculations of these various quantities for the isotopes of hydrogen (hydrogen [also called 'protium'] ^1H, deuterium ^2D, and tritium ^3T) and add in muonium as a light isotope of hydrogen and the results are shown in Table 3.1.

For all of the species listed in Table 3.1 the mass of the particle at the centre is much greater than that of the electron. They all have unit charge, and are essentially point particles with size much less than the wavefunctions of electrons bound to them. Therefore the chemical properties of all these species are very similar.[16] Therefore we can conclude that a positive muon at rest inside a sample will behave very much like

Property	Mu	H	D	T
Mass (m_H)	0.1131	1.000	1.998	2.993
Reduced Mass (m_e)	0.99519	0.99946	0.99973	0.99982
Ionization energy (eV)	13.540	13.598	13.602	13.603
Bohr Radius (Å)	0.53173	0.52946	0.52932	0.52927
Nuclear moment (μ_N)	8.8906	2.7928	0.8574	2.9788
Nuclear spin I	$\frac{1}{2}$	$\frac{1}{2}$	1	$\frac{1}{2}$
$\gamma/(2\pi)$ (MHz T^{-1})	135.53	42.576	6.536	45.413
Hyperfine constant (MHz)	4463.3	1420.4	327.4	1516.7

Table 3.1 Properties of muonium, hydrogen (protium), deuterium, and tritium. The first set of properties are related to the chemical properties (reduced mass, ionization energy, and Bohr radius) and are all rather similar. The second set of properties relate to the nuclear magnetism and here there are large differences.

hydrogen and attempt to reach chemical equilibrium with its surroundings. This equilibrium state may be a positive or negative "hydrogen" ion, a neutral atom, or chemically bound to a molecule or a lattice.

Where the hydrogen analogues differ strongly is in their nuclear magnetic properties. They have different magnetic moments $\mu = \gamma\hbar I$, where the nuclear spin I is $\frac{1}{2}$ (except for the case of deuterium where $I = 1$), and γ is the nuclear gyromagnetic ratio.[17] A quantity that will be very important in subsequent chapters is the contact hyperfine constant A which expresses the size of the interaction between the nuclear spin and the electron.[18] As shown in the following example, the most important contribution to the variation in A is the variation in γ (as well as the inclusion of a factor that goes like $I + \frac{1}{2}$ that is relevant for deuterium).

[17]The magnetic moment μ and the angular momentum $\hbar I$ are proportional to each other, and the proportionality constant is the gyromagnetic ratio. See Section 4.1.

[18]This interaction is only important for neutral muonium (Mu^0) because it contains an unpaired electron spin, and thus has the possibility of a hyperfine interaction. It is thus said to be **paramagnetic**. Both Mu$^+$ and Mu$^-$ are **diamagnetic** because neither contains any unpaired electron spins [Mu$^+$ has no electrons and Mu$^-$ has a pair of electrons (which in the ground state will exist in a $S = 0$ state)].

Example 3.2

The contact hyperfine interaction between a nuclear moment $\boldsymbol{\mu}_{\text{nuc}} = \hbar\gamma_N \boldsymbol{I}$ and an electron, where[19] $g_s\mu_B = \hbar\gamma_e$, can be written

$$\hat{\mathcal{H}} = -\boldsymbol{\mu}_{\text{nuc}} \cdot \boldsymbol{B}_e = A\boldsymbol{I} \cdot \boldsymbol{S}, \qquad (3.2)$$

where $\boldsymbol{B}_e$ is the magnetic field created by the electron. The electron is unpaired (there is just one electron in the hydrogen atom, not two) and it has a spin. It thus gives rise to a magnetization $\boldsymbol{M}(\boldsymbol{r}) = -g_s\mu_B\boldsymbol{S}|\psi(\boldsymbol{r})|^2$ and we are only interested in the value this takes inside the nucleus, which is $\boldsymbol{M}(0) = -g_s\mu_B\boldsymbol{S}|\psi(0)|^2$. We can consider the electron density inside the nucleus as a uniformly magnetized sphere and this gives rise to a magnetic induction $\boldsymbol{B}_e = \mu_0\boldsymbol{M}(0) - \mu_0\boldsymbol{M}(0)/3$, where the correction $-\mu_0\boldsymbol{M}(0)/3$ is due to the demagnetization field of a magnetized sphere.[20] Putting this all together yields

$$\boldsymbol{B}_e = -\frac{2}{3}\mu_0 g_s\mu_B|\psi(0)|^2\boldsymbol{S}. \qquad (3.3)$$

For a 1s electron, we have that $\psi(r) = e^{-r/a_0}/(\pi a_0^3)^{1/2}$ and hence $|\psi(0)|^2 = 1/(\pi a_0^3)$. We hence end up with an expression for the hyperfine constant as

$$A = \frac{2}{3}\mu_0 g_s\mu_B\hbar\gamma_N \frac{1}{\pi a_0^3}. \qquad (3.4)$$

[19]The g-factor of the electron $g_s \approx 2$. In this example, we take the nuclear spin operator $\boldsymbol{I}$ and the electron spin operator $\boldsymbol{S}$ to be dimensionless, so that A has the dimensions of energy.

[20]The demagnetization factor for a sphere is $N = \frac{1}{3}$ and the demagnetizing correction to the B-field is $-\mu_0 NM$. This is discussed further in Chapter 6.

[21]These are the levels $F = I - \frac{1}{2}$ and $F = I + \frac{1}{2}$.

[22]As is often the case in spectroscopy, we will use the same symbol A to refer to the energy gap between the two levels and for the frequency that corresponds to this energy gap (using the formula $E = hf$).

[23]See Exercise 3.2.

Coupling together an electron spin (for a 1s orbital $J = \frac{1}{2}$) and a nucleus (I) leads to a total angular momentum $\boldsymbol{F} = \boldsymbol{I} + \boldsymbol{J}$ and since $\boldsymbol{F}^2 = \boldsymbol{I}^2 + \boldsymbol{J}^2 + 2\boldsymbol{I} \cdot \boldsymbol{J}$ we have that the hyperfine energy levels are

$$E = A\boldsymbol{I} \cdot \boldsymbol{J} = \frac{A}{2}[F(F+1) - I(I+1) - J(J+1)]. \tag{3.5}$$

Since $J = \frac{1}{2}$, this gives two possible levels[21] separated by $A(I + \frac{1}{2})$, so for $I = \frac{1}{2}$ (relevant for hydrogen and muonium) the levels are separated by the hyperfine constant A. For hydrogen, this value is 5.87×10^{-6} eV. Dividing by h gives a frequency,[22] which is $A_H = 1420.4$ MHz, as written in Table 3.1. For muonium, we need to scale up A_H by the γ factor, so we might estimate $A_{Mu} \approx A_H \gamma_\mu / \gamma_p = 4521$ MHz which is within just over 1% of the right answer, but we can get pretty much spot on[23] by also including the small change in the Bohr radius between H and Mu.

In conclusion, we have seen that the chemical properties of muonium are very similar to hydrogen, but the nuclear magnetic properties are very different. Muonium will therefore have a unique spectroscopic signature which can be studied in experiments, and as we will see in Chapters 10 and 12 its environment within the solid state or within a chemical species will also modify the hyperfine interaction.

3.3 Measuring the states

For a practical experiment on a new sample, we need to know the charge and spin state(s) of the muon in order to interpret the data, especially if the muon is intended to observe properties such as magnetism or ionic motion. The simplest check is to observe the amplitude of the Larmor precession for the bare muon, at low transverse field (2 mT is typical) and, in magnets, above any ordering transition. Any missing amplitude must correspond to paramagnetic Mu^0 either as the free atom or bound in a radical, or a missing fraction initially formed as one of these but converted to another state.

In favourable cases, typically samples with few or no nuclear moments, a **triplet** muonium precession signal may be observed from interstitial Mu^0 in low transverse field.[24] If prompt conversions are suspected, a scan of low transverse fields may reveal amplitude and phase shifts. From this one can calculate the mean lifetime of the initial state. A longitudinal field scan gives the **repolarization**[25] of the radicals or muonium and will give the fraction in that state plus the approximate hyperfine constant. Repolarization also includes any missing fraction. The shape of the curve will be affected by dynamics or anisotropy.

Distinguishing between Mu^+ and Mu^- from the muon data alone is more difficult since neither has an electronic moment. The Mu^+ will be bound closer to other atoms so may show a larger nuclear dipole coupling in zero field. Unfortunately this is not a good guide in oxides where a Mu^+ is likely to have oxygen as its nearest neighbour and the fraction of magnetic ^{17}O is very small. Varying the doping of a semiconductor and observing a change in the diamagnetic fraction can imply which state is present. For example, a more n-type sample will tend to favour Mu^- over Mu^+.

[24]The principle behind transverse field measurements will be described in Section 4.4. Typically 0.2 mT is used at a pulsed muon source giving a frequency of approximately 2.8 MHz.

[25]Repolarization will be discussed in Section 4.4. Scans are performed typically from zero to around 0.2–0.5 T.

3.4 Influencing the states

Generally the charge states are fixed for a particular material and relatively similar for a series. In some cases it is possible to influence the balance of charge states to enhance the signal from a particular state.

Example 3.3

- A time-independent **electric field**[26] E applied to a sample will sweep the charge carriers from the muon track away from the stopped muon. In the usual $E = 0$ case the carriers form a cloud around the track and diffuse to the muon. Lower fields in the direction of implantation will sweep the carriers past the stopped muon and may enhance the interaction. Generally, a scan of E through zero is somewhat asymmetric.

- **Illumination** of an insulating or semiconducting sample can generate photocarriers which will interact with the muon. Often this causes rapid charge exchange cycles and therefore relaxation, rather than just forming a stable charge state.

- Semiconductor[27] samples can be **doped** n- or p-type and the resulting availability of electrons or holes will influence the prompt fractions.

- **Solution**[28] experiments, with the material of interest dissolved in a solvent which does not react with muons, will have initial charge state fractions dependent on the solvent. The radical fractions formed by reaction with the solute scale with the initial Mu^0 fraction in the solution. Similarly gas phase experiments often use an inert buffer gas to decrease the muon stopping range, either to study the concentration dependence of a reagent or simply that the vapour pressure of the sample alone is too low. Choice of the buffer gas can increase the fractions of the wanted charge state, usually Mu^0.

- **Low energy muon** experiments, with implantation energy of order 3–50 keV rather than 4 MeV, will have a much shorter track and fewer charge carrier pairs formed. The electron capture and stripping may not take place to the same extent. As a result the relative fractions of different species in a low energy experiment can be different to a muon experiment performed on a bulk sample of the same material, even without any surface-related phenomena in the sample.[29]

[26]See also Chapter 19.

[27]See also Chapter 10.

[28]See also Chapter 12.

[29]These will be considered in more detail in Chapter 18.

Now that we have a picture of the states in which the stopped muon is found, we shall turn to the magnetic interaction of its spin with the local environment. This is the subject of the next chapter, where we discuss how quantum mechanics can be used to describe the time-evolution of the muon spin.

Chapter summary

- The stopped muon can exist as the positively charged ion Mu^+, the neutral muonium Mu^0, or the negatively charged Mu^-.
- Neutral muonium Mu^0 is a paramagnetic species and has very similar chemical properties to atomic hydrogen H^0, but has a very different hyperfine constant.
- Both Mu^+ and Mu^-, the diamagnetic species, are also observed in experiments. In some cases, the various charged species can interconvert.

Further reading

- The following review contains a good discussion of the different charge states of the muon observed in semiconductors: S. F. J. Cox, Rep. Prog. Phys. **72**, 116501 (2009).

Exercises

(3.1) In the Bohr model of the hydrogen atom, the energy E of an electron is given by

$$E = \frac{L^2}{2m_e r^2} - \frac{Ze^2}{4\pi\epsilon_0 r}, \qquad (3.6)$$

where Z is the nuclear charge (equal to 1 for hydrogen). The first term is the angular kinetic energy and the second term is the electrostatic potential energy. By setting the angular momentum L to be $n\hbar$, minimize E and show that at the minimum value $r = a_0 n^2/Z$, where

$$a_0 = \frac{4\pi\epsilon_0 \hbar^2}{m_e e^2} = \frac{\hbar}{mc\alpha}, \qquad (3.7)$$

and the minimum energy is

$$E = -\frac{Z^2}{n^2}\left(\frac{1}{2}m_e c^2 \alpha^2\right), \qquad (3.8)$$

where $\alpha = e^2/(4\pi\epsilon_0 \hbar c)$ is the fine structure constant.

(3.2) Verify the scaling between the hyperfine constants in hydrogen, deuterium, tritium, and muonium using the values of Bohr radius and nuclear spin properties in Table 3.1.

The quantum muon

The muon is a spin-$\frac{1}{2}$ particle and is therefore a quantum object. To understand its properties, and the properties of muonium (described in the previous chapter), we need to use some *quantum mechanics*. This chapter provides a review of the quantum mechanics of angular momentum and works through the problem of a muon coupled to an electron in muonium, followed by the problem of the muon coupled to a nucleus. We begin with a discussion of magnetic moments and spin precession.

4.1 Larmor precession

Magnetic moment and angular momentum

The muon has a magnetic moment. What does magnetic moment mean? We can think of a classical magnetic moment as a current loop. If a charged particle goes around in a closed orbit then we have a current loop and the magnetic moment μ is defined by the current I multiplied by the area of the loop A. Note that a charged particle also has mass: not only does charge move round and round (meaning that we have a magnetic moment) but also mass moves round and round (meaning that we have angular momentum), as shown in Fig. 4.1. This means that the orbital magnetic moment $\boldsymbol{\mu}$ and the orbital angular momentum $\boldsymbol{L}$ are connected and should be proportional to each other, the proportionality constant being called the **gyromagnetic ratio** γ. Thus $\boldsymbol{\mu} = \gamma \boldsymbol{L}$.

But what if a particle doesn't go around in an orbit? Then $\boldsymbol{L} = 0$ but the particle still can have an *intrinsic* magnetic moment, connected to its *intrinsic* angular momentum (also called its **spin**). In this case the magnetic moment is given by[1]

$$\boldsymbol{\mu} = \gamma \boldsymbol{S}, \tag{4.1}$$

where $\boldsymbol{S}$ is the spin angular momentum. For a particle of mass m and charge q the gyromagnetic ratio γ is given by

$$\gamma = \frac{gq}{2m}, \tag{4.2}$$

where g is a constant known as the **g-factor** and for which for the electron and muon is very close[2] to 2. Putting the numbers in provides us with a very useful number:

$$\frac{\gamma_\mu}{2\pi} = 135.5\,\mathrm{MHz\,T^{-1}}. \tag{4.3}$$

Fig. 4.1 A massive charged particle in orbit gives rise to both a magnetic moment $\boldsymbol{\mu}$ and angular momentum $\boldsymbol{L}$.

[1]To make closer contact with undergraduate treatments of quantum mechanics, we will begin by defining $\boldsymbol{S}$ with the correct units of angular momentum (i.e. including the factor of $\hbar$); later in this chapter (on page 43), we will return to the convention adopted in more advanced work for which $\boldsymbol{S}$ is dimensionless.

[2]It is about 2.00233184 for the muon. Dirac's theory of the electron (and the muon is a heavy electron) gives $g = 2$ precisely, but quantum electrodynamics gives higher-order corrections to this.

Note that we have used a subscript on the gyromagnetic ratio γ_μ to remind us that we are talking about the muon. Also, we have chosen to express the result divided by a factor of 2π and in units of $\mathrm{MHz\,T^{-1}}$ for reasons that are about to become clear. Recall from Table 2.1 that the mass of the muon is intermediate between that of the electron and the proton, and thus so are its magnetic moment and gyromagnetic ratio. Because we are going to be dealing with the positive muon μ^+, its magnetic moment and gyromagnetic ratio are *positive*; note that the negatively charged electron e^- therefore has a magnetic moment and gyromagnetic ratio which are *negative*.

Classical treatment

A magnetic moment $\boldsymbol{\mu}$ in an applied magnetic field $\boldsymbol{B}$ has an energy E given by

$$E = -\boldsymbol{\mu} \cdot \boldsymbol{B}. \tag{4.4}$$

Thus we might think that a magnetic field would cause a magnetic moment to line up with it, so as to minimize its energy. However, because the magnetic moment is associated with angular momentum, there is a torque $\boldsymbol{G}$ given by

$$\boldsymbol{G} = \boldsymbol{\mu} \times \boldsymbol{B}. \tag{4.5}$$

Since torque is equal to the rate of change of angular momentum, eqn 4.5 can be rewritten as

$$\frac{\mathrm{d}\boldsymbol{\mu}}{\mathrm{d}t} = \gamma\boldsymbol{\mu} \times \boldsymbol{B}. \tag{4.6}$$

This means that the change in $\boldsymbol{\mu}$ is perpendicular to both $\boldsymbol{\mu}$ and to $\boldsymbol{B}$. Rather than turning $\boldsymbol{\mu}$ towards $\boldsymbol{B}$, the magnetic field causes the direction of $\boldsymbol{\mu}$ to precess around $\boldsymbol{B}$. Equation 4.6 also implies that $|\boldsymbol{\mu}|$ is time-independent.[3]

[3]See Exercise 4.1.

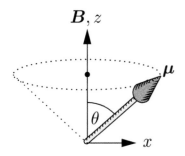

Fig. 4.2 A magnetic moment $\boldsymbol{\mu}$ in a magnetic field $\boldsymbol{B}$ which lies along the z-axis. The magnetic moment is initially in the xz-plane at an angle θ to $\boldsymbol{B}$. The magnetic moment precesses around a cone of semi-angle θ. The rotation is clockwise when viewed from above (from $+z$ back to the origin).

Example 4.1

Consider the case in which $\boldsymbol{B}$ is along the z-direction and $\boldsymbol{\mu}$ is initially at an angle of θ to $\boldsymbol{B}$ and in the xz-plane (see Fig. 4.2). Then

$$\dot{\mu}_x = \gamma B \mu_y$$
$$\dot{\mu}_y = -\gamma B \mu_x$$
$$\dot{\mu}_z = 0, \tag{4.7}$$

so that μ_z is constant with time and μ_x and μ_y both oscillate. Solving these differential equations leads to

$$\mu_x(t) = |\boldsymbol{\mu}| \sin\theta \cos\omega t$$
$$\mu_y(t) = -|\boldsymbol{\mu}| \sin\theta \sin\omega t$$
$$\mu_z(t) = |\boldsymbol{\mu}| \cos\theta, \tag{4.8}$$

where

$$\omega = \gamma B \tag{4.9}$$

is called the **Larmor precession (angular) frequency** [the frequency, measured in Hz, is of course $(\gamma/2\pi)B$] and the phenomenon is called **Larmor precession**.

Quantum mechanical treatment

The muon is a spin-$\frac{1}{2}$ particle, and so treating it as a classical magnetic moment cannot be the whole story.[4] Therefore we ought to treat the problem again but from a quantum mechanical perspective. We write the quantum state of the muon spin as $|\psi\rangle$, a state vector. For a spin-half particle, this state is a coherent superposition of a state with spin pointing up, $|\uparrow\rangle$, and a state with spin pointing down, $|\downarrow\rangle$. Thus in general we can write

$$|\psi\rangle = a\,|\uparrow\rangle + b\,|\downarrow\rangle = \begin{pmatrix} a \\ b \end{pmatrix}, \qquad (4.10)$$

where the two-component object on the right is called a **spinor** and a and b are complex numbers. To be normalized, we require[5] that

$$|a|^2 + |b|^2 = 1. \qquad (4.11)$$

In this notation, an up-spin is then denoted by

$$|\uparrow\rangle = \begin{pmatrix} 1 \\ 0 \end{pmatrix}, \qquad (4.12)$$

while a down-spin is denoted by

$$|\downarrow\rangle = \begin{pmatrix} 0 \\ 1 \end{pmatrix}. \qquad (4.13)$$

Similarly, a magnetic field $\boldsymbol{B}$ applied along z would give rise to the Zeeman Hamiltonian given by[6]

$$\hat{\mathcal{H}} = -\hat{\boldsymbol{\mu}} \cdot \boldsymbol{B} = -\frac{\hbar}{2}\gamma B \sigma_z = -\frac{\hbar\omega}{2}\begin{pmatrix} 1 & 0 \\ 0 & -1 \end{pmatrix}, \qquad (4.14)$$

so that the energy of the up and down spins is given by $-\hbar\omega/2$ or $+\hbar\omega/2$ respectively (remember $\omega = \gamma B$, and so the energy splitting $\hbar\omega$ between these two states depends on B) and the up and down spin states are both eigenstates of the $\hat{\mathcal{H}}$. The matrix in eqn 4.14 is the Pauli spin matrix σ_z. If we rotate the magnetic field so that it lies along a general direction $\hat{\boldsymbol{n}} = (\sin\theta\cos\phi, \sin\theta\sin\phi, \cos\theta)$, then $\hat{\mathcal{H}}$ would become

$$\hat{\mathcal{H}} = -\frac{\hbar\omega}{2}\hat{\boldsymbol{n}} \cdot \boldsymbol{\sigma} = -\frac{\hbar\omega}{2}(\sin\theta\cos\phi\,\sigma_x + \sin\theta\sin\phi\,\sigma_y + \cos\theta\,\sigma_z)$$
$$= -\frac{\hbar\omega}{2}\begin{pmatrix} \cos\theta & e^{-i\phi}\sin\theta \\ e^{i\phi}\sin\theta & -\cos\theta \end{pmatrix}, \qquad (4.15)$$

which again has eigenvalues $\mp\hbar\omega/2$ and the eigenstates are

$$|+\rangle = \begin{pmatrix} \cos\frac{\theta}{2} \\ e^{i\phi}\sin\frac{\theta}{2} \end{pmatrix} \quad \text{and} \quad |-\rangle = \begin{pmatrix} \sin\frac{\theta}{2} \\ -e^{i\phi}\cos\frac{\theta}{2} \end{pmatrix}, \qquad (4.16)$$

and this provides us with another method of deriving spin precession.

[4] However, the classical treatment in the previous section does nevertheless give the right answer.

[5] Because a and b are complex numbers, the state can be described by four real numbers (the real and imaginary parts of both a and b). However, the normalization condition, together with the fact that we can multiply $|\psi\rangle$ by an arbitary phase factor and still denote the same state, means that we only need two real numbers to describe the state. This is the same as a classical spin in which we need two angles (θ and ϕ) to describe the orientation of the magnetic moment.

[6] Here we use the fact that $\hat{\boldsymbol{\mu}} = \gamma\hat{\boldsymbol{S}}$ (eqn 4.1) and $\hat{\boldsymbol{S}} = \frac{\hbar}{2}\boldsymbol{\sigma}$ where $\boldsymbol{\sigma}$ is the vector of Pauli spin matrices:

$$\boldsymbol{\sigma} = (\hat{\sigma}_x, \hat{\sigma}_y, \hat{\sigma}_z).$$

The **Pauli spin matrices** are

$$\hat{\sigma}_x = \begin{pmatrix} 0 & 1 \\ 1 & 0 \end{pmatrix},$$

$$\hat{\sigma}_y = \begin{pmatrix} 0 & -i \\ i & 0 \end{pmatrix},$$

$$\hat{\sigma}_z = \begin{pmatrix} 1 & 0 \\ 0 & -1 \end{pmatrix}.$$

Example 4.2

If we keep the magnetic field aligned along z, the Hamiltonian is as in eqn 4.14. If the muon spin is initially in the xz-plane at an angle θ to the z-axis, then its state is

$$|\psi(0)\rangle = \begin{pmatrix} \cos\frac{\theta}{2} \\ \sin\frac{\theta}{2} \end{pmatrix} = \cos\frac{\theta}{2}|{\uparrow}\rangle + \sin\frac{\theta}{2}|{\downarrow}\rangle. \tag{4.17}$$

Its time dependence is then given by

$$|\psi(t)\rangle = e^{i\omega t/2}\cos\frac{\theta}{2}|{\uparrow}\rangle + e^{-i\omega t/2}\sin\frac{\theta}{2}|{\downarrow}\rangle = \begin{pmatrix} e^{i\omega t/2}\cos\frac{\theta}{2} \\ e^{-i\omega t/2}\sin\frac{\theta}{2} \end{pmatrix}, \tag{4.18}$$

and this corresponds to spin precession in the xy-plane. Working out the expectation values gives

$$\langle\psi(t)|\hat{\sigma}_x|\psi(t)\rangle = \sin\theta\cos\omega t$$
$$\langle\psi(t)|\hat{\sigma}_y|\psi(t)\rangle = -\sin\theta\sin\omega t$$
$$\langle\psi(t)|\hat{\sigma}_z|\psi(t)\rangle = \cos\theta, \tag{4.19}$$

which is just as we have had before in eqn 4.8.

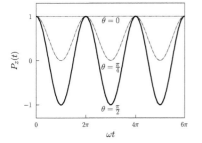

Fig. 4.3 The muon polarization $P_z(t)$ in eqn 4.20 plotted for three values of θ. When the field and initial muon polarization are aligned ($\theta = 0$) there is no spin precession. The amplitude of spin precession takes its maximum value when $\theta = \frac{\pi}{2}$. Note that, even though the *amplitude* varies as you change θ, the *frequency* of the precession is independent of θ.

[7]See Exercise 4.3.

Example 4.3

In the next chapter of the book we will need the result of a slightly different problem in which the muon spin is initially polarized along the z-direction and it is the magnetic field which is at angle of θ to the z-direction in the xz-plane (the opposite to what we are currently considering). This is worked out in Exercise 4.2 using the methods developed in this section, but here is another way to get it. We can rotate the expectation values in eqn 4.19 by an angle θ (which also will rotate the muon spin to lie along the z-axis, and the magnetic field will rotate to be at an angle θ from z). This gives immediately the time-dependent polarization $P_z(t)$ of the muon along the z-axis as

$$P_z(t) = \cos\theta\langle\sigma_z(t)\rangle + \sin\theta\langle\sigma_x(t)\rangle$$
$$= \cos^2\theta + \sin^2\theta\cos\omega t. \tag{4.20}$$

This result is plotted in Fig. 4.3 and will be very useful in the next few chapters.

We are now going to attack the problem worked out in Example 4.2 in a couple of different ways. Why bother doing this when we already have the answer? It is because finding the right formalism to attack a problem can play dividends when we have more complicated situations to deal with.

First, let us try to derive a time-evolution operator $\hat{U}(t)$ for the action of the magnetic field. This is related to the Hamiltonian $\hat{\mathcal{H}} = -\frac{\hbar\omega}{2}\hat{\boldsymbol{n}}\cdot\boldsymbol{\sigma}$ by the equation[7]

$$\hat{U}(t) = \exp\left(-\frac{i}{\hbar}\hat{\mathcal{H}}t\right) = \cos\left(\frac{\omega t}{2}\right)\hat{I} + i\sin\left(\frac{\omega t}{2}\right)\hat{\boldsymbol{n}}\cdot\boldsymbol{\sigma}. \tag{4.21}$$

Example 4.4

For our specific case of a muon initially polarized in the xz-plane, its state at time t is given by $\hat{U}(t)|\psi(0)\rangle$ and hence

$$
|\psi(t)\rangle = \hat{U}(t)|\psi(0)\rangle = \begin{pmatrix} \cos\frac{\omega t}{2} + \mathrm{i}\sin\frac{\omega t}{2} & 0 \\ 0 & \cos\frac{\omega t}{2} - \mathrm{i}\sin\frac{\omega t}{2} \end{pmatrix} \begin{pmatrix} \cos\frac{\theta}{2} \\ \sin\frac{\theta}{2} \end{pmatrix}
$$

$$
= \begin{pmatrix} \mathrm{e}^{\frac{\mathrm{i}\omega t}{2}} & 0 \\ 0 & \mathrm{e}^{-\frac{\mathrm{i}\omega t}{2}} \end{pmatrix} \begin{pmatrix} \cos\frac{\theta}{2} \\ \sin\frac{\theta}{2} \end{pmatrix} = \begin{pmatrix} \mathrm{e}^{\mathrm{i}\omega t/2}\cos\frac{\theta}{2} \\ \mathrm{e}^{-\mathrm{i}\omega t/2}\sin\frac{\theta}{2} \end{pmatrix}. \tag{4.22}
$$

The argument then proceeds as in the previous example.

This method hasn't made a lot of difference but has introduced the operator $\hat{U}(t)$ which will be important later.

4.2 Density matrices

The second method that we will introduce uses an object (formally an operator) which is called the **density matrix**. If[8] we know a quantum system is in state $|\psi\rangle$, the density matrix ρ is then given by

$$
\rho = |\psi\rangle\langle\psi|. \tag{4.23}
$$

This doesn't seem to be much of an advantage because we seem to have the same object, $|\psi\rangle$, just put in a different package. The real advantage of this approach will appear later when we make a more general definition of ρ. A rather useful result that follows from this is that the expected value of an operator $\hat{\mathcal{O}}$ is given by

$$
\langle\mathcal{O}\rangle = \langle\psi|\hat{\mathcal{O}}|\psi\rangle = \mathrm{Tr}(\hat{\mathcal{O}}|\psi\rangle\langle\psi|) = \mathrm{Tr}(\hat{\mathcal{O}}\rho). \tag{4.24}
$$

This is really useful because taking the trace of an operator is extremely easy.[9]

Example 4.5

A simple consequence of this is that if $\hat{\mathcal{O}} = \hat{I}$, where $\hat{I}$ is the identity operator (the operator that leaves states unchanged), then because $\langle I \rangle = 1$ we have immediately from eqn 4.24 that

$$
\mathrm{Tr}\rho = 1. \tag{4.25}
$$

We will also need the time dependence of ρ which we can work out by realising that $|\psi(t)\rangle = \hat{U}(t)|\psi(0)\rangle$ and $\langle\psi(t)| = (\hat{U}(t))^\dagger\langle\psi(0)|$, so that

$$
\rho(t) = \hat{U}(t)\rho(0)\hat{U}(t)^\dagger. \tag{4.26}
$$

Now let's solve our problem once more.

[8]This is a big 'if'! Because we know precisely which state our quantum system is in, the definition in eqn 4.23 applies only to what is known as a **pure state**. Our system is known to be $|\psi\rangle$, purely that and nothing else. A more general definition of the density matrix will be introduced later in eqn 4.38.

[9]The trace of an operator is obtained by taking the sum of the diagonal elements of the matrix representation of the operator. For example, the trace of any Pauli spin matrix is zero and the trace of a $n \times n$ identity matrix is n. The trace of a product of terms remains the same if those terms are cyclically permuted, a fact that we have used in eqn 4.24.

Example 4.6

The initial muon spin is $\begin{pmatrix} \cos\frac{\theta}{2} \\ \sin\frac{\theta}{2} \end{pmatrix}$, and the initial density matrix is

$$\rho(0) = |\psi(0)\rangle\langle\psi(0)| = \begin{pmatrix} \cos^2\frac{\theta}{2} & \cos\frac{\theta}{2}\sin\frac{\theta}{2} \\ \cos\frac{\theta}{2}\sin\frac{\theta}{2} & \sin^2\frac{\theta}{2} \end{pmatrix} = \frac{1}{2}\begin{pmatrix} 1+\cos\theta & \sin\theta \\ \sin\theta & 1-\cos\theta \end{pmatrix},$$
(4.27)

so that adding the time dependence produces

$$\rho(t) = \hat{U}(t)\rho(0)\hat{U}(t)^\dagger = \frac{1}{2}\begin{pmatrix} 1+\cos\theta & e^{i\omega t}\sin\theta \\ e^{-i\omega t}\sin\theta & 1-\cos\theta \end{pmatrix}.$$
(4.28)

Then the expected value of the spin is given by[10]

$$\langle\sigma_x(t)\rangle = \mathrm{Tr}(\sigma_x\rho(t)) = \sin\theta\cos\omega t$$
$$\langle\sigma_y(t)\rangle = \mathrm{Tr}(\sigma_y\rho(t)) = -\sin\theta\sin\omega t$$
$$\langle\sigma_z(t)\rangle = \mathrm{Tr}(\sigma_z\rho(t)) = \cos\theta.$$
(4.29)

[10] This agrees with eqn 4.19 and eqn 4.8.

[11] As is often the case in science, there's more than one way to get the same answer and having a variety of methods in our toolbox is extremely useful. This will become more important when we start coupling spins together. Though the case of a single spin interacting with a magnetic field can be treated classically or quantum mechanically, this problem has some very special symmetries which mean the answers come out the same. This is not the same with two interacting spins, and so from now on we will stick with quantum approaches.

[12] After a little algebra, see Exercise 4.4.

What is advantageous about using the density matrix approach is that all we have to do is to multiply various matrices together, and the results just tumble out. Fortunately, all these different approaches give the same answer.[11]

Before proceeding, let's derive an important formula for the density matrix for a spin-$\frac{1}{2}$ particle like the muon. For a general direction $\hat{\boldsymbol{n}} = (n_x, n_y, n_z) = (\sin\theta\cos\phi, \sin\theta\sin\phi, \cos\theta)$, the Pauli spin matrix for this direction is

$$\sigma_{\hat{\boldsymbol{n}}} = \boldsymbol{\sigma}\cdot\hat{\boldsymbol{n}} = \begin{pmatrix} n_z & n_x - in_y \\ n_x + in_y & -n_z \end{pmatrix},$$
(4.30)

so that an eigenstate for the state aligned with $\hat{\boldsymbol{n}}$ is[12]

$$|\hat{\boldsymbol{n}}\rangle = \begin{pmatrix} \sqrt{\frac{1+n_z}{2}} \\ \sqrt{\frac{1-n_z}{2}}\,e^{i\tan^{-1}\left(\frac{n_y}{n_x}\right)} \end{pmatrix},$$
(4.31)

and hence the density matrix $\rho = |\hat{\boldsymbol{n}}\rangle\langle\hat{\boldsymbol{n}}|$ is given by

[13] Quick shortcut: We often calculate expectation values of traceless spin operators σ_α, for which the product with the identity produces a zero trace and can be dropped. One can then replace ρ by $\rho' = \frac{1}{2}\boldsymbol{\sigma}\cdot\hat{\boldsymbol{n}}$, and $\mathrm{Tr}(\sigma_\alpha\rho')$ will give the same result as $\mathrm{Tr}(\sigma_\alpha\rho) \equiv \langle\sigma_\alpha\rangle$.

$$\rho = \frac{1}{2}\begin{pmatrix} 1+n_z & n_x - in_y \\ n_x + in_y & 1-n_z \end{pmatrix} = \frac{1}{2}\begin{pmatrix} 1+\cos\theta & \sin\theta\,e^{-i\phi} \\ \sin\theta\,e^{i\phi} & 1-\cos\theta \end{pmatrix},$$
(4.32)

which can be written compactly in the form

$$\rho = \frac{1}{2}(\hat{I} + \boldsymbol{\sigma}\cdot\hat{\boldsymbol{n}}).$$
(4.33)

These formulae[13] are given for special cases in Table 4.1.

4.3   Mixed states

Where density matrices come into their own is by being able to describe situations in which a spin state is not very well defined. This issue is explored in the following example.

Pauli spin matrices

σ_x	σ_y	σ_z
$\begin{pmatrix} 0 & 1 \\ 1 & 0 \end{pmatrix}$	$\begin{pmatrix} 0 & -i \\ i & 0 \end{pmatrix}$	$\begin{pmatrix} 1 & 0 \\ 0 & -1 \end{pmatrix}$

Eigenstates

$|{\uparrow_x}\rangle$	$|{\downarrow_x}\rangle$	$|{\uparrow_y}\rangle$	$|{\downarrow_y}\rangle$	$|{\uparrow_z}\rangle$	$|{\downarrow_z}\rangle$
$\frac{1}{\sqrt{2}}\begin{pmatrix} 1 \\ 1 \end{pmatrix}$	$\frac{1}{\sqrt{2}}\begin{pmatrix} 1 \\ -1 \end{pmatrix}$	$\frac{1}{\sqrt{2}}\begin{pmatrix} 1 \\ i \end{pmatrix}$	$\frac{1}{\sqrt{2}}\begin{pmatrix} 1 \\ -i \end{pmatrix}$	$\begin{pmatrix} 1 \\ 0 \end{pmatrix}$	$\begin{pmatrix} 0 \\ 1 \end{pmatrix}$

Density matrices

$|{\uparrow_x}\rangle\langle{\uparrow_x}|$	$|{\downarrow_x}\rangle\langle{\downarrow_x}|$	$|{\uparrow_y}\rangle\langle{\uparrow_y}|$	$|{\downarrow_y}\rangle\langle{\downarrow_y}|$	$|{\uparrow_z}\rangle\langle{\uparrow_z}|$	$|{\downarrow_z}\rangle\langle{\downarrow_z}|$
$\frac{1}{2}\begin{pmatrix} 1 & 1 \\ 1 & 1 \end{pmatrix}$	$\frac{1}{2}\begin{pmatrix} 1 & -1 \\ -1 & 1 \end{pmatrix}$	$\frac{1}{2}\begin{pmatrix} 1 & -i \\ i & 1 \end{pmatrix}$	$\frac{1}{2}\begin{pmatrix} 1 & i \\ -i & 1 \end{pmatrix}$	$\begin{pmatrix} 1 & 0 \\ 0 & 0 \end{pmatrix}$	$\begin{pmatrix} 0 & 0 \\ 0 & 1 \end{pmatrix}$

Table 4.1 The Pauli spin matrices and their eigenstates for the three principal Cartesian directions. The corresponding density matrices for pure states in these directions are also shown.

Example 4.7

A spin state for being aligned parallel to z is $\begin{pmatrix} 1 \\ 0 \end{pmatrix}$. If instead the spin is *antiparallel* to z, the state is written $\begin{pmatrix} 0 \\ 1 \end{pmatrix}$. If we make a **coherent superposition** of these two spin states, we get (after normalization) the state

$$\frac{1}{\sqrt{2}}\begin{pmatrix} 1 \\ 1 \end{pmatrix}. \tag{4.34}$$

This is *not* an unpolarized state but represents one polarized along the x-direction. In fact, the direction along which this state is polarized depends on the *phase relationship* between the terms in our superposition. Thus if we put a phase difference ϕ between the two terms in the superposition, i.e. we make the state

$$\frac{1}{\sqrt{2}}\begin{pmatrix} 1 \\ e^{i\phi} \end{pmatrix}, \tag{4.35}$$

then the spin will be polarized in the xy-plane with an angle ϕ from the x-axis.

If we want to describe an unpolarized collection of spins, in which the chance of a particular spin being polarized along z is $\frac{1}{2}$ and along $-z$ is $\frac{1}{2}$, so that a measurement of the z-component of the spin could equally go either way, then adding $|{\uparrow}\rangle$ to $|{\downarrow}\rangle$ is not going to help. This is where the density matrix comes into its own. An unpolarized state can be obtained by adding the density matrices $|{\uparrow}\rangle\langle{\uparrow}|$ and $|{\downarrow}\rangle\langle{\downarrow}|$ together, creating what is known as a **mixed state**. Thus

$$\rho_{\text{unpol}} = \frac{1}{2}\left(|{\uparrow}\rangle\langle{\uparrow}| + |{\downarrow}\rangle\langle{\downarrow}|\right) = \frac{1}{2}\begin{pmatrix} 1 & 0 \\ 0 & 1 \end{pmatrix}, \tag{4.36}$$

and in this case[14]

$$\langle\sigma_x\rangle = \text{Tr}(\sigma_x\rho_{\text{unpol}}) = 0$$
$$\langle\sigma_y\rangle = \text{Tr}(\sigma_y\rho_{\text{unpol}}) = 0$$
$$\langle\sigma_z\rangle = \text{Tr}(\sigma_z\rho_{\text{unpol}}) = 0. \tag{4.37}$$

[14] According to Note 13 one can work with $\rho'_{\text{unpol}} = 0$ which gives the answers very quickly.

A more general definition of ρ is

$$\rho = \sum_i p_i \left| \psi_i \right\rangle \left\langle \psi_i \right|. \tag{4.38}$$

This is a sum over pure states, with each term weighted by the probability p_i of that state occurring in the mixture. We require that $\sum_i p_i = 1$ so that the probabilities add up to unity. This means that even for a mixed state we will have (from eqn 4.25, which we only derived for pure states) that $\mathrm{Tr}\rho = 1$.

4.4   Two spins: muonium

Isotropic muonium

Having treated the precession of the muon in a magnetic field, we now turn to the behaviour of a muon interacting strongly with an electron: the muonium atom Mu^0. We introduced Mu^0 in Section 3.2 remarking its very similar chemical and very different magnetic properties with its hydrogen analogues. A muonium atom, like a hydrogen atom, is a neutral species (it has no net charge) but there are two spins, one on the muon and one on the electron. Using a basis where we consider spin along the z-axis, there are four possible states to consider: $\left| \uparrow^\mu \uparrow^e \right\rangle$, $\left| \uparrow^\mu \downarrow^e \right\rangle$, $\left| \downarrow^\mu \uparrow^e \right\rangle$ and $\left| \downarrow^\mu \downarrow^e \right\rangle$, where we denote the first arrow as representing the state of the muon and the second arrow as representing the state of the electron.[15]

Let's place the two spins in a magnetic field aligned along the z-axis. For now, we will ignore any interaction between the two spins so that they behave entirely independently. In this case we can write the Hamiltonian as the sum of two Zeeman terms, for the electron and muon $(-\hat{\boldsymbol{\mu}}_e \cdot \boldsymbol{B} - \hat{\boldsymbol{\mu}}_\mu \cdot \boldsymbol{B})$ and hence

$$\hat{\mathcal{H}} = \frac{\hbar \omega_e}{2} \sigma_e^z - \frac{\hbar \omega_\mu}{2} \sigma_\mu^z, \tag{4.39}$$

where both ω_e and ω_μ are positive quantities[16] defined by

$$\omega_\mu = \gamma_\mu B \quad \text{and} \quad \omega_e = -\gamma_e B, \tag{4.40}$$

with $|\gamma_e|/(2\pi) = 28024.21\,\mathrm{MHz\,T^{-1}}$ much larger than $\gamma_\mu/(2\pi) = 135.53\,\mathrm{MHz\,T^{-1}}$. The Hamiltonian in eqn 4.39 is simple to diagonalize because it is essentially two separate independent Hamiltonians added together. However, as we will shortly include interactions it is helpful to think about this in a more sophisticated way. If we consider our set of states $\{\left| \uparrow\uparrow \right\rangle, \left| \uparrow\downarrow \right\rangle, \left| \downarrow\uparrow \right\rangle, \left| \downarrow\downarrow \right\rangle\}$, then the operator σ_μ^z will only act on the first spin and the operator σ_μ^e will only act on the second spin. Thus in this basis the operators will be

$$\sigma_\mu^z = \begin{pmatrix} 1 & 0 & 0 & 0 \\ 0 & 1 & 0 & 0 \\ 0 & 0 & -1 & 0 \\ 0 & 0 & 0 & -1 \end{pmatrix} \quad \text{and} \quad \sigma_e^z = \begin{pmatrix} 1 & 0 & 0 & 0 \\ 0 & -1 & 0 & 0 \\ 0 & 0 & 1 & 0 \\ 0 & 0 & 0 & -1 \end{pmatrix}, \tag{4.41}$$

[15]To save writing, we will frequently abbreviate these states to $\left| \uparrow\uparrow \right\rangle$, $\left| \uparrow\downarrow \right\rangle$, $\left| \downarrow\uparrow \right\rangle$ and $\left| \downarrow\downarrow \right\rangle$, dropping the superscripts.

[16]Because the charge on the electron is negative, $\gamma_e < 0$, whereas the charge on the muon is positive and so $\gamma_\mu > 0$.

and hence the Hamiltonian becomes

$$\hat{\mathcal{H}}/\hbar = \begin{pmatrix} \omega_- & 0 & 0 & 0 \\ 0 & -\omega_+ & 0 & 0 \\ 0 & 0 & \omega_+ & 0 \\ 0 & 0 & 0 & -\omega_- \end{pmatrix}, \tag{4.42}$$

where we have defined (for convenience)

$$\omega_\pm = \frac{\omega_e \pm \omega_\mu}{2} = \begin{cases} 2\pi \times [14079.9\,\text{MHz} \times (B \text{ in T})] & \text{for } \omega_+ \\ 2\pi \times [13944.3\,\text{MHz} \times (B \text{ in T})] & \text{for } \omega_- \end{cases}. \tag{4.43}$$

These energy levels are plotted as a function of B (a plot known as a **Breit-Rabi diagram**) in Fig. 4.4, although in the diagram γ_μ has been increased by a factor of five (i.e. the muon has been made a factor of five lighter) so the very small splitting between the $\omega_\pm$ levels is visible.

Now let us look at the effect of switching on a hyperfine interaction. We have to add to our Hamiltonian the hyperfine term (see eqn 3.2) given by $(A/\hbar^2)\hat{S}_e \cdot \hat{S}_\mu$ where A is the hyperfine energy; writing $A = \hbar\omega_0$ the Hamiltonian becomes

$$\frac{\hat{\mathcal{H}}}{\hbar} = \frac{1}{2}\omega_e\sigma_e^z - \frac{1}{2}\omega_\mu\sigma_\mu^z + \frac{1}{4}\omega_0\boldsymbol{\sigma}_e \cdot \boldsymbol{\sigma}_\mu. \tag{4.44}$$

Example 4.8

Let us solve the simpler Hamiltonian in the case of zero applied magnetic field (so that $B = \omega_e = \omega_\mu = 0$ and the Hamiltonian in eqn 4.44 becomes

$$\frac{\hat{\mathcal{H}}}{\hbar} = \frac{1}{4}\omega_0\boldsymbol{\sigma}_e \cdot \boldsymbol{\sigma}_\mu. \tag{4.45}$$

In our basis $\{|\uparrow^\mu\uparrow^e\rangle, |\uparrow^\mu\downarrow^e\rangle, |\downarrow^\mu\uparrow^e\rangle, |\downarrow^\mu\downarrow^e\rangle\}$ we can evaluate the various matrices to show that

$$\boldsymbol{\sigma}_\mu \cdot \boldsymbol{\sigma}_e = \sigma_\mu^x\sigma_e^x + \sigma_\mu^y\sigma_e^y + \sigma_\mu^z\sigma_e^z = \begin{pmatrix} 1 & 0 & 0 & 0 \\ 0 & -1 & 2 & 0 \\ 0 & 2 & -1 & 0 \\ 0 & 0 & 0 & 1 \end{pmatrix}, \tag{4.46}$$

and hence that

$$\frac{\hat{\mathcal{H}}}{\hbar} = \begin{pmatrix} \frac{\omega_0}{4} & 0 & 0 & 0 \\ 0 & -\frac{\omega_0}{4} & \frac{\omega_0}{2} & 0 \\ 0 & \frac{\omega_0}{2} & -\frac{\omega_0}{4} & 0 \\ 0 & 0 & 0 & \frac{\omega_0}{4} \end{pmatrix}, \tag{4.47}$$

which has eigenvalues given by $\frac{\omega_0}{4}$, $\frac{\omega_0}{4}$, $\frac{\omega_0}{4}$, and $-\frac{3\omega_0}{4}$, the familiar triplet and singlet levels. The eigenstates for the triplet are $|\uparrow\uparrow\rangle$, $\frac{1}{\sqrt{2}}(|\uparrow\downarrow\rangle + |\downarrow\uparrow\rangle)$, $|\downarrow\downarrow\rangle$, while for the singlet the eigenstate is $\frac{1}{\sqrt{2}}(|\uparrow\downarrow\rangle - |\downarrow\uparrow\rangle)$.

Now including the Zeeman terms the Hamiltonian becomes

$$\frac{\hat{\mathcal{H}}}{\hbar} = \begin{pmatrix} \frac{\omega_0}{4} + \omega_- & 0 & 0 & 0 \\ 0 & -\frac{\omega_0}{4} - \omega_+ & \frac{\omega_0}{2} & 0 \\ 0 & \frac{\omega_0}{2} & -\frac{\omega_0}{4} + \omega_+ & 0 \\ 0 & 0 & 0 & \frac{\omega_0}{4} - \omega_- \end{pmatrix}, \tag{4.48}$$

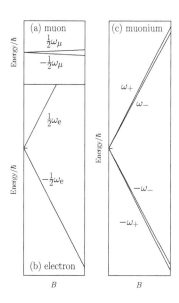

Fig. 4.4 The energy for (a) a muon and (b) an electron in a magnetic field, together with (c) the combined muon and electron system. No interactions between the muon and the electron are included. The gyromagnetic ratio of the muon has been increased by a factor of five so that the small splittings become visible. This plot of frequency (or energy) against magnetic field is known as a Breit-Rabi diagram (see also the note on page 180).

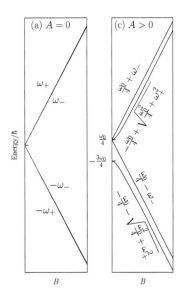

Fig. 4.5 The Breit–Rabi diagram for muonium (a) without interactions ($A = 0$) and (b) with a hyperfine interaction ($A > 0$). In this figure, the correct value of the gyromagnetic ratio has been used (in comparison with Fig. 4.4, where it was increased by a factor of five to make the splittings visible).

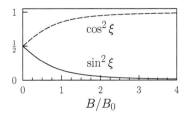

Fig. 4.6 The square of the amplitudes in eqn 4.49 (plotted as $\sin^2 \xi$ and $\cos^2 \xi$ as a function of $x = B/B_0$).

and the eigenvalues can be evaluated as

$$(1)\ \tfrac{\omega_0}{4} + \omega_-; \quad (2)\ \tfrac{\omega_0}{4} + \Omega; \quad (3)\ \tfrac{\omega_0}{4} - \omega_-; \quad (4)\ -\tfrac{3\omega_0}{4} - \Omega.$$

where we have defined $\Omega = \sqrt{(\frac{\omega_0}{2})^2 + \omega_+^2} - \frac{\omega_0}{2}$. These energy eigenvalues are plotted in Fig. 4.5. The eigenstates are given by $|\uparrow\uparrow\rangle$, $\sin\xi\,|\uparrow\downarrow\rangle + \cos\xi\,|\downarrow\uparrow\rangle$, $|\downarrow\downarrow\rangle$, and $\cos\xi\,|\uparrow\downarrow\rangle - \sin\xi\,|\downarrow\uparrow\rangle$, where

$$\sin\xi = \frac{1}{\sqrt{2}}\left(1 - \frac{x}{\sqrt{1+x^2}}\right)^{1/2}, \qquad \cos\xi = \frac{1}{\sqrt{2}}\left(1 + \frac{x}{\sqrt{1+x^2}}\right)^{1/2}.$$
$$(4.49)$$

and $x = 2\omega_+/\omega_0$. These amplitudes change as shown in Fig. 4.6 as a function of field in units of $B_0 = \omega_0/(|\gamma_e| + \gamma_\mu)$ (i.e. $x = B/B_0$).

Example 4.9

How large is B_0? For a muonium atom in vacuum (usually known as **vacuum muonium**), $A = 4.4633\,\mathrm{GHz}$ (see Table 3.1). This gives $B_0 = \omega_0/(2\gamma_+)$ [where $\omega_0 = 2\pi A$ and $\gamma_\pm = \frac{1}{2}(|\gamma_e| \pm \gamma_\mu)$] and hence $B_0 = 0.15853\,\mathrm{T}$.

Before proceeding, it is helpful to write these results out in terms of frequencies ($\frac{\omega}{2\pi}$), rather than angular frequencies (ω). Also, we will define a constant ζ by

$$\zeta = \frac{\gamma_-}{\gamma_+} = 0.99037, \tag{4.50}$$

which is very close to unity, but not precisely unity (the difference is given approximately by $1 - \zeta \approx 2\gamma_\mu/|\gamma_e|$). Hence, using $A = \frac{\omega_0}{2\pi}$, and recalling that $x = B/B_0 = 2\gamma_+ B/\omega_0$, we have the following equalities

$$\nu_+ \equiv \frac{\omega_+}{2\pi} = \frac{Ax}{2}$$
$$\nu_- \equiv \frac{\omega_-}{2\pi} = \frac{A\zeta x}{2}$$
$$\nu_e \equiv \frac{\omega_e}{2\pi} = \frac{A(1+\zeta)x}{2}$$
$$\nu_\mu \equiv \frac{\omega_\mu}{2\pi} = \frac{A(1-\zeta)x}{2}$$
$$\nu_\Omega \equiv \frac{\Omega}{2\pi} = \frac{A}{2}(\sqrt{1+x^2} - 1), \tag{4.51}$$

which are quick to derive. These yield the results in Table 4.2, which can then be plotted in the Breit–Rabi diagram, shown in Fig. 4.7 for the low-field region.

At very high field, the individual muon and electron spin values become good quantum numbers and one might think that there is nothing further of interest. However, when $x = B/B_0$ reaches a value of ≈ 103.34 there is a level crossing. This is explored in the following example.

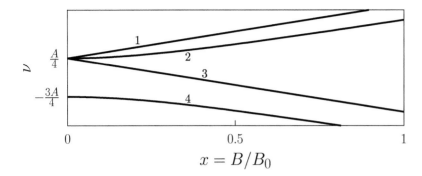

Fig. 4.7 A Breit-Rabi diagram (here shown as a plot of frequency against magnetic field) of the low-field energy levels of muonium.

Energy eigenstate	Frequency	Low-field ($x \ll 1$)
$\lvert 1 \rangle = \lvert \uparrow^\mu \uparrow^e \rangle$	$\nu_1 = \frac{A}{4}(1 + 2\zeta x)$	$\nu_1 = \frac{A}{4} + \frac{\zeta A x}{2}$
$\lvert 2 \rangle = s\lvert \uparrow^\mu \downarrow^e \rangle + c\lvert \downarrow^\mu \uparrow^e \rangle$	$\nu_2 = -\frac{A}{4}(1 - 2\sqrt{1+x^2})$	$\nu_2 = \frac{A}{4} + \frac{A x^2}{4} + \ldots$
$\lvert 3 \rangle = \lvert \downarrow^\mu \downarrow^e \rangle$	$\nu_3 = \frac{A}{4}(1 - 2\zeta x)$	$\nu_3 = \frac{A}{4} - \frac{\zeta A x}{2}$
$\lvert 4 \rangle = c\lvert \uparrow^\mu \downarrow^e \rangle - s\lvert \downarrow^\mu \uparrow^e \rangle$	$\nu_4 = -\frac{A}{4}(1 + 2\sqrt{1+x^2})$	$\nu_4 = -\frac{3A}{4} - \frac{A x^2}{4} + \ldots$

Table 4.2 Energy eigenstates and energy levels (displayed as frequencies, so $\nu_i = E_i/h$) in isotropic muonium. Here $s = \sin\xi$ and $c = \cos\xi$ (see eqn 4.49).

Example 4.10

There is a level crossing between states 1 and 2 because at sufficiently high field $\lvert \downarrow \uparrow \rangle$ should be higher in energy than $\lvert \uparrow \uparrow \rangle$, but at low-field the latter is the highest energy state. The crossing occurs when $E_1 = E_2$ and hence

$$1 + 2\zeta x = -1 + 2\sqrt{1+x^2}, \tag{4.52}$$

which has solutions when $x = 0$ (i.e. at zero-field, the trivial solution) or $x = 2\zeta/(1 - \zeta^2)$ which corresponds to a field

$$B = \frac{2\zeta}{1 - \zeta^2} B_0 \approx 103.34 B_0 = 16.38\,\text{T for vacuum muonium.} \tag{4.53}$$

This is, of course, a very large magnetic field and thus, for vacuum muonium, this level crossing is not very important. However, as we shall see, level crossings are very important in muonium states encountered in experiments which tend to be anisotropic and also have much smaller hyperfine constants.

Having worked out the energy levels in isotropic muonium, it is straightforward to work out the frequency of the transitions *between* these energy levels. These are given by $\nu_{mn} = (E_m - E_n)/h$ and are plotted[17] in Fig. 4.8, this time on a log-log plot.

[17]A simple calculation gives

$$\nu_{12} = \frac{A}{2}\left\lvert 1 + \zeta x - \sqrt{1 + x^2} \right\rvert$$

$$\nu_{14} = \frac{A}{2}(1 + \zeta x + \sqrt{1 + x^2})$$

$$\nu_{23} = \frac{A}{2}(-1 + \zeta x + \sqrt{1 + x^2})$$

$$\nu_{24} = A\sqrt{1 + x^2}$$

$$\nu_{34} = \frac{A}{2}(1 - \zeta x + \sqrt{1 + x^2}), \tag{4.54}$$

and we have omitted ν_{13} since this transition (involving flipping of both the muon and the electron) will not be allowed.

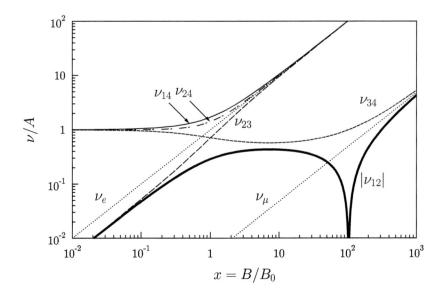

Fig. 4.8 The frequency of the transitions in isotropic muonium as a function of magnetic field. Note the level crossing at x=103.34 where $|\nu_{12}| \to 0$.

Example 4.11

- In the low-field limit ($x \ll 1$) the transitions ν_{14}, ν_{24}, and ν_{34} all $\to A$. The transitions ν_{12} and ν_{23} both $\to \nu_- = \gamma_- B/(2\pi) = \omega_-/(2\pi)$ and this will be 13.9 GHz T^{-1} (or 1.39 MHz G^{-1} in old-fashioned units).

- In the high-field limit ($x \gg 1$) the ν_{14}, ν_{24}, and ν_{23} all $\to \nu_e$ which will be 28.0 GHz T^{-1}. The transitions ν_{12} and ν_{34} both $\to \nu_\mu$ which will be 135.5 MHz T^{-1}.

We have worked out the *frequencies* of all the possible transitions, but that doesn't mean that a particular transition will actually occur. We now need to evaluate what happens in a real experiment and calculate the *amplitudes* of the transitions.

Example 4.12

The muon is in a well-defined state of angular momentum (a pure state) but the electron is unpolarized (a mixed state), and so we can now work out the density matrix and time-evolve it. The density matrix for the combined system is simply the tensor product[18] of the two relevant density matrices, that is of the spin-polarized muon $\rho = \frac{1}{2}(1 + \boldsymbol{\sigma} \cdot \hat{\boldsymbol{n}}) = \begin{pmatrix} 1 & 0 \\ 0 & 0 \end{pmatrix}$ and the unpolarized electron $\frac{1}{2}\begin{pmatrix} 1 & 0 \\ 0 & 1 \end{pmatrix}$, and the result is

$$\rho = \frac{1}{4}(1 + \boldsymbol{\sigma} \cdot \hat{\boldsymbol{n}}) = \frac{1}{2}\begin{pmatrix} 1 & 0 & 0 & 0 \\ 0 & 1 & 0 & 0 \\ 0 & 0 & 0 & 0 \\ 0 & 0 & 0 & 0 \end{pmatrix}. \tag{4.55}$$

[18]If $\underline{\boldsymbol{A}}$ is a $m \times m$ matrix and $\underline{\boldsymbol{B}}$ is a $n \times n$ matrix, the tensor product $\underline{\boldsymbol{A}} \otimes \underline{\boldsymbol{B}}$ is the $mn \times mn$ matrix

$$\begin{pmatrix} A_{11}\underline{\boldsymbol{B}} & \cdots & A_{1n}\underline{\boldsymbol{B}} \\ A_{21}\underline{\boldsymbol{B}} & \cdots & A_{2n}\underline{\boldsymbol{B}} \\ \vdots & \ddots & \vdots \\ A_{n1}\underline{\boldsymbol{B}} & \cdots & A_{nn}\underline{\boldsymbol{B}} \end{pmatrix}.$$

To work out the polarization $P_z(t)$ in a longitudinal-field experiment, in which the muon spin is implanted in the z-direction and measured in the z-direction, we need to evaluate

$$P_z(t) = \mathrm{Tr}(\sigma_\mu^z \rho(t)) = \frac{1}{4}\mathrm{Tr}(\sigma_\mu^z \hat{U}(t)(1 + \sigma_\mu^z)\hat{U}(t)^\dagger). \qquad (4.56)$$

The term involving the factor 1 gives zero trace, and so this simplifies to

$$P_z(t) = \frac{1}{4}\mathrm{Tr}(\sigma_\mu^z \hat{U}(t)\sigma_\mu^z \hat{U}(t)^\dagger). \qquad (4.57)$$

Using the resolution of the identity,[19] and working in the basis of energy eigenstates, we obtain

$$P_z(t) = \frac{1}{4}\sum_{m,n}\langle m|\sigma_\mu^z|n\rangle\langle n|\sigma_\mu^z|m\rangle e^{i\omega_{mn}t} = \frac{1}{4}\sum_{m,n}|\langle m|\sigma_\mu^z|n\rangle|^2 e^{i\omega_{mn}t}, \qquad (4.58)$$

which is of the general form

$$P_z(t) = \sum_{mn} a_{mn}\cos\omega_{mn}t, \qquad (4.59)$$

in other words a sum of terms oscillating at the transition frequencies $\nu_{mn} = \omega_{mn}/(2\pi)$, each with amplitude given by a_{mn}.

For a transverse-field experiment, the muon is implanted again with its spin along the z-direction, but its polarization is monitored along the x-direction. Thus

$$P_x(t) = \mathrm{Tr}(\sigma_\mu^x \rho(t)) = \frac{1}{4}\mathrm{Tr}(\sigma_\mu^x \hat{U}(t)\sigma_\mu^z \hat{U}(t)^\dagger), \qquad (4.60)$$

and we will once again recover an equation of the form $P_x(t) = \sum_{mn} a_{mn}\cos\omega_{mn}t$, but with different amplitudes. This procedure is rather complicated and fortunately the main results have already been worked out and in the next section we will summarize them without deriving them explicitly.

In a longitudinal field experiment, the initial muon polarization is parallel to B. Only the transition $2 \leftrightarrow 4$ is possible (see Fig. 4.9) because with the field along z the spin-polarized muon initially along z involves a superposition of the 2 and 4 states $\sin\xi\,|{\uparrow}{\downarrow}\rangle + \cos\xi\,|{\downarrow}{\uparrow}\rangle$ and $\cos\xi\,|{\uparrow}{\downarrow}\rangle - \sin\xi\,|{\downarrow}{\uparrow}\rangle$. In this case, the muon polarization is given by

$$P_z(t) = (1 - a_{24}) + a_{24}\cos\omega_{24}t, \qquad (4.61)$$

with $a_{24} = \frac{1}{2}\sin^2 2\xi = 1/[2(1+x^2)]$. The oscillatory term is very high frequency and usually hard to observe directly. Therefore, one simply observes the constant $1 - a_{24}$ component and this gives rise to what is known as **repolarization**, with

$$1 - a_{24} = \frac{\frac{1}{2} + x^2}{1 + x^2}, \qquad (4.62)$$

as shown in Fig. 4.10.

Example 4.13

This makes sense physically. The muon is implanted in a state $|{\uparrow}\rangle$ but the electron is unpolarized. In zero-field, half of the time the muonium is formed in a state $|{\uparrow}{\uparrow}\rangle$, which is an eigenstate, giving rise to a polarization of $\frac{1}{2}$, and the other half of the time the muonium is an state $|{\uparrow}{\downarrow}\rangle$ which rapidly oscillates back and forth as $|{\uparrow}{\downarrow}\rangle \leftrightarrow |{\downarrow}{\uparrow}\rangle$ and does not contribute. Thus the measured polarization is 0.5. In high field, both $|{\uparrow}{\uparrow}\rangle$ and $|{\uparrow}{\downarrow}\rangle$ are eigenstates, so whichever way the muon is implanted it keeps its polarization. Thus the measured polarization is 1. This behaviour is shown in Fig. 4.10 and the idea is illustrated further in Fig. 4.11.

[19] $\sum_n |n\rangle\langle n| = 1$. We also use the expression for the trace of a matrix, that $\mathrm{Tr}X = \sum_m \langle m|X|m\rangle$.

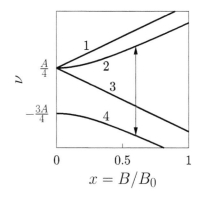

Fig. 4.9 The $2 \leftrightarrow 4$ transition in isotropic muonium.

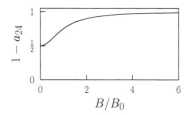

Fig. 4.10 Repolarization of isotropic muonium is described by the function $1 - a_{24}$.

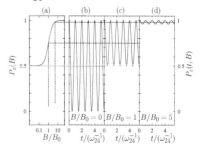

Fig. 4.11 Effect of *repolarization* of the initial muon polarization, shown in (a), the same plot as in Fig. 4.10 but with a logarithmic horizontal axis, is the result of the oscillations plotted in (b), (c), and (d) being averaged out when the hyperfine frequency ω_{24} is too fast to be observed.

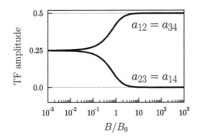

Fig. 4.12 The amplitudes in a transverse-field experiment on isotropic muonium.

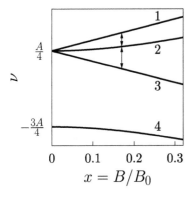

Fig. 4.13 At low field, the $1 \leftrightarrow 2$ and $2 \leftrightarrow 3$ are the only transitions which are easily observable.

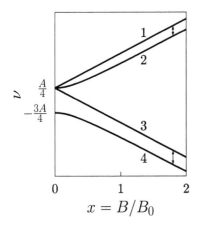

Fig. 4.14 At high field, only the $1 \leftrightarrow 2$ and $3 \leftrightarrow 4$ transitions can be observed.

[20] See Exercise 4.6.

[21] See Exercise 4.6.

In the transverse geometry, the initial muon polarization is perpendicular to B and there are four possible transitions. In this case, the observed polarization is

$$P_x(t) = \sum_{mn} a_{mn} \cos \omega_{mn} t, \tag{4.63}$$

where the four non-zero components are given as follows:

a_{12}	a_{34}	a_{14}	a_{23}
$\frac{1}{2}\cos^2\xi =$	$\frac{1}{2}\cos^2\xi =$	$\frac{1}{2}\sin^2\xi =$	$\frac{1}{2}\sin^2\xi =$
$\frac{1}{4}\left(1 + \frac{x}{\sqrt{1+x^2}}\right)$	$\frac{1}{4}\left(1 + \frac{x}{\sqrt{1+x^2}}\right)$	$\frac{1}{4}\left(1 - \frac{x}{\sqrt{1+x^2}}\right)$	$\frac{1}{4}\left(1 - \frac{x}{\sqrt{1+x^2}}\right)$

The first two components persist to high field, while the final two components disappear at high field (see Fig. 4.12). We can therefore distinguish[20] between two distinct regimes for experiments on isotropic muonium:

- At low transverse magnetic field, all four components are present but only $1 \leftrightarrow 2$ and $2 \leftrightarrow 3$ are at sufficiently low frequency to be observed (see Fig. 4.13), the other two being close to A which is usually too high to easily detect. Those two observable frequency components can be used to estimate A using the following equation:

$$A = \frac{1}{2}\left[\frac{(\nu_{12} + \nu_{23} + 2\nu_\mu)^2}{\nu_{23} - \nu_{12}} + \nu_{12} - \nu_{23}\right]. \tag{4.64}$$

- At high transverse magnetic field, only $1 \leftrightarrow 2$ and $3 \leftrightarrow 4$ are observed since the other transitions have zero amplitude (see Fig. 4.12). These two transitions have different precession frequencies (see Fig. 4.14), but a simple calculation[21] shows that

$$A = \nu_{12} + \nu_{34}. \tag{4.65}$$

An important conclusion can be drawn concerning the high field states in Fig. 4.14. At high field, $\cos\xi \approx 1$ and $\sin\xi \approx 0$ (see eqn 4.49 or Fig. 4.6) and inserting these expressions into those in Table 4.2 tells us that $|2\rangle \approx |{\downarrow}{\uparrow}\rangle$ and $|3\rangle \approx |{\uparrow}{\downarrow}\rangle$ at high field (we know already that $|1\rangle = |{\uparrow}{\uparrow}\rangle$ and $|4\rangle = |{\downarrow}{\downarrow}\rangle$). Thus we note that (in the absence of level crossings, a case we will deal with later) **high field transitions are pure muon spin flip transitions**, i.e. high-field transitions only occur between states of equal electron spin orientation.

In all the calculations performed so far we have assumed that the hyperfine interaction is isotropic. This should be a sensible assumption in the case of vacuum muonium, or for muonium in an approximately spherically symmetric environment, such as the centre of a cage in silicon (see Chapter 10). But what if this assumption does not hold? In the case of muonium forming in the middle of a chemical bond between two atoms, one would expect the hyperfine interaction to be highly anisotropic.

Anisotropic muonium

The muonium Hamiltonian should now be written as[22]

$$\frac{\hat{\mathcal{H}}}{\hbar} = -\gamma_e \mathbf{S}_e \cdot \mathbf{B} - \gamma_\mu \mathbf{S}_\mu \cdot \mathbf{B} + \mathbf{S}_\mu \cdot (2\pi \underline{\mathbf{A}}) \cdot \mathbf{S}_e, \qquad (4.66)$$

where the crucial difference is that the hyperfine constant is now written as a **tensor** $\underline{\mathbf{A}}$. In this equation, we have also chosen the units of the hyperfine interaction to be frequency (hence the factor of 2π) since that will be most useful experimentally. Also, we have now defined the spin operators in units of $\hbar$, so that $\mathbf{S} = \frac{1}{2}\boldsymbol{\sigma}$. We will here just consider the case in which the hyperfine interaction has axial symmetry, so that

$$\underline{\mathbf{A}} = \begin{pmatrix} A_\perp & 0 & 0 \\ 0 & A_\perp & 0 \\ 0 & 0 & A_\parallel \end{pmatrix} = \begin{pmatrix} \nu_0 & 0 & 0 \\ 0 & \nu_0 & 0 \\ 0 & 0 & \nu_0 + \nu^* \end{pmatrix}, \qquad (4.67)$$

where we have chosen the axial direction as the z-direction. Equation 4.67 also shows two different possible ways of writing the parameters of the hyperfine tensor.[23] Other conventions are possible and, for example, it can be useful to write the hyperfine tensor as the sum of an isotropic part and a traceless part, so that[24]

$$\underline{\mathbf{A}} = \begin{pmatrix} A_{\rm iso} & 0 & 0 \\ 0 & A_{\rm iso} & 0 \\ 0 & 0 & A_{\rm iso} \end{pmatrix} + \begin{pmatrix} -D/2 & 0 & 0 \\ 0 & -D/2 & 0 \\ 0 & 0 & D \end{pmatrix}. \qquad (4.68)$$

These parameters are all related to each other so that $\nu_0 = A_\perp = A_{\rm iso} - D/2$ and $\nu^* = A_\parallel - A_\perp = 3D/2$.

[22]Recall that $\omega_e = -\gamma_e B$ and $\omega_\mu = \gamma_\mu B$, the signs chosen because $\gamma_e < 0$.

[23]If angular frequency units are preferred then one can use $\omega_0 = 2\pi\nu_0$ and $\omega^* = 2\pi\nu^*$, thereby absorbing the 2π factor in eqn 4.66.

[24]Yet another convention found in the literature defines $A_s = A_{\rm iso}$ and $A_p = D/2$.

Example 4.14

To solve this model in detail, one can write eqn 4.66 in terms of Pauli spin matrices (as in eqn 4.44). Since the magnetic field is aligned along the z-direction, to keep things completely general we will put the axial direction of the hyperfine tensor along an arbitrary unit vector $\hat{\boldsymbol{n}}$. Hence

$$\frac{\hat{\mathcal{H}}}{\hbar} = \frac{1}{2}\omega_e \sigma_e^z - \frac{1}{2}\omega_\mu \sigma_\mu^z + \frac{1}{4}\omega_0 \boldsymbol{\sigma}_e \cdot \boldsymbol{\sigma}_\mu + \frac{1}{4}\omega^*(\boldsymbol{\sigma}_e \cdot \hat{\boldsymbol{n}})(\boldsymbol{\sigma}_\mu \cdot \hat{\boldsymbol{n}}). \qquad (4.69)$$

This problem can be solved analytically in exactly the same way as we did with the isotropic muonium Hamiltonian. This is just some tedious algebra, so we will not write it out in full; fortunately the solution is available using a variety of computer programs. Suffice to say that the big difference we have when the hyperfine interaction is anisotropic is that the eigenvalues will depend on the *angle* between the magnetic field and the axial direction of the hyperfine tensor. This means that the transverse-field signal from such a state will depend on the angle of the magnetic field and the sample. This will be important when we study semiconductors (see Chapter 10).

(a)   (b)

Isotropic muonium [see Fig. 4.15(a)] is spherically symmetric, while anisotropic muonium [see Fig. 4.15(b)] is not. We will expect to find anisotropic muonium states in semiconductor "bond-centre" states (see

Fig. 4.15 (a) Isotropic muonium. (b) Anisotropic muonium. In each case the positive muon is shown at the centre of the electron cloud.

Chapter 10) or when muonium is attached to a chemical species (see Chapter 12). Sometimes, in addition to the hyperfine interaction, there is a dipolar interaction. The dipolar interactions is a traceless interaction and so in this case the separation of the hyperfine interaction into an isotropic component and a traceless component (as in eqn 4.68) makes a lot of sense.

4.5   Multiple spins

Three spins: muonium plus a proton

Isotropic MuH$^+$ is the prototype of more complex paramagnetic centers covered in Chapter 12, although this simple molecular ion is never observed. Notice that the isotropic condition, instead, is not unfrequent. For instance it is obtained by rapid molecular reorientations in solutions. Once again, the general analytic solution is rather tedious and better performed by computer symbolic manipulation. However, there are a few lessons to learn just from the approximate solution that we sketch here.

The Hamiltonian is (in units of frequency)

$$\frac{\hat{\mathcal{H}}}{h} = \frac{\nu_e}{2}\sigma_e^z - \frac{\nu_\mu}{2}\sigma_\mu^z - \frac{\nu_p}{2}\sigma_p^z + \frac{A_\mu}{4}\boldsymbol{\sigma}_e \cdot \boldsymbol{\sigma}_\mu + \frac{A_p}{4}\boldsymbol{\sigma}_e \cdot \boldsymbol{\sigma}_p, \qquad (4.70)$$

with $\nu_e = -\gamma_e B/(2\pi)$, $\nu_\mu = \gamma_\mu B/(2\pi)$, and $\nu_p = \gamma_p B/(2\pi)$. In the following example we sketch how such a system can be treated with the methods we have developed.

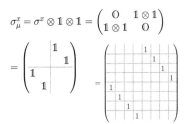

$$\sigma_\mu^x = \sigma^x \otimes 1 \otimes 1 = \begin{pmatrix} 0 & 1 \otimes 1 \\ 1 \otimes 1 & 0 \end{pmatrix}$$

$$\sigma_e^x = 1 \otimes \sigma^x \otimes 1 = \begin{pmatrix} \sigma^x \otimes 1 & 0 \\ 0 & \sigma^x \otimes 1 \end{pmatrix}$$

$$\sigma_p^x = 1 \otimes 1 \otimes \sigma^x = \begin{pmatrix} 1 \otimes \sigma^x & 0 \\ 0 & 1 \otimes \sigma^x \end{pmatrix}$$

Fig. 4.16 Construction of the spin matrices σ_μ^x, σ_e^x, and σ_p^x using the tensor product. They are all $8{\times}8$ matrices (all missing elements are zeros).

Example 4.15

Let us follow the first step of Section 4.4: the addition of three spin-$\frac{1}{2}$ components produces one state with total spin $F = \frac{3}{2}$ (a quadruplet) and two states with total spin $F = \frac{1}{2}$ (two doublets). A quadruplet and two doublets sums to eight levels in total, which is what you would expect for the combination of three spin-$\frac{1}{2}$ angular momenta ($2^3 = 8$). We can also check this by building the tensor product spin matrices by the procedure sketched in Fig. 4.16 (see Exercise 4.9). The resulting Hamiltonian is then separated in blocks corresponding to the three eigenvalues given by

$$\frac{A_\mu + A_p}{4}, \quad -\frac{A_\mu + A_p}{4} \pm \frac{1}{2}\sqrt{A_\mu^2 - A_p^2 - A_\mu A_p}.$$

This simple case is rarely encountered in experiments, because the anisotropy of the coupling discussed in the previous section mixes these states. However, the LF and high TF cases prove to be more robust and the following discussion is important also for more realistic situations.

Let us now add the Zeeman terms $\nu_{\pm\pm} = \nu_e \pm \nu_\mu \pm \nu_p$ and thereby obtain a Hamiltonian, equivalent to eqn 4.48, with two $1{\times}1$ blocks, providing directly two eigenvalues $\pm\frac{A_+}{2} \pm \nu_{--}$ for the states $|\sigma_\mu^z \sigma_e^z \sigma_p^z\rangle = |\uparrow\uparrow\uparrow\rangle, |\downarrow\downarrow\downarrow\rangle$ and a central $6{\times}6$ block. The central block can be simplified greatly by performing linear combinations of its rows and columns: if we simply swap rows 3 and 4, and columns 3 and 4 as well, (in any order, this is globally an even permutation), we can separate the central block of the Hamiltonian into two $3{\times}3$ blocks as follows:

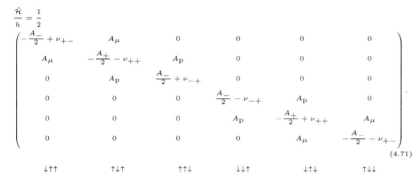

$$(4.71)$$

In eqn 4.71 we have defined $A_\pm = A_\mu \pm A_p$. The bottom labels indicate the states corresponding to each column after we have done the swapping described above. Each of the 3×3 blocks can be diagonalized separately. This is still an tedious task, but it can be accomplished either by means of linear symbolic algebra programmes, or numerically. The result is shown in Fig. 4.17.

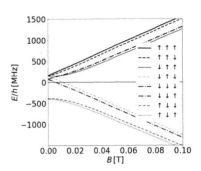

Fig. 4.17 Breit-Rabi diagram (frequency eigenvalues as a function of applied field) for eqn 4.71. The hyperfine frequencies are close to those of the cyclohexadienyl radical (see Chapter 12). In the high-field regime the spin eigenstates corresponds almost exactly to those shown by the arrows. For larger B the four upper levels must cross: eventually the two states where I_μ^z is $\uparrow$ are at higher energy than the two wihere it is $\downarrow$.

[25] The high field regime is known as the **Paschen-Back regime** in atomic physics.

In the high field[25] regime $A_\mu, A_p \ll \nu_{\pm\pm}$, and the off-diagonal matrix elements in eqn 4.71 may be neglected, so that the eigenvalues are closely approximated by the diagonal elements. The high field states in the figure are directly assigned on the basis of the field dependence of their $\nu_{\pm\pm}$ terms.

Under this approximation, whose limits we discuss in Section 4.5, we can now draw two simple conclusions. The first regards the LF geometry. Once again, in high longitudinal field all initial states, whatever the electron and proton spin orientation, are eigenstates. This justifies full repolarization when $\gamma_\mu B$ greatly exceeds A_μ and A_p. The second consideration addresses the high field TF transitions, that are muon spin-flip transitions between states of equal electron and proton spin.[26] In this regime the two upper transitions are (nearly) equal to the Mu^0 transition ν_{12} and the two lower ones to ν_{34} (with $\omega_0 = 2\pi A_\mu$), as it can be checked directly with the difference of the appropriate diagonal elements of the matrix in eqn 4.71. The neglect of the off-diagonal terms is possible also in the presence of more nuclei, and even anisotropies, and therefore both conclusions are widely applicable.

[26] As we pointed out in Section 4.4, high-field transitions observed in μSR are those in which the muon spin flips but the other spins (here, those of the electron and the proton) remain the same.

The F–μ–F state

An interesting effect was discovered[27] in the 1980's concerning muons implanted in non-magnetic inorganic fluorides such as NaF (which contains Na^+ cations and F^- anions). Despite the absence of electronic magnetism in these compounds, a clear and somewhat complex precession signal was observed. The fluorine ion is rather unique in that (i) fluorine is extremely electronegative and thus attractive to the positively-charged muon, (ii) F^- has a very small ionic radius (so that the muon sits very close), (iii) the fluorine nucleus has spin one-half (so that any relaxation retains a simple, yet characteristic, time dependence), and (iv) the fluorine nucleus has a large nuclear moment, associated with an

[27] See J. H. Brewer *et al.*, Phys. Rev. B **33**, 7813(R) (1986).

[28]In these equations, $\boldsymbol{S}_\mu$ is the muon spin operator, $\boldsymbol{S}_{\text{F}i}$ is the spin operator for the ith fluorine, $\boldsymbol{r}_i$ is the vector from μ^+ to the ith fluorine nucleus, and $\boldsymbol{r}_{\text{FF}}$ is the vector between the two fluorine nuclei, and once again we are defining our spin operators in units of $\hbar$. The results described in this section will also apply to a muon coupled with any spin-$\frac{1}{2}$ nucleus (the protons in the hydride ion or an OH^- group would also do, but the dipolar interaction tends to be weaker and the oscillations less clear than in the fluorine case).

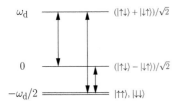

Fig. 4.18 The energy levels and eigenstates for a muon coupled to a single fluorine. The left-hand transition is excited when the muon is polarized along the F–μ^+ bond; the other two transitions are excited when the muon polarization and F–μ^+ bond are perpendicular. A polycrystalline average will give the polarization in eqn 4.80, plotted in Fig. 4.19.

[29]Of course, in a real experiment the muon spin would be held fixed and the direction of $\boldsymbol{r}$ would be varied.

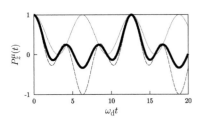

Fig. 4.19 The time-dependent muon polarization for a muon coupled to a single fluorine. The dotted line is for when the muon is polarized along the F–μ^+ bond, the dashed line is for the perpendicular case, and the thick solid line is the polycrystalline average (eqn 4.80).

isotope which occurs with 100% abundance (so that all states behave the same). It was recognised that the effect could be explained if a muon stops between two fluorine ions to form what is known as an F–μ–F state.

This state can be described by the Hamiltonian $\hat{\mathcal{H}}$ given by[28]

$$\hat{\mathcal{H}} = \hat{\mathcal{H}}_{\mu\text{F}} + \hat{\mathcal{H}}_{\text{FF}}, \tag{4.72}$$

the sum of two terms expressing the dipolar interaction between the muon and the fluorine nuclei

$$\hat{\mathcal{H}}_{\mu\text{F}} = \frac{\mu_0}{4\pi} \sum_{i=1}^{2} \frac{\hbar^2 \gamma_\mu \gamma_\text{F}}{r_i^3} \left[\boldsymbol{S}_\mu \cdot \boldsymbol{S}_{\text{F}i} - \frac{3(\boldsymbol{S}_\mu \cdot \boldsymbol{r}_i)(\boldsymbol{S}_{\text{F}i} \cdot \boldsymbol{r}_i)}{r_i^2} \right], \tag{4.73}$$

and the dipolar interaction between the fluorine nuclei

$$\hat{\mathcal{H}}_{\text{FF}} = \frac{\mu_0}{4\pi} \frac{\hbar^2 \gamma_\text{F}^2}{r_{\text{FF}}^3} \left[\boldsymbol{S}_{\text{F}1} \cdot \boldsymbol{r}_{\text{FF}})(\boldsymbol{S}_{\text{F}2} \cdot \boldsymbol{r}_{\text{FF}}) \right]. \tag{4.74}$$

Example 4.16

Consider a muon coupled to a *single fluorine*. The Hamiltonian is then simply

$$\hat{\mathcal{H}}_{\mu\text{F}} = \frac{\mu_0 \hbar^2 \gamma_\mu \gamma_\text{F}}{4\pi r^3} [\boldsymbol{S}_\mu \cdot \boldsymbol{S}_\text{F} - 3(\boldsymbol{S}_\mu \cdot \hat{\boldsymbol{r}})(\boldsymbol{S}_\text{F} \cdot \hat{\boldsymbol{r}})]$$

$$= \frac{\mu_0 \hbar^2 \gamma_\mu \gamma_\text{F}}{4\pi r^3} \frac{1}{4} [\boldsymbol{\sigma}_\mu \cdot \boldsymbol{\sigma}_\text{F} - 3(\boldsymbol{\sigma}_\mu \cdot \hat{\boldsymbol{r}})(\boldsymbol{\sigma}_\text{F} \cdot \hat{\boldsymbol{r}})], \tag{4.75}$$

where $\boldsymbol{r}$ is the vector separating the fluorine from the muon and $\hat{\boldsymbol{r}} = \boldsymbol{r}/|\boldsymbol{r}|$. Setting the characteristic energy $\hbar\omega_\text{d}$ as

$$\hbar\omega_\text{d} = \frac{\mu_0 \hbar^2 \gamma_\mu \gamma_\text{F}}{4\pi r^3}, \tag{4.76}$$

we can write the Hamiltonian, with $\hat{\boldsymbol{r}} = \hat{\boldsymbol{z}}$, as

$$\hat{\mathcal{H}} = \frac{\hbar\omega_\text{d}}{2} \begin{pmatrix} -1 & 0 & 0 & 0 \\ 0 & 1 & 1 & 0 \\ 0 & 1 & 1 & 0 \\ 0 & 0 & 0 & -1 \end{pmatrix}, \tag{4.77}$$

which has energy eigenvalues $-1, -1, 0, 2$ multiplied by $\hbar\omega_\text{d}/2$ as shown in Fig. 4.18. We can now follow through the analysis used for muonium to show that if the muon is implanted with its spin along the z-direction then its polarization will be

$$P_z(t) = \frac{1}{2}(1 + \cos\omega_\text{d}t), \tag{4.78}$$

whereas if its spin is along the x-direction then

$$P_x(t) = \frac{1}{2}\left(\cos\frac{\omega_\text{d}t}{2} + \cos\frac{3\omega_\text{d}t}{2}\right). \tag{4.79}$$

In a real experiment on a polycrystalline material one would need to average over all angles.[29] The result is

$$\langle P_z(t) \rangle = \frac{1}{3}P_z(t) + \frac{2}{3}P_x(t) = \frac{1}{6}(1 + \cos\omega_\text{d}t) + \frac{1}{3}\left(\cos\frac{\omega_\text{d}t}{2} + \cos\frac{3\omega_\text{d}t}{2}\right), \tag{4.80}$$

and is plotted in Fig. 4.19.

A similar analysis can be performed for the F–μ–F state, but with an extra spin-$\frac{1}{2}$ species the Hilbert space increases in size and the Hamiltonian is an 8×8 matrix, rather than a 4×4 matrix.

Example 4.17

If we ignore the interaction between the two fluorine moments, then eigenvalues are -1, $\frac{1-\sqrt{3}}{2}$, 0, and $\frac{1+\sqrt{3}}{2}$, all in units of $-\omega_d$ and twice-repeated, so the energy levels are four doublets, as shown in Fig. 4.20. The polycrystalline average polarization is then (see Exercise 4.13)

$$\langle P_z(t) \rangle = \frac{1}{6} \left\{ 3 + \cos(\sqrt{3}\omega_d t) + \left(1 - \frac{1}{\sqrt{3}}\right) \cos\left[\left(\frac{3-\sqrt{3}}{2}\right)\omega_d t\right] \right.$$
$$\left. + \left(1 + \frac{1}{\sqrt{3}}\right) \cos\left[\left(\frac{3+\sqrt{3}}{2}\right)\omega_d t\right] \right\}, \tag{4.81}$$

If the interaction between the two fluorine moments is also included then the factors in this expression change a little bit (and the expression doesn't look as pretty). Ignoring the fluorine-fluorine interaction is not too bad an approximation since the fluorine nuclear moment is smaller than the muon moment, and also there is an additional factor of $1/8$ due to the fact that for F–μ–F we have $r_{FF} = r_{F\mu}/2$ and there is a $1/r^3$ factor in the dipolar interaction. The final result is plotted in Fig. 4.21.

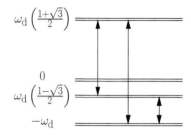

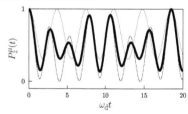

Fig. 4.20 The four doublet energy levels for the F–μ–F state, together with the possible transitions. These eigenvalues ignore the effect of the fluorine-fluorine interaction.

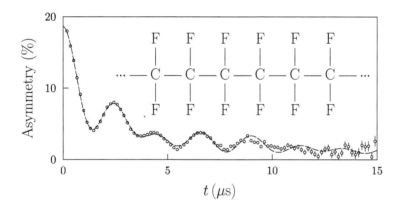

Fig. 4.21 The time-dependent muon polarization for a F–μ–F state. The dotted line is for when the muon is polarized along the F–μ–F bond, the dashed line is for the perpendicular case, and the thick solid line is the polycrystalline average (eqn 4.81).

Fig. 4.22 Low-temperature muon spectrum in PTFE showing a characteristic F–μ–F signal.

The characteristic F–μ–F signal shown in Fig. 4.21 is found in many inorganic fluoride compounds, as well as in teflon (polytetrafluoroethylene, or PTFE) which is a fluoropolymer. An example of the latter is shown in Fig. 4.22 which shows the oscillatory signal expected from an F–μ–F state (even though of course there is no magnetic order in this material). Teflon consists of long helical chains of repeating CF_2 units (with successive CF_2 units rotated slightly due to the steric interference of adjacent fluorine atoms) and a typical piece of teflon is a mixture of crystalline and amorphous regions. Nevertheless, the muon is able to form states in which it is positioned between two adjacent fluorines in

[30]See J. M. Wilkinson and S. J. Blundell, Phys. Rev. Lett. **125**, 087201 (2020).

the material. Note that in data on real materials (and in contrast to the simulations in Fig. 4.21) there is also an additional relaxation of the F–μ–F signal which is attributed to the effects of additional couplings that have been neglected. The coupling between the muon and the two nearest-neighbour fluorines is by far the largest coupling in the problem, but the coupling to slightly more distant nuclei is not zero.[30]

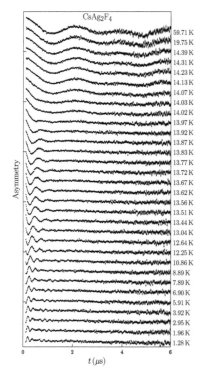

Fig. 4.23 Zero-field muon-spin rotation data on Cs_2AgF_4 [from T. Lancaster *et al.*, Phys. Rev. B **75**, 220408(R) (2007)].

Example 4.18

High-T_c cuprates share a layered structure of $[CuO_2]$ planes with strong antiferromagnetic interactions between $S = 1/2$ $3d^9$ Cu^{2+} ions. It is interesting to test whether Cu can be replaced with another similar element. Moving down the periodic table one step leads to silver, the $4d$ analogue of Cu^{2+}, namely $S = \frac{1}{2}$ $4d^9$ Ag^{2+}. This motivated the synthesis of the layered fluoride Cs_2AgF_4 which contains silver in the unusual divalent oxidation state. It looks rather similar to the superconducting parent compound La_2CuO_4; it is comprised of planes of $[AgF_2]$ instead of $[CuO_2]$ separated by planes of $[CsF]$ instead of $[LaO]$. Both Cs and Ag strongly absorb neutrons, resulting in limited resolution and a poor signal-to-noise ratio in such experiments. In contrast, spin-polarized muons are ideal probes for studying this material.

μSR data on this compound are shown in Fig. 4.23 and demonstrate that the sample is magnetic below about 14 K. The magnetism gives rise to the rapid oscillations which are visible in the data below 14 K. We will return to the magnetic properties of this material in Chapter 6. However, here notice that there are some low-frequency oscillations visible at higher temperature, the frequency of which does not seem to change much with temperature. These arise from the F–μ–F state which is formed in this compound and gives information about the muon site.

Note that the F–μ–F oscillations are only visible when the long-range order from the magnetic sites has disappeared, i.e. at high temperature. This is because the internal field of electron magnetism is much stronger than the weak nuclear interaction which gives rise to the F–μ–F state. The muon therefore responds to the internal field of the magnetic order rather than to the weak dipole-dipole interaction with the nearby fluorine ions.

Avoided level crossings

After solving a complicated Hamiltonian where the muon couples with the nearby nuclei, or unpaired electrons, and all these particles couple with the applied magnetic field, we end up with many different energy levels which move in complicated ways with the magnetic field. The energy versus field plot can sometimes resemble the pattern of railway lines at the entrance to a major railway station! As these energy levels weave in and out of each other, they sometimes cross each other. At these crossing points, interesting things may happen. To study this, let's just focus on two energy levels which, as a function of magnetic field, cross at some particular magnetic field which we will call B^* corresponding to energy E_0. Therefore, we can write the Hamiltonian near this special point (focusing on just these two energy levels) as a 2×2 matrix, so that

$$\hat{\mathcal{H}} = \begin{pmatrix} E_0 + \alpha(B - B^*) & 0 \\ 0 & E_0 + \beta(B^* - B) \end{pmatrix}, \tag{4.82}$$

where α and β are some positive real constants. This Hamiltonian obviously has energy eigenvalues $E = E_0 + \alpha(B - B^*)$ and $E = E_0 + \beta(B^* - B)$. These two energies become equal at $B = B^*$, where they both take the value E_0, as shown in Fig. 4.24(a), where there is a **level crossing**. However, the two energy levels just pass through each other and nothing interesting happens.[31]

However, we have not treated the most general case since the general form of a Hermitian 2×2 matrix can also contain off-diagonal elements.[32] These can result from other terms in the Hamiltonian, perhaps relating to couplings with some of the other levels which we have ignored by focusing on just two levels. These additional interactions lead to a mixing between our two states. Therefore, a more general form of our Hamiltonian is

$$\hat{\mathcal{H}} = \begin{pmatrix} E_0 + \alpha(B - B^*) & \delta \\ \delta^* & E_0 + \beta(B^* - B) \end{pmatrix}, \qquad (4.83)$$

where δ is some complex number. The energy eigenvalues are now

$$E_\pm = E_0 + \frac{\alpha - \beta}{2}(B - B^*) \pm \sqrt{\left(\frac{\alpha + \beta}{2}(B - B^*)\right)^2 + |\delta|^2}, \qquad (4.84)$$

and in particular, at $B = B^*$ this reduces to

$$E_\pm = E_0 \pm |\delta|. \qquad (4.85)$$

This is now an **avoided level crossing**, as shown in Fig. 4.24(b), and the two energy levels now appear to repel each other close to the crossing.

An important feature of an avoided level crossing is that the nature of the state changes near the avoided crossing.

[31] In particular, note that the eigenvectors in this basis remain as $\begin{pmatrix} 1 \\ 0 \end{pmatrix}$ and $\begin{pmatrix} 0 \\ 1 \end{pmatrix}$ and so these two states remain independent as the levels cross.

[32] The off-diagonal elements must be complex conjugates of each other, to make sure the Hamiltonian is Hermitian.

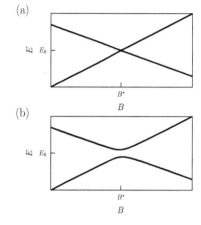

Fig. 4.24 (a) Two levels crossing as a function of field at the position $B = B^*$ and $E = E_0$. (b) An avoided level crossing.

Example 4.19

Far from the level crossing, the two energy levels are independent and thus have their separate identities. The eigenstates will be very close to $|a\rangle = \begin{pmatrix} 1 \\ 0 \end{pmatrix}$ and $|b\rangle = \begin{pmatrix} 0 \\ 1 \end{pmatrix}$, just as we had when there was no off-diagonal coupling term. However, because of the avoided crossing, the two curves now have to swap identities. Thus near the crossing, the states will become mixtures of the two states $|a\rangle$ and $|b\rangle$ (see Fig. 4.25).

For example, let us assume that the muon is coupled to a number of nearby nuclear spins, but that the field is high enough that all of the individual spin states are good quantum numbers. Therefore the energy levels of the system will be of the form such as $|\uparrow\downarrow\downarrow\uparrow\ldots\rangle$, i.e. the spin of each species is well defined. We will call the first spin in this state the muon spin, and let's look at a case where the two levels that cross are

$$|a\rangle = |\uparrow\ldots\downarrow\ldots\rangle \quad \text{and} \quad |b\rangle = |\downarrow\ldots\uparrow\ldots\rangle, \qquad (4.86)$$

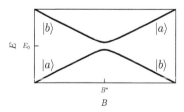

Fig. 4.25 At an avoided level crossing, the two states become mixtures of the states $|a\rangle$ and $|b\rangle$ which are well-defined away from the crossing.

where we are focusing our attention on the muon spin and some other spin, and assuming that all others do not change. Far from the level crossing, the muon spin is a good quantum number and so a longitudinal-field experiment would show the muon retains its full polarization (we implant the muon in the $|\uparrow\rangle$ state, which is associated with $|a\rangle$ and not $|b\rangle$, and this is a stationary state[33] of the system). However, near the level crossing, the stationary states will be mixtures of $|a\rangle$ and $|b\rangle$, and thus the muon polarization will be reduced. This is because even though we implant the muon spin in the $|\uparrow\rangle$ state, this is no longer associated with a stationary state of the system. Transitions between $|a\rangle$ and $|b\rangle$ are now possible, which in this case are flip-flop transitions involving the other nuclear spin which flips from down to up when the muon flops from up to down, and *vice versa*. This leads to an **avoided level crossing resonance**, a topic which we will explore further for chemical systems in Chapter 12.

A very good example of a level crossing resonance involves a quadrupolar nucleus. Nuclei with a spin quantum number I which is larger than $\frac{1}{2}$ have a non-spherical charge distribution of charge and this gives rise to a **quadrupole moment**. An electric charge interacts with an electric field, but an electric quadrupole moment interacts with an electric field gradient. An implanted μ^+ will produce an electric field together with an electric field gradient around itself and this can interact with the quadrupolar moments of nearby nuclei.

Example 4.20

An example of a quadrupolar level crossing resonance occurs in elemental copper (Cu). Muons sit at the centre of an octahedron inside the face-centred cubic lattice of Cu and the Cu atoms each have a nucleus with a spin-3/2 moment. (69% of the atoms are ^{63}Cu and 31% are ^{65}Cu, but both isotopes are spin-3/2 with a very similar magnetic moment.) The implanted muon causes an electric field gradient to act outwards on the six nearest-neighbouring Cu atoms and interacts with their quadrupolar moments. In this example, we simplify the calculation to deal with a single Cu atom.

The energy levels of a spin-3/2 nucleus in a magnetic field are shown in Fig. 4.26(a), illustrating the Zeeman splitting into four separate levels. In the presence of an electric field gradient these levels are split into two doublets, as shown in Fig. 4.26(b). The quadrupolar Hamiltonian is given by

$$\hat{\mathcal{H}}_Q = \hbar\omega_Q \left[(\boldsymbol{n} \cdot \boldsymbol{I})^2 - \frac{I(I+1)}{3} \right], \tag{4.87}$$

where $\boldsymbol{n}$ is the direction of the electric-field gradient eq and the quadrupolar frequency ω_Q is given by $\omega_Q = 3e^2qQ/4\hbar I(2I-1)$ where Q is the quadrupole moment. The quadrupole moment in Cu is negative, and so the $\pm\frac{3}{2}$ levels are lowered in energy and the $\pm\frac{1}{2}$ levels are raised in energy.

If we consider the muon by itself, then its energy levels would be as shown in Fig. 4.26(c). There is now a magnetic field, given by $B_{\text{cross}} = 2|\omega_Q|/(\gamma_\mu - \gamma_{\text{Cu}}) \approx 0.0085$ T at which the gap between the two muon levels precisely match the gap between the $|3/2\rangle$ and $|1/2\rangle$ levels of the Cu nucleus. This means that the two spins can now "talk to each other"; we can imagine flipping the muon spin and using the energy released to cause the Cu nucleus to change state, and this joint process would involve no net input or output of energy. This means that at this magnetic field we would expect the energy levels of the joint system to cross.

Now let's consider this joint system, containing the Cu nucleus and the muon in a magnetic field, including the quadrupolar splitting for the Cu nucleus which is caused by the electric field gradient from the muon. This will consist of eight energy levels coming from the four Cu nuclear levels multiplied by the two muon levels (for each Cu energy level we have two possible states for the muon). However, there will also be an additional term in the Hamiltonian which comes from the dipolar coupling $\hat{\mathcal{H}}_{\mathrm{dip}}$ between the muon and the Cu nucleus given by

$$\hat{\mathcal{H}}_{\mathrm{dip}} = \hbar\omega_{\mathrm{d}}\left[\boldsymbol{S}_\mu \cdot \boldsymbol{I} - 3(\boldsymbol{S}_\mu \cdot \boldsymbol{n})(\boldsymbol{I} \cdot \boldsymbol{n})\right]. \qquad (4.88)$$

This is a very small effect and so it is negligible almost everywhere. Although $\hat{\mathcal{H}}_{\mathrm{dip}}$ breaks some degeneracies near $B = 0$, at higher field we expect the system to be in the same eigenstates, with the muon and Cu spins along the direction of the applied field, apart from a small range of fields around B_{cross}.

The energy levels of this joint system are plotted in Fig. 4.26(d) (for this plot, we have chosen the applied field to be along the z-direction and the muon and Cu nucleus to be separated also along the z-direction) and indeed, as expected, there is a crossing of energy levels at the field value ≈ 0.0085 T. However, close inspection of Fig. 4.26(d) reveals that it is an avoided crossing due to the additional dipolar interaction in the problem. This dipolar interaction contains flip-flop terms which remove the degeneracy between the $|-\frac{1}{2}; +\frac{3}{2}\rangle$ and $|+\frac{1}{2}; +\frac{1}{2}\rangle$ states (there are other level crossings in the energy level scheme, but for no others are the degeneracies removed by the dipolar coupling; see Exercise 4.10). The avoided level crossing resonance that occurs at this field can then be detected in experiments on Cu at this field.[34]

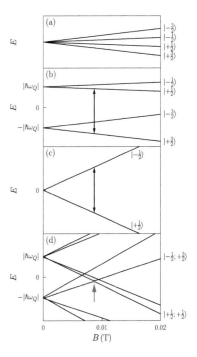

Fig. 4.26 (a) The energy levels in an $I = 3/2$ nucleus in a magnetic field. (b) Including the quadrupolar interaction (with a negative quadrupole moment) the $\pm 1/2$ levels are raised in energy and the $\pm 3/2$ levels are lowered in energy. (c) The muon energy levels in a magnetic field are shown on the same scale. (d) The energy levels for a dipolar coupled $I = \frac{3}{2}$ nucleus and the muon. The arrow in panel (d) points at the avoided level crossing at $B_{\mathrm{cross}} \approx 0.0085$ T and possible transitions at this field are shown in panels (b) and (c).

[34]See G. M. Luke *et al.*, Phys. Rev. B **43**, 3284 (1991).

Chapter summary

- A muon spin in a magnetic field undergoes spin precession at an angular frequency $\omega = \gamma_\mu B$. This can be treated classically or quantum mechanically by a variety of methods.

- Muonium has to be treated quantum mechanically and the energy levels and transitions have been worked out as a function of magnetic field. In anisotropic muonium these energy levels and transitions depend on the angle between the magnetic field and the axis of the hyperfine tensor.

- Another manifestation of the quantum nature of the muon is revealed when the muon couples to two nearby fluorine nuclei in an F–μ–F state, resulting in a characteristic oscillatory polarization.

- When two energy levels cross as a function of magnetic field, it is possible to get an avoided level crossing due to other interactions in the system.

Further reading

- A good review of the quantum mechanical behaviour of isotropic and anisotropic muonium states can be found in B. D. Patterson, Rev. Mod. Phys. **60**, 69 (1988), and see also E. Roduner and H. Fischer, Chem. Phys. **54**, 261 (1981) for an excellent treatment of the quantum behaviour of muonium states in free radicals.

• Many aspects of the quantum-mechanical treatment of relaxation functions are reviewed in Chapters 3, 4, 7, and 8 of A. Yaouanc and P. Dalmas de Réotier, *Muon Spin Rotation, Relaxation, and Resonance: Applications to Condensed Matter*, OUP (2011).

• The theory of avoided level crossing resonances can be found in M. Heming *et al.* Chem. Phys. Lett. **128**, 100 (1986) and S. R. Kreitzman and E. Roduner, Chem. Phys. **192**, 189 (1995).

Exercises

(4.1) By evaluating $d|\boldsymbol{\mu}|^2/dt$ and $d(\boldsymbol{\mu} \cdot \boldsymbol{B})/dt$, show that eqn 4.6 implies that both $|\boldsymbol{\mu}|$ and $\boldsymbol{\mu}\cdot\boldsymbol{B}$ must be time-independent. These conditions are satisfied by the solution in eqn 4.8.

(4.2) A muon is implanted with its spin parallel to the z-axis. A magnetic field B is applied at an angle θ to the z-axis so that $\boldsymbol{B} = (B\sin\theta, 0, B\cos\theta)$. Show that the Hamiltonian is

$$\hat{\mathcal{H}} = -\frac{\gamma_\mu B}{2}\begin{pmatrix} \cos\theta & \sin\theta \\ \sin\theta & -\cos\theta \end{pmatrix}, \qquad (4.89)$$

and hence that the time-evolution operator $\hat{U}(t)$ is given by

$$\begin{pmatrix} \cos\frac{\omega t}{2} + i\cos\theta\sin\frac{\omega t}{2} & i\sin\theta\sin\frac{\omega t}{2} \\ i\sin\theta\sin\frac{\omega t}{2} & \cos\frac{\omega t}{2} - i\cos\theta\sin\frac{\omega t}{2} \end{pmatrix}, \qquad (4.90)$$

where $\omega = \gamma_\mu B$. Thus deduce that

$$|\psi(t)\rangle = \begin{pmatrix} \cos\frac{\omega t}{2} + i\cos\theta\sin\frac{\omega t}{2} \\ i\sin\theta\sin\frac{\omega t}{2} \end{pmatrix}, \qquad (4.91)$$

and hence that

$$\langle\sigma_z(t)\rangle = \cos^2\theta + \sin^2\theta\cos\omega t. \qquad (4.92)$$

This result is equivalent to eqn 4.20 and will end up being very useful in the next chapter. Try proving this result again, but this time using density matrices.

(4.3) Show that even powers of a general Pauli spin matrix are given by the identity matrix, and hence that odd powers of a general Pauli spin matrix are given by the same Pauli spin matrix. Use this result to prove eqn 4.21.

(4.4) Diagonalise the matrix in eqn 4.30 and show that the eigenstate associated with eigenvalue $+1$ is that given by eqn 4.31.

(4.5) Show that the transition frequencies for isotropic muonium are given by

$$\begin{aligned} \omega_{12} &= \omega_- - \Omega \\ \omega_{23} &= \omega_- + \Omega \\ \omega_{14} &= \omega_0 + \omega_- + \Omega \\ \omega_{34} &= \omega_0 - \omega_- + \Omega. \end{aligned} \qquad (4.93)$$

(4.6) At low applied field, the only transitions that we are likely to observe in isotropic muonium are ω_{12} and ω_{23}. Show that using these we can derive the value of the hyperfine constant expressed as a frequency $A = \omega_0/(2\pi)$ using the frequencies $\nu_{12} = \omega_{12}/(2\pi)$ and $\nu_{23} = \omega_{23}/(2\pi)$ according to the equation

$$A = \frac{1}{2}\left[\frac{(\nu_{12}+\nu_{23}+2\nu_\mu)^2}{\nu_{23}-\nu_{12}} + \nu_{12} - \nu_{23}\right]. \qquad (4.94)$$

At high magnetic field the only transitions that we are likely to observe are ω_{12} and ω_{34}. Show that these are related by

$$A = \nu_{12} + \nu_{34}. \qquad (4.95)$$

Note that in the very high field regime, levels 1 and 2 swap round above the level crossing, and so experimentally one sometimes needs to take the difference between the frequencies to obtain the hyperfine constant.

(4.7) Show that for isotropic muonium $a_{24} = \frac{1}{2}\sin^2 2\xi = 1/[2(1+x^2)]$ and hence derive eqn 4.62.

(4.8) Write the unit vector in spherical polars so that $\boldsymbol{n} = (\sin\theta\cos\phi, \sin\theta\sin\phi, \cos\theta)$ and hence show that

$$\boldsymbol{n}\cdot\underline{\boldsymbol{A}}\cdot\boldsymbol{n} = A_{\text{iso}} + \frac{D}{2}(3\cos^2\theta - 1), \qquad (4.96)$$

where $\underline{\boldsymbol{A}}$ is as defined in eqn 4.68.

(4.9) (a) Show that adding $I_\mu = \frac{1}{2}$, $S_e = \frac{1}{2}$, $I_p = \frac{1}{2}$ one obtains $F = \frac{3}{2}, \frac{1}{2}, \frac{1}{2}$.

(b) Build the Hamiltonian $\frac{\hat{\mathcal{H}}}{h}$ of eqn 4.71 for $B = 0$ starting from eqn 4.70, with the convention of Fig. 4.16 for the spin operators. These are tensor spin operators. For instance, the z terms in the muon hyperfine coupling is often written $\frac{A_\mu}{4} \sigma_p^z \otimes \sigma_e^z \otimes \hat{I}_p$. Check that the three tensor σ_z operators are 8×8 diagonal matrices with elements $(1, -1, 1, -1, 1, -1, 1, -1)$, $(1, 1, -1, -1, 1, 1, -1, -1)$, $(1, 1, 1, 1, -1, -1, -1, -1)$ for the muon, the electron, and the proton, respectively. Obtain the eigenvalues.

(c) Obtain eqn 4.71 for $B > 0$ with the tensor spin operators.

(4.10) Label the energy levels in Fig. 4.26(d) (only two are labelled in the figure) and explain why only the crossing between the $|-\frac{1}{2}; +\frac{3}{2}\rangle$ and $|+\frac{1}{2}; +\frac{1}{2}\rangle$ levels is avoided. Show that this crossing takes place at a field given by $B = 2|\omega_Q|/(\gamma_\mu - \gamma_{\mathrm{Cu}})$.

(4.11) If a muon, initially polarized along the x-direction, is coupled to various other spins then show that its density matrix can be written $\rho = (\mathbb{1} + \sigma_\mu^x)/\mathrm{Tr}\mathbb{1}$, where $\mathbb{1}$ is the identity matrix in the appropriate Hilbert space. By using

$$P_z(t) = \langle \sigma_\mu^z(t) \rangle = \mathrm{Tr}(\sigma_\mu^z \rho(t)), \qquad (4.97)$$

show that

$$P_z(t) = \frac{\mathrm{Tr}(\sigma_\mu^z \sigma_\mu^x(t))}{\mathrm{Tr}(\mathbb{1})} = \frac{\mathrm{Tr}(\sigma_\mu^z e^{i\hat{\mathcal{H}}t/\hbar} \sigma_\mu^x e^{-i\hat{\mathcal{H}}t/\hbar})}{\mathrm{Tr}(\mathbb{1})}, \qquad (4.98)$$

and explain the possible relevance to a transverse-field measurement.

(4.12) For a zero- or longitudinal-field experiment, the muon is initially polarized along the z-direction. Using eqn 4.97 find an equation analogous to eqn 4.98 relevant to such a measurement.

(4.13) Derive eqn 4.81. Be warned – this is quite an involved calculation! For enthusiasts only.

5

Polarization functions

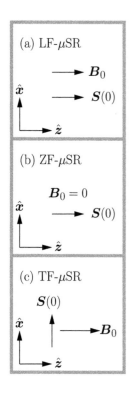

Fig. 5.1 The geometries of the experiments showing the direction of the applied field B_0 and initial muon spin direction $S(t = 0)$.

In the last chapter we saw how a muon at rest interacts magnetically with its local environment. Our next task is to use this insight to predict how the average spin polarization of a large number of stopped muons evolves with time. This is important as: (i) the **polarization function** that describes this time evolution depends sensitively on the nature of the material into which we've implanted our muons, and (ii) the asymmetry $A(t)$ that we measure in a μSR experiment is directly proportional to the polarization function. In short, if we can describe our data using a polarization function, we can extract information on the material in question.

More formally, the polarization function $P_\alpha(t)$ describes the evolution of the average muon spin projected along the α-axis, or

$$P_\alpha(t) = \langle \sigma_\alpha(t) \rangle, \tag{5.1}$$

where σ_α is the Pauli spin operator for the α direction. The average here is performed over the ensemble of muons implanted during the experiment. The underlying reason for the time evolution is, classically speaking, that an individual spin will precess in a perpendicular magnetic field, so that the polarization of the muon spin oscillates. The process of averaging over many muon precessions can wash out the oscillatory signal leading to relaxation of the polarization. In this chapter we describe the derivations of the key polarization functions that are commonly used to fit data. The models we shall discuss apply to diamagnetic muon states. Our task here is to provide the mathematical basis for computing the relaxation functions. In the following chapters, we will discuss the physical basis for the models.

As we have seen, two geometries are used in experiments: the **transverse-field** (TF) geometry and the **zero-field** (ZF) or (equivalently) the **longitudinal-field** (LF) geometry. These are shown in Fig. 5.1. The **laboratory reference frame** is defined such that the z-axis is parallel to the externally applied magnetic field B_0. In the transverse-field geometry [Fig. 5.1(c)] the direction of the muon polarization at an initial time $t = 0$, which is by definition perpendicular to $\hat{z}$, fixes the direction of x-axis. The polarization function here specifies the evolution of the muon-spin polarization projected along this x-axis and is denoted $P_x(t)$. In the zero- [Fig. 5.1(b)] or longitudinal-field geometry [Fig. 5.1(a)] the muon polarization is initially along the z-axis, and the function $P_z(t)$, describing the projection along the z-axis at time t, is of interest.

In this book we shall discuss two approaches to describing the average muon-spin polarization. The first, which is the subject of this chapter, is a stochastic approach: the ensemble of muons implanted in the material under study is assumed to probe a distribution of magnetic fields, and the evolution of the muon spin is computed using (classical) statistical physics. The fields can remain constant, or they can evolve during the time spent by a given muon in the system. The second approach is quantum mechanical, where the starting point is the Hamiltonian describing the spin degrees of freedom of the system and their interactions with the muon spin. We postpone the discussion of the quantum-mechanical approach until Chapter 7 and Appendix F.

5.1 Static fields

Our task is to compute the polarization function assuming that the ensemble of muons experiences a distribution of magnetic fields. The magnetic field[1] $\boldsymbol{B}$ at the muon position results from the vector sum of any applied field $\boldsymbol{B}_0$ added to the field produced by the system in which the muon is implanted, which stems from nuclei and unpaired electrons. If the magnetic field remains static during the time spent by the muon in the system, the evolution of the muon spin is given by the solution of the equations of motion discussed in Chapter 4. In fact, for the computation of $P_\alpha(t)$ (where $\alpha = x$ or z), we are interested in the component $\sigma_\alpha(t)$ of the muon spin polarization. Defining the polar angle between $\boldsymbol{B}$ and the axis α as θ (see Fig. 5.2), eqn 4.20 results in an expectation value for a single muon in a field $\boldsymbol{B}$ of

$$\langle \sigma_\alpha(t) \rangle = \cos^2\theta + \sin^2\theta \cos(\gamma_\mu Bt), \tag{5.2}$$

where the muon's spin vector has components $S_\alpha(t) = \frac{\hbar}{2}\sigma_\alpha(t)$ (see Chapter 4). In the stochastic approach, the polarization function $P_\alpha(t)$ is given by this expectation value, averaged over the distribution of magnetic fields experienced by the ensemble of muons, and written

$$P_\alpha(t) = \int \mathrm{d}^3 B \, p(\boldsymbol{B}) \langle \sigma_\alpha(t) \rangle, \tag{5.3}$$

where, in Cartesian coordinates, $\mathrm{d}^3 B = \mathrm{d}B_x \mathrm{d}B_y \mathrm{d}B_z$. The quantity $p(\boldsymbol{B})$ is the **probability density function**, i.e. $p(\boldsymbol{B}) \, \mathrm{d}^3 B$ is the probability that the local field at the muon position takes the vector value $\boldsymbol{B}$ in range $\mathrm{d}^3 B$, or, in components, the probability of the field vector being found between (B_x, B_y, B_z) and $(B_x + \mathrm{d}B_x, B_y + \mathrm{d}B_y, B_z + \mathrm{d}B_z)$. Naturally the probability density function is normalized such that $\int p(\boldsymbol{B}) \, \mathrm{d}^3 B = 1$.

[1] Notation: we will write the magnitude of a vector field $\boldsymbol{A}$ (such as the B-field $\boldsymbol{B}$ or magnetization $\boldsymbol{M}$) as

$$|\boldsymbol{A}| = A,$$

a unit vector as $\hat{\boldsymbol{x}}$ and the components of a tensor $\underline{\boldsymbol{D}}$ as D_{ij}.

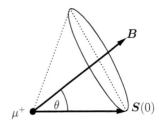

Fig. 5.2 The muon spin $\boldsymbol{S} = \frac{\hbar}{2}\boldsymbol{\sigma}$ initially lies at an angle θ to the direction of the local magnetic field $\boldsymbol{B}$. In experiments, we are usually interested in muon spins initially directed along $\alpha = z$ or x.

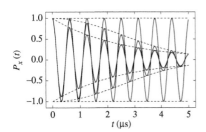

Fig. 5.3 Transverse-field polarization functions corresponding to eqns 5.4, 5.10, and 5.13. The envelopes are shown as dashed lines for convenience. The plots are drawn for $\gamma_\mu B_0$ = 10 μs^{-1}, corresponding to B_0 = 11.74 mT, and $\Delta = a = 0.4$ μs^{-1}.

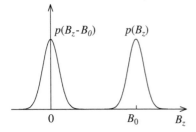

Fig. 5.4 Relation between $p(B_z)$ and $p(B_z - B_0)$.

[2]Note that eqn 5.7 is valid provided that $p(X)$ is an even function of X. Otherwise an additional term, proportional to $\sin(\gamma_\mu B_0 t)$, is present.

Computing the polarization function from this distribution relies on carrying out Gaussian integrals. The basic Gaussian integral is

$$\int_{-\infty}^{\infty} dx\, e^{-\frac{ax^2}{2}+bx} = \sqrt{\frac{2\pi}{a}}\, e^{\frac{b^2}{2a}}. \quad (5.7)$$

Another useful integral is

$$\int_{-\infty}^{\infty} dx\, x^2 e^{-\frac{ax^2}{2}} = \sqrt{\frac{2\pi}{a^3}}, \quad (5.8)$$

and another is

$$\int_{-\infty}^{\infty} dx\, x^2 e^{-\frac{ax^2}{2}+bx}$$
$$= \sqrt{\frac{2\pi}{a^3}} \left(1 + \frac{b^2}{a}\right) e^{\frac{b^2}{2a}}. \quad (5.9)$$

Example 5.1

As a first application of eqns 5.2 and 5.3, we examine the case where the field produced by the system is negligible compared to B_0. We then have a probability density $p(B) = \delta^{(3)}(B - B_0)$. As shown in Fig. 5.1, the transverse-field geometry corresponds to $S(0) \parallel \hat{x}$ and $B_0 \parallel \hat{z}$, and hence to setting $\theta = \pi/2$ in $\sigma_\alpha(t)$. The longitudinal geometry ($S(0) \parallel \hat{z}$ and $B_0 \parallel \hat{z}$) corresponds to $\theta = 0$. For the transverse-field polarization we find

$$P_x(t) = \int_{-\infty}^{\infty} d^3B\, \delta^{(3)}(B - B_0)\cos(\gamma_\mu Bt) = \cos(\gamma_\mu B_0 t). \quad (5.4)$$

Figure 5.3 shows $P_x(t)$. For the longitudinal-field function we set $\theta = 0$ and we find

$$P_z(t) = \int_{-\infty}^{\infty} d^3B\, \delta^{(3)}(B - B_0) = 1. \quad (5.5)$$

As might be expected, a transverse field gives oscillations; a longitudinal field does not cause any evolution of the polarization.

An important simplification to eqn 5.3 occurs for transverse-field ($\theta = \pi/2$) experiments with a large enough applied field B_0. In this case, among the values of B with a sizeable weight in the distribution, components B_x and B_y are negligible compared to $B_z \approx B_0$. The triple integral in eqn 5.3 reduces to the single integral

$$P_x(t) = \int_{-\infty}^{\infty} dB_z\, p(B_z)\cos(\gamma_\mu B_z t). \quad (5.6)$$

Here, $p(B_z)\,dB_z$ is the probability that the z-component of the local field takes the value B_z within an interval dB_z.

We shall assume that the field originating from the nuclei and electrons of the system under study (the internal field) is not affected by B_0. Since the probability density function $p(B_z)$ in eqn 5.6 depends on the magnitudes of the internal and external fields, it is useful to introduce a function shifted by $-B_0$, as shown in Fig. 5.4. The shift is obtained using the substitution $B_z = B_0 + X$ in eqn 5.6, which then reads[2]

$$P_x(t) = \left[\int_{-\infty}^{\infty} dX\, \cos(\gamma_\mu Xt)p(X)\right]\cos(\gamma_\mu B_0 t) = G_x(t)\cos(\gamma_\mu B_0 t).$$

This polarization function is the product of a factor oscillating with a frequency corresponding to B_0 and a function $G_x(t)$ inside the square bracket, which is the **transverse relaxation function** introduced in Chapter 2, and acts as a damping factor. Notice that $G_x(t)$, which is the Fourier cosine transform of $p(X)$, is intrinsic to the system since it is independent of B_0.

Example 5.2

Figure 5.4 shows the probability densities for the case where $p(B_z)$ corresponds to a Gaussian distribution with standard deviation Δ/γ_μ. The shifted part $p(X)$ is written

$$p(X) = \frac{\gamma_\mu}{\sqrt{2\pi}\Delta}\exp\left(-\frac{\gamma_\mu^2 X^2}{2\Delta^2}\right). \quad (5.10)$$

We compute a transverse relaxation function

$$G_x(t) = \left(\frac{\gamma_\mu}{\sqrt{2\pi}\Delta}\right) \int_{-\infty}^{\infty} dX\, e^{-\frac{\gamma_\mu^2 X^2}{2\Delta^2}} \cos(\gamma_\mu X t)$$

$$= \frac{1}{2}\left(\frac{\gamma_\mu}{\sqrt{2\pi}\Delta}\right) \int_{-\infty}^{\infty} dX\, e^{-\frac{\gamma_\mu^2 X^2}{2\Delta^2}} \left(e^{i\gamma_\mu X t} + e^{-i\gamma_\mu X t}\right)$$

$$= \left(\frac{\gamma_\mu}{\sqrt{2\pi}\Delta}\right) \int_{-\infty}^{\infty} dX\, e^{-\frac{\gamma_\mu^2 X^2}{2\Delta^2} + i\gamma_\mu X t}. \tag{5.11}$$

Using the formulae in the margin, we obtain a Gaussian relaxation function

$$G_x(t) = e^{-\frac{\Delta^2 t^2}{2}}. \tag{5.12}$$

We can also examine the case of $p(B_z)$ corresponding to a Lorentzian distribution with half-width at half maximum a/γ_μ, with shifted distribution written

$$p(X) = \frac{1}{\pi} \frac{\gamma_\mu a}{(a^2 + \gamma_\mu^2 X^2)}. \tag{5.13}$$

A cosine Fourier transform in this case results in an exponential function

$$G_x(t) = e^{-at}. \tag{5.14}$$

The relaxation functions resulting from the distributions in eqns 5.10 and 5.13 are plotted in Fig. 5.3.

One strength of μSR is its ability to probe systems in zero external field. It is often appropriate to consider the field distribution as being isotropic, for example when the sample being studied is a powder. In this case, eqn 5.3 is most conveniently evaluated in spherical polar coordinates, with polar and azimuthal angles θ and ϕ. (The picture to keep in mind is Fig. 5.2.)

Example 5.3

We start again with

$$P_z(t) = \int d^3B\, p(\boldsymbol{B})\langle\sigma_z(t)\rangle. \tag{5.15}$$

In spherical polar coordinates we have $\langle\sigma_z(t)\rangle = \cos^2\theta + \sin^2\theta\cos(\gamma_\mu Bt)$ and so we need to convert $p(\boldsymbol{B})$ to spherical polars. We use the rule that[3]

$$\begin{aligned} p(\boldsymbol{B})\,d^3B &= g(B_x)dB_x\, g(B_y)dB_y\, g(B_z)dB_z \\ &= g(B_x)g(B_y)g(B_z)B^2\sin\theta\, dB\, d\theta\, d\phi \\ &= f(\theta,\phi)p(B)\, dB\, d\theta\, d\phi, \end{aligned} \tag{5.16}$$

where the isotropy allows us to say that the distribution of magnetic field magnitudes $p(B)$ is independent of orientation. Since the distribution of angles is uniform over solid angle, we have a normalized distribution $f(\theta,\phi) = \sin\theta/4\pi$. We then compute the polarization function

$$P_z(t) = \frac{1}{4\pi}\int_{B=0}^{\infty}\int_{\theta=0}^{\pi}\int_{\phi=0}^{2\pi} dB\, d\theta\, d\phi\, p(B)\sin\theta\left[\cos^2\theta + \sin^2\theta\cos(\gamma_\mu Bt)\right]. \tag{5.17}$$

The azimuthal (ϕ) part is straightforward[4] and contributes a factor of 2π. Doing the θ part of the integral, we find

$$\begin{aligned} P_z(t) &= \int_{B=0}^{\infty} dB\, p(B)\int_{\theta=0}^{\pi} d\theta\, \frac{\sin\theta}{2}\left[\cos^2\theta + \sin^2\theta\cos\gamma_\mu Bt\right] \\ &= \int_{B=0}^{\infty} dB\, p(B)\left[\frac{1}{3} + \frac{2}{3}\cos(\gamma_\mu Bt)\right]. \end{aligned} \tag{5.19}$$

[3]Here we call the distribution function for field components $g(B_i)$, that for field magnitudes $p(B)$ and that for angles $f(\theta,\phi)$. The $B^2\sin\theta$ factor in the second line of eqn 5.16 originates from the usual Jacobian for spherical coordinates. We define the normalized distributions via

$$\int_0^{\infty} dB\, p(B) = 1,$$

where $p(B)$ includes the factor B^2 from the Jacobian, and

$$\int d\theta\, d\phi\, f(\theta,\phi) = 1,$$

where $f(\theta,\phi)$ includes the factor $\sin\theta$ from the Jacobian.

[4]We integrate over the azimuthal angle using

$$\frac{1}{4\pi}\int_{\phi=0}^{2\pi} d\phi\, d\theta\, \sin\theta = \frac{1}{2}\sin\theta d\theta. \tag{5.18}$$

Using the normalization of $p(B)$, we arrive at the simplified form of the last equation:

$$P_z(t) = \frac{1}{3} + \frac{2}{3} \int_0^\infty p(B) \cos(\gamma_\mu B t)\, dB. \tag{5.20}$$

Equation 5.20 consists of time-independent and dependent terms, respectively accounting for $1/3$ and $2/3$ of the full polarization. These directly arise from the isotropic nature of the distribution and can be thought of as describing an ensemble of muons with $1/3$ of the local magnetic field components pointing parallel to the initial muon spin (and are therefore unchanged by the magnetic field) and $2/3$ of the components in a perpendicular direction. *The perpendicular components lead to a time dependence given by the Fourier cosine transform of the field distribution* $p(B)$.

Example 5.4

The zero-field response in an experiment performed on a powder sample of a magnetic compound can be straightforwardly predicted from eqn 5.20 in the case where all the muons are subject to a spontaneous field $\boldsymbol{B}_1$ of random orientation but uniform modulus, i.e. $p(B) = \delta(B - B_1)$. We find

$$P_z(t) = \frac{1}{3} + \frac{2}{3} \int \delta(B - B_1) \cos(\gamma_\mu B t)\, dB$$

$$= \frac{1}{3} + \frac{2}{3} \cos(\gamma_\mu B_1 t), \tag{5.21}$$

which is the sum of the $1/3$ term, typical of the isotropic distribution, and an oscillating term with a frequency determined by B_1.

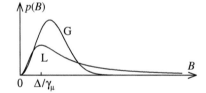

Fig. 5.5 The distribution $p(B)$ vs B for the Gaussian distribution (G) in eqn 5.24. For comparison the Lorentzian distribution (L) from eqn 5.46 is also shown.

Of particular interest is the case of a Gaussian distribution for the three Cartesian components of the field, since this provides a reasonable model for a magnetically disordered material. The distribution for B becomes a Maxwell-Boltzmann distribution (see Fig. 5.5) which is well known in the kinetic theory of ideal gases, and the remaining integration in eqn 5.20 can be performed analytically as we show in the next example.

Example 5.5

Assuming we have an independent Gaussian distribution in each of the Cartesian components of the magnetic field vector $\boldsymbol{B} = (B_x, B_y, B_z)$, gives us

$$p(\boldsymbol{B})\mathrm{d}^3 B = \left(\frac{\gamma_\mu^2}{2\pi\Delta^2} \right)^{\frac{3}{2}} \mathrm{e}^{-\frac{\gamma_\mu^2 (B_x^2 + B_y^2 + B_z^2)}{2\Delta^2}} \mathrm{d}^3 B. \tag{5.22}$$

Swapping to spherical coordinates (B, θ, ϕ) we find a distribution

$$[p(B)\mathrm{d}B]\,[f(\theta,\phi)\mathrm{d}\theta\mathrm{d}\phi] = \left[\left(\frac{\gamma_\mu^2}{2\pi\Delta^2} \right)^{\frac{3}{2}} (4\pi B^2) \mathrm{e}^{-\frac{\gamma_\mu^2 B^2}{2\Delta^2}} \mathrm{d}B \right] \left[\frac{\sin\theta}{4\pi} \mathrm{d}\theta\mathrm{d}\phi \right]. \tag{5.23}$$

We conclude that the field magnitude is distributed according to the Maxwell-Boltzmann-type distribution

$$p(B) = \left(\frac{\gamma_\mu^2}{2\pi\Delta^2}\right)^{\frac{3}{2}} (4\pi B^2) e^{-\frac{\gamma_\mu^2 B^2}{2\Delta^2}}. \tag{5.24}$$

We can now integrate the muon spin component $\langle\sigma_z(t)\rangle = \cos^2\theta + \sin^2\theta\cos(\gamma_\mu Bt)$ over θ and B. The angular average is identical to that in Example 5.3, and reduces the integrand to $\left[\frac{1}{3} + \frac{2}{3}p(B)\cos(\gamma_\mu Bt)\right]$. After that, averaging over the distribution $p(B)$, we obtain

$$\begin{aligned}P_z(t) &= \frac{1}{3} + \frac{2}{3}\left(\frac{\gamma_\mu^2}{2\pi\Delta^2}\right)^{\frac{3}{2}} \int_0^\infty dB\,(4\pi B^2)e^{-\frac{\gamma_\mu^2 B^2}{2\Delta^2}}\cos(\gamma_\mu Bt)\\ &= \frac{1}{3} + \frac{2}{3}(1-\Delta^2 t^2)e^{-\frac{\Delta^2 t^2}{2}},\end{aligned} \tag{5.25}$$

where use was again made of the Gaussian integrals. The resulting function is shown in Fig. 5.6. Expanding the function at early times we find that it initially decays according to $P_z(t) \approx 1 - \Delta^2 t^2 + ...$, and so resembles a Gaussian function for the early part of the decay.

The result from the last example is an important one and is known as the **Kubo-Toyabe function**

$$P_z(t) = \frac{1}{3} + \frac{2}{3}\left(1 - \Delta^2 t^2\right)\exp\left(-\frac{\Delta^2 t^2}{2}\right). \tag{5.26}$$

As in eqn 5.10, Δ/γ_μ is the standard deviation of the distributions and the three Cartesian field components have the same Δ in line with the assumed isotropy.[5] It can be thought of as a damped oscillation around the value $1/3$. The minimum of the function is approximately 0.03583 at $t = \sqrt{3}/\Delta$.

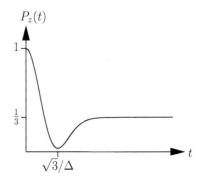

Fig. 5.6 The Kubo-Toyabe function from eqn 5.26.

[5]While the function shown in eqn 5.26 was originally the result of a *Gedanken-Experiment* devised by Kubo and Toyabe in 1967 for zero-field nuclear magnetic resonance, it later turned to be a powerful model for the interpretation of μSR spectra.

Example 5.6

One way to think about the Kubo-Toyabe function is that, in a disordered material, $1/3$ of the magnetic field components will point along the direction of the initial muon spin (i.e. the z-direction) and so will not lead to any precession or, consequently, any relaxation. The remaining $2/3$ all point perpendicular to the muon spin so causes precession at different frequencies. Since the muon-spins are initially polarized, the oscillations all start from the same value. However, the distribution in fields leads to their dephasing, washing out the oscillations so that we obtain the characteristic minimum and recovery. This is shown in Fig. 5.7.

A common way of gaining information on a system is to observe the effect of an externally applied magnetic field on the measured spectra. We again assume that this field only shifts the distribution probed by the muons, i.e. it does not affect the system under study. In transverse-field, provided the large transverse-field limit applies, a change in B_0 is not very instructive since only the frequency of the oscillations is changed while the envelope function remains the same. The situation is

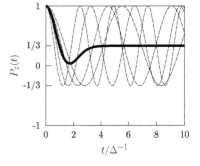

Fig. 5.7 Muon-spin polarization for precession in different local B-fields. The weighted average over these yields the Kubo-Toyabe relaxation function with its characteristic dip and recovery to a value of $1/3$.

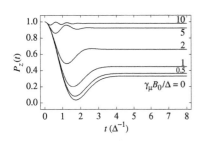

Fig. 5.8 Longitudinal-field polarization functions computed for the probability density function shown in eqn 5.27. The external field values are normalized to the standard deviation Δ/γ_μ of the distribution. The function for $B_0 = 0$ is analytical (eqn 5.26), giving the Kubo-Toyabe function.

[6]The evolution of the field can stem from different origins: (i) in a magnetic system it is the magnetic fluctuations or excitations; (ii) in ionic systems it's the motion of ions bearing magnetic moments of nuclear origin; (iii) the muon can also move from a site to another.

more interesting in a longitudinal field geometry where the probability density function becomes

$$p(\boldsymbol{B}) = \left(\frac{\gamma_\mu}{\sqrt{2\pi}\Delta} \right)^3 \exp \left\{ -\frac{\gamma_\mu^2 [B_x^2 + B_y^2 + (B_z - B_0)^2]}{2\Delta^2} \right\}. \quad (5.27)$$

As usual the associated polarization is computed from eqn 5.3. The result is shown in Fig. 5.8 for different values of B_0. When the field is large enough, oscillations of small amplitude with a frequency corresponding to B_0 are seen. For even larger fields, $P_z(t)$ tends to be time independent and approaches unity, as expected from eqn 5.5. The muon spin is said to be **decoupled** from the system. The static Kubo-Toyabe function represented in Fig. 5.8 can be written as $G_{\text{KT}}^{\text{Stat}}$. It will be used in examples of data analysis in Chapters 11, 15, and 22.

5.2 Dynamical fields

In many instances the magnetic field at the muon site evolves during the time spent by the muon in the system. In many cases of interest to us, the evolution of the field[6] at the muon site is a random process. The derivation of the polarization function in the presence of dynamics is usually made assuming that the dynamic process is a stationary **Gaussian-Markovian process**. This implies (i) that the properties of the dynamics are independent of the origin in time (ii) that the probability of occurrence of a future state does not depend on its history, and (iii) that the field distribution accessible at any instant is Gaussian. Two models are often used for the derivation of $P_\alpha(t)$. In the first, $\boldsymbol{B}(t)$ evolves through a series of sudden changes; in the second the field is assumed to vary continuously with time. Both methods lead to very similar results and can be equally applied for the derivation of $P_x(t)$ and $P_z(t)$.

The first model for the computation of $P_\alpha(t)$ is often called the **strong-collision model**, since $B(t)$ suddenly changes from one value to another. In this model, the muon ensemble evolves as a function of time in a static field. However, after an average time τ, the local field suffers a discontinuous fluctuation event that changes its value. It chooses a new value subject to the range of fields available via the reconfiguration of the magnetic moments. The muon ensemble then precesses in the new static configuration until it suffers another discontinuous change. This represents a regular effective resetting of the muon precession process. The routine to take these processes into account is known as **dynamicization** and can be carried out mathematically. It is shown graphically in Fig. 5.9 for a muon ensemble in a disordered environment (in zero field) where the underlying relaxation follows a Kubo-Toyabe function. If the fluctuation time τ is long compared to the precession of muons in the local field, such that muon spins can precess through a number of complete cycles (that is, the fluctuating field has the property $\gamma B\tau \gg 1$, putting us in the slow-fluctuation limit, described in more detail below) we have the situation shown in Fig. 5.9(a). Here the initial part of the polarization is completely unrelaxed by the dynamical process, and only the tail of the relaxation decays away exponentially. Often in magnetic materials we find ourselves in the fast-fluctuation limit, where $\gamma B\tau \ll 1$, and we have the situation shown in Fig. 5.9(b), where the rapid resetting of the Kubo-Toyabe muon relaxation leads to an exponential relaxation.

Mathematically, the polarization function can be decomposed in a sum of contributions $R_{\alpha,\ell}(t)$ describing muons that have experienced ℓ field changes between instants 0 and t. For $P_z(t)$ we can then write

$$P_z(t) = \sum_{\ell=0}^{\infty} R_{z,\ell}(t). \qquad (5.28)$$

Since the probability that $\boldsymbol{B}$ has remained constant between 0 and t is $\exp(-\nu t)$ we must have $R_{z,0}(t) = P_z^{\mathrm{s}}(t)\exp(-\nu t)$, where $P_z^{\mathrm{s}}(t)$ is the static relaxation function. Given these hypotheses, the recursion formula

$$R_{z,\ell+1}(t) = \nu \int_0^t R_{z,\ell}(t-t')R_{z,0}(t')\,\mathrm{d}t', \qquad (5.29)$$

can then be derived. In words, the $\ell + 1$ collisions imply no collisions up to time t', when one occurs with probability ν, followed by ℓ further identical collisions.[7] Combining eqns 5.28 and 5.29 we find the following equation for $P_z(t)$:

$$P_z(t) = P_z^{\mathrm{s}}(t)\exp(-\nu t) + \nu \int_0^t P_z(t-t')P_z^{\mathrm{s}}(t')\exp(-\nu t')\,\mathrm{d}t'. \qquad (5.30)$$

This integral equation relates the dynamic to the static polarization functions and, in principle, can be applied to any static polarization function $P_z^{\mathrm{s}}(t)$. Figure 5.10 illustrates $P_z(t)$ for the case where $P_z^{\mathrm{s}}(t)$ is the zero field Kubo-Toyabe function (eqn 5.26). The behaviour of this **dynamicized-Kubo-Toyabe function** can be computed numerically from the above equation and is very frequently used to model experimental data. It can be written as $G_{\mathrm{KT}}^{\mathrm{Dyn}}$ and this function will be used for the analysis of data in Chapters 11 and 22.

The second approach to describing dynamics is to treat the magnetic field as evolving continuously, as outlined in the next example, for the case of the transverse polarization $P_x(t)$.

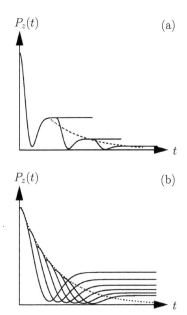

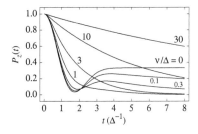

Fig. 5.9 The strong-collision model in pictures in the (a) slow and (b) fast limits. The dashed line shows the observed relaxation that results from the dynamics.

[7]This is examined further in the exercises.

Fig. 5.10 Dynamical zero-field polarization function computed for an isotropic Gaussian distribution of field. The curves correspond to different values of ν measured in units of Δ, as indicated in the graph. These useful functions are generally known as dynamicized-Kubo-Toyabe functions.

Example 5.7

We shall examine the large transverse field limit for which only component $B_z(t)$ matters. To make progress here, we introduce a total precession angle $\phi = \int_0^t \gamma_\mu B_z(t')\,\mathrm{d}t'$ undergone by the muon spin between 0 and t. The point here is that since the field $B_z(t)$ varies in time, we can only evaluate the precession through the small angle $\mathrm{d}\phi$ that occurs in a short time interval $\mathrm{d}t$, since after this interval the field will have changed. The integral allows us to add up all of these contributions to find the total precession that has taken place in the limit that $\mathrm{d}t \to 0$, so that $B_z(t)$ is a continuous function. To evaluate the relaxation we then assume

$$P_x(t) = \mathcal{R}e\left\langle \exp\left[\mathrm{i}\int_0^t \gamma_\mu B_z(t')\,\mathrm{d}t'\right]\right\rangle, \qquad (5.31)$$

[9]This functional form is a consequence of **Doob's theorem**, which says that in a stochastic process one can have random variables selected from a Gaussian distribution using a Markov chain (a sequence of events that has no history dependence, so each selection is independent of the previous one) and this will give rise to a correlation function which decays exponentially with time.

[10]Anatole Abragam (1914–2011) made a number of significant contributions to nuclear magnetic resonance and wrote the influential book *Principles of Nuclear Magnetism*.

where we have introduced the complex exponential, and $\mathcal{R}eA$ stands for the real part of complex number A. Next we make the substitution $B_z(t) \longrightarrow \langle B_z \rangle + \delta B_z(t)$. The idea here being that the field varies around a time-independent average $\langle B_z \rangle$ by an amount $\delta B_z(t)$, that varies with time. For our problem we set $\langle B_z \rangle = B_0$, that is, the field fluctuates around the applied field. With the prescription that the stochastic process is Gaussian-Markovian. We assume, by analogy to our strong collision model,[8] that the dynamics of the field dynamics are exponentially distributed, which implies that the fields obey the equation[9]

$$\langle \delta B_\alpha(t_0) \delta B_\alpha(t_0 + t) \rangle = \langle \delta B_\alpha^2 \rangle \exp\left(-\nu|t|\right), \qquad (5.32)$$

where $\nu^{-1} = \tau$ is the field **correlation time** (i.e. the average time between fluctuations). The average $\langle \delta B_\alpha(t_0) \delta B_\alpha(t_0 + t) \rangle$ is an example of a *correlation function* and is central to the more general description of dynamics discussed in Chapter 7. With these ingredients the integral in eqn 5.31 can be computed.

The result of the integration discussed in the previous example is that the transverse polarization $P_x(t)$ is given by

$$P_x(t) = \exp\left\{-\frac{\Delta^2}{\nu^2}\left[\exp(-\nu t) - 1 + \nu t\right]\right\} \cos(\gamma_\mu B_0 t), \qquad (5.33)$$

where $\Delta^2/\gamma_\mu^2 \equiv \langle (\delta B_z)^2 \rangle$.

Equation (5.33) is known as the **Abragam function**.[10] It describes the transverse polarization in the presence of dynamics with fluctuation rate ν. In this context Δ^2/γ_μ^2 represents the average square amplitude of the fluctuating field. When considering dynamics it is often useful to examine the limiting behaviour for **slow** and **fast fluctuations**.

• Slow fluctuations imply (i) that the fluctuation rate ν is small and so $\nu t \to 0$, and (ii) that the fluctuating amplitude Δ and the correlation time $\tau = 1/\nu$ obey $\Delta\tau \gg 1$. This latter equation says that a muon spin is expected to make many precessions in the field Δ/γ_μ before a fluctuation would be expected to take place.

• Fast fluctuations imply (i) that the fluctuation rate is large ($\nu t \gg 1$), and (ii) that $\Delta\tau \ll 1$. The latter expression means that a muon spin is unlikely to make a precession through 2π in the field Δ/γ_μ before a fluctuation process takes place.

Example 5.8

The damping factor in the Abragam function provides an interpolation between Gaussian and exponential envelopes. In the limit of a static field ($\nu \to 0$) we have

$$P_x(t) = \exp\left(-\frac{\Delta^2 t^2}{2}\right) \cos(\gamma_\mu B_0 t), \qquad (5.34)$$

while in the fast fluctuating limit ($\nu t \gg 1$)

$$P_x(t) = \exp(-\lambda t) \cos(\gamma_\mu B_0 t) \quad \text{with} \quad \lambda = \frac{\Delta^2}{\nu} = \Delta^2 \tau. \qquad (5.35)$$

The latter behaviour is referred to as the **motional narrowing limit**.[11]

Let's now return to the strong collision model. While, in general, eqn 5.30 can usually be solved numerically using a computer, we can also make some progress into solving it analytically by considering its **Laplace transform**. We'll see that providing an analytical solution valid in limiting cases will help provide some physical insight into $P_z(t)$. (The analytical solutions also greatly speeds up the numerical computation for certain intervals of parameters.) Recalling the definition of a Laplace transform, $\tilde{f}(s) = \int_0^\infty f(t) \exp(-st) \, dt$, eqn 5.30 takes the simple form

$$\tilde{P}_z(s) = \frac{\tilde{P}_z^s(s+\nu)}{1 - \nu \tilde{P}_z^s(s+\nu)}. \tag{5.36}$$

We see how this can be used in the next example.

Example 5.9

We shall apply the last equation to the physical case examined before i.e. an isotropic Gaussian distribution for the three components of $\boldsymbol{B}$, which results in the Kubo-Toyabe function in the static case. For zero-field measurements, we start from the Laplace transform of eqn 5.26 and first assume extremely slow dynamics. From eqn 5.36 we find $\tilde{P}_z(s) = (s+2\nu/3)/3$ in the small s limit, i.e. $P_z(t) = \exp(-2\nu t/3)/3$ in the large time limit. Therefore we deduce

$$P_z(t) \approx \frac{1}{3} \exp\left(-\frac{2}{3}\nu t\right) + \frac{2}{3}\left(1 - \Delta^2 t^2\right) \exp\left(-\frac{\Delta^2 t^2}{2}\right), \tag{5.37}$$

for $\nu/\Delta \ll 1$. Interestingly, only the so-called '1/3 tail' is damped, with a damping rate equal to 2/3 of ν. This may be viewed as the result of muon-spin precession with dephasing induced by only two (x and y) of the three Cartesian components of $\boldsymbol{B}$.

In the fast fluctuation limit, i.e. $\nu/\Delta \gg 1$, using a similar procedure we arrive at

$$P_z(t) = \exp\left\{-\frac{2\Delta^2}{\nu^2}\left[\exp(-\nu t) - 1 + \nu t\right]\right\}. \tag{5.38}$$

This expression is very reminiscent of the Abragam formula [eqn 5.33], except for a factor of 2. The origin of the extra factor in the former case is that the two components of $\boldsymbol{B}$ perpendicular to z contribute to the relaxation in zero field, instead of one in transverse field (B_z). A numerical study shows that eqn 5.38 provides an excellent approximation to $P_z(t)$ for $\nu/\Delta \gtrsim 3$. In the same spirit as for the transverse field case, eqn 5.38 reduces, in the fast fluctuation limit, to

$$P_z(t) = \exp(-\lambda t) \quad \text{with} \quad \lambda = \frac{2\Delta^2}{\nu}, \tag{5.39}$$

when $\nu t \gg 1$.[12]

The dynamical polarization function can generally be derived from the static function using eqn 5.30. This is also the case for the dynamical polarization function observed in a longitudinal external field B_0. The function $P_z(t)$ now depends on two parameters: ν and B_0. There is an appealing limit form that holds in the case of the Gaussian distribution. For sufficiently fast fluctuations and external field, $P_z(t)$ becomes an exponential function

$$P_z(t) = \exp(-\lambda t) \quad \text{with} \quad \lambda = \frac{2\Delta^2\nu}{\nu^2 + \gamma^2 B_0^2}. \tag{5.40}$$

[12]Interestingly, by comparing the behaviour of the 1/3 tail in eqn 5.37 and the expression of λ in eqn 5.39, it is seen that the relaxation rate starts to increase before decreasing as dynamics gets faster and faster. The intermediate maximum can be seen in Fig. 5.10.

[13]The Lorentzian form of eqn 5.40 is often used for the determination of ν. Indeed, in a zero-field experiment only a combination of Δ and ν is determined (eqn 5.39). From the field $B_0^{(1/2)}$ at which λ is divided by a factor 2, we get $\nu = \gamma_\mu B_0^{(1/2)}$, from which Δ can also be obtained.

[14]The method described below works equally well for the description of the field distribution arising from electronic or nuclear magnetic moments.

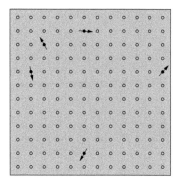

Fig. 5.11 Sketch of a lattice of atoms where magnetic atoms (arrows) randomly substitute non-magnetic atoms (open circles).

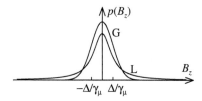

Fig. 5.12 Shifted density distribution functions of interest for large transverse-field experiments in the case of small (L, eqns 5.44 or 5.13) and large concentration of magnetic atoms (G, eqn 5.10).

[15]See Dalmas de Réotier and Yaouanc for details.

This expression is known as the **Redfield formula**, or alternatively the **BPP model** (referring to the three authors Bloembergen, Purcell, and Pound). It corresponds to a Lorentzian decay of λ as a function of B_0.[13] We discuss it further in Chapter 8 where we give some examples, and also in Chapter 7 where we revisit dynamics from an approach based on correlation functions. We mention the expression here because its limit for vanishing magnetic field is exactly the relaxation rate of eqn 5.39, the fast fluctuation limit of the dynamicized Kubo-Toyabe polarization.

5.3   Disordered systems

So far we have considered two distributions of disordered magnetic fields: the Gaussian and the Lorentzian. Here we meet the origin of these distributions in disordered magnetic materials.[14]

Let's consider the case of magnetic atoms randomly distributed at substitutional sites in a lattice of non-magnetic atoms (Fig. 5.11). We denote as c_{imp} the probability of occupation of a site by a magnetic impurity atom. The case $c_{\mathrm{imp}} = 1$ corresponds to a lattice fully occupied by magnetic atoms. As usual we investigate the large transverse and zero field cases in turn.

Example 5.10

Once again, if the transverse field is large enough, the expression of $P_x(t)$ depends only on the distribution of the z-component of the local field. Its probability density function is given by the formula

$$p(B_z) = \int \cdots \int \delta \left(B_z - \sum_{j=1}^{n_c} B_{z,j} \right) \times \tag{5.41}$$

$$\prod_{j=1}^{n_c} \left[(1 - c_{\mathrm{imp}}) \delta(B_{z,j}) + c_{\mathrm{imp}} w_j(B_{z,j}) \right] \mathrm{d}B_{z,1} \ldots \mathrm{d}B_{z,n_c},$$

where the index j runs on the lattice atoms, magnetic or not, which give rise to field contributions $B_{z,j}$ at the muon site. If a magnetic atom is present at site j its contribution to B_z is given by the density $w_j(B_{z,j})$. If no magnetic moment is present, this is $\delta(B_{z,j})$. The factorization of the quantities in the square brackets stems from our hypothesis of the absence of correlations between the magnetic moments.

To proceed, we introduce the **characteristic function** of $p(B_z)$, defined as

$$C(t) = \int p(B_z) \exp(\mathrm{i}\gamma_\mu B_z t) \, \mathrm{d}B_z. \tag{5.42}$$

This quantity is very closely related to the Fourier transform of the probability density function that we encountered before. The calculation involves doing this integral[15] for the $p(B_z)$ given in eqn 5.41. The result is that $C(t)$ takes simple forms in two cases. In the large dilution limit (i.e. $c_{\mathrm{imp}} \lesssim 1\%$) we obtain

$$C(t) = \exp(-a|t|), \tag{5.43}$$

The density function is derived from eqn 5.42 using the standard Fourier transform formula, and yields

$$p(B_z) = \frac{1}{\pi} \frac{\gamma_\mu a}{a^2 + \gamma_\mu^2 B_z^2}. \tag{5.44}$$

It corresponds to the Lorentzian (or Cauchy) distribution that was discussed in Example 5.2. It is plotted in Fig. 5.12.

In the case of concentrated magnetic atoms, i.e. $c_{\text{imp}} \approx 1$, an analytical function valid at relatively short times is found for $C(t)$:

$$C(t) = \exp\left(-\frac{\Delta^2 t^2}{2}\right). \qquad (5.45)$$

The probability density function corresponding to eqn 5.45 is a Gaussian function (eqn 5.10) plotted in Fig. 5.12. The resulting polarization function has therefore a Gaussian envelope (see, e.g. eqn 5.34).

We conclude that the Gaussian distribution describes a dense array of disordered fields, while the Lorentzian one describes a dilute array of disordered fields.

A procedure similar to the one just described can be used for the determination of the field distribution relevant for zero-field measurements. It starts with an expression for $p(\boldsymbol{B})$, in the spirit of eqn 5.41. Then the characteristic function $C(\boldsymbol{t})$ is evaluated.[16] Considering a system in the large dilution limit, it is found that $C(\boldsymbol{t}) = \exp(-at)$, where the expression of a is similar to that valid in large transverse field, although with a different numerical factor. Interestingly, $C(\boldsymbol{t})$ depends only on the modulus $|\boldsymbol{t}|$, implying that this discussion can only be applied to isotropic field distributions. We eventually find that the distribution of field magnitudes is given by

$$p(B) = \frac{\gamma_\mu^3}{\pi^2}\left(4\pi B^2\right)\frac{a}{\left(a^2 + \gamma_\mu^2 B^2\right)^2}, \qquad (5.46)$$

i.e. a Maxwell-Boltzmann function (see Fig. 5.5). Inserting this expression into eqn 5.20, we arrive at the formula

$$P_z(t) = \frac{1}{3} + \frac{2}{3}(1 - at)\exp(-at), \qquad (5.47)$$

sometimes referred to as the **Lorentzian Kubo-Toyabe** function (Fig. 5.13). This function will be further discussed in Chapter 7.

Finally, we note that in a system with concentrated magnetic atoms, $C(\boldsymbol{t})$ is found to only depend on t^2 and the three Cartesian components of the local field are Gaussian distributed, leading to the recovery of the ordinary, Gaussian, Kubo-Toyabe function (eqn 5.26) for $P_z(t)$.

5.4   The stretched exponential

A cursory inspection of the μSR literature, especially that related to magnetism, shows the frequent use of the **stretched exponential** for modelling experimental data. We dedicate the last section of the chapter to this function. The stretched exponential function defined as

$$P_z(t) = \exp\left[-(\lambda t)^\beta\right], \qquad (5.48)$$

with $0 < \beta < 2$ is sometimes used to fit μSR spectra.[17]

[16]Here the (important) vector-nature of parameter $\boldsymbol{t}$ arises from the three dimensional character of field $\boldsymbol{B}$.

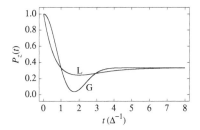

Fig. 5.13 Comparison of the zero-field polarization functions resulting from the probability density functions sketched in Fig. 5.5. Their analytical expression is given in eqns 5.26 (Kubo-Toyabe function, G) and 5.47 (Lorentzian Kubo-Toyabe function, L), respectively.

[17]If the exponent β is allowed to take values above 1, strictly speaking, the function should be called a compressed exponential function. The words stretched and compressed characterize the shape of the graph representing $\log[P(t)]$ versus t, by comparison with the case $\beta = 1$.

In physics, stretched exponential functions are generally associated with relaxation phenomena in disordered or inhomogeneous systems. They are then viewed as resulting from a collection of exponential decays with variable relaxation rates. The shape of the density probability of the relaxation rates depends on β, with lower values of $\beta < 1$ implying a broader distribution. As far as μSR is concerned, the stretched exponential function with $\beta = 1/2$ is physically and mathematically justified for the case of a dilute magnetic system in the limit of fast dynamics. (This is discussed in Chapter 7.) In some cases, detailed studies of specific physical situations have shown $P_z(t)$ to closely approach stretched exponential functions; e.g. in the case of anisotropy in the magnetic response or magnetic segregation.[18] Otherwise, this function should be treated simply as a mathematical model accounting for the evolution of μSR spectra without direct physical significance.

[18]A notable example is the moderately concentrated spin glass, where relaxation with $\beta = 1/3$ has been observed as the glass transition is approached in temperature from above. This value of β can be justified geometrically by describing the realization of the microstates of the spin glass in terms of random walks near a percolation threshold. [See the article by I. A. Campbell in *Muon Science: Muons in Physics, Chemistry and Materials*, CRC Press (1999) and also A. T. Ogielski, Phys. Rev. B **32**, 7384 (1985) and I. A. Campbell *et al.* Phys. Rev. Lett. **72**, 1291 (1994).]

Chapter summary

- The polarization function is computed in the stochastic approach using

$$P_\alpha(t) = \int d^3B\ p(\boldsymbol{B})\langle\sigma_\alpha(t)\rangle. \tag{5.49}$$

- For isotropically distributed moments the zero field response is

$$P_z(t) = \frac{1}{3} + \frac{2}{3}\int_0^\infty p(B)\cos(\gamma_\mu Bt)\,dB. \tag{5.50}$$

- Magnetic disorder results in the Kubo-Toyabe function for zero field

$$P_z(t) = \frac{1}{3} + \frac{2}{3}\left(1 - \Delta^2 t^2\right)\exp\left(-\frac{\Delta^2 t^2}{2}\right). \tag{5.51}$$

- The Kubo-Toyabe function becomes modified in applied field and in the presence of dynamics. In both these cases it must be computed numerically.

- Dynamics in the transverse field case lead to the Abragam function

$$P_x(t) = \exp\left\{-\frac{\Delta^2}{\nu^2}\left[\exp(-\nu t) - 1 + \nu t\right]\right\}\cos(\gamma_\mu B_0 t). \tag{5.52}$$

- Fast dynamics in the longitudinal geometry give

$$P_z(t) = \exp(-\lambda t) \quad\text{with}\quad \lambda = \frac{2\Delta^2\nu}{\nu^2 + \gamma^2 B_0^2}. \tag{5.53}$$

Further reading

- R. S. Hayano *et al.*, Phys. Rev. B **20**, 850 (1979).

- R. Kubo and T. Toyabe in *Magnetic Resonance and Relaxation,* ed. R. Blinc, p810, North-Holland (1967).
- Chapter 6 of A. Yaouanc and P. Dalmas de Réotier, *Muon Spin Rotation, Relaxation, and Resonance: Applications to Condensed Matter*, OUP (2011).

Exercises

(5.1) Verify eqn 5.14.

(5.2) By performing an appropriate angular average, show that $\langle \cos^2 \theta \rangle = 1/3$ and $\langle \sin^2 \theta \rangle = 2/3$. Using the same approach, evaluate the following:
(a) $\langle 5 - 3\cos^2 \theta \rangle$,
(b) $\langle \cos^4 \theta \rangle$, and
(c) $\langle (3\cos^2 \theta - 1)^2 \rangle$.

(5.3) (a) Verify the expression for the Gaussian Kubo-Toyabe function (eqn 5.25).
(b) Show that the early-time behaviour of the Kubo-Toyabe function looks like $P_z(t) = 1 - \Delta^2 t^2$.

(5.4) Justify (a) eqn 5.29 and verify (b) eqn 5.30 and (c) eqn 5.36.

(5.5) By inserting eqn 5.46 into eqn 5.20, derive the Lorentzian Kubo-Toyabe function (eqn 5.47).

Part II

Science with μSR

Some major areas of application for μSR within science are covered in this part of the book.

- One major application of μSR is in the science of magnetism, since the muon is sensitive to static magnetic fields. The physics of magnetism in this context is explored in Chapter 6

- In many cases, the local magnetic field is not static but fluctuates. In this case, as explored in Chapter 7, the muon is a powerful probe of magnetic dynamics.

- This ability to follow fluctuating magnetic fields allows the muon to measure many other dynamical processes in the sample, or even the motion of the muon itself, as shown in Chapter 8.

- Superconductivity is a major research field in which the use of muons has made an important contribution. This area of application is discussed in Chapter 9.

- The muon can act as an analogue of hydrogen and this idea has allowed it to be an effective probe of defect states in semiconductors, as described in Chapter 10.

- μSR can be applied to a particular dynamical process: ionic motion. This is highly topical because of the importance of energy storage technologies such as lithium-ion batteries and the use of μSR in such systems is described in Chapter 11.

- Chapter 12 describes the use of muons in chemistry. By focusing on the formation of muoniated radicals and the study of their spectroscopic and dynamical properties, much useful information has been obtained, which has made substantial contributions to research in chemistry.

6 Magnetism

An important use of muon spectroscopy is as a probe of magnetism in condensed matter. The magnitude of the muon gyromagnetic ratio, $\gamma_\mu = 2\pi \times 135.5$ MHz T^{-1}, makes the muon spin a sensitive probe of small magnetic fields. The muon lifetime of 2.2 μs is long enough to allow the muon spin to respond to the presence of typical magnetic fields in magnetically ordered materials. In this chapter, we discuss the application of these ideas to the study of magnetic materials.

6.1 The basics

The magnetic flux-density field B (known as the B-field) may be written

$$B = \mu_0 \left(H + M \right), \tag{6.1}$$

where H is the magnetic field strength (known as the H-field) in the material and the magnetization M is the average magnetic moment m per unit volume. One of the most important magnetic properties of a solid is the **magnetic susceptibility** χ, defined for a linear material via

$$M = \chi H. \tag{6.2}$$

The susceptibility χ is sometimes known as the response, since it describes a response in the magnetization M for a given stimulus field H. It is an average quantity that can be measured by **magnetometry**, which is an experimental technique that probes a sample as a whole and allows one to extract M/H_0, where H_0 is the experimentally applied field.[1] When M is a large contribution to eqn 6.1,[2] the field B inside the material will be rather different to the applied field $B_0 = \mu_0 H_0$ in free space, far from the sample, because of the demagnetization field. However, in the frequently encountered case when the magnetic susceptibility is small ($\chi \ll 1$), the two quantities B and B_0 will be very similar and, moreover, $B \approx \mu_0 H$ (since $M \ll H$). Hence, we can use these approximations to write a useful expression for the susceptibility in terms of field magnitudes which is given by

$$\chi \approx \frac{\mu_0 M}{B}. \tag{6.3}$$

An insulating magnetic material is formed by an arrangement of ions, each carrying a magnetic moment m, proportional to the total angular

---

[1] The applied field can be written $B_0 = \mu_0 H_0$ and may be different from the internal field because of the demagnetization field, a contribution that depends on the shape of the sample.

[2] This is the case, for example, for ferro- and ferrimagnets, which have large magnetic susceptibility and magnetic charges (poles) on the surface of the sample, leading to a demagnetization field.

momentum $\boldsymbol{J}$ of the ion.[3] In the case that these moments do not interact, their behaviour is purely paramagnetic and is described by the Hamiltonian $\mathcal{H} = -\sum_i \boldsymbol{m}_i \cdot \boldsymbol{B}$. This leads to a **Curie Law** for the susceptibility as a function of temperature T, given by

$$\chi(T) = \frac{C}{T}, \tag{6.4}$$

where C is the Curie constant. Magnetic ions can interact with each other in different ways, giving rise to a number of more complicated Hamiltonians. A common one for magnetic insulators is the Heisenberg interaction between nearest neighbour (nn) spins:

$$\mathcal{H} = -J \sum_{ij \in \text{nn}} \boldsymbol{S}_i \cdot \boldsymbol{S}_j, \tag{6.5}$$

where, in typical inorganic materials, the **exchange coupling constant** J can reach strengths of up to 100 meV. In the presence of interactions the susceptibility will differ from Curie's law. For example, in a ferromagnet with ordering temperature T_c, an approximation to the form of the susceptibility for $T > T_c$ is given by the Curie-Weiss law $\chi = C/(T - T_c)$.

Our task in this chapter is to investigate how the muon can be used to probe magnetism in materials, allowing us to measure the internal field. We exploit the fact that the muon spin in a field $\boldsymbol{B}(\boldsymbol{r}_\mu)$, assumed not to be aligned with the initial muon spin direction, undergoes a precession at an angular frequency

$$\omega = \gamma_\mu B. \tag{6.6}$$

This feature provides the basis for the muon's sensitivity to magnetism in materials.

The fields in eqn 6.1 might be microscopic fields, $\boldsymbol{B}(\boldsymbol{r}), \boldsymbol{M}(\boldsymbol{r}), \boldsymbol{H}(\boldsymbol{r})$, varying with the position $\boldsymbol{r}$ inside the material, or they might be **thermodynamic variables**, which is the name given to volume averages of the microscopic fields, that result in a smooth dependence on $\boldsymbol{r}$. For the purpose of magnetometry, only these thermodynamic variables are relevant. However, the internal field that each muon experiences at its stopping site $\boldsymbol{r}_\mu$ is the microscopic **local field** $\boldsymbol{B}(\boldsymbol{r}_\mu)$. This depends strongly on the muon position[4] and on the details of the microscopic magnetization $\boldsymbol{M}(\boldsymbol{r})$, reflecting the arrangement of the magnetic ions and the nature of their magnetic moments $\boldsymbol{m}_i$.

As we shall discuss in the next section, all muons may feel the same field, and in this case their coherent spin precessions produce oscillations in the time-dependent asymmetry. However, when they experience a distribution of field values[5] the superposition of the corresponding out-of-phase spin precessions of the muon ensemble leads to a decaying asymmetry oscillation. The decay is governed by the variance of the distribution

$$\frac{\Delta^2}{\gamma_\mu^2} = \langle (B - \langle B \rangle)^2 \rangle = \langle B^2 \rangle - \langle B \rangle^2. \tag{6.7}$$

[3]We will often be dealing with cases when the orbital moment is quenched, so that $\boldsymbol{J} = \boldsymbol{S}$, where $\boldsymbol{S}$ is the spin of the ion. We often say 'total spin', when what we really mean is 'total angular momentum'.

[4]This is still true after the average over an ensemble of millions of individual muons, since they will usually stop at structurally equivalent interstitial sites.

[5]For simplicity, let us imagine that the local magnetic field $\boldsymbol{B} = B\hat{\boldsymbol{n}}$ at the muon site is aligned along some direction $\hat{\boldsymbol{n}}$, but that B can vary between different muon sites with probability distribution $p(B)$. The average field is

$$\langle B \rangle = \int \mathrm{d}B \, p(B) B.$$

By the same token, we can compute

$$\langle B^n \rangle = \int \mathrm{d}B \, p(B) B^n.$$

[6]It corresponds to a **fluctuation rate** $\nu = 1/\tau$, telling us how frequently the field changes.

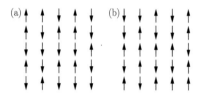

Fig. 6.1 (a) A disordered magnet, with spins arranged randomly. The magnetization $M = 0$. Turning all of the spins through 180° results in (b), which also has $M = 0$.

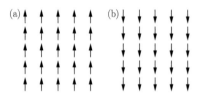

Fig. 6.2 (a) An ordered magnetic configuration with $M = M_0$. Turning the spins through 180° results in (b), which has magnetization $M = -M_0$.

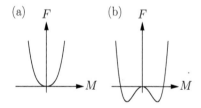

Fig. 6.3 (a) Landau free energy F for temperatures $T > T_c$. (b) The function F for $T < T_c$.

[7]This is because the Landau model assumes that every magnetic moment experiences the same environment and the magnetization is simply determined by a global minimization of the total free energy. Mean field theories treat every site to be the same, all exhibiting precisely the same values of each physical quantity. No deviations from average behaviour are permitted.

[8]For example, a rigorous treatment of the Heisenberg model (eqn 6.5) in three dimensions predicts $\beta = 0.367$.

When the average field vanishes, $\langle B \rangle = 0$, the variance simplifies to $\Delta^2/\gamma_\mu^2 = \langle B^2 \rangle$. When the B-field at the muon site jumps as a function of time to a different value, extracted from the same field distribution, the precession of the muon ensemble resulting from this dynamics encodes information both about the distribution and about the correlation time τ. This additional variable represents the average time that elapses between magnetic field value jumps.[6]

6.2 Static magnetic order

A snapshot of a paramagnet is shown in Fig. 6.1(a). This disordered system can be understood via its symmetry: the magnetization is zero in panel (a) and if we turn each of the moments through 180° then we get panel (b), which also has $M = 0$. An ordered **ferromagnet** with $M = M_0 \neq 0$ is shown in Fig. 6.2. The system has lost its symmetry since now a rotation of all of the moments leads to a reversal of the non-zero magnetization, that is $M_0 \to -M_0$. We say that the symmetry was broken (or lowered) on magnetic ordering. The ordering takes place at a magnetic phase transition at a temperature T_c.

A simple description of the magnetic phase transition is provided by the **Landau model**, where a system is described by a free energy

$$F(T, M) = F_0 + a_0 (T - T_c) M^2 + bM^4, \tag{6.8}$$

where a_0 and b are positive constants. Here M is the magnetization, which is also known as the **order parameter** in this context. This free energy is plotted for $T > T_c$ and $T < T_c$ in Fig. 6.3. The equilibrium state of the system is found by minimizing this free energy function. It is minimized by $M = 0$ for $T > T_c$, which corresponds to a magnetically disordered state. For $T < T_c$ it has minima at $M = \pm M_0$, with $M_0 = \pm \frac{a_0}{2b}(T_c - T)^{\frac{1}{2}}$, corresponding to the system aligning spins up ($M = +M_0$) or down ($M = -M_0$). The model is intended to apply only at temperatures close to the phase transition at T_c.

From the Landau model we obtain the prediction that, in the ordered magnetic state, $M(T) \propto (T_c - T)^\beta$ with $\beta = 1/2$. Important here is that this behaviour is solely governed by a *power law*, with no underlying length or time scale explicitly required. The Landau model provides a very powerful example of a so-called **mean field model**.[7] More refined models of the magnetic behaviour describe not only the mean field but also its correlations and fluctuations. Close to the phase transition all of these models predict similar power-law, scale-free behaviour, although the predicted values of β depend on details of the dimensionality of the model.[8] Furthermore, β is just one of a class of **critical exponents** which describe power-law behaviour of thermodynamic quantities near a phase transition. They are important because they classify fundamental aspects of the interactions. As we shall see, in favourable cases μSR can be very good at measuring critical exponents; locking on to a precession signal and measuring its frequency as $M \to 0$ as $T \to T_c$ is often better than attempting to measure the amplitude of a weak magnetic diffraction peak as it disappears into the noisy background.

In addition to the ferromagnet, in this chapter we shall also discuss the **antiferromagnetically ordered** state, shown schematically in Fig. 6.4. It can be thought of as two interpenetrating ferromagnetic sublattices, one with all spins pointing up and one with all spins pointing down.[9] For the antiferromagnet, the thermodynamic bulk magnetization vanishes, but the sublattice (or staggered) magnetization does not, and it is this that defines the order parameter in this case.

The simplest μSR experiment on a magnetically ordered material takes place in zero applied field. The local magnetic field at the muon site, determined mainly by the dipolar field of the ionic magnetic moments (as we shall discuss later), is then proportional to the local microscopic spontaneous magnetization M. Equation 6.6 then implies that measuring the muon precession frequency at different temperatures gives access to $M(T)$, and hence to the exponent β.[10] We are left with a general rule:

In a magnetically ordered material the zero-field muon precession frequency $\omega = \gamma_\mu B$ can often be taken as an order parameter.

Let's examine the muon-spin polarization function in a magnetically ordered material. Consider the geometry in Fig. 6.5, given in spherical polar coodinates, with a local field $\boldsymbol{B}$ oriented at an angle θ to the initial muon spin direction (assumed parallel to $\hat{\boldsymbol{z}}$). As discussed in Chapter 5, the muon spin expectation value is given by

$$\langle \sigma_z(t) \rangle = \cos^2\theta + \sin^2\theta\cos(\gamma_\mu B t). \tag{6.9}$$

The $\cos^2\theta$ term captures the contribution of the component of muon spin parallel to the field, which does not precess. The $\sin^2\theta$ term captures the component of the muon spin that lies perpendicular to the local field, which must precess at a frequency $\omega = \gamma_\mu B$, determined by the *magnitude* of the local magnetic field. We therefore have access to the magnitude of the local field at the muon site via this precession, while information about the direction of the field is encoded in the amplitudes of the two contributing components.

In general, in an experiment we measure the polarization of many muons as a function of time, as an average over the appropriate distributions of (i) the field direction $f(\theta, \phi)$ and (ii) the field magnitude $p(B)$ that describe the magnetic environment of the muon ensemble. This is to say we measure the time dependence of the average polarization $P_z(t)$ of the muon ensemble.

One useful example is the case where a magnetic field points, with equal probability, along all possible directions with respect to the initial muon spin direction. This might describe a polycrystalline, perfectly magnetically ordered material, in which muons always see the same magnitude of the local magnetic field, but the direction of the magnetization is different in each of the large numbers of crystallites in which the muons implant. In this case we must average the muon polarization over a random distribution of angles θ as discussed in Chapter 5. The result is

[9]It is worth noting that when we take quantum mechanics into account, the antiferromagnet is rather more complicated than the simple classical cartoon would suggest. It turns out that the classical antiferromagnet is not an eigenstate of the Heisenberg Hamiltonian. However, it can be constructed as a superposition of low-energy eigenstates, which form a narrow and stable wavepacket.

[10]Notice that an external field requires additional terms to the Landau free energy in eqn 6.8, which may modify (or remove altogether) the magnetic transition. In this context it is particularly important that μSR can be performed in zero field.

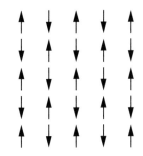

Fig. 6.4 In a two sub-lattice antiferromagnet, the magnetic moments alternatively point in opposite directions. This is known as the Néel state.

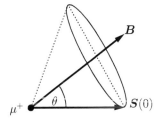

Fig. 6.5 The muon spin initially lies at an angle θ to the direction of the local magnetic field $\boldsymbol{B}$.

that for an environment featuring randomly oriented magnetic fields of constant magnitude B_1 we measure a muon-spin polarization

$$P_z(t) = \frac{1}{3} + \frac{2}{3}\cos(\gamma_\mu B_1 t). \qquad (6.10)$$

This can be thought of as describing an ensemble of muons with $1/3$ of the local magnetic field components pointing parallel to the initial muon spin and $2/3$ of the components in a perpendicular direction. In all real magnetic materials there will be some amount of variation between the magnetic field at the different muon sites, which will lead to some relaxation. However, if this variation is small we will observe a precession about the average local field at the muon site. The simple principle of using the frequency of the oscillations of the muon polarization to access the local magnetic field B is a powerful one that we can use to understand the evolution of the magnetization of a magnetic material.

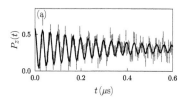

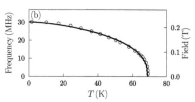

Fig. 6.6 (a) Oscillations in the muon precession frequency in ferromagnetic EuO are seen below T_c. (b) The temperature evolution of the muon precession frequency. [Adapted from S. J. Blundell *et al.*, Phys. Rev. B **81**, 092407 (2010).]

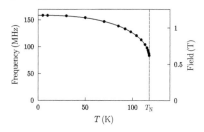

Fig. 6.7 Frequency of the oscillations observed in the antiferromagnet MnO below the ordering temperature T_N. In this particular example, the frequency appears to fall to a non-zero value as the order temperature (indicated by the vertical dashed line) is reached. [Data from S. J. Blundell *et al.* in preparation; MnO first measured using muons in Y. J. Uemura *et al.*, Hyp. Int. **17**, 339 (1984).]

Example 6.1

EuO is a cubic ferromagnet with localized moments that order below a critical temperature $T_c = 69$ K. Oscillations are seen in the muon polarization below this temperature [Fig. 6.6(a)] which enable the precession frequency to be followed as a function of temperature [Fig. 6.6(b)] allowing a parameter $\beta = 0.32(2)$ to be estimated.

The same principle applies to an antiferromagnet, since its microscopic magnetization does not vanish (as its thermodynamic average, measured by a magnetometer, does), but rather alternates between the two oppositely oriented sublattices. The local field at the muon site will also, generally speaking, not vanish. In antiferromagnetically ordered MnO we again see oscillations, allowing the precession frequency to be followed (Fig. 6.7).

It is often the case that the oscillations are observed at a number of frequencies. This reflects the fact that there might be a number of possible muon sites that are magnetically inequivalent in the crystal (that they can be either crystallographically equivalent or not). In this case, we can simply add their contributions to the polarization, weighted by their occupation probability.

Example 6.2

Consider a polycrystalline magnet in which there are a range of possible muon sites and let's say that there is a probability p_i of an implanted muon landing in the ith muon site and hence experiencing a local magnetic field of B_i. Then, by a simple process of averaging, we have that

$$P_z(t) = \sum_i p_i \left[\frac{1}{3} + \frac{2}{3}\cos(\gamma_\mu B_i t)\right] = \frac{1}{3} + \frac{2}{3}\sum_i p_i \cos(\gamma_\mu B_i t), \qquad (6.11)$$

where we have used that $\sum_i p_i = 1$, so that the probabilities sum to unity. This situation can arise if there are a number of magnetically inequivalent muon sites in the unit cell (which might or might not be structurally equivalent) and/or there are different magnetic regions within the same sample.

As a demonstration that useful information can be obtained when many muon sites occur, we can consider Cu(pyz)$_2$(ClO$_4$)$_2$. This compound is a molecular antiferromagnet (with the magnetic moments on the Cu^{2+} ions) that shows a complicated time-dependent muon spectrum at low temperature [Fig. 6.8 (top)]. These data can be Fourier transformed, revealing three different frequency components. Fitting the data in the time domain using three frequency components gives a best fit, shown as the solid line through the data. This process is repeated for data measured at other temperatures and the resulting temperature dependence of the three frequencies is shown in the lower part of Fig. 6.8. These three curves follow the same temperature dependence, albeit scaled by different amounts. One can deduce that each of the frequencies is proportional to the same order parameter (the staggered magnetization of the molecular antiferromagnet) but with distinct constants of proportionality, reflecting the location of the three different muon sites with respect to the copper moments.

Note that there are even more complicated cases where the frequencies at different muon sites can show different temperature dependences.

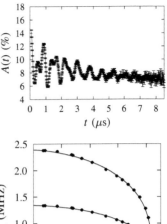

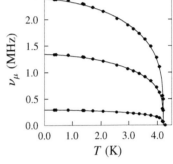

Fig. 6.8 Oscillations at three frequencies are seen in the molecular antiferromagnet Cu(pyz)$_2$(ClO$_4$)$_2$ below an ordering temperature of $T_N = 4.3$ K. [Adapted from T. Lancaster *et al.* Phys. Rev. B **75**, 094421 (2007).]

Having extracted the temperature dependence of the order parameter, the next task is to model it. This can be a delicate business since theoretical expressions are not available in the general case. We can however expect that (i) at very low temperature power-law behaviour is likely due to magnetic excitations reducing the order [a good example of which is the famous Bloch-$T^{3/2}$ law that is expected in three-dimensional ferromagnets, leading to $M(T) = M(0)(1 - aT^{3/2})$ where the constant a is proportional to an average Bose factor]; (ii) at temperatures near to the transition temperature, we also expect power-law behaviour (though with a different scaling exponent) due to critical scaling. A good first step in extracting the general form of the temperature dependence of the precession frequency $\nu(T)$ (and hence the temperature dependence of the order parameter) is provided by the phenomenological function

$$\nu = \nu_0 \left[1 - \left(\frac{T}{T_c} \right)^\alpha \right]^\beta, \tag{6.12}$$

where $\alpha > 1$ and $\beta < 1$ are numbers. The properties of this function are explored in the following example.

Example 6.3

At temperatures just below the transition we have $T/T_c = 1 - \varepsilon$, where ε is a small parameter, and so the expression in eqn 6.12 becomes $\nu \approx \nu_0(\alpha\varepsilon)^\beta$, resulting in

$$\nu \approx \nu_0 \alpha^\beta \left(1 - \frac{T}{T_c} \right)^\beta \quad (T \to T_c^-), \tag{6.13}$$

and hence $\nu(T) \propto (T_c - T)^\beta$ just below the transition, as expected for critical scaling, and yielding the critical exponent β. In order to get a good fit to this behaviour near the transition, it can be very useful to use a **scaling plot** such as the one shown inset in Fig. 6.9. This is a plot of $\nu(T)/\nu(0)$ against $(1 - T/T_c)$ on a log-log scale in the region close to $T = T_c$. This is very sensitive to the choice of T_c (which is not known a priori), but can be repeated for different values of T_c until it leads a straight line behaviour (at least close to the transition). Once this is achieved, it can yield fairly reliable values of β. At low temperatures we can take a limit that yields

$$\nu \approx \nu_0 \left[1 - \beta \left(\frac{T}{T_c} \right)^\alpha \right] \quad (T \to 0). \tag{6.14}$$

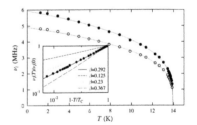

Fig. 6.9 Two frequencies are observed in Cs$_2$AgF$_4$. Critical parameters can be extracted using a scaling plot as described in the text. [Adapted from T. Lancaster *et al.* Phys. Rev. B **75**, R220408 (2007).]

[11]The Bloch–$T^{3/2}$ law is made quantitatively equivalent to eqn 6.14 (at low T) by shifting the critical temperature to $(T_c^*)^{\frac{3}{2}} = \beta/a$.

This behaviour approximates the true reduction in magnetization $\Delta M(T)/M(0)$ due to spin waves. For isotropic ferromagnets $\Delta M(T)/M(0) = (1 - aT^{\frac{3}{2}})$ is given by the Bloch–$T^{3/2}$ law.[11] Magnetic anisotropy gives rise to a gap in spin wave excitations, $k_{\mathrm{B}} T_{\mathrm{g}}$, for which $\Delta M(T)/M(0) \approx [1 - b \exp(-2T_{\mathrm{g}}/T)]$ at low temperatures, and this is roughly mimicked by values of $\alpha > \frac{3}{2}$.

Muon data for both ferromagnets and antiferromagnets look rather similar, since in both cases we observe a precession frequency proportional to the order parameter. Thus in general, muons cannot be used to straightforwardly and reliably distinguish between ferromagnets and antiferromagnets. However, μSR is not a technique in isolation and it is always helpful to have access to the temperature-dependent static magnetic susceptibility $\chi(T)$, which will usually give a clear indication of whether a material hosts ferromagnetic or antiferromagnetic interactions. Susceptibility is, however, a bulk thermodynamic technique that actually measures the total magnetic moment of a sample, i.e. $\int \boldsymbol{M} \, \mathrm{d}V$, where V is the volume, in the presence of a field $\boldsymbol{H}$ applied to the entire sample. It only therefore gives an unambiguous response if $\boldsymbol{M}$ is uniform across the sample and so is unable to distinguish between a sample with a uniform, weakly magnetic response, and one with a small amount of a strongly magnetic impurity. The muon asymmetry, on the other hand, will locally probe the presence of different phases in a material. They will appear as distinct components of the spectrum, with the proportions of the magnetic phases present reflected in their relative amplitudes. Moreover, the μSR examples in this section all used zero-field data and so no field was applied to the sample (so that we can be sure that the state of the sample was not affected by the applied field). More detail on this combination of techniques can be found in Chapter 23.

6.3   The local magnetic field

The muon is a spin $1/2$ particle, so all muon spin interactions can be written in the form of a Zeeman Hamiltonian, with the muon polarization entirely determined by the local field at the muon site, $\boldsymbol{B}(\boldsymbol{r}_\mu)$. However, inside matter there are too many sources of $\boldsymbol{B}$ to calculate such a local field as a solution of Maxwell's equations, as eqn 6.1 might imply. It is easier to classify the sources, grouping them in several (somewhat arbitrary) categories, and to calculate the local field as a sum of these terms, $\boldsymbol{B}(\boldsymbol{r}_\mu) = \sum_i \boldsymbol{B}_i(\boldsymbol{r}_\mu)$.

We divide the field into four classes of contribution: (i) the external field, (ii) the dipolar field (the main subject of this section), (iii) the local hyperfine field, and the (iv) diamagnetic contribution of superconducting currents. To simplify things further we shall subdivide these classes into a total of six contributions as

[12]Notice that, although we need just the magnitude B to compute the muon precession frequency, we must compute the individual vector field terms, since they are not necessarily collinear.

$$\boldsymbol{B} = \boldsymbol{B}_0 + \boldsymbol{B}_{\mathrm{dip}} + \boldsymbol{B}_{\mathrm{L}} + \boldsymbol{B}_{\mathrm{dem}} + \boldsymbol{B}_{\mathrm{hf}} + \boldsymbol{B}_{\mathrm{dia}}. \tag{6.15}$$

Our task in this section is to examine each of these in turn.[12] We start with the first two of these.

- If a field is applied externally, we have $B_0 = \mu_0 H_0$, where H_0 is the externally applied magnetic field, usually due to currents in a solenoid, or a pair of Helmholtz coils, at the centre of which is the sample.

- There is also a magnetic field due to the **dipolar interaction** between the magnetic moments in the sample and the magnetic moment of the muon.[13] Here we consider only magnetic moments arising from ions that are sufficiently far from the muon that there is no appreciable overlap between the muon wave function and that of the magnetic moments.

The dipolar interaction is a rich source of important physics, so we pause to examine it further. Let's start with a single magnetic moment vector m at the origin of a coordinate system. This gives rise to a vector dipole field $B(r)$ at position r, described by[14]

$$B(r) = \frac{\mu_0}{4\pi r^3}\left[\frac{3(m\cdot r)r}{r^2} - m\right]. \tag{6.16}$$

Although this expression looks rather complicated, it simply represents the familiar field pattern around a bar magnet, as shown in Fig. 6.10. The equation can be more easily understood when rewritten as a tensor equation

$$B(r) = \underline{D}(r)\cdot m, \tag{6.17}$$

where the **dipolar tensor** $\underline{D}(r)$ (think of it as a 3×3 matrix), which is a function of $r = (x, y, z)$, can be written as

$$\underline{D}(r) = \frac{\mu_0}{4\pi r^3}\begin{pmatrix} -1+\frac{3x^2}{r^2} & \frac{3xy}{r^2} & \frac{3xz}{r^2} \\ \frac{3yx}{r^2} & -1+\frac{3y^2}{r^2} & \frac{3yz}{r^2} \\ \frac{3zx}{r^2} & \frac{3zy}{r^2} & -1+\frac{3z^2}{r^2} \end{pmatrix}. \tag{6.18}$$

Since the dipolar field is linearly related to the size of the ordered magnetic moment, the magnitude of which is a measure of the order parameter, the dipolar field gives a direct measure of the order parameter.

[13]We avoid describing this field by a specific symbol, since below we want to reserve B_{dip} just for a part of this dipolar term.

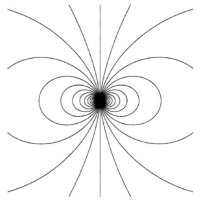

Fig. 6.10 An example of dipolar magnetic field (eqn 6.16) is that around a common bar magnet.

[14]This equation has an unphysical singularity at the origin, which will be addressed later (using eqn 6.25).

Example 6.4

We can explore the dipolar field in some simple geometries.

(a) The local field $B(r)$ at position $r = (0, 0, a)$, due to one moment oriented in a generic direction, requires the following product of the dipolar tensor and the vector moment:

$$B(r) = \frac{\mu_0}{4\pi a^3}\begin{pmatrix} -1 & 0 & 0 \\ 0 & -1 & 0 \\ 0 & 0 & 2 \end{pmatrix}\begin{pmatrix} m_x \\ m_y \\ m_z \end{pmatrix}. \tag{6.19}$$

The result is

$$B = -\frac{\mu_0}{4\pi a^3}\begin{pmatrix} m_x \\ m_y \\ -2m_z \end{pmatrix}. \tag{6.20}$$

We can also map the local field at different locations on a sphere of radius a, fixing the magnetic moment along $\hat{z}$, with components $(0, 0, m)$. We obtain

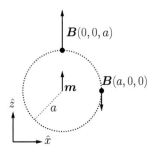

Fig. 6.11 A moment m directed along z. The dipole field vector at $(0,0,a)$ is twice the magnitude and opposite to that at $(a,0,0)$.

(b) $\boldsymbol{B} = \mu_0 m/2\pi a^3 \hat{\boldsymbol{z}}$ at the position $(0,0,a)$, at the pole.

(c) $\boldsymbol{B} = -\mu_0 m/4\pi a^3 \hat{\boldsymbol{z}}$ at the positions $(a,0,0)$, and, by symmetry, at all positions $(a\cos\phi, a\sin\phi, 0)$, with any ϕ; this is the field value at the equator.

These two example show that the field at the poles, i.e. along the moment direction, is twice as large and oppositely oriented than in the equatorial plane. This is shown in Fig. 6.11.

Usually there are many magnetic moments contributing to the field at $\boldsymbol{r}_\mu$, and so we modify eqn 6.17 as follows: if the ith moment is at position $\boldsymbol{r}_i$ the field at position $\boldsymbol{r}_\mu$ is simply given by $\boldsymbol{B}(\boldsymbol{r}_\mu) = \underline{\boldsymbol{D}}(\boldsymbol{r}_\mu - \boldsymbol{r}_i) \cdot \boldsymbol{m}(\boldsymbol{r}_i)$. In this case the total local field is given by the sum

$$\boldsymbol{B}(\boldsymbol{r}_\mu) = \sum_i \underline{\boldsymbol{D}}(\boldsymbol{r}_\mu - \boldsymbol{r}_i) \cdot \boldsymbol{m}(\boldsymbol{r}_i). \tag{6.21}$$

This sum is extended to an extremely large number of moments in the material. We note that, far away, individual moments contribute much less than those close to the muon, since the field dies off as r^{-3}. However, their number also grows with distance as r^2, and the calculation must be dealt with carefully.[15] In fact, we can calculate the dipolar field for a magnetic sample by using a trick invented by Lorentz and draw a large sphere (known as the Lorentz sphere) around the point $\boldsymbol{r}_\mu$ where we seek to evaluate the field. We then split the sum in two parts, as shown in Fig. 6.12. The two parts are

(i) sites within the Lorentz sphere, close to the muon, where we need the microscopic details of the dipole fields, and

(ii) sites outside the Lorentz sphere, far from the muon, where it is convenient to use the continuum limit, and so we replace the individual discrete magnetic moments by the continuous magnetization field (the macroscopic bulk average). Here we need to remember that divergences in the magnetization act as sinks of the $\boldsymbol{H}$-field (because $\nabla \cdot \boldsymbol{B} = 0$ implies $\nabla \cdot \boldsymbol{H} = -\nabla \cdot \boldsymbol{M}$) which you can think of as effective magnetic poles (or charges).

In (ii) we conventionally further distinguish two contributions to $\boldsymbol{B}$: one from the surface of the spherical cavity obtained by carving out the Lorentz sphere and one from the sample's external surface.[16] Hence, we can subdivide the full distant dipole field into:

- $\boldsymbol{B}_{\mathrm{dip}}$, the lattice sum of dipole fields inside the Lorentz sphere.
- $\boldsymbol{B}_{\mathrm{L}}$, the Lorentz field, i.e. the contribution from the poles on the Lorentz cavity. In the next example we show that it is given by $\boldsymbol{B}_{\mathrm{L}} = \mu_0 \boldsymbol{M}/3$, where $\boldsymbol{M}$ is the bulk magnetization of the material.
- $\boldsymbol{B}_{\mathrm{dem}}$, the demagnetizing field, arising from poles on the surfaces of the magnetic sample. In general this produces a non-uniform contribution to the field at the muon site, and is larger for sites close to the sharp edges of the sample under study. Some simple shapes produce a uniform demagnetizing field, whose αth Cartesian component is given by $(\boldsymbol{B}_{\mathrm{dem}})_\alpha = -\mu_0 N_{\alpha\beta}(\boldsymbol{M})_\beta$, where $(\boldsymbol{M})_\beta$ is the

[15]The sum therefore has terms which grow like $1/r$; such a series is said to be conditionally convergent, which means you have to be careful in finding the limit. As we shall see, this implies that the surface of the sample therefore plays an important role and can't simply be assumed to be sufficiently far away that it doesn't matter.

[16]These two terms are proportional to the bulk magnetization. Therefore they vanish for both antiferromagnets and unmagnetized ferro-, ferrimagnetic samples, which have $M = 0$.

βth component of the bulk magnetization and $N_{\alpha\beta}$ are the components of the demagnetizing tensor $\underline{N}$, entirely determined by sample geometry.[17] Example 6.5 shows that the infinite slab with axis along $\hat{z} \parallel M$ has $N_{zz} = 1$ (with all other components vanishing).

[17]It turns out (as shown first by Maxwell) that ellipsoids, including their extremal cases, the infinite plane and the infinitely long cylinder, produce uniform fields. The components of $\underline{N}$ for these cases along with the numerical methods for the more general inhomogeneous case can be found in the literature.

Example 6.5

Consider a slab of magnetized material containing a spherical cavity (the Lorentz sphere) as shown in Fig. 6.12. A basic result of electromagnetism is that the field B due to an assembly of magnetic dipoles confined to some volume is equivalent to that produced by two contributions (i) a bound current density $\nabla \times M$ inside the volume and (ii) a contribution from the surface. In this problem, the magnetization is uniform inside the volume and so we only have to consider the effect of the surface. This is just the effect of a surface magnetic charge density σ_m given by

$$\sigma_m = M \cdot \hat{n}, \tag{6.22}$$

where $\hat{n}$ is a unit vector normal to the surface, pointing out of the surface.[18] Using this result we can show how we obtain (1) the demagnetizing field from the flat surfaces of our slab and (2) the Lorentz field from the surface of the spherical cavity.

(1) Let's do the easy surface first, which is the boundary of the slab, i.e. the top and bottom flat surfaces. As shown in Fig. 6.12, we have positive magnetic charges on the top surface (because there M is *parallel* to $\hat{n}$) and negative magnetic charges on the bottom surface (because there M is *antiparallel* to $\hat{n}$). Thus there is a magnetic charge density $\sigma_m = M = |M|$ on the top surface and $\sigma_m = -M$ on the bottom surface. This is just like the parallel plate capacitor problem, for which the electric field would be $E = \sigma/\epsilon_0$, pointing from positive to negative (equivalently $D = \sigma$). By analogy, the magnetic field $H = \sigma_m = -M$ and we can write this contribution as the demagnetizing field $B_{\mathrm{dem}} = -\mu_0 M$ (and in terms of the demagnetizing tensor, this is equivalent to showing that $N_{zz} = 1$).

(2) Now we can tackle the spherical surface and work out the field at the origin due just to the surface density $\sigma_m(\theta)$ on the Lorentz sphere. This is given by

$$\sigma_m(\theta) = -M \cdot \hat{n} = -M \cos\theta. \tag{6.23}$$

The calculation of the field inside this surface distribution is discussed in Exercise 6.2. Here we evaluate it by direct integration, taking the radius of the Lorentz sphere as r. Between θ and $\theta + \mathrm{d}\theta$ there is a ring of charge equal to $\mathrm{d}q_m = \sigma(\theta) \times 2\pi r \sin\theta \times r\,\mathrm{d}\theta$ and this will produce a field of magnitude $\mathrm{d}H_z = \mathrm{d}q_m \cos\theta / 4\pi r^2$ along the z-axis, the factor of $\cos\theta$ representing the component along z and all the transverse components cancel by symmetry. Thus

$$H_z = \int_0^\pi \frac{|\sigma(\theta)| \times 2\pi r^2 \sin\theta \, \cos\theta \, \mathrm{d}\theta}{4\pi r^2} = \int_0^\pi \frac{1}{2}\cos^2\theta \, \sin\theta \, \mathrm{d}\theta = \frac{M}{3}. \tag{6.24}$$

In general, therefore, in the cavity, the sample outside of the Lorentz sphere contributes an extra field at the muon site of $B_L = \mu_0 H_L = \mu_0 M/3$. Note that both B_{dem} and B_L contribute in ferromagnets, but B_{dem} depends on the shape of the sample. (For the case of a spherical sample $B_{\mathrm{dem}} = -\mu_0 M/3$ and hence $B_{\mathrm{dem}} + B_L = 0$.)

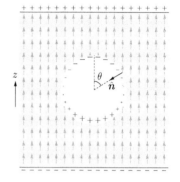

Fig. 6.12 The Lorentz sphere cut into a slab of magnetized material, showing the normal unit vector $\hat{n}(\theta)$ at its surface. The z-axis is vertical.

[18]Although a magnetic dipole is really a loop of current, you can think of it here just like an electric dipole: each arrow can thus be replaced a positive and negative magnetic charge at the head and tail of the arrow respectively. Inside the volume of the uniformly magnetized material all the positive and negative magnetic charges cancel out where the head of each arrow touches the tail of the next arrow. But, just considering the slab in Fig. 6.12, then at the top surface, where all the arrow heads poke out, you accumulate positive magnetic charge; at the bottom surface, all the arrow tails lead to negative magnetic charge.

Here are the last local-field terms, B_{hf} and B_{dia}:

- The **hyperfine field** B_{hf}, is the local field due the electronic structure in the immediate surrounding of the muon, at distances comparable to the extent of the electron wavefunctions.[19] Under these

[19]Be aware that our definition is not universally adopted; other texts might include the distant dipole contributions in this term.

conditions we must consider the case when the positions of the electron, r, and of the muon, r_μ coincide. This is particularly important in metals, where there is finite electronic probability density at the muon site. Quantitatively the magnitude of this interaction is difficult to assess accurately, not least because of the perturbation of the electronic density caused by the presence of the muon's charge. In practice the formula

$$B(r_\mu) = \frac{\mu_0}{4\pi} \sum_{j=1}^{N'} \frac{A_j m_j}{V_c}, \quad (6.25)$$

provides a fair description of the field. In eqn 6.25 the sum is restricted to the N' nearest (and possibly next-nearest) neighbours of the muon and V_c is the volume per magnetic atom. The quantity A_j is a dimensionless parameter,[20] which takes the same value for all the neighbours of the same type. A simple classical evaluation of the sum shows that the term can be written

$$B(r) = \frac{2\mu_0}{3} m \delta^{(3)}(r - r_\mu). \quad (6.26)$$

This expression can then be used to replace the singularity in the dipolar sum (eqn 6.16). Furthermore, treating eqn 6.16 with this additional part as an operator, and averaging over the electronic wavefunction $\psi(r)$, allows us to identify two contributions to the hyperfine field (see Fig. 6.13):

(i) The **contact-hyperfine field**[21] B_c, proportional to the local electron spin density at the muon site

$$\langle \psi | \hat{\sigma} \delta^{(3)}(r - r_\mu) | \psi \rangle = |\psi_\uparrow(r_\mu)|^2 - |\psi_\downarrow(r_\mu)|^2. \quad (6.27)$$

(ii) The **transferred-hyperfine field B_{trans}**. It includes the so-called pseudo-dipolar term, the average of eqn 6.16 over the higher orbital momentum components centered at r_μ. In magnetic insulators the transferred-hyperfine term can also include contributions from nearest-neighbour magnetic ions, either directly or mediated by the hybridization of the ligand wave functions (known as super-hyperfine couplings). In metals the interaction yielding this field, is the **Ruderman-Kittel-Kasuya-Yosida (RKKY) interaction** and arises due to the extended electron wave functions.[22]

- The diamagnetic field B_{dia} is due to spin-paired electron currents and is generally very small. It is relevant in μSR only for superconducting materials, where it arises from their large supercurrent density, as discussed in Chapter 9.

To summarize, in the most general case we may write the total magnetic field at the muon site as a sum of seven terms. However, some good news is that these terms are not all present at once. For instance B_{dia} and the previous five terms typically exclude each other. For an antiferromagnet we have that the bulk magnetization is $M = 0$; therefore the

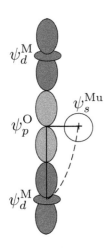

Fig. 6.13 Sketch of a muon (cross) bound to an M-O-M group, where M is a $3d$ atom, O is an oxygen atom and valence atomic orbitals are shown. The contact-hyperfine field is $B_c \propto |\psi_s^{\text{Mu}}(0)|^2$, the dashed and solid paths indicate the overlap contributing to the direct- and transferred-hyperfine fields, respectively.

[21]This is sometimes called the **Fermi-contact field**. Projecting the total electron ψ onto the local atomic wave function centred at r_μ, only the s-wave projection ψ_s^{Mu} contributes, since higher orbital momentum components have nodes at the muon site.

[22]The RKKY mechanism originally refers to the indirect interaction among nuclear spins mediated by conduction electrons.

total field at the muon site is given just by $\boldsymbol{B} = \boldsymbol{B}_0 + \boldsymbol{B}_{\mathrm{dip}} + \boldsymbol{B}_{\mathrm{c}} + \boldsymbol{B}_{\mathrm{trans}}$. Some more good news is that $\boldsymbol{B}_{\mathrm{trans}}$ and $\boldsymbol{B}_{\mathrm{c}}$ are often small in non-metals. Therefore, for the commonly encountered case of an antiferro-magnetic insulator, the applied field $\boldsymbol{B}_0$ and the dipolar field $\boldsymbol{B}_{\mathrm{dip}}$ are sufficient for an order of magnitude evaluation, and often even for an accurate computation of the local field.

Example 6.6

Although we have concentrated on electronic moments, the muon spin also interacts with nuclei. For a muon localized at an interstitial site, there is no overlap between the nuclei and muon wavefunctions and therefore no direct interaction. Only the classical dipolar interaction must be accounted for, and therefore eqn 6.16 applies, in which the magnetic moment $\boldsymbol{m}_j$ is generally written in terms of the nucleus spin $\boldsymbol{I}_j$ and gyromagnetic ratio γ_j through the relation $\boldsymbol{m}_j = \gamma_j \hbar \boldsymbol{I}_j$. Of course, only non-spinless nuclei contribute to this field.[23] Finally, we note that in contrast to the case of nuclear techniques like NMR, Mössbauer spectroscopy, perturbed angular correlations etc., no quadrupole interaction between the muon spin and its environment need be considered. This results from the muon's spin 1/2. Sometimes, however, the effect of an electric field gradient acting on the system nuclei, including the muon-induced electric field gradient, must be accounted for since it affects the nuclear magnetic field acting on the muon.

[23]Tables of nuclear spins and moments are given in Appendix B.

6.4   Static field distributions

We frequently find muon data which result from a distribution of magnetic fields because there is a different value of $\boldsymbol{B}(\boldsymbol{r}_\mu)$ experienced at each muon site. Formally, assuming a distribution function $p(\boldsymbol{B})$, our task is to compute the muon polarization, given by

$$P_z(t, \theta) = \int \mathrm{d}^3 B \, p(\boldsymbol{B}) \left[\cos^2 \theta + \sin^2 \theta \cos(\gamma_\mu B t) \right]. \qquad (6.28)$$

Here the $\cos^2 \theta$ factor tells us about the component of field directed parallel to the muon spin, whereas the $\sin^2 \theta$ factor multiplying the precessing term tells us about the perpendicular component.

We saw in Chapter 5 and again in eqn 6.10 that for the case of randomly directed fields, all with the same magnitude B_1, we had the result $P_z(t) = \frac{1}{3} + \frac{2}{3} \cos \gamma_\mu B_1 t$. We interpreted this in terms of 1/3 of the field components being directed parallel to the initial muon spin and 2/3 directed perpendicular. In the general case of a distribution of the magnitudes of the randomly oriented fields, we need to consider the spread in field components perpendicular to the initial muon-spin direction since they induce precessions at different frequencies, whose dephasing cause the decay of the time dependent signal. The field components directed parallel to the muon spin do not induce precessions, nor decay, giving rise to the constant 1/3 term.[24] If the field along the ith Cartesian direction

[24]Known as the 1/3 tail.

is distributed normally about $B_i = 0$ then the normalized distribution $p(B_i)$ takes the form

$$p(B_i) = \sqrt{\frac{\gamma_\mu^2}{2\pi\Delta^2}}\, e^{-\frac{\gamma_\mu^2 B_i^2}{2\Delta^2}}, \qquad (6.29)$$

where Δ^2/γ_μ^2 is proportional to the variance of the distribution, that is, $\Delta^2 = \gamma_\mu^2\langle B_i^2\rangle$. Owing to the central limit theorem, independent random variables will have a distribution tending towards a normal distribution, so this is frequently a good approximation.

Example 6.7

[25] Often the case for crystals with random ion substitutions.

Consider an inhomogeneous magnetically ordered system,[25] where the three magnetic field components at the muon site have a Gaussian distribution, narrow compared to the average field magnitude $B_1 = |\langle \boldsymbol{B}\rangle|$. In this case, the direction of the local field is always very close to that of the average, since the small deviations in the two orthogonal directions contribute negligibly to the total vector, and the local field magnitude itself is to a good approximation distributed according to a one-dimensional Gaussian probability

$$p(B) = \left(\frac{\gamma_\mu^2}{2\pi\Delta^2}\right)^{\frac{1}{2}} e^{-\frac{\gamma_\mu^2 (B-B_1)^2}{2\Delta^2}}, \qquad (6.30)$$

where $\Delta/\gamma_\mu^2 = \langle B^2\rangle$ is the width of the distribution of the components parallel to the direction of the average local field. We can then compute $P_z(t) = \frac{1}{3} + \frac{2}{3}\int dB\, p(B)\cos(\gamma_\mu B t)$. The Fourier cosine transform in this expression can be computed using the shifting theorem, and the polarization becomes

$$P_z(t) = \frac{1}{3} + \frac{2}{3} e^{-\left(\frac{\Delta^2 t^2}{2}\right)} \cos\gamma_\mu B_1 t. \qquad (6.31)$$

That is, the corresponding non-monochromatic oscillations dephase and decay with a Gaussian envelope. Expanding the envelope function we see that at early times the decay varies as $1 - \Delta^2 t^2/2$, which is half the decay rate of the Kubo-Toyabe function.[26] This is explained by noting that here, in high field, *only* the static deviations *along* the field direction dephase the muons, and the deviations in the two orthogonal directions are ineffective because they do not change the field magnitude. By contrast, in the Kubo-Toyabe case, the relaxation is caused by deviations along the *two* directions perpendicular to the muon spin.

[26] See Chapter 5.

The static local dipolar fields arise from non-interacting static magnetic moments surrounding the muon. Next we shall investigate how the magnetic field distribution at the muon site can be related back to the electronic spins in the material.

Example 6.8

In the same way that crystalline systems are formed by a periodic arrangement of atoms, the magnetic moments in a magnetically ordered compound form a periodic array. Because of the periodic character of magnetic moment arrangement, they can be described with a spatial Fourier transform. The equations

$$\boldsymbol{S_q} = \frac{1}{\sqrt{N_c}}\sum_{j=1}^{N_c} \exp(-i\boldsymbol{q}\cdot\boldsymbol{j})\boldsymbol{S_j} \quad \text{and} \quad \boldsymbol{S_j} = \frac{1}{\sqrt{N_c}}\sum_q \exp(i\boldsymbol{q}\cdot\boldsymbol{j})\boldsymbol{S_q}, \qquad (6.32)$$

relate the spins $\boldsymbol{S}_j$, where j spans the set of the N_c magnetic atoms, to their Fourier components $\boldsymbol{S_q}$. The jth magnetic atom is located at vector position $\boldsymbol{j}$ in the crystal and the $\boldsymbol{q}$ vectors belong to the first Brillouin zone.

We shall consider two types of magnet: the ferromagnet for which $\boldsymbol{S_q} = 0$ for all $\boldsymbol{q}$ vectors with the exception of $\boldsymbol{q} = 0$, and the antiferromagnet for which $\boldsymbol{S_q} = 0$, except for $\boldsymbol{q} = \pm\boldsymbol{k}$. The vector $\boldsymbol{k}$ characterizes the spatial periodicity of the magnetic moment arrangement: it is called the magnetic structure propagation wavevector. A ferromagnet is then viewed as a magnetic structure with a $\boldsymbol{k} = 0$ propagation wavevector.

At this stage it is convenient to express $\boldsymbol{B}_{\mathrm{dip}}$ and $\boldsymbol{B}_c$ as a function of the Fourier components of $\boldsymbol{S}_j$ (eqn 6.32). Since both $\boldsymbol{B}_{\mathrm{dip}}$ and $\boldsymbol{B}_c$ linearly depend on $\boldsymbol{S}_j$, we define the dimensionless tensor $\boldsymbol{C_{r_j}}$ such that

$$\boldsymbol{B}_{\mathrm{loc}} = \boldsymbol{B}_{\mathrm{dip}} + \boldsymbol{B}_c = -\frac{\mu_0}{4\pi}\frac{g\mu_{\mathrm{B}}}{\mathcal{V}_c}\sum_{j=1}^{N_c}\boldsymbol{C_{r_j}}\boldsymbol{S}_j, \tag{6.33}$$

where $\mathcal{V}_c$ is the volume per ion. Inspired by the Fourier transform, we introduce[27]

$$\boldsymbol{C_q} = \sum_j \exp(\mathrm{i}\boldsymbol{q}\cdot\boldsymbol{r}_j)\boldsymbol{C_{r_j}}, \text{ and accordingly } \boldsymbol{C_{r_j}} = \frac{1}{N_c}\sum_{\boldsymbol{q}}\exp(-\mathrm{i}\boldsymbol{q}\cdot\boldsymbol{r}_j)\boldsymbol{C_q}, \tag{6.34}$$

which leads to the following expression for $\boldsymbol{B}_{\mathrm{loc}}$,

$$\boldsymbol{B}_{\mathrm{loc}} = -\frac{\mu_0}{4\pi}\frac{g\mu_{\mathrm{B}}}{\sqrt{N_c}\mathcal{V}_c}\sum_{\boldsymbol{q}}\exp(-\mathrm{i}\boldsymbol{q}\cdot\boldsymbol{r}_0)\boldsymbol{C_q}\boldsymbol{S_q}. \tag{6.35}$$

Here, vector $\boldsymbol{r}_0$ denotes the vector distance from the muon to the origin of the lattice.[28] Returning to the expression for $\boldsymbol{B}_{\mathrm{loc}}$ in the ordered phase of a magnet, we have

$$\boldsymbol{B}_{\mathrm{loc}} = -\frac{\mu_0}{4\pi}\frac{g\mu_{\mathrm{B}}}{\sqrt{N_c}\mathcal{V}_c}\boldsymbol{C}_{\boldsymbol{q}=0}\boldsymbol{S}_{\boldsymbol{q}=0}, \tag{6.36}$$

$$\boldsymbol{B}_{\mathrm{loc}} = -\frac{\mu_0}{4\pi}\frac{g\mu_{\mathrm{B}}}{\sqrt{N_c}\mathcal{V}_c}\sum_{\boldsymbol{q}=\pm\boldsymbol{k}}\exp(-\mathrm{i}\boldsymbol{q}\cdot\boldsymbol{r}_0)\boldsymbol{C_q}\boldsymbol{S_q}, \tag{6.37}$$

respectively, for a ferro- and an antiferromagnet.

Let's assume that in a ferromagnet $\boldsymbol{B}_{\mathrm{loc}}$ takes a single value[29] denoted $\boldsymbol{B}_1$. The associated field probability density function is

$$p(\boldsymbol{B}) = \delta^{(3)}(\boldsymbol{B} - \boldsymbol{B}_1), \tag{6.38}$$

from which $P_z(t)$ is computed according to eqn 5.3. For a powder sample an orientation average is needed (see eqn 5.21).

A vanishing bulk magnetization is not exclusive to Néel antiferromagnets. Two types of $M = 0$ order can be distinguished: these are magnetic structures that are (i) commensurate with the underlying lattice and (ii) incommensurate. The former structure is fairly common; the latter less so, with incommensurate magnetic field distributions found in materials unstable towards the formation of a spin density wave (SDW).[30] Examples of the two sorts of antiferromagnet are shown in Fig. 6.14. The simple antiferromagnet is a wave of spin density commensurate with the lattice [Fig. 6.14(a)] whose distribution of local fields is discrete. A more complicated distribution arises when the wave of spin density is incommensurate with the underlying lattice [see Fig. 6.14(b)].

[27] For the computation of the dipole field contribution to $\boldsymbol{C_q}$, the use of the Ewald summation technique is particularly recommended. This is a fast and exact way to handle lattice sums.

[28] We have used the notation $\boldsymbol{C_q}$ which suggests that this quantity does not depend on the muon site $\boldsymbol{r}_0$, although $\boldsymbol{C_q}$ depends on $\boldsymbol{r}_j \equiv \boldsymbol{r}_0+\boldsymbol{j}$. While $\boldsymbol{C_q}$ does depend on $\boldsymbol{r}_0$, it takes the same value for two $\boldsymbol{r}_0$ vectors differing by a direct lattice vector, i.e. for two muons sitting at the same location in different crystal unit cells. This means the evolution of the field at the muon site from one unit cell to another is handled by the $\exp(-\mathrm{i}\boldsymbol{q}\cdot\boldsymbol{r}_0)$ factor in eqn 6.35, irrespective of the magnetic structure.

[29] When the multiplicity of the crystal structure position occupied by the muon is a number m greater than 1, $\boldsymbol{B}_1$ can take up to m different values.

[30] They typically arise from nesting instabilities of the Fermi surface, i.e. several of its points, with a large density of states, connected by the same wave vector. Often these structures emerge when interactions are low-dimensional.

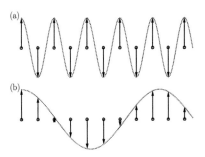

Fig. 6.14 Spin density wave configurations. (a) A simple antiferromagnet with a wavelength of twice the underlying lattice constant. (b) An incommensurate structure with wavelength much longer than the lattice constant.

[31] A useful related formula is

$$\delta[g(x)] = \sum_n \frac{1}{|g'(x_n)|}\delta(x - x_n),$$

(6.41)

where $g(x_n) = 0$.

[32] This example was originally discussed by Albert Overhauser (1925–2011).

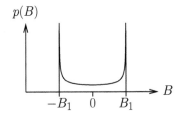

Fig. 6.15 The probability distribution $p(B)$ for the SDW state.

Example 6.9

We first examine the case of a commensurate magnetic structure, i.e. $\boldsymbol{k} = r\boldsymbol{Q}$, where r is a rational number and $\boldsymbol{Q}$ is a vector of the reciprocal lattice. Then the factor $\exp(\mp i\boldsymbol{k}\cdot\boldsymbol{r}_0)$ in eqn 6.37 takes a unique value (or, at most, a finite number of values when $\boldsymbol{r}_0$ spans the different muon positions). It follows that the probability density function is given by eqn 6.38, or a sum of Dirac delta functions, and $P_z(t)$ involves the sum of cosine functions.

For an incommensurate structure the real number r is irrational and $\exp(\mp i\boldsymbol{k}\cdot\boldsymbol{r}_0)$ uniformly spans the circle of radius 1 centered at the origin of the complex plane. As a result it can be shown that in general $\boldsymbol{B}_{\text{loc}}$ as given by eqn 6.37 describes an ellipse, i.e.

$$\boldsymbol{B}_{\text{loc}} = \cos\alpha\,\boldsymbol{B}_{1,\text{M}} + \sin\alpha\,\boldsymbol{B}_{1,\text{m}}, \quad \text{with } \alpha \in [0, 2\pi],$$

(6.39)

and $\boldsymbol{B}_{1,\text{M}}$ and $\boldsymbol{B}_{1,\text{m}}$ are respectively the semimajor and semiminor axes of the ellipse.

Let's consider an incommensurate magnetic structure, where the field magnitude, $B(\boldsymbol{r})$, varies with position $\boldsymbol{r}$. The probability distribution may be defined as

$$p(B) = \frac{\int \mathrm{d}^3 r\,\delta^{(3)}\left[B(\boldsymbol{r}) - B\right]}{\int \mathrm{d}^3 r},$$

(6.40)

where the δ-function here simply picks out[31] the field-magnitude of interest from the field function $B(\boldsymbol{r})$. Since the ratio of the wavelength in B to the underlying unit cell is irrational, implanted muons almost continuously sample the field distribution, reproducing the calculation of the following example.[32]

Example 6.10

Consider the case where the B-field is spatially varying in one dimension according to $B(x) = B_1 \cos kx$. Call $kx = \phi$ and we have

$$p(B) = \frac{\int_0^{2\pi} \mathrm{d}\phi\,\delta\left[B_1 \cos\phi - B\right]}{\int_0^{2\pi} \mathrm{d}\phi}.$$

(6.42)

Using eqn 6.41 with $g(\phi) = B_1 \cos\phi - B$, we have

$$\delta\left(B_1 \cos\phi - B\right) = \frac{2}{|B_1 \sin\phi_n|}\delta(\phi - \phi_n),$$

(6.43)

where the factor of 2 arises as there are two places where $B_1 \cos\phi_n - B = 0$. Using $\cos\phi_n = B/B_1$, we have $\sin\phi_n = \sqrt{1 - (B/B_1)^2}$, and so

$$p(B) = \frac{1}{\pi}\frac{1}{\left(B_1^2 - B^2\right)^{\frac{1}{2}}}.$$

(6.44)

This is the Overhauser spectrum, shown in Fig. 6.15. For simplicity, let's set $\theta = \pi/2$ in the polarization of eqn 6.28 (this means that the local fields are perpendicular to the initial muon spin), and compute $P_z(t) = \int \mathrm{d}B\,p(B)\cos(\gamma_\mu B t)$ within the field range of the distribution, $-B_1 \leq B \leq B_1$. We obtain

$$P_z(t) = \frac{1}{\pi}\int_{-B_1}^{B_1} \mathrm{d}B\,\frac{\cos(\gamma_\mu B t)}{\left(B_1^2 - B^2\right)^{\frac{1}{2}}}.$$

(6.45)

The integral in this expression can be related to the zero-order Bessel function,[33] giving us the answer

$$P_z(t) = J_0(\gamma_\mu B_1 t).$$ (6.47)

A property of this Bessel function is that for $x \gg 1$, $J_0(x) \approx \left(\frac{2}{\pi x}\right)^{\frac{1}{2}} \cos\left(x - \frac{\pi}{4}\right)$. So we may approximate the polarization after the first oscillations as

$$P_z(t) = J_0(\gamma_\mu B_1 t) \approx \left(\frac{2}{\pi \gamma_\mu B_1 t}\right)^{\frac{1}{2}} \cos\left(\gamma_\mu B_1 t - \frac{\pi}{4}\right).$$ (6.48)

We conclude that the polarization resembles a damped cosine with a $-45°$ phase shift.

The Bessel function behaviour has been observed in magnetic materials showing spin density wave ordered states, where it is often spotted from the characteristic phase shift.[34] An example is shown in Fig. 6.16 for the molecular charge transfer salt (TMTSF)(PF$_6$). This material is based on one-dimensional stacks of molecules, leading to strongly overlapping electron spin density that undergoes a spin-density wave transition below $T \approx 12$ K.

Chapter summary

- Spin precession allows the muon to act as a sensitive probe of local magnetic fields, with the precession frequency being proportional to the order parameter in a magnetically ordered material.

- The local magnetic field at the muon site is a sum of several terms, including contributions from the dipole and hyperfine fields.

- The time-dependent relaxation from static magnetic field distributions can be computed from the field distribution function $p(B)$ via a cosine Fourier transform $\int_0^\infty \mathrm{d}B\, p(B) \cos\gamma_\mu Bt$.

Further reading

- An introduction to magnetism can be found in S. J. Blundell, *Magnetism in Condensed Matter*, OUP (2001).

- A more advanced treatment of symmetry breaking and its consequences is given in P. M. Chaikin and T. C. Lubensky, *Principles of Condensed Matter Physics*, CUP (1995).

- Magnetic structures can be explored via the Bilbao Crystallographic Server at http://cryst.ehu.es.

- Spin density waves are discussed in G. Grüner, *Density Waves in Solids*, CRC Press (2000).

[33]The first-kind zero-order Bessel function is defined as

$$J_0(ax) = \frac{2}{\pi} \int_0^x \mathrm{d}y\, \frac{\cos(ay)}{(x^2 - y^2)^{\frac{1}{2}}}$$ (6.46)

for $a, x \in \mathbb{R}, x > 0$.

[34]Often the low-amplitude low-field tail of the distribution, corresponding to nodes in the SDW, is hidden within the noise of the Fourier spectrum. The peaks, corresponding to loci where the field changes slowly, such as the crests and troughs of the SDW, always stand out in the spectrum. As a result, there is sometimes some ambiguity in assigning a measured spectrum to SDW order.

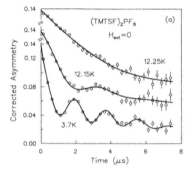

Fig. 6.16 Asymmetry spectra measured for the SDW material (TMTSF)$_2$(PF$_6$). [Taken from L. P. Le *et al.*, Phys. Rev. B **48**, 7284 (1993).]

Exercises

(6.1) Show that in the Landau model, the critical exponent $\beta = 1/2$.

(6.2) Show by Gauss' theorem ($\int \boldsymbol{H} \cdot \mathrm{d}\boldsymbol{S} = q_{\mathrm{m}}$ where q_{m} is the magnetic charge) that the field $\boldsymbol{H}$ at radius r from a uniformly magnetically charged ball is $\boldsymbol{H} = \rho_{\mathrm{m}}\boldsymbol{r}/3$. Consider two such charged balls with radius R, one with charge density $+\rho_{\mathrm{m}}$ and the other with charge density $-\rho_{\mathrm{m}}$ and displace their centres by a small vector $\boldsymbol{d}$ with $d \ll R$. Argue

that the field inside this sphere can be written as

$$\boldsymbol{H} = \frac{\rho_{\mathrm{m}}\boldsymbol{r}_{+}}{3} - \frac{\rho_{\mathrm{m}}\boldsymbol{r}_{-}}{3} = -\frac{\rho_{\mathrm{m}}\boldsymbol{d}}{3} \qquad (6.49)$$

where $\boldsymbol{r}_{\pm}$ is the vector from the centre of each sphere to the point at which $\boldsymbol{H}$ is measured and $\boldsymbol{r}_{+} + \boldsymbol{d} = \boldsymbol{r}_{-}$. Now, for the spherical cavity in Fig. 6.12, argue that this gives the Lorentz field $\boldsymbol{B}_{\mathrm{L}} = \mu_{0}\boldsymbol{M}/3$.

(6.3) Verify eqn 6.31.

(6.4) Verify the steps leading to eqn 6.47.

Dynamic effects in magnetism

7

If the microscopic magnetization in a material is time-dependent, we say that the material exhibits **magnetic dynamics**. The muon polarization in magnetic materials is sensitive to these dynamics. We often discuss two closely linked concepts: (i) dynamic fluctuations, meaning the time-dependent deviations of the microscopic magnetization from its average value and (ii) dissipation, meaning the loss of energy in a system owing to the creation or destruction of excitations. These both result in the relaxation of the time-dependent muon polarization.

We start by introducing the key quantities in measuring dynamics. There are two experimentally accessible functions that are of interest: the correlation function and the dynamic susceptibility. A useful rule of thumb is that correlation functions describe the fluctuations, while (part of) the dynamic susceptibility is proportional to the dissipation. Actually, these two quantities measure essentially the same thing since the two concepts are closely linked by the **fluctuation-dissipation theorem**, as we shall see in this chapter.

7.1 Correlation functions

The **correlation function** of two components $i, j (= x, y, z)$ of a field $\boldsymbol{M}(\boldsymbol{x}, t)$, is defined as the average[1] of the product of the two components at different coordinates $\boldsymbol{x}, \boldsymbol{x}'$ and times t, t':

$$G_{ij}(\boldsymbol{x}, t, \boldsymbol{x}', t') = \langle M_i(\boldsymbol{x}, t) M_j(\boldsymbol{x}', t') \rangle. \tag{7.1}$$

In most physical systems, either after a characteristic time, or at sufficiently large distance between the coordinates, the correlation decays. In a paramagnet, for instance, the average magnetization $\langle M_i(\boldsymbol{x}, t) \rangle$ is zero for all $i, \boldsymbol{x}$, and t.[2] However, the field will have an instantaneous non-zero value and its correlation function can be non-zero for a short time and over a short distance. The quantity $\langle M_i(\boldsymbol{x}, t) M_j(\boldsymbol{x}', t') \rangle$ therefore measures for how long and how far away M_j at $(\boldsymbol{x}', t')$ is nearly equal to M_i at $(\boldsymbol{x}, t)$. The correlation function can then be thought of as providing a measure of how similar the field is (or how well *correlated* it is) through intervals in space and time. In homogeneous condensed matter, the correlation function will only depend on the difference in coordinates. For simplicity, we start by considering a one-dimensional,

[1] In thermal equilibrium, the probability of finding a state with energy E_λ is given by

$$p_\lambda = \frac{1}{Z} e^{-\frac{E_\lambda}{k_B T}},$$

where $Z = \sum_\lambda e^{-\frac{E_\lambda}{k_B T}}$. A **thermal average** of a quantity $O^{(\lambda)}$ is then given by

$$\langle O \rangle = \sum_\lambda O^{(\lambda)} p_\lambda = \frac{1}{Z} \sum_\lambda O^{(\lambda)} e^{-\frac{E_\lambda}{k_B T}}.$$

[2] In a ferromagnet, which has $\langle M \rangle \neq 0$ we are interested in the fluctuations that cause the system to vary from the magnetic order. In that case, the quantity of interest is the **connected correlation function**, defined as

$$\begin{aligned} G_{ij}(\boldsymbol{x}, t, \boldsymbol{x}', t') &= \langle M_i(\boldsymbol{x}, t) M_j(\boldsymbol{x}', t') \rangle \\ &\quad - \langle M_i(\boldsymbol{x}, t) \rangle \langle M_j(\boldsymbol{x}', t') \rangle. \end{aligned} \tag{7.2}$$

[3]We define the Fourier transforms of
the field ϕ in three dimensions, omit-
ting $\pm\infty$ integral limits for simplicity:

$$\tilde{\phi}_i(\boldsymbol{p}, t) = \int d^3x\, e^{-i\boldsymbol{p}\cdot\boldsymbol{x}} \phi_i(\boldsymbol{x}, t) \quad (7.3)$$

$$\phi_i(\boldsymbol{x}, t) = \frac{1}{V} \sum_{\boldsymbol{p}} e^{i\boldsymbol{p}\cdot\boldsymbol{x}} \tilde{\phi}_i(\boldsymbol{p}, t) \quad (7.4)$$

$$\tilde{\phi}_i(\boldsymbol{x}, \omega) = \int dt\, e^{i\omega t} \phi_i(\boldsymbol{x}, t) \quad (7.5)$$

$$\phi_i(\boldsymbol{x}, t) = \int \frac{d\omega}{2\pi} e^{-i\omega t} \tilde{\phi}_i(\boldsymbol{x}, \omega), \quad (7.6)$$

where V is the volume of the sys-
tem and where $\tilde{\phi}(\boldsymbol{p}, t)$ in eqn 7.3 and
$\tilde{\phi}_i(\boldsymbol{x}, \omega)$ in eqn 7.5 are different functions. We also have the useful relations

$$\int d^3x\, e^{-i(\boldsymbol{p}-\boldsymbol{p}')\cdot\boldsymbol{x}} = V\delta_{\boldsymbol{p},\boldsymbol{p}'} \quad (7.7)$$

$$\frac{1}{V} \sum_{\boldsymbol{p}} e^{i\boldsymbol{p}\cdot(\boldsymbol{x}-\boldsymbol{x}')} = \delta^{(3)}(\boldsymbol{x} - \boldsymbol{x}').$$
$$(7.8)$$

space-only correlation function $G(z)$, with $z = x - x'$. We investigate
this quantity and its Fourier transform[3] in the following example.

Example 7.1

The correlation function of the field $\phi(x)$ in one spatial dimension is $G(x - x') = \langle\phi(x)\phi(x')\rangle$, and, since $x = z + x'$, we have

$$G(z) = \int dx'\, \langle\phi(z + x')\phi(x')\rangle. \quad (7.9)$$

In words, (i) we pick a point x', (ii) ask how well correlated a point is that lies a
distance z from it, and then (iii) average over our original choice of x'. The Fourier
transform of $G(z)$ is given by

$$\tilde{G}(p) = \int dz\, G(z)e^{-ipz}$$
$$= \int dz\, dx'\, \langle\phi(z + x')\phi(x')\rangle e^{-ipz}. \quad (7.10)$$

If we insert the Fourier transform of the fields, we have

$$\tilde{G}(p) = \frac{1}{V^2} \sum_{q,s} \int dz\, dx'\, \langle\tilde{\phi}(q)\tilde{\phi}(s)\rangle e^{-ipz} e^{iq(z+x')} e^{isx'}, \quad (7.11)$$

where V is the (one-dimensional) volume of the system. Integration over x' and z,
recalling eqn 7.7, yields

$$\tilde{G}(p) = \sum_{q,s} \langle\tilde{\phi}(q)\tilde{\phi}(s)\rangle \delta_{q,p}\delta_{q,-s}, \quad (7.12)$$

thence

$$\tilde{G}(p) = \langle\tilde{\phi}(p)\tilde{\phi}(-p)\rangle. \quad (7.13)$$

We conclude that the Fourier transform of a correlation function that depends only on the distance between two points has a wavevector dependence
of the form $\langle\tilde{\phi}(p)\tilde{\phi}(-p)\rangle$. This is an important result for computing dynamics.

A complementary approach to correlation functions is to measure the
system's response through a **dynamic susceptibility** $\chi_{ij}(\boldsymbol{x}, t, \boldsymbol{x}', t')$,
defined via

$$\delta M_i(\boldsymbol{x}, t) = \int dt'd^3x'\, \chi_{ij}(\boldsymbol{x}, t, \boldsymbol{x}', t')\delta H_j(\boldsymbol{x}', t'). \quad (7.14)$$

[4]Here i and j label different components: a stimulus along x may result
in an effect along y.

This says we apply a stimulus (a small change δH_j) and measure a
response (a small change δM_i).[4] The susceptibility χ_{ij} tells us about
the dynamic response of the system. If, as in the next example, we
just consider the time variable then (assuming the susceptibility only
depends on time differences) we have

$$\delta M_i(t) = \int dt'\, \chi_{ij}(t - t')\delta H_j(t'), \quad (7.15)$$

and, by the convolution theorem, in frequency space we obtain the product of the Fourier transforms $\tilde{\chi}_{ij}(\omega)$ and $\delta\tilde{H}_j(\omega)$

$$\delta\tilde{M}_i(\omega) = \tilde{\chi}_{ij}(\omega)\delta\tilde{H}_j(\omega). \quad (7.16)$$

This formulation is useful as it allows us to work in the complex plane distinguishing the real and the imaginary parts[5] of the susceptibility $\tilde{\chi}_{ij}(\omega) = \chi'_{ij}(\omega) + i\chi''_{ij}(\omega)$.

The description of a system in terms of correlation and response is admittedly rather abstract. An insight can be provided by using the same language to describe a simple harmonic oscillator, as we do in the next example.

Example 7.2

Consider a simple harmonic oscillator with spring constant $m\omega_0^2$ under an external force $f(t)$, with equation of motion

$$\ddot{x}(t) + \gamma\dot{x}(t) + \omega_0^2 x(t) = \frac{f(t)}{m}. \tag{7.19}$$

For a periodic forcing function $f(t) = f_0 e^{-i\omega t}$, we can solve this equation using the ansatz $x(t) = A e^{-i\omega t}$. The oscillator responds to driving frequencies ω with amplitudes proportional to its susceptibility, which can be defined as

$$\tilde{\chi}(\omega) = \frac{\tilde{x}(\omega)}{\tilde{f}(\omega)} = \frac{1}{m}\frac{1}{(-\omega^2 + \omega_0^2 - i\omega\gamma)}. \tag{7.20}$$

The static ($\omega = 0$) response is given by $\chi = 1/m\omega_0^2$, and so we can rewrite the susceptibility

$$\tilde{\chi}(\omega) = \frac{\chi}{1 - \left(\frac{\omega}{\omega_0}\right)^2 - i\omega\frac{\gamma}{\omega_0^2}}. \tag{7.21}$$

This latter function has real and imaginary parts given respectively by

$$\chi'(\omega) = \frac{\chi(\omega_0^2 - \omega^2)}{\omega_0^2\left[1 - \left(\frac{\omega}{\omega_0}\right)^2\right]^2 + \left(\frac{\omega}{\omega_0}\gamma\right)^2} \qquad \chi''(\omega) = \frac{\chi\omega\gamma}{\omega_0^2\left[1 - \left(\frac{\omega}{\omega_0}\right)^2\right]^2 + \left(\frac{\omega}{\omega_0}\gamma\right)^2}. \tag{7.22}$$

The response to a forcing function $f_0 \cos\omega t$ is then

$$x(t) = f_0 |\tilde{\chi}(\omega)| \cos[\omega t - \phi(\omega)], \tag{7.23}$$

where $\tan\phi(\omega) = \chi''(\omega)/\chi'(\omega)$.

Useful for our purposes is the extremely overdamped limit of the oscillator, which occurs when $\frac{\gamma}{2} \gg \omega_0$, hence $\omega_0 \gg \omega$. We identify $\tau_s = \gamma/\omega_0^2$ as the characteristic decay time in the frequency response,[6] and we obtain an expression for the overdamped susceptibility

$$\tilde{\chi}(\omega) = \frac{\chi}{1 - i\omega\tau_s}. \tag{7.25}$$

As we shall see, this expression also describes the dynamic susceptibility of a magnetic system interacting with the broad excitation spectrum of a thermal reservoir. The imaginary part of this expression can be written in the form

$$\frac{\chi''(\omega)}{\omega} = \frac{\chi\tau_s}{1 + \omega^2\tau_s^2}. \tag{7.26}$$

To calculate the work done per unit time by the force, i.e. the rate of energy dissipation averaged over one cycle with period $T = 2\pi/\omega$, we compute[7]

$$P = \left\langle \frac{dW}{dt} \right\rangle = \frac{1}{T}\int_0^T dt\, f(t)\frac{dx(t)}{dt}$$

$$= \frac{1}{2}f_0^2\omega\chi''(\omega). \tag{7.27}$$

This justifies the fact that χ'' is sometimes known as the dissipation.

---

[5] The real and imaginary parts are not independent, but obey sum rules, which are useful to know. We have a sum rule

$$\lim_{\omega\to 0}\tilde{\chi}(\omega) = \int_{-\infty}^{\infty}\frac{d\omega}{\pi}\frac{\chi''(\omega)}{\omega} = \chi, \tag{7.17}$$

where χ is the static susceptibility. Real and imaginary parts are linked by the Kramers-Kronig relation

$$\chi'(\omega) = -P\int_{-\infty}^{\infty}\frac{d\nu}{\pi}\frac{\chi''(\nu)}{\nu - \omega}, \tag{7.18}$$

where P is the instruction to take the principal part of the integral.

[6] The transient response also features a faster decay rate, neglected here. The two characteristic decay rates are

$$\tau_f^{-1} = \frac{\gamma}{2}\left[1 + (1 - 4\omega_0^2\gamma^{-2})^{\frac{1}{2}}\right]$$

$$\tau_s^{-1} = \frac{\gamma}{2}\left[1 - (1 - 4\omega_0^2\gamma^{-2})^{\frac{1}{2}}\right], \tag{7.24}$$

where the subscripts stand for fast (f) and slow (s).

[7] See Exercise 7.2.

[8]Compared with eqn 7.14 we simplify to a single dimension here, so we don't have the complication of the components of the susceptibility.

The fluctuation-dissipation theorem tells us that the same information is to be found in dissipation, measured via[8] $\chi''(\omega)$, as is to be found in fluctuations, measured by the correlation function $\tilde{G}(\omega) = \langle \tilde{M}(\omega)\tilde{M}(-\omega)\rangle$. Specifically, the theorem says that

$$\frac{1}{2\hbar}\left(1 - e^{-\frac{\hbar\omega}{k_{\mathrm{B}}T}}\right)\tilde{G}(\omega) = \chi''(\omega). \tag{7.28}$$

We will often work in the limit of small ω, where this expression becomes

$$2k_{\mathrm{B}}T\lim_{\omega\to 0}\frac{\chi''(\omega)}{\omega} = \tilde{G}(\omega\to 0). \tag{7.29}$$

[9]This is the key ingredient in the calculation of dynamic relaxation rates (as seen below in eqn 7.47). The general definition of the autocorrelation function, which accounts for correlations between different components, is

$$\Phi_{ij}(t_1, t_2) = \frac{\gamma_\mu^2}{2}\langle \delta B_i(t_1)\delta B_j(t_2)$$
$$+ \delta B_j(t_2)\delta B_i(t_1)\rangle. \tag{7.30}$$

For a fluctuating field, its correlations die away into the future but also into the past, so for example $\Phi_{ii}(t) = \Phi_{ii}(-t)$ and so $\Phi_{ii}(t)$ can be written as a function of $|t|$.

Example 7.3

In μSR we will be interested in the (local) **field-field autocorrelation function**[9]

$$\Phi_{ii}(t) = \gamma_\mu^2\langle \delta B_i(t)\delta B_i(0)\rangle, \tag{7.31}$$

where $\delta B_i(t)$ is the fluctuating magnetic field. This tells us to measure the ith component of the local magnetic field at time zero and then measure it at a later time t and take an average. This quantity only depends on the time difference $|t|$. If $\Phi_{ii}(t) \approx \Phi_{ii}(0)$, the fields remain correlated up to time t. If $\Phi_{ii}(t) \ll \Phi_{ii}(0)$, then correlations have died away after time t. If we take a Fourier transform of the autocorrelation function we obtain $\tilde{J}(\omega)$, known as the **spectral density** of the field distribution. This function tells us the range of frequencies over which we should expect fluctuations. An important point is that

$$\int_{-\infty}^{\infty}\frac{d\omega}{2\pi}\tilde{J}(\omega) = \gamma_\mu^2\langle \delta B_i(0)^2\rangle, \tag{7.32}$$

and is therefore a constant, so the areas under curves representing $\tilde{J}(\omega)$ are the same for correlations that have the same $\langle \delta B_i(t=0)^2\rangle$. A sharp peak at $\omega = 0$ [Fig. 7.1(a)] tells us that the slowly varying fields remain correlated for long periods. A sharp peak at non-zero $\omega = \omega_0$ [Fig. 7.1(b)] tells us that the correlations are periodic. A broad peak at $\omega = 0$ [Fig. 7.1(c)] tells us about the frequency spectrum of correlations that die away with time.

An example of the latter is the correlation function that we met in Chapter 5: $\Phi_{ii}(t) = \Delta^2 e^{-|t|/\tau}$. This is the typical form for the correlation function expected for a two-state system with an energy barrier separating the states. It will be convenient for our later discussion to give the spectral density in terms of a Fourier transform

$$\tilde{J}(\omega) = \int_{-\infty}^{\infty}dt\,\Delta^2 e^{-|t|/\tau}e^{i\omega t} = \frac{2\Delta^2\tau}{1 + \omega^2\tau^2}. \tag{7.33}$$

It is interesting to note the resemblance of this spectral density function to χ''/ω in eqn 7.26. It implies that we can also think of these fluctuations as transient and random forced oscillations.

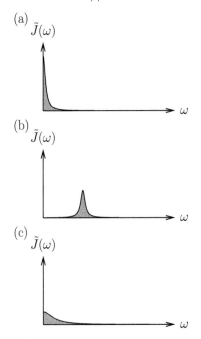

Fig. 7.1 Spectral density $\tilde{J}(\omega)$ for different correlations that die away in time. (a) Long-lived correlations; (b) long-lived periodic correlations in time; and (c) short-lived correlations.

We shall return to the behaviour discussed in the last example a little later in this chapter.

7.2 Dynamics in magnets

The scale of the muon response time in μSR is of order $1/\omega_0 = 1/(\gamma_\mu B)$, set by the muon gyromagnetic ratio and the local field B. The energy required to flip a spin in an interacting magnet is set by the exchange energy J, which corresponds[10] to a timescale $\hbar/J$, typically very short on the scale of μSR. As a result, the muon is often insensitive (directly, at least) to the high-frequency dynamics at the exchange energy scale and is, instead, sensitive to lower-frequency excitations. Where do such low frequency excitations come from in a magnet?

Consider setting up a disturbance in time and space, with frequency ω and wavevector q. In general, different excitations in solids will have their characteristic microscopic fluctuation time τ_{micro} and mean free path λ_{micro} at the atomic scale.[11] However, processes that involve the cooperative behaviour of large numbers of atoms will take place at longer timescales than τ_{micro}, and over longer length scales than λ_{micro}. These are known as **hydrodynamic** processes[12] and for $q \to 0$ their frequencies vanish ($\omega \to 0$).

In a magnetic material, if you set up an excitation (e.g. a patch of parallel spins of width ξ, with local $M > 0$, against a disordered background, with $M = 0$) there are potentially many ways the spins can flip back to a disordered configuration, with a characteristic short time constant τ_{micro}. However, in an interacting magnet, the magnetization M is often a conserved quantity and, as such, must obey a **continuity equation**

$$\nabla \cdot \boldsymbol{J} = -\frac{\partial M}{\partial t}, \tag{7.34}$$

where $\boldsymbol{J}$ is a magnetization (or spin) current, telling us the rate at which net local magnetization flows through a unit area perpendicular to the direction of $\boldsymbol{J}$.[13] The result of this conservation law is that any local excesses of magnetization will spread out slowly over the whole system, and magnetization can therefore be described as a hydrodynamic variable. For instance in a three-dimensional ferromagnet there are four conserved quantities: the energy density E and the three components of the total magnetization M_x, M_y, and M_z. In the disordered phase each of these gives rise to a **diffusive mode** in the material.[14] These are excitations that obey a **diffusion** equation of the form $\partial n/\partial t = D_n \nabla^2 n$, where $n = E, M_x, M_y$ or M_z, and D_n are diffusion constants. This justifies the slow variation of all conserved variables.

Symmetry breaking changes the excitation spectrum. This is because new slowly varying quantities (known as broken-symmetry variables) emerge. Once symmetry is broken in a ferromagnet on magnetic ordering, a new excitation in the magnetization, the spin wave, can be supported. Spin waves, or magnons, are well-defined eigenstates of the magnet in the limit $T \to 0$, each representing a **propagating mode**, which is a wave-like excitation in the magnetization that is qualitatively different from a diffusing mode. Magnons are less well defined for larger T where multi-magnon processes take place, eventually degenerating into the diffusive behaviour described above.

By conservation of energy, any coupling between an excitation of lifetime $1/\omega$ and the muon spin depends crucially on how closely the characteristic frequency ω matches the muon precession frequency $\omega_0 = \gamma_\mu B$. The frequency range of spin waves has an upper limit at the Brillouin

[10]We make use of the relationship $\Delta E \Delta t \approx \hbar$ here.

[11] Typically another length scale characterizes excitations, their correlation (or coherence) length, ξ, discussed in the next chapter. Here we assume $\xi \lesssim \lambda_{\text{micro}}$.

[12]Hydrodynamics, named after the study of the processes that take place in liquid water, and other liquids, describes low-frequency, long wavelength excitations. These have frequencies ω and wavevectors q such that $\omega\tau_{\text{micro}} \ll 1$ and $q\lambda_{\text{micro}} \ll 1$.

[13]We have considered a single component of the magnetization in this discussion so far. For a three-dimensional magnetization vector $\boldsymbol{M}$, the magnetization current is a rank 2 tensor.

[14]A mode is a solution to the equations of motion of the system's dynamic variables, telling us how these evolve in time.

[15]This is guaranteed by **Goldstone's theorem** for all variables that emerge on breaking a continuous symmetry.

[16]This effect is also called magnetic Raman scattering, and it involves two (virtual) magnons, one emitted and one absorbed.

[17]We might also remark that, whereas spin waves contribute very little directly to muon-spin relaxation, the reduction of the magnetic order parameter with temperature is due to the dynamic excitations we have described. So, in this sense, the determination of the static local field in a magnet using μSR is sensitive to magnetic dynamics.

zone boundary, set by the exchange, $J/\hbar$, generally much larger than ω_0. In isotropic magnets the lower limit is $\omega = 0$ for acoustic modes with $q \to 0$.[15] Since this is a very broad range compared to ω_0, we conclude that muons typically have only a very limited sensitivity to the dynamics of spin waves. Apart from acoustic modes at $\omega \to 0$, only a two-magnon process[16] can exchange magnetization with the muon and conserve energy. This is a second order process which requires a large magnon population, hence high temperatures, which is where magnons are less well defined and diffusive excitations are dominant.[17]

Example 7.4

In the text above we discussed the four diffusive modes of the isotropic ferromagnet in its paramagnetic, disordered phase. In the ferromagnetically ordered phase, assuming one magnetic ion per cell, symmetry is broken along one direction (for example, z) and there is one pair of spin wave modes (for M_x and M_y), while two diffusive modes are retained for E and M_z, leading to four modes overall.

An isotropic *antiferromagnet* also has four conserved quantities E, M_x, M_y, and M_z. There will therefore be four diffusive modes in the paramagnetic, disordered phase: one for each conserved variable. In the antiferromagnetically ordered phase things are different to the ferromagnet case. We still find two diffusive modes corresponding to E and M_z, but now we have *two pairs* of propagating spin-wave-like modes, leading to six modes overall.

The difference is due to the magnetization in the ferromagnet playing the dual role of both a conserved quantity *and* the broken-symmetry variable. In the antiferromagnet, although the magnetization is conserved, the broken-symmetry variable is now the sublattice magnetization. The consequence is that there are fewer constraints for the modes to be expressed as excitations, and so we find more possible modes in the system.

---

[18]This can be seen in the harmonic oscillator Example 7.2. As the spring constant $m\omega_0^2$ tends to zero, we see (eqn 7.24) that the slow rate τ_s^{-1} tends to zero, while the fast rate becomes $\tau_f^{-1} \to \gamma$. In this case the spring constant controls the separation of the timescales.

[19]This is often not the case and necessitates a more sophisticated treatment of critical dynamics. See the next chapter and also P. C. Hohenberg and B. I. Halperin, Rev. Mod. Phys. **49**, 435 (1977).

This hydrodynamic description of the behaviour can be extended to other slowly relaxing variables with characteristic timescales τ_s. This is possible as long as there is a **division of timescales** between variables determined by microscopic collision frequencies τ_{micro}^{-1} and those[18] for which the rate τ_s^{-1} obeys $\tau_{\text{micro}}/\tau_s \ll 1$. It is exactly this state of affairs that occurs near a second order phase transition. According to the conventional theory of critical dynamics (formulated by van Hove in 1954) there will be a dissipative mode for the order parameter with $\tau_s^{-1} \approx c/\chi$, where c is a constant coefficient and χ is the susceptibility. We might expect[19] that c approaches a constant in the limit $T \to T_c$ and so τ_s diverges as $|T - T_c|^{-\gamma}$. This phenomenon is known as **critical slowing down** and is discussed further in the following chapter.

After this round up of the possible sources of excitation in a magnet, we are ready to see how muons probe them.

7.3   Dynamics with muons

We showed in Chapter 4 that in the presence of an applied field B_0 along $\hat{z}$, the eigenstates of the muon spin Hamiltonian $\hat{\mathcal{H}}_0 = -\gamma_\mu B_0 \hat{S}_z$, denoted[20] by $|\uparrow\rangle$ and $|\downarrow\rangle$, have eigenenergies $E = -\hbar\gamma_\mu B_0\, m_s = \pm\hbar\omega_0/2$. The two states are degenerate in zero field and the applied field generates the Zeeman energy gap $\Delta E = \hbar\omega_0 = \hbar\gamma B_0$ between them. If we prepare the muon in one of these eigenstates (say, $|\downarrow\rangle$), then the field is along the initial muon spin polarization direction and the state won't change unless it is perturbed by some additional interaction. In this way we have designed a means to measure any perturbation that can induce a transition to the other muon eigenstate. In order to change the muon's state, the perturbation must develop, at some future time, a component of the local magnetic field along the x- or y-direction. Such a time-dependent perturbation can scatter a muon from the $|\downarrow\rangle$ state, reducing the polarization of the muon ensemble. The resulting decay is the basis of a method to probe dynamics with the muon.

Example 7.5

Quantum mechanically, the time evolution of a spin can be understood by considering Heisenberg's equation of motion

$$\frac{\mathrm{d}}{\mathrm{d}t}\hat{\boldsymbol{S}}(t) = \frac{1}{\mathrm{i}\hbar}\left[\hat{\boldsymbol{S}}(t), \hat{\mathcal{H}}\right], \tag{7.35}$$

where the spin operators are given time dependence within the Heisenberg picture. We therefore have the rule of thumb that to cause a change in the value of the muon spin, we need a perturbation that does not commute with the operator for the initial spin direction. For a muon spin oriented along z in a longitudinal field, the S_z component will therefore acquire some time dependence if the Hamiltonian $\mathcal{H} = \mathcal{H}_0 + \mathcal{H}_1(t) \propto \boldsymbol{S} \cdot \boldsymbol{B}(t)$ contains any contributions from S_x and S_y, because $[S_z, S_i] \neq 0$ for $i = x, y$. This will be the case if the instantaneous local field $\boldsymbol{B}(t)$ is not oriented along z. In general, integrating equations of motion such as that in eqn 7.35 is not a simple matter, as it involves the commutator of a given operator, say S_z, with a time-dependent Hamiltonian $\mathcal{H}_1(t)$. This commutator will change with time, making the theory rather more complicated than that for a stationary perturbation.

The basis of our measurement of dynamics is to apply a magnetic field to form a known gap in energy between muon-spin-up and spin-down states, as shown in Fig. 7.2. If we observe relaxation then it must be that the (non-commuting) fluctuating field is large enough to cause transitions between the up and down states. To see how this works we can split up the field such that we have time-dependent fluctuations in the local field described by a Hamiltonian

$$\hat{\mathcal{H}}(t) = -\gamma_\mu \hat{\boldsymbol{S}} \cdot [\boldsymbol{B}_0 + \delta\boldsymbol{B}(t)]. \tag{7.36}$$

The point here is to treat the small field $\delta\boldsymbol{B}(t)$, as a fluctuation about the average, time-independent field $\boldsymbol{B}_0$ and perform an expansion in terms of this small fluctuating field.

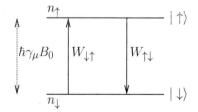

Fig. 7.2 Dynamics as a two level system with energy gap $\Delta E = \hbar\gamma_\mu B_0$.

Example 7.6

In order to get an idea of what to expect from dynamics we can consider a simple picture where the perturbation scatters the muon spin at a rate given by Fermi's golden rule

$$W_{fi} \propto |\langle f|\gamma_\mu \hat{\boldsymbol{S}} \cdot \delta \boldsymbol{B}(t)|i\rangle|^2, \tag{7.37}$$

where $|i\rangle$ and $|f\rangle$ are initial and final muon-spin states respectively (which can be $|\uparrow\rangle$ or $|\downarrow\rangle$). Using the resulting transition rates we can write rate equations for the population of up and down muon spins as

$$\dot{n}_\uparrow = -W_{\uparrow\downarrow}n_\uparrow + W_{\downarrow\uparrow}n_\downarrow$$
$$\dot{n}_\downarrow = W_{\uparrow\downarrow}n_\uparrow - W_{\downarrow\uparrow}n_\downarrow, \tag{7.38}$$

where n_i is the population of the state $|i\rangle$. Subtracting, we have

$$\frac{\mathrm{d}}{\mathrm{d}t}(n_\uparrow - n_\downarrow) = -(W_{\uparrow\downarrow} + W_{\downarrow\uparrow})(n_\uparrow - n_\downarrow), \tag{7.39}$$

which can be solved to find that $P_z(t) \propto (n_\uparrow - n_\downarrow)$ can be written

$$P_z(t) = P_z(0)\mathrm{e}^{-\lambda t}, \tag{7.40}$$

where $\lambda = W_{\uparrow\downarrow} + W_{\downarrow\uparrow}$.

We therefore use an applied field to probe dynamics: it opens up a gap between the muon up and down states. We then test the dynamics to see if they have the strength to cause the muon spins to flip. The transition rate is measured via an exponential relaxation of the muon spin polarization.

In the final part of this section we are going to make progress beyond the LF μSR experiment, by introducing two different classes of relaxation process, following a simple model formulated by Felix Bloch.[21] The idea is that we apply a field along $\hat{z}$, but we won't assume the muon spin is parallel to this direction. We then make a distinction between (i) relaxation of the spin components longitudinal to the field and (ii) relaxation of the spin components perpendicular to the field.

[21] Felix Bloch (1905–1983), Nobel Prize 1952. The model was originally devised for nuclear and electronic magnetization but is also widely used in coherent laser experiments, quantum computation, etc.

Example 7.7

We return to the example of simple Larmor precession, where a field B_0 is applied along $\hat{z}$ and the initial muon spin lies in a direction, $(S_x(0), S_y(0), S_z(0))$. We have equations of motion

$$\dot{S}_x(t) = \gamma_\mu B_0 S_y(t)$$
$$\dot{S}_y(t) = -\gamma_\mu B_0 S_x(t)$$
$$\dot{S}_z(t) = 0. \tag{7.41}$$

We know, however, that the muon-spin precession is damped exponentially when we have dynamics. We also know that any spread in the field experienced by a spin leads to relaxation, but only of those spin components perpendicular to the field. In Bloch's model, the relaxation is introduced phenomenologically via two distinct decay constants, which in this context are known as the T_1 **and** T_2 **relaxation times**.[22] The **Bloch equations** are then given by

$$\dot{S}_x(t) = \gamma_\mu B_0 S_y(t) - \frac{1}{T_2}S_x$$
$$\dot{S}_y(t) = -\gamma_\mu B_0 S_x(t) - \frac{1}{T_2}S_y$$
$$\dot{S}_z(t) = -\frac{1}{T_1}S_z. \tag{7.42}$$

[22] The time T_1 is often referred to as a spin-lattice relaxation time, while T_2 is called the spin-spin relaxation time.

The reason for two distinct relaxation rates is that there are different processes leading to the decay of the longitudinal (z) and transverse (x, y) components, with rates T_1^{-1} and T_2^{-1} respectively. These rates have different physical origins. For the first, T_1^{-1} is the recovery of the equilibrium value of the component $S_z(t \to \infty)$ aligned along the external field. This T_1 process is due to dissipative interactions with a thermal reservoir, therefore it does not conserve angular momentum $\boldsymbol{S}$ nor total energy $-\gamma_\mu \boldsymbol{B} \cdot \boldsymbol{S}$. In contrast the T_2 process conserves both quantities, and the decay corresponds to dephasing of the average spin precession.[23]

The model avoids a microscopic description of these two processes, which are due to inelastic and elastic scattering events, respectively, but also, for the transverse T_2 decay, to simple static inhomogeneity in the field distribution. The solution of Bloch's equations is

$$S_z(t) = S_z(\infty) - [S_z(\infty) - S_z(0)]\, \mathrm{e}^{-t/T_1}$$
$$S_\perp(t) = S_\perp(0)\mathrm{e}^{-t/T_2} \cos(\gamma_\mu B_0 t). \tag{7.43}$$

This description is not very widely used in μSR, probably because it does not apply to one of the most important muon experiments: zero-field μSR. In the Bloch equations the field direction is necessary for the definitions, and moreover, for the distinction between energy-conserving and non-conserving phenomena. Still, whenever a large static field is present (either external, or spontaneous), the nomenclature helps in reminding us of the two different physical origins of relaxation.

[23]The first process requires that the spin system eventually shares the temperature of the reservoir, whereas the second process can't be described in terms of a temperature, being determined instead by quantum dynamics.

We conclude that the LF μSR experiment we introduced at the start of this section probes $1/T_1$ processes. In the next section we discuss the role of the applied field in picking out a particular part of the fluctuation spectrum.

[24]Specific examples of muon-spin resonance experiments are described in more detail in Chapter 19.

7.4 Relaxation as resonance

A very important concept, that justifies **resonance** being part of the μSR abbreviation, is that our measurement of dynamics is intrinsically **resonant**.[24] We illustrate this through the following two examples that apply to resonance experiments, but whose importance goes beyond that specific application.

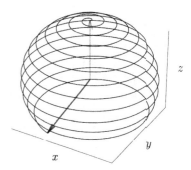

Fig. 7.3 Trajectory of the tip of the muon spin on the surface of a sphere during its nutation at resonance. Here the effective field is $B_1 \hat{\boldsymbol{x}}$ and the motion stops at $\omega_1 \tau = \pi/2$.

Example 7.8

Imagine that we apply the static field $\boldsymbol{B}_0 = B_0 \hat{\boldsymbol{z}}$ and additionally a small orthogonal rotating field $\boldsymbol{B}_1 = B_1(\hat{\boldsymbol{x}} \cos \omega_0 t + \hat{\boldsymbol{y}} \sin \omega_0 t)$, with $B_1 \ll B_0$. We choose the oscillation rate so that $\omega_0 = \gamma_\mu B_0$. This is known as the **resonance condition** and implies that the small field oscillates at a frequency that corresponds to the energy gap between up and down spins caused by the application of the large field B_0.

We can visualize the resulting spin dynamics by placing ourselves in a **rotating reference frame**, so that we turn around $\hat{\boldsymbol{z}}$ at the same frequency as the rotating field, such that the field $\boldsymbol{B}_1$ is static. However, we are now in an non-inertial, accelerating reference frame and so, if we want to use classical mechanics, we must also include the presence of a fictitious field $-\omega_0/\gamma_\mu = -B_0$ directed along $\hat{\boldsymbol{z}}$. This exactly cancels the applied field such that the **effective field**, the only field left in the rotating frame, is simply the small field $\boldsymbol{B}_1$. As a result, the muon spin, as observed in the rotating frame, will precess (slowly) around the small field at a low frequency $\omega_1 = \gamma_\mu B_1$. Transforming back into the laboratory frame, the fast precession around $\boldsymbol{B}_0$, taking place at $\omega_0 \gg \omega_1$, is superimposed on the much slower rotating-frame precession. The composite motion that results is a spiral called a **nutation**,[25] shown in Fig. 7.3.

[25]This is the condition produced by a $\pi/2$ pulse in magnetic resonance experiments, achieved by switching on the B_1-field just for a time $\tau = 1/4\omega_1$. The spin is left orthogonal to $\boldsymbol{B}_0$ and, from there on, its free motion will be a maximum amplitude precession.

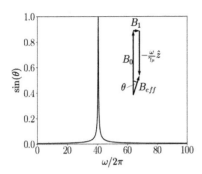

Fig. 7.4 Precession cone aperture for $B_0/B_1 = 500$. Inset: the effective field, sum of the external, fictitious, and rotating field.

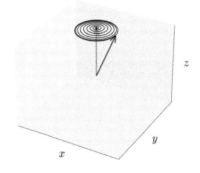

Fig. 7.5 Trajectory of the tip of the muon spin arrow on the surface of a sphere during an out of resonance nutation. Here $\boldsymbol{B}_{\mathrm{eff}} = (\omega - \omega_0)/\gamma_\mu \hat{\boldsymbol{z}} + B_1 \hat{\boldsymbol{x}}$, with $(\omega - \omega_0)/\omega_1 = 0.1$.

We might have expected the large field B_0 in the previous example to pin the spin along z. However, the example demonstrates how a small field, if chosen to have dynamics at a resonant frequency, can cause a dramatic change to the direction of a spin. It is worth stressing that this example is intrinsically resonant, even in cases when we are not carrying out a resonance experiment. The following example explains why.

Example 7.9

Now consider an out-of-resonance rotating field, at a some frequency ω. Even small deviations $\omega - \omega_0$ result in a large z-component of the effective field, $B_{\mathrm{eff}} = (\omega - \omega_0)/\gamma_\mu \gg B_1$, as shown in the inset of Fig. 7.4. The precession in the rotating frame takes place around the direction of the effective field $\boldsymbol{B}_{\mathrm{eff}} = (\omega - \omega_0)/\gamma_\mu \hat{\boldsymbol{z}} + B_1 \hat{\boldsymbol{x}}$, which is almost aligned along $\hat{\boldsymbol{z}}$. The motion is now described by a very small aperture cone and the maximum transverse spin deviation that it can produce (which in the previous example was $\pi/2$) reduces here to the precession cone aperture, which is a very small quantity.

When we transform back into the laboratory frame, we see only a very partial nutation, like that of Fig. 7.5. This results in very low-amplitude transverse components.

The result of the last example is that *out-of-resonance rotating fields are ineffective in causing full nutations and thus leave the spin along its initial direction.* So when do we encounter this condition outside a specific magnetic resonance experiment? Time-dependent, random interactions with a solid can result in a fluctuating field at the muon site and transverse components of this field can potentially tilt the muon spin by inducing precessions. However, they have a broad Fourier spectrum of frequencies $\tilde{J}(\omega)$ and we now know that only a very narrow component of this Fourier spectrum, centred at the Larmor frequency $\omega_0 = \gamma_\mu B_0$, can resonantly contribute to the muon spin dynamics. All the other components can often be ignored since they have little influence.

7.5 Dynamic magnetism

After warming up with the classical notions of spin relaxation due to magnetic dynamics, we turn to the results of a quantum mechanical treatment. Recall that in Section 5.1, the derivation of the polarization functions was performed within the framework of classical statistical physics, which led us to a relatively straightforward treatment of several complex physical situations. However, in cases such as relaxation due to dynamics, the quantum character of the muon and the system under study must be taken into account. In Chapter 4, the general formula

$$P_\alpha(t) = \mathrm{Tr}\left(\sigma_\alpha \rho(t)\right), \qquad (7.44)$$

was given for the polarization function, where $\hat{\rho}$ is the density operator of the muon system under study and σ_α is a Pauli matrix. In a limited number of physical situations, $P_\alpha(t)$ can be computed exactly.

However, our interest is typically in situations where the number of interacting spins is too large to be tractable without approximations and a quantum-mechanical perturbation expansion must be carried out in order to understand the relaxation in detail. This is rather lengthy and so is covered in Appendix F. Here we sketch out the main results appropriate to dynamics in magnetic materials. The idea of the perturbation expansion is that the Hamiltonian describing the muon-system is split into a static and dynamic part. The static part is the Hamiltonian representing the system only and the Zeeman interaction of the muon in the time-averaged field $\langle \boldsymbol{B} \rangle$ at the muon site. That is, we write, $\boldsymbol{B} = \langle \boldsymbol{B} \rangle + \delta \boldsymbol{B}(t)$. (Note that $\langle \boldsymbol{B} \rangle$ might include an externally applied field $\boldsymbol{B}_0$.) The perturbation Hamiltonian then corresponds to the muon Zeeman interaction in the fluctuating field $\delta \boldsymbol{B}(t)$.

Dynamics in a paramagnet

We start by considering a disordered magnet in the LF geometry, where the muon spin is originally oriented along z, which is also the direction of the applied field B_0. The fluctuations that dephase the muon spin are encoded in the correlation function of the local field at the muon site, defined as

$$\Phi_{ii}(t) = \gamma_\mu^2 \langle \delta B_i(t) \delta B_i(0) \rangle, \tag{7.45}$$

where $i = x, y$ and we make the assumption that there are no cross correlations.[26] If we want a picture of what the fluctuations do we can recall the *strong-collision model* of Chapter 5. There the fluctuations occurred discontinuously and had the effect of resetting the muon polarization. Appendix F shows that for the longitudinal field configuration we expect a polarization function

$$P_z(t) = \exp(-\lambda t), \tag{7.46}$$

where

$$\lambda t = \int_0^t \mathrm{d}t' (t - t') \cos \omega_0 t' \left[\Phi_{xx}(t') + \Phi_{yy}(t') \right]. \tag{7.47}$$

Here, $\omega_0 = \gamma_\mu \langle B \rangle$ is the Larmor frequency in the applied field, which is the average field around which the fluctuations are assumed to take place (since we assume $\langle \boldsymbol{B} \rangle$ in a paramagnet). Recall that there are two key effects in the measurement of dynamics: (i) the resetting of the muon relaxation by the fluctuating field and (ii) our ability to select the fluctuations that influence the measurement by matching the applied field, via $\hbar \omega_0$, to the energy of the dynamic processes that we wish to probe. Both of these features are contained in eqn 7.47.

We usually work in the limit of relatively long times ($t \gg t'$), which is to say that we are assuming that a measurement takes place over a long time compared to the dynamic processes. In that case we have

$$\lambda = \int_0^\infty \mathrm{d}t' \cos \omega_0 t' \left[\Phi_{xx}(t') + \Phi_{yy}(t') \right]. \tag{7.48}$$

[26] Cross correlations are described by functions like $\langle \delta B_x \delta B_y \rangle$. These should be expected to be very small in the absence of magnetic order. When order is present, the rigidity of the spin structure means that we might expect cross correlations to be present, although they usually remain relatively small.

This equation means that the longitudinal relaxation rate is governed by the Fourier transform of the correlations (i.e. the spectral density) of fields perpendicular to the average field. Returning briefly to the simplified classical picture, we concluded that only a very narrow slice of the spectral density, close to the Larmor frequency ω_0, can induce spin nutation. The classical picture also applies to the ensemble average in Heisenberg time evolution in the quantum case, where short-lived, fluctuating transverse-field components can induce nutations, and hence relaxation, only in the resonant case. This is why, in the presence of a static field, the relaxation rate is proportional to the Fourier transform of the spectral density evaluated *at* the corresponding Larmor frequency ω_0.

Example 7.10

Let's consider the case that $\delta B_x = \delta B_y = \delta B_\perp$ and $\langle \delta B_\perp(t) \delta B_\perp(0) \rangle = \langle \delta B_\perp^2 \rangle e^{-t/\tau}$, where $1/\tau$ is the field fluctuation rate. We therefore have

$$\Phi_{xx}(t) + \Phi_{yy}(t) = 2\gamma_\mu^2 \langle \delta B_\perp^2 \rangle e^{-t/\tau}, \tag{7.49}$$

and we find a relaxation rate

$$\lambda = \frac{2\Delta^2 \tau}{1 + \omega_0^2 \tau^2}, \tag{7.50}$$

where $\Delta^2 = \gamma_\mu^2 \langle \delta B_\perp^2 \rangle$. This is the Redfield equation that we met in Chapter 5.

If we don't make the simplifying assumption that $t \gg t'$ then the computation is a little more involved.

Example 7.11

Again using $\gamma_\mu^2 \langle \delta B_\perp^2 \rangle = \Delta^2$, we substitute eqn (7.49) into eqn (7.47). We can perform the integrals to obtain

$$\lambda(t)t = 2\Delta^2 \tau^2 \frac{(e^{-\frac{t}{\tau}} \cos \omega_0 t - 1)(1 - \omega_0^2 \tau^2) + \frac{t}{\tau}(\omega_0^2 \tau^2 + 1) - 2\omega_0 \tau e^{-\frac{t}{\tau}} \sin \omega_0 t}{(1 + \omega_0^2 \tau^2)^2}. \tag{7.51}$$

This very useful expression, known as the[27] **Keren function**, allows us to describe analytically the influence of exponentially correlated dynamics across the range of fluctuation rates $\nu = 1/\tau$. The limiting behaviour of this expression provides some familiar expressions.

In the *zero*-field limit, we set $\omega_0 = 0$, corresponding to small fluctuations around zero. We obtain the Abragam function, eqn 5.33, with an additional factor[28] of 2:

$$P_z(t) = \exp\left[-2\Delta^2 \tau^2 \left(\frac{t}{\tau} - 1 + e^{-\frac{t}{\tau}}\right)\right]. \tag{7.52}$$

In the fast fluctuation regime we have both $\omega_0 \tau \ll 1$ (i.e. lots of relaxation events in the time taken for a complete oscillation in the average field) and $t/\tau \gg 1$ (i.e. lots of events compared to the measurement time), and so eqn 7.51 becomes

$$P_z(t) = \exp\left[-\frac{2\Delta^2 \tau}{1 + \omega_0^2 \tau^2} t\right], \tag{7.53}$$

where we recognise the Redfield formula once again. This reduces, at zero fields ($\omega_0 = 0$), to

$$P_z(t) = \exp(-2\Delta^2 \tau t), \tag{7.54}$$

[27] A. Keren, Phys. Rev. B **50**, 10039 (1994).

[28] Note that the Abragam function is obtained in the *high-field* limit of the TF geometry, that attributes relaxation solely to fluctuations along z. In contrast, the LF relaxation is due to fluctuations in two directions, x and y, which justifies the factor of 2 in eqn 7.51. This simplified argument would not apply, strictly speaking, to the ZF case, which breaks the distinction between transverse and longitudinal components, and involves all three contributions, Φ_{xx}, Φ_{yy}, and Φ_{zz}.

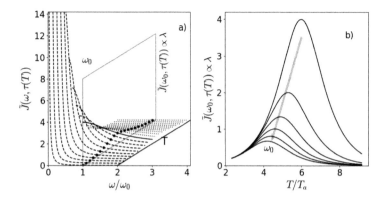

Fig. 7.7 a) The spectral densities $\tilde{J}(\omega, \tau)$ with an Arrhenius correlation time, at different temperatures. The slice at the Larmor frequency $\omega = \omega_0$ shows a peak in the muon rate $\lambda \propto J(\omega_0, \tau)$. b) The peaks in spectral density $J(\omega_0, \tau)$ for increasing values of ω_0.

i.e. an exponential, as predicted by our simplified approach, and shown in Fig. 7.6.

Finally, in the low-field, static limit of eqn 7.51 at early times ($t \ll \tau, \Delta^{-1}, \omega_0^{-1}$), we again obtain the Gaussian function

$$P_z(t) \approx 1 - \Delta^2 t^2 + \cdots \approx \exp(-\Delta^2 t^2), \tag{7.55}$$

resembling the early-time behaviour of the Kubo-Toyabe function that we saw before for the case of static disorder.

In the next example we shall gain some physical insight into the typical *temperature-dependent* behaviour of the relaxation rates through which muons probe a system's excitations.

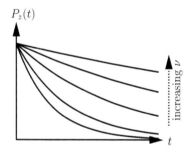

Fig. 7.6 Exponential relaxation in the fast fluctuation limit. The relaxation rate decreases as the fluctuation rate $\nu = \tau^{-1}$ increases.

Example 7.12

We assume the spectral density is that discussed in Example 7.3, where we had $\tilde{J}(\omega, \tau) = \frac{2\Delta^2 \tau}{1 + \omega^2 \tau^2}$. Here τ is the characteristic time of the thermal reservoir with which the muon is coupled.[29] In other words, a particular excitation modulates the local field at the muon site, taking place on average with rate τ^{-1}. The temperature dependence of $\tau(T)$ depends on the details of these excitation, but in most cases it will have two standard limits.

[29]Remember that the area $\int d\omega\, \tilde{J}(\omega) \propto \langle \delta B(0)^2 \rangle$ is a constant, independent of temperature.

- It will freeze out, [$\tau(T) \to \infty$], for $T \to 0$ so that $\tilde{J}(\omega_0, \tau) = 2(\Delta/\omega_0)^2 \tau^{-1} \to 0$.

- It will become shortest, and reach the minimum value $\tau(T) \to \tau_\infty \ll \omega_0^{-1}$, for $T \to \infty$ so that $\tilde{J}(\omega_0, \tau) = 2\Delta^2 \tau_\infty$, generally small.

In these two limits the relaxation rate is low, but in between them there is a temperature T_p where $\omega_0 \tau(T_p) = 1$, the denominator is minimum, and the spectral density peaks at its maximum value.

To be more specific let's assume that $\tau(T)$ follows an Arrhenius, activated behaviour $\tau(T) = \tau_\infty e^{T_a/T}$, producing the spectral densities in Fig. 7.7(a). Notice that these spectral densities are very high and narrow at low T, and their value at ω_0 is small. They decrease in height and increase in width with temperature, so that initially their value at ω_0 also increases. However, they become so broad that, in order to conserve the area, the value at ω_0 starts decreasing again. The full process produces the peak in $\tilde{J}(\omega_0, \tau)$ as a function of temperature[30] shown in the slanted slice of Fig. 7.7(a). Figure 7.7(b) shows how this peak shifts to lower temperature, and decreases as $\tilde{J}(\omega_0, \tau(T_p)) = \Delta^2 \tau(T_p) = \Delta^2/\omega_0$ with increasing ω_0, i.e. with increasing applied field. This feature actually allows a direct measure of $\tau(T)$ in some circumstances. The peak at T_p in the last example discriminates between two regimes of the relaxation: (i) the freezing regime, for $T < T_p$, where the rate goes to zero because the dynamics are slower than spin precessions; and (ii) the motional narrowing regime, for $T > T_p$ where the dynamics is faster than precessions, and the rate also decreases because the amplitude of the local fields Δ is reduced by the fraction $\Delta/\nu = \Delta\tau$ sampled by the dynamic process.

The most common state of affairs for a magnetic material above the ordering temperature is to be in the fast-fluctuation regime such that, in zero field, the polarization along z decays according to $P_z(t) = e^{-\lambda t}$, with $\lambda = 2\Delta^2\tau$. This is similar to spin lattice relaxation in magnetic resonance where the relaxation rate is called $1/T_1$. Since the fluctuation timescale of electronic moments in many magnetic systems is very fast compared to the muon timescale, we can have very large values of $\nu(= 1/\tau)$ compared to Δ^2, reducing the relaxation rate to very low values (i.e. *motionally narrowing* the magnetic contribution from the spectra, as discussed in the example). This is the situation shown in Fig. 7.6. However, when extreme motional narrowing occurs, we don't, in reality, end up with an unrelaxed muon ensemble, since the magnetic nuclear spins are often themselves unaffected by the dynamics, and still available to relax the muons.

Dynamics in an ordered magnet

Finally, we move on to the dynamics expected in an *ordered* magnetic material. In a static, perfectly ordered, polycrystalline magnet we expect a polarization function $P_z(t) = \frac{1}{3} + \frac{2}{3}\cos(\gamma_\mu B t)$. This corresponds to a coordinate system in which the initial muon spin points parallel to z along with $1/3$ of the local magnetic field components, while $2/3$ of the local field components point in the x-y plane. This separation of longitudinal and transverse components therefore allows us to employ the Bloch model discussed in Section 7.3. Earlier we discussed the example of a disordered material, using the geometry with initial muon spins (and any applied field) oriented along z and decided that the relaxation rate λ we considered is analogous to $1/T_1$ discussed in Section 7.3. For an ordered magnet we can assume that this sort of relaxation applies to the $1/3$ part of the muon polarization, describing magnetic fields initially parallel to muon spins. For the remaining field components, which give rise to the oscillations, we have behaviour that reflects a field transverse to the initial muon spin. The relaxation of this part is described by the $1/T_2$ relaxation processes of Section 7.3. A subtlety here is that $1/T_2$ needs to include not only dynamics, but also the possibility of static disorder, which contributed a relaxation $e^{-\Delta^2 t^2/2}$ where Δ/γ_μ is the standard deviation of the static field distribution about the average. We call the relaxation rate that encompasses both of these effects $1/T_2^*$. The relaxation rate $1/T_2^*$ is then the sum of the rate due to fluctuating components (usually called the **secular part**) and of the second moment of the static distribution of the field. We conclude then, that if the initial muon spin is along $\hat{z}$, we have

$$P_z(t) = \frac{1}{3}e^{-t/T_1} + \frac{2}{3}e^{-t/T_2^*}\cos(\gamma_\mu B t), \qquad (7.56)$$

where B is the average field at the muon site.

Example 7.13

The expression $P_z(t) = \frac{1}{3} + \frac{2}{3}e^{-(\Delta^2 t^2/2)}\cos(\gamma_\mu B t)$, appropriate for a static, polycrystalline, magnetically ordered sample (with a little disorder), also provides us with another means of finding a magnet transition temperature: the constant 1/3-tail. This arises in the magnetically ordered state since, at a local level, the ordering picks out a unique direction and those muon spin components pointing along that direction don't, in the absence of dynamics, relax. The basic idea is that, if we ignore relaxation, the 2/3 transverse component disappears at T_c, where the longitudinal component makes a sudden jump from 1/3 to 1.

When dynamics are present, as is usually the case, the 1/3 tail will relax exponentially following the $e^{-\lambda t}$ form discussed above. The remaining 2/3 of the muons could be relaxed by a mixture of static and dynamic processes, which dephase the oscillations more rapidly than the 1/3 tail is relaxed. If there are dynamics in the magnetically disordered regime (for $T > T_c$) then these should relax *all* of the muon spins to zero polarization. In the fast fluctuation regime these give rise to an exponential relaxation with $\lambda = 2\Delta^2/\nu$. If we compare the behaviour of the relaxation in the ordered and disordered phases of the magnet, it is notable that we often observe the disappearance of the 1/3 tail on warming from $T < T_c$ to $T > T_c$. In short: even in the presence of relaxation, the longitudinal component is easy to identify and its amplitude jumps at the transition. This effect is often an accurate means of locating the magnetic ordering temperature T_c.

7.6   Coupling tensors

The correlation function $\Phi_{ii}(t)$ describes the dynamics in the local field at the muon site. How does this relate to the underlying magnetic moments in the material under investigation? This question is important because many theoretical descriptions of a magnetic system are given in terms of spin-spin correlation functions. If we assume that the magnetic fields arise from the dipole fields of the spin moments, then we can come up with an answer that relates the correlation function of the spins to the relaxation rate.

Let's examine a zero-field experiment on a magnetically disordered material (so $\omega_0 = \gamma_\mu \langle B \rangle = 0$), in the long-time limit of Example 7.11, where we predict exponential relaxation with a decay rate

$$\lambda = g \sum_{i=x,y} \int_0^\infty \mathrm{d}t \, \langle \delta B_i(t) \delta B_i(0) \rangle, \tag{7.57}$$

where g is a constant. We can now link the magnetic fields to their sources, the spins, and come up with an expression in terms of the correlation function of the spins themselves.[31]

[31] We could use the dipolar coupling tensor, for example, although here we'll keep things general.

Example 7.14

For a muon site at position r, the $\alpha = x, y, z$ components of the magnetic field arising from the β component of a spin at position r' is given by

$$B_\alpha(r) = \sum_\beta D_{\alpha\beta}(r - r')S_\beta(r'), \tag{7.58}$$

where $D_{\alpha\beta}(r_i - r_j)$ encodes the coupling. Assuming no cross correlations among the spins, the relaxation rate becomes

$$\lambda = g \sum_{\substack{\alpha=x,y \\ \beta=x,y,z}} \sum_{r'r''} D_{\alpha\beta}(r - r')D_{\alpha\beta}(r - r'') \int_0^\infty \mathrm{d}t \, \langle S_\beta(r',t)S_\beta(r'',0) \rangle, \tag{7.59}$$

where we sum over the positions of the spins r and r'. Substituting Fourier transforms[32], we obtain

[32] The Fourier decompositions we need here are

$$S_\alpha(r) = \frac{1}{\sqrt{N}} \sum_q S_\alpha(q)\mathrm{e}^{\mathrm{i}q \cdot x}, \quad (7.60)$$

and

$$D_{\alpha\beta}(r) = \frac{1}{N} \sum_q D_{\alpha\beta}(q)\mathrm{e}^{\mathrm{i}q \cdot x}, \quad (7.61)$$

as we used in the previous chapter. Recall also the rule for taking the Fourier transform of a correlation function that depends only on the separation of coordinates.

$$\lambda = \frac{g}{N^3} \sum_{kqp} \sum_{\alpha\beta} \sum_{r'r''} D_{\alpha\beta}(k)\mathrm{e}^{\mathrm{i}k \cdot (r-r')} D_{\alpha\beta}(q)\mathrm{e}^{\mathrm{i}q \cdot (r-r'')}$$
$$\times \int_0^\infty \mathrm{d}t \, \langle S_\beta(-p,t)S_\beta(p,0) \rangle \mathrm{e}^{-\mathrm{i}p \cdot r'}\mathrm{e}^{\mathrm{i}p \cdot r''}, \tag{7.62}$$

where N is the number of spins. Collecting terms

$$\lambda = \frac{g}{N^3} \sum_{kqp} \sum_{\alpha\beta} \sum_{r'r''} D_{\alpha\beta}(k)D_{\alpha\beta}(q) \int_0^\infty \mathrm{d}t \, \langle S_\beta(-p,t)S_\beta(p,0) \rangle$$
$$\times \mathrm{e}^{\mathrm{i}(k+q) \cdot r}\mathrm{e}^{\mathrm{i}(-p-k) \cdot r'}\mathrm{e}^{\mathrm{i}(-q+p) \cdot r''}. \tag{7.63}$$

Sum over r' and r'' using the rule that $\sum_r \mathrm{e}^{\mathrm{i}(p+q) \cdot r} = N\delta_{-p,q}$, to get

$$\lambda = \frac{g}{N^3} \sum_{kqp} \sum_{\alpha\beta} D_{\alpha\beta}(k)D_{\alpha\beta}(q) \int_0^\infty \mathrm{d}t \, \langle S_\beta(-p,t)S_\beta(p,0) \rangle$$
$$\times \mathrm{e}^{\mathrm{i}(k+q) \cdot r} N\delta_{p,-k} N\delta_{p,q}. \tag{7.64}$$

Sum over k and q to find

$$\lambda = \frac{g}{N} \sum_p \sum_{\alpha\beta} D_{\alpha\beta}(-p)D_{\alpha\beta}(p) \int_0^\infty \mathrm{d}t \, \langle S_\beta(-p,t)S_\beta(p,0) \rangle \mathrm{e}^{\mathrm{i}(-p+p) \cdot r} \tag{7.65}$$

Tidying up we obtain

$$\lambda = \frac{g}{N} \sum_{\alpha\beta} \sum_p |D_{\alpha\beta}(p)|^2 \int_0^\infty \mathrm{d}t \, \langle S_\beta(-p,t)S_\beta(p,0) \rangle. \tag{7.66}$$

The result is a relatively simple expression in terms of the Fourier transform of the coupling (e.g. dipolar) and the spin-spin correlation function.

One rule of thumb to take away from this last example is that the muon response is not resolved in momentum space. That is to say, we lose most of the spatial information as a result of the sum over p. This can be traced back to the fact that the muon is a local probe.

Doing the integral over time[33], we obtain $\lambda = g'\sum_{\boldsymbol{q}}|D(\boldsymbol{q})|^2 G(\boldsymbol{q},0)$, where the momentum is $\boldsymbol{q}$. We have also suppressed the α and β indices[34] and rescaled the constant g. We see that the response involves a sum over the coupling and a factor of the spin-spin correlation function evaluated at zero frequency and summed over all $\boldsymbol{q}$. If we now use the fluctuation-dissipation theorem (eqn 7.29), we can rewrite

$$\lambda = 2g'k_{\mathrm{B}}T\lim_{\omega\to 0}\sum_{\boldsymbol{q}}|D(\boldsymbol{q})|^2\frac{\chi''(\boldsymbol{q},0)}{\omega}, \qquad (7.67)$$

which gives us an expression in terms of the susceptibility.

Example 7.15

Recall our example of the harmonic oscillator susceptibility. This is very similar to the response in a magnetic system. In the overdamped limit of a magnetic excitation we have

$$\frac{\chi''(\boldsymbol{q},\omega)}{\omega} = \frac{\chi(\boldsymbol{q})\tau_s(\boldsymbol{q})}{1+\omega^2\tau_s(\boldsymbol{q})^2}, \qquad (7.68)$$

and so we can write

$$\lambda = 2g'k_{\mathrm{B}}T\lim_{\omega\to 0}\sum_{\boldsymbol{q}}|D(\boldsymbol{q})|^2\frac{\chi(\boldsymbol{q})\tau_s(\boldsymbol{q})}{(1+\omega^2\tau_s(\boldsymbol{q})^2)} = 2g'k_{\mathrm{B}}T\sum_{\boldsymbol{q}}|D(\boldsymbol{q})|^2\frac{\chi(\boldsymbol{q})}{\Gamma(\boldsymbol{q})}, \qquad (7.69)$$

where $\Gamma(\boldsymbol{q}) = 1/\tau_s(\boldsymbol{q})$ is known as the linewidth.

One use for the approach described in the last example is in describing the muon response to dynamics in metals. We postpone a discussion of this to the next chapter. Instead, we now turn to the important problem of how dynamics affects the muon response to a dilute and disordered array of spins.

7.7 Dilute spins

A dilute spin system is one where there are relatively few magnetic moments compared to lattice sites. These moments are arranged randomly on the lattice and are magnetically disordered (i.e. the spin direction is randomized), as shown in Fig. 7.8. We allow the disordered moments to fluctuate as a function of time. This is the situation that occurs in **spin glass** materials comprising a small concentration of magnetic ions disolved in a non-magnetic, metallic host material. Examples are Fe disolved in Au, or Mn disolved in Cu, at around a 1% concentration. The RKKY interaction mediates magnetic exchange between the magnetic constituents. Owing to the oscillatory nature of RKKY exchange in space, however, the exchange interaction between the ions is highly randomized, varying in magnitude and sign depending on the separation of the magnetic ions. We can try to compute the muon response in such a system. However, going through the motions as we've done previously leads to a problem.

[33]Here we write $G_{\alpha\alpha}(\boldsymbol{q},t) = \langle S_\alpha(-\boldsymbol{q},t)S_\alpha(\boldsymbol{q},0)\rangle$ for $\alpha = x,y,$ or z. Recall that from our definition of the Fourier transform

$$G(\boldsymbol{q},\omega) = \int \mathrm{d}t\,\mathrm{e}^{i\omega t}G(\boldsymbol{q},t),$$

and so, in the limit that $\omega \to 0$ we can write

$$G(\boldsymbol{q},0) = \int \mathrm{d}t\,G(\boldsymbol{q},t).$$

[34]The result of the sum over α and β indices will depend on the situation being considered. We are often justified in assuming the spin fluctuations are isotropic, so that $G_{xx} = G_{yy} = G_{zz}$. To describe hyperfine coupling we set $D_{xx} = D_{yy} = D_{zz}$ and imagine the other components vanishing. The non-zero coupling constants are $\boldsymbol{q}$ independent in this case. This is the approach taken in the next chapter where this expression is used to describe the relaxation in a metal. When describing dipolar coupling the non-zero values of $D_{\alpha\beta}(\boldsymbol{q})$ are $\boldsymbol{q}$ dependent. Therefore, the loss of spatial information alluded to above is not absolute, with $|D_{\alpha\beta}(\boldsymbol{q})|^2$ acting as a structure factor that can provide a partial sensitivity to wavevector.

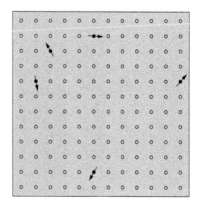

Fig. 7.8 A cartoon of a spin glass.

Key to this section is an exact result for the field distribution of a dilute set of randomized moments with a magnetic interaction falling off as $1/r^3$ (e.g. a dipolar interaction). The result, as examined in Chapter 5, is that averaging over the distribution that results from a dilute array of such moments gives a *Lorentzian* probability distribution of field components

$$p^{\mathrm{L}}(B_i) = \frac{\gamma_\mu}{\pi} \left(\frac{a}{a^2 + \gamma_\mu^2 B_i^2} \right), \tag{7.70}$$

with $i = x, y, z$ and where a is proportional to the concentration of the moments. Compared to the Gaussian function (appropriate for a dense distribution of spins) we might expect the probability distribution for the dilute distribution has a broader shape. The Lorentzian, with its long tails at large B, fits the bill, capturing the property that a point picked at random could be near to, or (more likely) far from, dipolar spins. We saw further in Chapter 5 that, in terms of the field magnitude B, this distribution is given by

$$p^{\mathrm{L}}(B) = \frac{\gamma_\mu^3}{\pi^2} \frac{a}{\left(a^2 + \gamma_\mu^2 B^2\right)^2} 4\pi B^2. \tag{7.71}$$

[35]In this section we shall describe the fluctuations in terms of the fluctuation rate $\nu = 1/\tau$, where τ is the fluctuation time used earlier in the chapter.

To capture the observed behaviour we will treat the dynamics as having the usual exponential distribution, such that[35] $\langle \delta B_i(t) \delta B_i(0) \rangle = \langle \delta B_i^2(0) \rangle \mathrm{e}^{-\nu t}$. As in the previous section, dynamicization using this distribution can be thought of in terms of a static configuration of magnetic spins that, once every $1/\nu$ seconds, jumps to a new configuration. The field at a given point will therefore take another value from that range of possible fields that are generated by the different possible orientations of the magnetic spins.

Example 7.16

The distribution in eqn 7.71, when combined with our polarization function $P_z(t, B) = \frac{1}{3} + \frac{2}{3} \cos(\gamma_\mu B t)$, gives rise to a (so-called) Lorentzian-Kubo-Toyabe function for the zero-field polarization

$$P_z(t) = \frac{1}{3} + \frac{2}{3}(1 - at)\mathrm{e}^{-at}. \tag{7.72}$$

This describes the muon polarization for muons implanted at random into the static array of dilute moments. It has a 1/3 tail, much like a conventional, Gaussian-Kubo-Toyabe function, reflecting the behaviour of those muons with their spins directed along the local magnetic field direction. This distribution can be dynamicized to account for fluctuations with fluctuation rate ν. The result turns out to be that there is not much dependence on ν at the start of the decay with the dynamics merely relaxing the late-time 1/3 tail of the distribution.

The relaxation described in the last example is *not* what is observed experimentally in spin glasses. The problem with our previous approach is that it involves (i) the average over a static distribution followed by (ii) a dynamicization. However, this ordering [(i) followed by (ii)] turns out to be incorrect.

To understand this problem we take the point of view of a muon, as shown in Fig. 7.9. The muon close to the magnetic moments can, depending on the directions of the local spins, experience a wide range of magnitudes of the local magnetic field. The muon far from the moments will, whatever the orientations of the magnetic spins, experience a far more narrow possible range of fields. In this way, we think of *different muon sites as having access to different distributions of magnetic field*. If each of the disordered moments are assumed to be fluctuating then, after each fluctuation event, a typical local field seen by a muon essentially chooses a new value from the available distribution, which is different at each muon site. Such a process can't be captured by the dynamicization of the Lorentzian distribution, since that assumes the muon ensemble is relaxed according to the same distribution at every step.

The correct way to take dynamics into account, therefore, is (i) to dynamicize the muon polarization function for each muon site, according to the range of fields available at that specific site, and then (ii) use the distribution of field ranges available across the sample to produce the spatial average. In short, we dynamicize and then spatially average.

Let's do this. We say that some muon site (labeled j) will have access to a range of random local fields with a Gaussian distribution[36] $p^{\mathrm{G}}(B, \Delta_j)$ of width Δ_j. When the spins fluctuate, they pick a field from within this distribution, causing a muon at site j to precess in this new field.

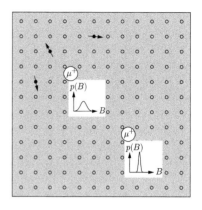

Fig. 7.9 The field distributions at different muon sites. A site close to magnetic ions can select fields from a broader distribution than one far from the magnetic ions.

[36]We have, from before, that
$$p^{\mathrm{G}}(B_i, \Delta_j) = \frac{\gamma_\mu}{\sqrt{2\pi}\Delta_j} e^{-\gamma_\mu^2 B_i^2 / 2\Delta_j^2}.$$

Example 7.17

If we imagine an ensemble of muons at site j, then their instantaneous polarization will be given by a (Gaussian) Kubo–Toyabe polarization function $P_z^{\mathrm{G}}(t, \Delta_j) = \frac{1}{3} + \frac{2}{3}(1 - \Delta_j^2 t^2) e^{-\frac{\Delta_j^2 t^2}{2}}$. We can dynamicize this function as usual, to give, in the fast-fluctuation limit for example, $P_z^{\mathrm{G}}(t, \Delta_j, \nu) = e^{-\frac{2\Delta_j^2}{\nu} t}$.

The potentially tricky thing now is how to find the distribution of Δ_j (i.e. the distribution of Gaussian widths across the sample). However, we are saved because we know the one key fact that *in the static case, the total distribution of magnetic fields must be Lorentzian*, as in eqn (7.70)! Therefore, in order to recover the Lorentzian distribution for each component we must find the distribution function $\rho(\Delta_j)$ such that

$$\int_0^\infty \mathrm{d}\Delta_j \, p^{\mathrm{G}}(B_i, \Delta_j)\rho(\Delta_j) = p^{\mathrm{L}}(B_i). \tag{7.73}$$

The distribution of Δ_j that, on integration, gives the required Lorentzian distribution turns out to be

$$\rho(\Delta_j) = \left(\frac{2}{\pi}\right)^{\frac{1}{2}} \left(\frac{a}{\Delta_j^2}\right) e^{-\frac{a^2}{2\Delta_j^2}}. \tag{7.74}$$

Now that all of the parts are in place, we can compute the final relaxation function by averaging $P_z^G(t, \Delta_j, \nu)$, the dynamicized polarization function for each j, over the distribution $\rho(\Delta_j)$:

$$P_z(t, a, \nu) = \int_0^\infty \mathrm{d}\Delta \, P_z^G(t, \Delta_j, \nu) \rho(\Delta_j). \qquad (7.75)$$

Example 7.18

Analytical forms can be obtained for $P_z(t)$ in eqn 7.75 in limiting cases. Within the same framework as in Section 5.2, in the limit of slow fluctuations we have

$$P_z(t, a, \nu) = \frac{1}{3} \exp\left(-\frac{2}{3}\nu t\right) + \frac{2}{3}(1 - at) \exp(-at), \qquad (7.76)$$

since the 1/3 recovery value of the Kubo–Toyabe function is independent of Δ. We assume that the fluctuations are exponentially correlated, taking the familiar form $\langle \delta B(t) \delta B(0) \rangle = \langle \delta B(0)^2 \rangle e^{-\nu t}$, where $\delta B(0)$ is the local field at some particular point in the system. In the fast fluctuation limit, i.e. $\nu/a \gg 1$, we obtain

$$P_z(t, a, \nu) = \exp\left\{ -\sqrt{\frac{4a^2}{\nu^2}[\exp(-\nu t) - 1 + \nu t]} \right\}, \qquad (7.77)$$

which reduces to

$$P_z(t, a, \nu) = \exp\left(-\sqrt{\frac{4a^2 t}{\nu}}\right), \qquad (7.78)$$

when $\nu t \gg 1$. This latter polarization function is a square-root exponential function. It illustrates a rare case where fast dynamics does not lead to an exponential relaxation.

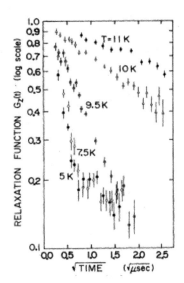

Fig. 7.10 ZF μSR data measured for the spin glass material *Au*Fe showing root exponential behaviour at several temperatures. [Taken from Y. J. Uemura *et al.*, Phys. Rev. B **31**, 546 (1985).]

In the fast-fluctuation limit the last example showed that we have $P_z(t, a, \nu) = \mathrm{e}^{-\left(\frac{4a^2 t}{\nu}\right)^{\frac{1}{2}}}$, which is stretched-exponential behaviour, i.e. the relaxation following $\mathrm{e}^{-(\lambda t)^\beta}$ with $\beta = 1/2$. As shown in Fig. 7.10, the muon-spin relaxation in a spin glass material, when plotted on a log scale against $\sqrt{t}$, gives approximately straight lines demonstrating this so-called *root exponential* (i.e. $\beta = 1/2$) behaviour. The model, which assumed rapid fluctuations, works well above the glass freezing temperature T_g. Below the $T_g (\approx 10$ K for the data shown in Fig. 7.10), the model only captures the initial relaxation, with a slowly relaxing tail to the spectra reflecting more complicated behaviour in this regime. The incorporation of this freezing into our model is the subject of the next example.

Example 7.19

When a ferromagnet undergoes a transition to static long-range magnetic order, we expect a correlation function for the field at a point to be $\langle B(\tau)B(0) \rangle = \langle B^2 \rangle$, which is independent of time. The value of B should be proportional to the order parameter of the magnetic system. If the ferromagnet hosts fluctuations, then we expect that the correlation function eventually relaxes to just such a non-zero value, reflecting the presence of the long-range order and a non-zero order parameter.

When a glass freezes it rarely becomes completely static. Each spin will have a preferred direction that biases its orientation, but it will likely continue to fluctuate about this axis. This is a complicated problem which was addressed by Sam Edwards and Philip Anderson, who suggested that a simple description could be well motivated using a technique known as replica theory. According to the Edwards-Anderson approach, if neighbouring spins are uncorrelated, then it is permissible to treat the average magnetic field at the muon site as being formed from a static component $\sqrt{Q}\delta B$ and a dynamic one $\sqrt{(1-Q)}\delta B$ that fluctuates with a rate ν, where $0 \leq Q \leq 1$. In terms of these components, the correlation function is written

$$\left\{ \frac{\langle \delta B(\tau)\delta B(0)\rangle}{\langle \delta B(0)^2\rangle} \right\} = (1-Q)e^{-\nu\tau} + Q, \qquad (7.79)$$

where $\langle\rangle$ denotes the usual thermal average and, here, $\{\}$ denotes a spatial average over the system. After long intervals in τ, we see that the correlation function, averaged across the system, relaxes to a non-zero value determined by Q, reflecting the presence of the frozen parts of the spins after spatial averaging. By analogy with the order parameter in a conventional magnet, Q is often called the **Edwards-Anderson order parameter**.

Using these ideas, the computation of the muon relaxation function proceeds in a similar manner as we saw previously for the purely dynamic spin glass. So, in the presence of a static component and in the fast fluctuation limit, the spin relaxation at each muon site j becomes $P_z^G(t, \Delta_j, \nu, Q)$, which can be approximated as a product of the static and dynamic parts:[37]

$$P_z^G(t, \Delta_j, \nu, Q) = \left[\frac{1}{3} + \frac{2}{3}\left(1 - \Delta_{sj}^2 t^2\right)e^{-\frac{\Delta_{sj}^2 t^2}{2}} \right] \times e^{-\frac{2\Delta_{dj}^2 t}{\nu}}, \qquad (7.80)$$

where $\Delta_{sj} = \sqrt{Q}\Delta_j$ is the static part of the width and $\Delta_{dj} = \sqrt{1-Q}\Delta_j$ is the dynamic part. We can then perform the spatial average analytically using the function $\rho(\Delta_j)$, from which we obtain

$$P_z(t, a, \nu) = \frac{1}{3}e^{-(\lambda_d t)^{\frac{1}{2}}} + \frac{2}{3}\left[1 - \frac{a_s^2 t^2}{(\lambda_d t + a_s^2 t^2)^{\frac{1}{2}}}\right]e^{-(\lambda_d t + a_s^2 t^2)^{\frac{1}{2}}}, \qquad (7.81)$$

where here $a_s = \sqrt{Q}a$ is the static part of the Lorentzian width, $\lambda_d = 4a_d^2/\nu$ is the dynamic relaxation rate, and $a_d = \sqrt{1-Q}a$ is the dynamic part of the width. Just as for the relaxation functions for ordered magnets in the presence of dynamics, we have 1/3 of the polarization that reflects local field components longitudinal to the initial muon spin and is purely relaxed by dynamic processes that (as in the last example) give a root-exponential response. We also have 2/3 of the polarization, reflecting fields transverse to the initial muon polarization direction, that is more complicated owing to the combination of dynamic and static effects. Equation 7.81 is graphed in Fig. 7.11 for several values of $\sqrt{Q} = a_s/a$, where we can see how it interpolates between dynamic, root-exponential behaviour for $Q = 0$ to Lorentzian Kubo-Toyabe behaviour for $Q = 1$.

[37]The use of a product here is similar in spirit to the Born-Oppenheimer approximation, and is permissible owing to the division of timescales between the two contributions.

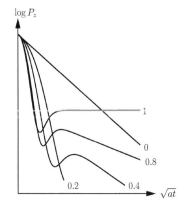

Fig. 7.11 Equation 7.81 graphed for given values of $\sqrt{Q} = a_s/a$ for $\nu/a = 100$. [Based on Y. J. Uemura *et al.*, Phys. Rev. B **31**, 546 (1985).]

The ideas presented in this chapter are applicable to a wide variety of physical problems both in the field of magnetism and beyond. We examine some further examples in the next chapter.

Chapter summary

- The μSR relaxation rate in the presence of dynamics in a field B_0 can be understood as reflecting the resetting of the field distribution after a correlation time τ via dynamic processes with a characteristic energy $\hbar\omega_0 = \hbar\gamma_\mu B_0$.

- The muon relaxation can be related to the correlations or response functions governing the dynamics in a magnetic material using

$$\lambda = \int_0^\infty dt' \cos\omega_0 t' \left[\Phi_{xx}(t') + \Phi_{yy}(t')\right], \qquad (7.82)$$

where $\Phi_{ii}(t')$ is the autocorrelation function for the ith component of the local magnetic field.

- The longitudinal-field polarization function in the presence of fast fluctuations is given by

$$P_z(t) = \exp\left[-\frac{2\Delta^2\tau}{1 + \omega_0^2\tau^2}t\right]. \qquad (7.83)$$

Further reading

- Dynamic susceptibility is introduced in S. J. Blundell, *Magnetism in Condensed Matter*, OUP (2001).

- A full treatment of dynamic effects can be found in P. M. Chaikin and T. C. Lubensky, *Principles of Condensed Matter Physics*, CUP (1995).

- The response of muons in a spin glass is presented in Y. J. Uemura *et al.*, Phys. Rev. B **31**, 546 (1985).

Exercises

(7.1) Take a Fourier tranform of the susceptibility equation

$$\delta M(t) = \int dt'\, \chi(t - t')\delta H(t'), \qquad (7.84)$$

to reveal its form in frequency space.

(7.2) (a) By considering the Fourier transform of the equation of motion, show that the susceptibility for the simple harmonic oscillator is given by

$$\tilde{\chi}(\omega) = \frac{\tilde{x}(\omega)}{\tilde{f}(\omega)} = \frac{1}{m}\frac{1}{(-\omega^2 + \omega_0^2 - i\omega\gamma)}. \qquad (7.85)$$

(b) Show that, in the overdamped limit, we obtain

$$\tilde{\chi}(\omega) = \frac{\chi}{1 - i\omega\tau_s}. \qquad (7.86)$$

(c) Show that the power dissipated by the overdamped oscillator is given by

$$P = \frac{1}{2}f_0^2\omega\chi''(\omega). \qquad (7.87)$$

(7.3) In Appendix F we show that, for dynamics in the transverse geometry, we obtain a second order correction to the polarization function of

$$\langle \sigma_x(t) \rangle_{(2)} = -\int_0^t dt' \left\{ (t-t')\Phi_{zz}(t') \cos\omega_0 t \right.$$
$$+ \frac{1}{2}(t-t') \left[\Phi_{xx}(t') + \Phi_{yy}(t') \right] \cos\omega_0(t-t')$$
$$\left. - \frac{1}{2\omega_0} \left[\Phi_{xx}(t') - \Phi_{yy}(t') \right] \sin\omega_0(t-t') \right\}. \quad (7.88)$$

(a) Verify that at time t longer than any correlation time, and assuming $t \gg \omega_0^{-1}$, and ignoring any out of phase component, eqn 7.88 predicts a relaxation rate

$$\lambda = \int_0^\infty dt' \times$$
$$\left\{ \Phi_{zz}(t') + \frac{1}{2}\cos\omega_0 t' \left[\Phi_{xx}(t') + \Phi_{yy}(t') \right] \right\}. \quad (7.89)$$

(b) Derivation of the Abragam function only necessitates considering the first term in the integral in eqn 7.88. Assuming correlations of the form $\Phi_{zz}(t') = \gamma_\mu^2 \langle \delta B_z^2 \rangle e^{-\nu t'}$, show that

$$P_x(t) = \exp\left[-\frac{\Delta^2}{\nu^2}(\nu t - 1 + e^{-\nu t}) \right] \cos\omega_0 t. \quad (7.90)$$

(7.4) The relaxing part of the Abragam function, $G_x(t)$, is defined such that

$$P_x(t) = G_x(t)\cos\omega_0 t. \quad (7.91)$$

(a) Show that, in the static limit ($\nu t \to 0$), we recover the Gaussian form

$$P_x(t) = \exp\left(-\frac{\Delta^2 t^2}{2} \right)\cos\omega_0 t. \quad (7.92)$$

(b) In the fast fluctuation limit ($\nu t \to \infty$ and $\nu/\Delta \gg 1$) show that we obtain

$$P_x(t) = \exp\left(-\frac{\Delta^2 t}{\nu} \right)\cos\omega_0 t, \quad (7.93)$$

i.e. exponential relaxation.

(7.5) By direct integration, show that the distribution function

$$\rho(\Delta_j) = \left(\frac{2}{\pi} \right)^{\frac{1}{2}} \left(\frac{a}{\Delta_j^2} \right) e^{-\frac{a^2}{2\Delta_j^2}}, \quad (7.94)$$

recreates the Lorentzian distribution, as described in the text.

Measuring dynamic processes

We saw in the Chapter 7 how the time dependence of a distribution of local magnetic fields relaxes the muon-spin ensemble and how, using applied longitudinal magnetic fields, we are able to use the muon to probe the spectrum of fluctuations. The principle of this measurement can be applied to a wide range of dynamic phenomena, both in magnetism and more widely. In this chapter we discuss some examples of the different ways in which muons can be used to probe dynamics, along with the quantities that can be extracted from the measurements.

8.1 Critical dynamics

At temperatures close to the ordering temperature T_c of a magnet, it is found experimentally that several physical properties of a magnet described above follow *power law* behaviour[1] as a function of temperature, applied magnetic field or distance. This motivates the definition of critical exponents to describe the phase transition. The usefulness of these parameters comes from the hypothesis of the **universality** of critical behaviour, which says that the exponents characterizing a symmetry-breaking phase transition don't depend on the type of phase transition (e.g. magnetic, structural, superconducting etc.), but rather only on the dimensionality d of the system, the dimensionality $\mathcal{D}$ of the order parameter and the range of the interactions.[2] In this way, a set of critical exponents characterizes a class of phase transitions seen in nature.

We discussed in Chapter 6 how the exponent β, defined via $M(T) = M(0)(1 - T/T_c)^\beta$ could be extracted from the temperature dependence of the muon precession frequency $\nu(T)$, measured in the magnetically ordered state. This β parameter is one of the set of critical exponents that characterize phase transitions; we meet some of the others in the next example.

[1]Unlike, for example, exponential functions such as $e^{-x/\xi}$, power laws are free from any dependence on a length scale. This scale-free property is a hallmark of critical behaviour, where the characteristic fluctuations occur over all length scales.

[2]Recall from Chapter 6 that the order parameter in a magnet is the magnetization $\boldsymbol{M}$. The dimensionality $\mathcal{D}$ of $\boldsymbol{M}$ tells us the number of dimensions in which magnetic moments are free to point while the dimensionality of the system d tells us the number of dimensions in which the moments interact, often called the dimensionality of the lattice. We might expect $d = 3$ in isotropic magnets, although cases with $d = 2$ (isolated layers) and $d = 1$ (isolated chains) are well known. Idealized cases with $\mathcal{D} = 3$ are known as Heisenberg models; $\mathcal{D} = 2$ cases are X-Y models; and $\mathcal{D} = 1$ cases are Ising models.

Example 8.1

Defining the reduced temperature as $t = (T - T_c)/T_c$, the critical exponents $\alpha, \beta, \gamma, \delta$, and ν are related to physical properties as follows:

- Heat capacity: $C \sim |t|^{-\alpha}$,
- Magnetization: $M \sim (-t)^\beta$, for $B \to 0$, $T < T_c$,

- Magnetic susceptibility: $\chi \sim |t|^{-\gamma}$,
- Field dependence of χ at $T = T_c$: $\chi \sim |B|^{1/\delta}$.

In addition to measuring β, we can use muons to access the exponent γ for $T > T_c$. By applying a small transverse field in the paramagnetic phase, a measurement of the temperature dependence of the local magnetic susceptibility may be obtained from the shift of the muon spin rotation frequency caused by the total field at the muon site. Specifically we write

$$\left(\frac{\nu(T)}{\nu(0)} - 1 \right) \propto \chi(T) \propto |t|^{-\gamma}. \tag{8.1}$$

An example of the use of this method is shown in Fig. 8.1 for the antiferromagnet cobalt glycerolate $[Co(C_3H_6O_3)$ or Co(gly)]. In this system μSR was used to extract $\beta = 0.228(9)$ via zero field measurements of muon-spin precession and $\gamma = 1.05(4)$ was extracted from the measurements made in weak transverse field.

To the thermodynamic quantities in the last example, we can add the spatial correlation function,[3] which depends on the **correlation length** ξ, and tells us how well correlated the *fluctuations* are in space (Fig. 8.2). For example, in a magnet above T_c, despite the spins being disordered they may save energy by pointing in the same direction, and will tend to do so within a region of size ξ. The correlation between spins at position x and position y decreases with increasing $|x-y|$ because thermal fluctuations have a tendency to randomize the spins. Thus we find that[4]

$$G_c(x, y) \sim e^{-|x-y|/\xi}, \tag{8.3}$$

for $|x-y| \gg \xi$. The subtraction of the average fields in our definition of G_c means that this is also the behaviour below T_c and represents the fluctuations away from perfect order.

Example 8.2

We can add to our list of critical exponents an exponent for the correlation length ξ and an exponent η that described the short-range behaviour of the correlation function.

- Correlation length: $\xi \sim |t|^{-\nu}$.
- The correlation function $G(r)$ behaves like

$$G(r) \sim \begin{cases} \frac{1}{|r|^{d-2+\eta}} & |r| \ll \xi \\ e^{-\frac{|r|}{\xi}} & |r| \gg \xi, \end{cases} \tag{8.4}$$

where r is distance and d is the dimensionality of the system. The penultimate expression gives us the definition of η.

The set of critical exponents we have defined describe the static behaviour of the magnet close to the phase transition. These critical exponents are not independent and so, although we have only described the measurement of two of the exponents using μSR, we already have enough information to determine the others.

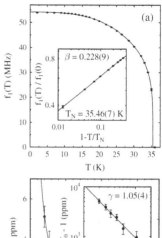

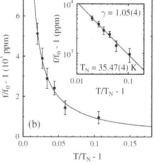

Fig. 8.1 (a) Precession frequencies measured in Co(gly) from ZF measurements. (b) The shift in the frequency of weak TF measurements made in the paramagnetic regime. The insets show scaling plots whose gradients give the critical parameters. [Taken from F. L. Pratt *et al.*, Phys. Rev. Lett. **99**, 017202 (2007).]

[3] A modification of the correlation function $G(x, y) = \langle M(x)M(y) \rangle$ is useful when studying phase transitions. This is the **connected correlation function** $G_c(x, y)$, given by

$$G_c(x, y) = \langle M(x)M(y) \rangle \tag{8.2}$$
$$- \langle M(x) \rangle \langle M(y) \rangle,$$

and differs from $G(x, y)$ by the subtraction of a product of thermally averaged fields $\langle M(x) \rangle \langle M(y) \rangle$. At temperatures above the transition temperature T_c we have $\langle M(x) \rangle = 0$ and the definition is identical to the usual form of the correlation function. At temperatures below T_c, subtracting off the mean fields allows us to just see the fluctuations away from average behaviour.

[4] Closer to the disturbance the correlations are found to decay more slowly, as described below.

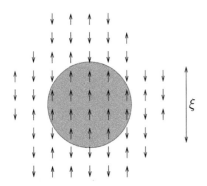

Fig. 8.2 The correlation length in a paramagnet showing a ferromagnetic fluctuation of size ξ.

[5]Recall in Chapter 6 that we briefly mentioned van Hove's prediction that the characteristic fluctuation time τ should diverge as $\tau \propto \chi \propto |t|^{-\gamma} = |t|^{-\nu(2-\eta)}$, but that this is often not observed. Here we extend that discussion by including the parameter z to correct the latter power law.

[6]For example, in an isotropic ferromagnet we expect $z = \frac{1}{2}(d + 2 - \eta)$; for an isotropic antiferromagnet, we have instead that $z = \frac{d}{2}$.

Example 8.3

There are only two independent static critical exponents. Further static exponents can be derived using the **scaling relations**. We can extract α from

$$\alpha = 2 - 2\beta - \gamma, \tag{8.5}$$

and ν from

$$\nu = \frac{(2 - \alpha)}{d} = \frac{(2\beta + \gamma)}{d}, \tag{8.6}$$

where d is the dimensionality of the spin system. We also have an expression

$$2 - \eta = \frac{\gamma}{\nu} = d\left(\frac{\delta - 1}{\delta + 1}\right). \tag{8.7}$$

Using these expressions and a knowledge of the dimensionality of the system near the phase transition, the full range of parameters is therefore accessible.

So far, all of the exponents relate to the static behaviour close to the phase transitions. However, there is also a rich range of **dynamic critical behaviour** that describes the behaviour of the dynamic susceptibility $\chi(\boldsymbol{q}, \omega)$ and the characteristic frequency $\omega(\boldsymbol{q})$ of fluctuations, close to the phase transition.[5] This gives rise to a **dynamic critical exponent** z. Although the parameter z is not independent of the other critical exponents and the dimensionality, the details of its relationship with the other parameters depends on the nature of the dynamics under examination.[6] This sensitivity makes z a useful way of understanding the nature of the dynamics in a material.

The good news is that the dynamic exponent z can be accessed experimentally using LF μSR measurements close to the phase transition. This is because, when the muon response is coupled linearly to the magnetization (as is the case for dipole fields, for example), the dynamic correlation time τ describing the field-field correlation function, $\Phi(\tau) = \gamma_\mu^2 \langle \delta B(\tau) \delta B(0) \rangle$, is expected to diverge according to

$$\tau = \left(\frac{1}{\xi}\right)^{-2+d+\eta-z} = |t|^{-\nu(2-d-\eta+z)}. \tag{8.8}$$

As a result, the longitudinal muon-spin relaxation rate λ should be expected to be sensitive to dynamic critical behaviour. If, for example, we work at zero applied magnetic field in the fast-fluctuation regime, where the relaxation rate is given by $\lambda \propto \Delta^2 \tau$, then if the fluctuation amplitude Δ isn't strongly dependent on temperature we should expect that close to the transition

$$\lambda \propto \tau \propto |T - T_c|^{-w}, \tag{8.9}$$

with

$$w = \nu(2 - d - \eta + z). \tag{8.10}$$

In terms of the measured μSR exponents β and γ, we can then write an expression for the dynamic exponent

$$z = \frac{d(2\beta + 2)}{2\beta + \gamma}. \tag{8.11}$$

In Fig. 8.3 the dynamic exponent for Co(gly) is extracted from scaling plots where a value of $w = 0.20(1)$ is found close to the transition, enabling the relationship between exponents to be established.

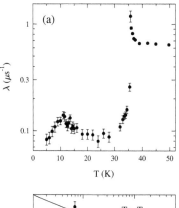

Example 8.4

The prototypical Heisenberg antiferromagnetic insulator MnF_2 (with isotropic exchange coupling J) is a tetragonal crystal with $T_N = 67.336$ K and its staggered magnetization lies along the c-axis. As a consequence, magnetic anisotropy must eventually dominate close to T_N, where the system is expected to crossover to a regime of behaviour described by an Ising-type model. This material offers an example of critical relaxation that can be detected in a TF-μSR experiment.

In the TF geometry the dynamical relaxation is due to fluctuations of the secular component of the local field (see discussion above eqn 7.56), which, in a paramagnetic single crystal, is defined by the direction of the applied field (e.g. either along, or perpendicular to c). Figure 8.4 shows that the relaxation rate diverges at T_N, following a power law (inset).[7] The figure shows two sets of data, with fields along two distinct crystal axes. Their ratio inverts at the crossover between two regimes, one close to T_N (i.e. in the critical regime), and a second further away. This is understood if we remember that in the TF case the relaxation rate in eqn 7.48 is replaced by

$$\lambda_z = \int_0^\infty d\tau \left\{ \Phi_{zz}(\tau) + \frac{1}{2} \cos\omega_0\tau \left[\Phi_{xx}(\tau) + \Phi_{yy}(\tau) \right] \right\}, \quad (8.12)$$

(see eqn F.34 and Exercise 7.3) where $z = a, c$ for the two data sets of Fig. 8.4. Far from T_N fluctuations are isotropic, but the ratio λ_c/λ_a reflects the dipolar origin of the field at the muon, with two equivalent nearest neighbour (nn) Mn ions along a, as sketched in the figure. The result depends on the exact dipolar sums of the MnF_2 lattice that predict $\lambda_c/\lambda_a = 0.7$ (a stronger dipolar field when the moment points towards the muon), in agreement with the data. In the critical regime $\lambda_c/\lambda_a = 2$ can be directly read from eqn 8.12, assuming that $\Phi_{cc}(\tau) \gg \Phi_{aa}(\tau), \Phi_{bb}(\tau)$.[8]

The anisotropy of relaxation rates can only be measured in single crystal experiments. However, the divergence of transverse relaxation rates due to critical slowing down of spin fluctuations is a more common phenomenon, often observed below T_c, where the role of the applied field is played by the local spontaneous field. This implies that zero-field oscillations in the muon polarization become critically damped at T_c.

8.2 Magnetism in metals

A metal can be characterized as a gas of electrons interacting via the Coulomb interaction. Metals have a rich structure of excitations and properties that result from this apparently simple interaction, including a magnetic response. If we have access to the dynamic magnetic susceptibility of a metal we can relate it to the muon dynamics using the ideas from the previous chapter.

The simplest model of a metal sets aside electron-electron interactions completely. This is the Fermi-gas model which has a momentum distribution

$$n_{\boldsymbol{p}} = \left[1 + e^{\frac{(E_{\boldsymbol{p}} - \mu)}{k_B T}} \right]^{-1}, \quad (8.13)$$

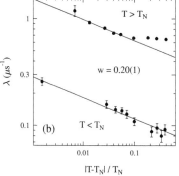

Fig. 8.3 (a) Longitudinal relaxation rate in Co(gly) which shows a critical divergence at the 35 K antiferromagnetic transition. (b) Scaling plots near the transition show the critical behaviour characterized by $w = 0.20(1)$. [Taken from F. L. Pratt *et al.*, Phys. Rev. Lett. **99**, 017202 (2007).]

[7]The value of the critical exponent $w = 0.8(2)$ is in agreement with both Heisenberg and Ising models and cannot distinguish between them.

[8]It is crucial that the two nearest-neighbour Mn ions belong to the same sublattice. In the opposite case the instantaneous local field of an antiferromagnetic fluctuation would cancel at the muon site. This shows that the symmetry of the site drastically influences how the muon probes critical fluctuations.

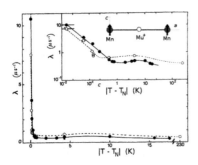

Fig. 8.4 The behaviour of the transverse relaxation rate in MnF$_2$ just above the antiferromagnetic transition. The field is applied both along c (filled symbols) and along a (open symbols). Inset: scaling plot. [Taken from R. De Renzi *et al.*, Phys. Rev. B **30**, 197 (1984).]

[9]In a Hubbard model described by the Hamiltonian

$$\mathcal{H} = T + U \sum_i n_{i\uparrow} n_{i\downarrow}, \qquad (8.16)$$

we have $I = 2U/(g\mu_B)^2$.

[10]Notice that taking $\Gamma(\mathbf{q})^{-1} = \tau$, we can rewrite this as

$$\chi(\mathbf{q}, \omega) = \frac{\chi(\mathbf{q})}{1 - i\omega\tau}.$$

The complex susceptibility can be rewritten

$$\chi(\mathbf{q}, \omega) = \chi'(\mathbf{q}, \omega) + i\chi''(\mathbf{q}, \omega)$$
$$= \frac{\chi(\mathbf{q})}{1 + \omega^2\tau^2} + i\frac{\omega\tau\chi(\mathbf{q})}{1 + \omega^2\tau^2},$$

and hence

$$\frac{\chi''(\mathbf{q}, \omega)}{\omega} = \frac{\tau\chi(\mathbf{q})}{1 + \omega^2\tau^2},$$

which we recognise from the simple harmonic oscillator example in the previous chapter. We also have

$$\lim_{\omega \to 0}\left(\frac{\chi''(\mathbf{q}, \omega)}{\omega}\right) = \tau\chi(\mathbf{q}) = \frac{\chi(\mathbf{q})}{\Gamma(\mathbf{q})}.$$

where $E_\mathbf{p} = \mathbf{p}^2/2m_e$, where $\mathbf{p}$ is a momentum and μ is the chemical potential. The susceptibility of the electron gas is given by

$$\chi_0(\mathbf{q}, \omega) = \sum_\mathbf{p} \frac{n_{\mathbf{p}+\mathbf{q}} - n_\mathbf{p}}{E_{\mathbf{p}+\mathbf{q}} - E_\mathbf{p} - \hbar\omega - i\varepsilon}, \qquad (8.14)$$

with ε as small parameter. The electron gas is certainly interesting in its own right, but will not undergo a magnetic transition without the inclusion of interactions.

Interactions in metals can be included at many levels of complexity. A relatively straightforward treatment involves the random phase approximation result for the dynamic susceptibility, given by

$$\chi(\mathbf{q}, \omega) = \frac{\chi_0(\mathbf{q}, \omega)}{1 - I\chi_0(\mathbf{q}, \omega)}, \qquad (8.15)$$

where I describes the interaction strength.[9] This result is a part of the so-called *Stoner theory* which suffices to describe some metals. However, in cases where there are very strong correlations, Stoner theory does not model the behaviour successfully. A more sophisticated model, developed by Toru Moriya and Sebastian Doniach, is self-consistent renormalization. This approach, which involves considering the coupling of excitation modes, gives rise to a complex-valued magnetic susceptibility

$$\chi(\mathbf{q}, \omega) = \frac{\chi(\mathbf{q})\Gamma(\mathbf{q})}{\Gamma(\mathbf{q}) - i\omega}, \qquad (8.17)$$

where, $\Gamma(\mathbf{q})$ is the linewidth (i.e. an inverse lifetime) for an excitation with wavevector $\mathbf{q}$.[10] We can use this susceptibility to predict the response of the muon to fluctuations in a correlated metal.

Example 8.5

Consider a metallic ferromagnet above its magnetic ordering temperature T_c. To make progress here, some observations are in order:
(i) Near a phase transition the static, $\mathbf{q}$-dependent part of the susceptibility is often well approximated by the so-called Orstein-Zernike form

$$\chi(\mathbf{q}) = \frac{\chi_0}{1 + q^2/\kappa^2}, \qquad (8.18)$$

where κ is an inverse correlation length with temperature dependence $\kappa = \kappa_0 \left(|T - T_c|/T_c\right)^{\frac{1}{2}}$;
(ii) Close to the phase transition, but not so close that critical fluctuations dominate the behaviour, we find that the uniform susceptibility follows the Curie-Weiss law, such that $\chi_0 = C/(T - T_c)$;
(iii) In this regime the linewidth (or inverse lifetime) for a ferromagnetic metal can be shown to follow $\Gamma(\mathbf{q}) = kq/\chi(\mathbf{q})$, where k is a parameter. These three facts give us what we need to compute the muon response.

In zero applied field, the muon relaxation rate λ is given by the expression from Section 7.6, which has the form

$$\lambda \propto gk_B T \sum_\mathbf{q} \frac{\chi(\mathbf{q})}{\Gamma(\mathbf{q})}, \qquad (8.19)$$

where we assume isotropic, q-independent (hyperfine) coupling between the muons and the electronic spins. Substituting for $\Gamma(q)$ and $\chi(q)$, we have[11]

$$\lambda \propto gk_B T \sum_q \frac{1}{kq} \frac{\chi_0^2}{(1 + q^2/\kappa^2)^2}$$

$$\propto gk_B T \int dq \frac{q\chi_0^2}{(1 + q^2/\kappa^2)^2}. \qquad (8.21)$$

This integral can be done if we assume that there is some high-momentum cutoff in q at $q = q_{max}$. Then we find[12]

$$\int_0^{q_{max}} dq \frac{q\chi_0^2}{(1 + q^2/\kappa^2)^2} = \left[\frac{-\chi_0^2\kappa^2}{2(1 + q^2/\kappa^2)} \right]_0^{q_{max}}$$

$$= \frac{\chi_0^2 q_{max}^2}{2(1 + q_{max}^2/\kappa^2)}. \qquad (8.22)$$

Since χ_0 follows a Curie-Weiss form $\chi_0 = C/(T - T_c)$ then in the limit $q_{max} \gg \kappa$ we expect

$$\lambda \propto gk_B T\chi_0^2\kappa^2 \propto T \left(\frac{1}{T - T_c} \right)^2 \left(\frac{|T - T_c|}{T_c} \right) \qquad (8.23)$$

and finally

$$\lambda \propto \frac{T}{T - T_c}. \qquad (8.24)$$

The prediction of this example is found to describe measurements of the temperature dependence of λ in the ferromagnetic itinerant (i.e. metallic) system MnSi. As shown in Fig. 8.5 the temperature dependence is well described by the model above $T_c = 29.5$ K.

8.3 BPP relaxation

In the previous examples, the time-dependent part of the local magnetic field of eqn 7.36 is modulated over a wide range of characteristic times τ_c, thanks to the slowing down of critical fluctuations. In Example 7.12 we discussed the contrasting case of activated dynamics, where τ_c obeys an Arrhenius law and produces a peak in the relaxation rates (Fig. 7.7). Examples of this kind are relevant for muons experiencing slow molecular motion. Unfortunately small molecules in liquid phases have dynamics that are too fast, while large viscous molecules that have more than one muon site often complicate the picture. However, a simpler example is offered by small molecules in zeolites, whose relaxation can described with the BPP (Redfield) formula introduced in Chapter 5.

Example 8.6

Zeolites are frameworks formed by empty connected cages. The zeolite geometry and counterion (Li, Na, K, etc.) hinder the small-molecule mobility in different ways. The cyclohexadienyl radical, discussed in Chapter 12 and obtained by muonium addition to benzene (Fig. 12.2), is one of the first muoniated organic radicals to be investigated by μSR.

The left panel of Fig. 8.6 shows the peak of the relaxation rate λ of the $\nu_{12} = 103$ MHz transition of the radical versus temperature in a TF experiment at $B = 0.01$ T. The curves represent the BPP model described in Example 7.12, corresponding more precisely to

$$\lambda = \lambda_0 + \frac{(\pi D)^2 \tau}{1 + (2\pi\nu_{12}\tau)^2}, \qquad (8.25)$$

[11] In the second step, we need to use the rule that, in three dimensions,

$$\sum_q \to \int V \frac{d^3q}{(2\pi)^3} \qquad (8.20)$$

and also that $d^3q = 4\pi q^2 \, dq$.

[12] The standard integral we need here is

$$\int \frac{dx \, x}{(ax^2 + c)^2} = \frac{-1}{2a(ax^2 + c)} + \text{const.}$$

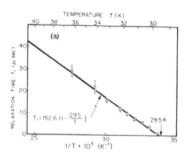

Fig. 8.5 The dependence of $T_1 \equiv 1/\lambda$ in MnSi, plotted against $1/T$. The straight line form demonstrates the $\lambda \propto T/(T - T_c)$ behaviour derived in the example. [Taken from R. S. Hayano et al., Phys. Rev. Lett. **41**, 1743 (1978).]

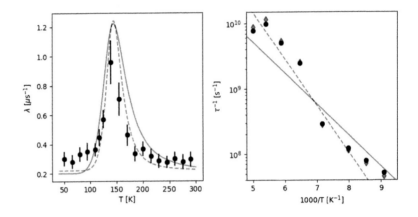

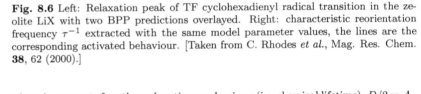

[13]This is a T_2^{*-1} rate (eqn F.34). Since presumably the zeolite is polycrystalline, we should really take a powder average of $T_2^{*-1}(\theta)$ with $D/2 = A_p(3\cos^2\theta - 1)$.

Fig. 8.6 Left: Relaxation peak of TF cyclohexadienyl radical transition in the zeolite LiX with two BPP predictions overlayed. Right: characteristic reorientation frequency τ^{-1} extracted with the same model parameter values, the lines are the corresponding activated behaviour. [Taken from C. Rhodes *et al.*, Mag. Res. Chem. **38**, 62 (2000).]

where λ_0 accounts for other relaxation mechanisms (i.e. chemical lifetime), $D/2 = A_p$ is the anisotropic hyperfine coupling of Section 4.4[13] and $\tau = \tau_\infty \exp(T_a/T)$ is the activated reorientation. The solid curve has the activation energy quoted by the authors, the better-fitting dashed curve has slightly different parameter values. These parameters are rather correlated (to get a feeling for this see Exercise 8.1). The model is a bit rough, but it provides satisfactory qualitative agreement with the data.

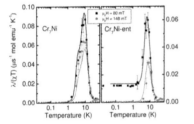

Fig. 8.7 Temperature dependence of the slow LF relaxation rate in single molecule magnets, at two field values. Rates are T_1^{-1}, scaled by χT to account for the paramagnetic susceptibility. [Taken from S. Sanna *et al.*, Phys. Rev. B **96**, 184403 (2017).]

The best strategy to reduce correlations among parameters, such as those found in Example 8.6, is to seek additional constraints such as those provided on molecular reorientation by avoided level crossing spectroscopy (Section 12.5). Alternatively one can perform additional experiments varying the frequency in the denominator of eqn 8.25. An example is provided by the ring-structured single molecule magnet, Cr$_7$Ni, and its Cu bridged dimer Cr$_7$Ni-ent. They are medium-sized molecules that include tens of oxygens and many more hydrogen atoms. They yield several muon sites at different distances from the magnetic ions. The muon sites further from the magnetic ions can be singled out by their slow LF relaxation rate. The molecules form crystals with negligible intermolecular exchange. Intramolecular exchange produces a magnetic ground state and several excited states, subject to more than one relaxation channels for the ensemble magnetization (possibly via both thermally activated and quantum coherent paths). The slow muon relaxation rate is shown in Fig. 8.7. Both molecules show sharp BPP-like peaks, and the clear field dependence allows a self-consistent evaluation of the parameters in the model (solid curves). However, the presence of more relaxation channels forces the use of more complex models with distributions of activation temperatures. This is quite common in the practical use of the BPP mechanism.

8.4   Mobile excitations

So far the dynamics we have described are assumed to be due to a dense array of magnetic moments that are assumed to fluctuate randomly, producing dynamics in the local field at the muon sites. However, we saw in the last chapter that propagating and diffusive modes are the expected excitations of many magnets. These **mobile excitations** are also found more generally in condensed matter and their properties can be investigated directly using μSR, particularly in low-dimensional systems.[14]

A general approach to understanding how static implanted muons can be used to probe mobile excitations in a material is provided by a simple anisotropic lattice model (Fig. 8.8). The hopping rates of the mobile excitations along the three primary axes of a cubic lattice are taken to be D_1, D_2 and D_3 and a system with reduced effective dimensionality can be treated by tuning the anisotropy of these rates. A muon sitting at an interstitial site in the lattice will couple to fluctuating fields associated with the motion of the excitations. In general there may be both a contact hyperfine coupling A and a dipolar coupling D, that act when the excitation comes within range of the muon site, thus providing a direct interaction with the mobile excitation [Fig. 8.8(a)]. Alternatively, if a localized electron is associated with the muon in a muonium-like probe state, a more indirect mechanism is possible, where the localized electron interacts with the mobile electron spins via an intermittent spin exchange process [Fig. 8.8(b)]. The internal hyperfine coupling within the muonium state described by A and D will then transfer the external electron spin fluctuations to the muon.

The types of excitation we consider here will mainly reflect unpaired electron spins and we will show examples later for solitons in conducting polymers and spinons in quantum spin chains. Other types of excitation with associated magnetic fields could also be considered in a similar way, e.g. charge carriers and triplet excitons in semiconductors or domain walls and other topological excitations in magnets. For the muonium-like probe state of Fig. 8.8(b), it is not even strictly necessary for the mobile excitations to have a magnetic character, since a spinless structural excitation that modulates the muonium hyperfine coupling (A, D) will also produce an effective modulation of the local field at the muon site, leading to muon spin relaxation. One example of this could be the conformational excitations of a polymer chain.

The muon spin relaxation rate derived from perturbation theory can in general be resolved into sums of contributing terms. Each contributing term can be further resolved into the product of a field-independent coupling term and a field-dependent spectral density term $J(\omega)$, i.e.

$$\lambda(B) = \sum_{\alpha,\beta,q} \Omega^q_{\alpha\beta} \, J(\omega_{\alpha\beta}), \tag{8.26}$$

where $\omega_{\alpha\beta}$ is a function of B and q, and reflects different types of relaxation mechanism. The field-independent terms $\Omega^q_{\alpha\beta}$ reflect the coupling mechanism and geometrical factors that translate the dynamics into a

[14]These mobile excitations can be thought of as localized in real space, in contrast to wave-like excitations like phonons and magnons, which are localized in momentum space and extended in real space.

(a)

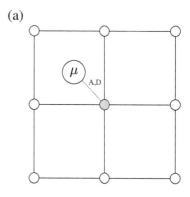

(b)

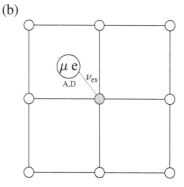

Fig. 8.8 The muon coupling to mobile excitations diffusing on a simple cubic lattice with anisotropic hopping rates. (a) The muon intermittently interacts with a spin whenever it is present at an adjacent site. The coupling is via a generic hyperfine coupling (A, D). (b) The muon probe state is muonium-like with a localized electron, coupled to it via an internal coupling (A, D). Mobile spin excitations couple to the electron via a spin exchange mechanism ν_{ex}. Modulation of (A, D) provides an alternative coupling mechanism for non-magnetic excitations.

fluctuating field at the muon site, whereas the field dependent terms $J(\omega_{\alpha\beta})$ reflect the spectral density of the fluctuations for the frequency $\omega_{\alpha\beta}$ corresponding to the level difference for the α to β state transition that relaxes the muon. Note that the spectral density we are using here is the Fourier transform of a normalized autocorrelation function[15] $S(t)$ with $S(0) = 1$, so that it only contains information about the spectral distribution of the fluctuations, amplitude information being provided by the $\Omega_{\alpha\beta}^{q}$ terms.

[15]In terms of the quantities in the previous chapter, we have $S(t) = \Phi(t)/\Phi(0)$, where $\Phi(t) = \gamma_{\mu}^{2}\langle\delta B(t)\delta B(0)\rangle$. The corresponding spectral density $J(\omega)$ has units of inverse frequency and its sum rule is $\int_{-\infty}^{\infty}\frac{d\omega}{2\pi}J(\omega) = 1$.

Example 8.7

As an example of this approach we return to the Redfield expression for the longitudinal relaxation covered in Chapter 5 (eqn 5.40) and also in Chapter 7. Here the normalized autocorrelation functions of the field components in the transverse x- and y-directions are taken to be simple exponentials $\exp(-\nu t)$, so that the corresponding normalized spectral density function $J(\omega)$ is given by

$$J(\omega) = \frac{2\nu}{\nu^{2} + \omega^{2}}. \tag{8.27}$$

The factor $\Delta^{2} = (\gamma_{\mu}B_{x})^{2} = (\gamma_{\mu}B_{y})^{2}$ describes the coupling amplitude of each transverse fluctuation component. These contribute two independent terms to eqn 8.26 so that the relaxation becomes

$$\lambda(B) = \frac{1}{2}\left[(\gamma_{\mu}B_{x})^{2}J(\omega_{\mu}) + (\gamma_{\mu}B_{y})^{2}J(\omega_{\mu})\right] = \frac{2\Delta^{2}\nu}{\nu^{2} + (\gamma_{\mu}B)^{2}}. \tag{8.28}$$

For the type of coupling described in Fig. 8.8(a), which has both contact hyperfine and dipolar terms, the following powder averaged expression can be used[16]

[16]F. Devreux, J.-P. Boucher, and M. Nechtschein, J. Physique **35**, 19 (1974).

$$\lambda(B) = \frac{A^{2}}{4}J(\omega_{e}) + \frac{D^{2}}{4}\left[\frac{7J(\omega_{e}) + 3J(\omega_{\mu})}{5}\right]. \tag{8.29}$$

If $D \ll A$ then the second term in eqn 8.29 can be ignored. To obtain the muon spin relaxation for the anisotropic lattice diffusion model, it now just remains to determine the spectral density function. For the model represented in Fig. 8.8(a), where the presence of an excitation located at a specific lattice point is probed by the muon, the autocorrelation function is represented by an anisotropic random walk, taking the form of a product of terms for each lattice direction[17]

[17]M. A. Butler, L. R. Walker, and Z. G. Soos, J. Chem. Phys. **64**, 3592 (1976).

$$S(t) = \prod_{i=1}^{3}e^{-2D_{i}t}I_{0}(2D_{i}t), \tag{8.30}$$

where $I_{0}(x)$ is the zeroth order modified Bessel function. The spectral density $J(\omega)$ of the fluctuations associated with this motion is then given by the Fourier transform[18] of eqn 8.30. Systems of different dimensionality d can simply be represented in terms of fast and slow rates $D_{\parallel}$ and $D_{\perp}$:

[18]We write $J(\omega)$, the Fourier transform of $S(t)$, as follows:

$$J(\omega) = \int_{-\infty}^{\infty}S(|t|)e^{-i\omega t}\,dt$$

$$= 2\int_{0}^{\infty}S(t)\cos\omega t\,dt. \tag{8.31}$$

$$\begin{array}{lll} D_{1} = D_{\parallel}, & D_{2}, D_{3} = D_{\perp} & (d = 1) \\ D_{1}, D_{2} = D_{\parallel}, & D_{3} = D_{\perp} & (d = 2) \\ D_{1}, D_{2}, D_{3} = D_{\parallel} & & (d = 3), \end{array} \tag{8.32}$$

and when $1/D_{\parallel} \ll t \ll 1/D_{\perp}$, the time dependence of eqn 8.30 follows the power law $t^{-d/2}$. This is illustrated in Fig. 8.9(a) for the cases $d = 1$ to 3, with the corresponding spectral density $J(\omega)$ shown in Fig. 8.9(b). For the $d = 1$ case a characteristic $\omega^{-1/2}$ behaviour is obtained, so that

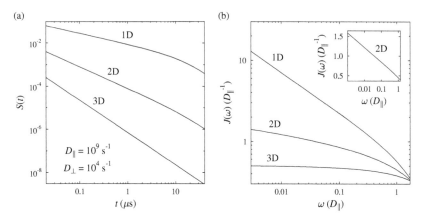

Fig. 8.9 (a) The random-walk autocorrelation function for diffusing excitations versus effective dimensionality. (b) The corresponding spectral density of the fluctuations sensed by a local muon probe. For the $d = 1$ case a power law $\omega^{-1/2}$ is obtained. The inset plot shows that the $d = 2$ case has a $\log(1/\omega)$ dependence. In the $d = 3$ case there is very little ω dependence below the cutoff, similar to case of the BPP model.

the LF relaxation rate will follow $B^{-1/2}$. The spectral density for the $d = 2$ case shows a much weaker logarithmic dependence [Fig. 8.9(b), inset]. When the diffusion has a $d = 3$ character the spectral density hardly changes until ω approaches $D_{\parallel}$. This behaviour is very similar to that of the BPP model with its exponential autocorrelation function. Thus in $d = 3$ systems, relaxation originating from diffusion is very hard to distinguish experimentally from that produced by localized fluctuations.

In general the spectral density must be obtained from eqn 8.30 by numerical integration, however in the $d = 1$ case a useful approximate expression is available[19] for the region where $\omega \ll D_{\parallel}$, given by

$$J(\omega) = \frac{1}{\sqrt{4D_{\parallel}D_{\perp}}} \sqrt{\frac{1 + \sqrt{1 + (\omega/2D_{\perp})^2}}{1 + (\omega/2D_{\perp})^2}}. \tag{8.33}$$

This expression gives the expected $\omega^{-1/2}$ dependence at high frequencies and low frequency divergence is avoided by a cutoff when ω becomes smaller than $D_{\perp}$.

Solitons in polyacetylene

One of the first reports of using μSR to study diffusing one-dimensional excitations was by Nagamine and co-workers[20] in the conducting poly-

[19]K. Mizoguchi, Jpn. J. Appl. Phys. **34**, 1 (1995).

[20]K. Nagamine *et al.*, Phys. Rev. Lett. **53**, 1763 (1984).

[21]In one dimension, spin and charge excitations can be separate entities, with distinct dynamics.

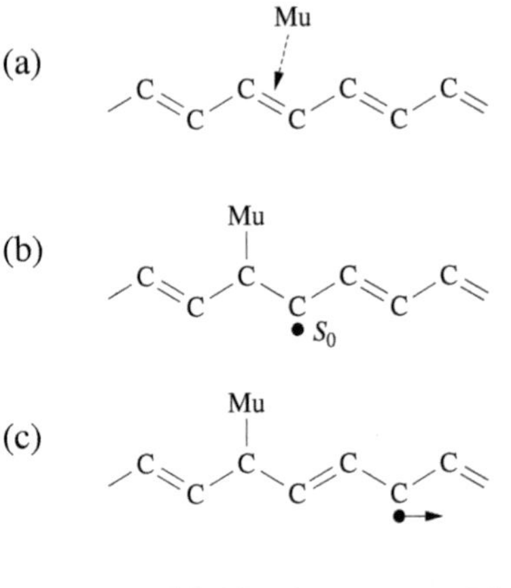

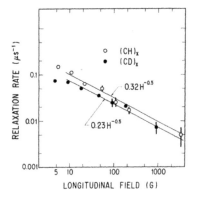

Fig. 8.10   (a) Muonium reacts with one of the double bonds on the polyacetylene chain, leaving the muonium bonded to one of the carbon sites and an unpaired electron spin. (b) The unpaired spin takes the form of a charge-neutral soliton, S_0. (c) The soliton is free to diffuse along the 1D chain of the polymer. Note that the soliton spin density is actually delocalized across 20–30 sites.

Fig. 8.11   Field dependent relaxation from 1D diffusing solitons in *trans*-polyacetylene.   [Taken from K. Nagamine *et al.*, Phys. Rev. Lett. **53**, 1763 (1984).]

[22]R. Risch and K. W. Kehr, Phys. Rev. B **46**, 5246 (1992).

mer *trans*-polyacetylene. This polymer is the simplest form of conducting polymer and it supports mobile soliton excitations, which take the form of charge-neutral, spin-1/2 quasiparticles.[21] In these studies the longitudinal muon spin relaxation was measured and the main source of muon relaxation was attributed to hyperfine coupling between muons bonded to the polymer and the mobile solitons which primarily move along a single polymer chain. The scenario starts off with the formation of muonium which reacts with the chain to form a neutral soliton, which is free to move away from the muonium-bonded site on one side of the chain (Fig. 8.10).   The data were first analysed using an exponential relaxation function and the field dependence of the relaxation rate was found to follow the expected $B^{-1/2}$ dependence for 1D diffusion (Fig. 8.11).

Although an exponential was used initially for fitting the relaxation, Nagamine *et al.* noticed that the relaxation rate appeared to slow down with time, i.e. the time evolution of the relaxation was non-exponential. This was originally interpreted in terms of the interchain motion of solitons becoming more important at longer times. However we saw in the previous section that the correlation function for 1D diffusion is not an exponential, and hence the muon spin relaxation is not expected to be a simple exponential either.

The theorists Risch and Kehr (RK) specifically looked at this issue and pointed out that since the correlation time for the return to the origin of a particle diffusing in 1D is divergent, the standard NMR theory used for the earlier data analysis is not strictly valid, as it assumes the existence of a finite correlation time that is short compared to all other timescales in the problem. RK then developed a stochastic diffusion theory for the model of a static muon interacting through an intermittent hyperfine coupling with an electron that is randomly diffusing along a 1D chain. This model provides a natural explanation for the non-exponential relaxation and this RK theory will be described in more detail in the next section.

Risch–Kehr relaxation

The model used by RK[22] for describing the relaxation of the muon spin is illustrated in Fig. 8.12. There are just three model parameters, the inter-site hopping rate of the mobile soliton $D_{\parallel}$, the intermittent contact hyperfine coupling A between the muon spin and the soliton spin, which is only present when the soliton is located at the site next to the muon, and finally an electron spin relaxation rate ν resulting from interaction of the diffusing soliton with spin defects or lattice excitations. This model corresponds to a 1D version of the case shown in Fig. 8.8(a), with the dipolar coupling D set to zero.

Despite the simplicity of the model, the full expression for the polarization evaluated in the Laplace transform domain is quite complicated. An inverse Laplace transform can be made to give the full time dependence and examples in zero applied field are plotted in Fig. 8.13. Fortunately, the time dependence at all but the earliest times is described well by the following relatively straightforward relaxation function

$$G_z(t) = \phi_{\mathrm{RK}}(\Gamma_0 t), \tag{8.34}$$

where the RK function $\phi_{\mathrm{RK}}(x)$ is defined as

$$\phi_{\mathrm{RK}}(x) = \exp(x)\,\mathrm{erfc}(\sqrt{x}), \tag{8.35}$$

with erfc being the complementary error function

$$\mathrm{erfc}(x) = \frac{2}{\sqrt{\pi}} \int_x^\infty \mathrm{d}t\, e^{-t^2}. \tag{8.36}$$

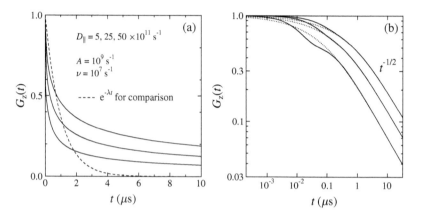

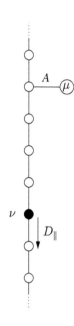

Fig. 8.13 (a) The RK model relaxation for the longitudinal muon polarization versus diffusion rate in zero applied field. An exponential relaxation is shown for comparison. (b) Log-log plot of the same RK relaxation curves showing the power law long time behaviour. The dotted lines show the asymptotic RK relaxation function defined by eqns 8.34 to 8.37.

Fig. 8.12 The RK model for describing the field dependent relaxation from electronic spin excitations diffusing on a single chain with hopping rate $D_{\parallel}$. The muon interacts with the spin on one adjacent site via a contact hyperfine coupling A. The relaxation rate of the mobile electronic spin is ν.

The RK relaxation function is shown as the dotted lines in Fig. 8.13(b). It can be seen that at long times the RK function mirrors the $t^{-1/2}$ behaviour of the 1D random walk autocorrelation function. In comparison with an exponential relaxation function, the RK relaxation starts faster and then becomes slower [Fig. 8.13(a)]. This could tempt the unwary into trying to fit this type of relaxation in an overly complex way using two separate exponential components or a single stretched exponential component, both requiring additional fitting parameters, whereas a single RK component only requires one relaxation parameter. The zero-field RK relaxation parameter Γ_0 in eqn 8.34 is given by

$$\Gamma_0 = \frac{\nu}{\left(1 + 2\nu D_{\parallel}\omega_0^{-2}\right)^2}, \tag{8.37}$$

where $\omega_0 = 2\pi A$. At finite LF the relaxation also follows the RK functional form $\phi_{\mathrm{RK}}(\Gamma t)$, with the general field-dependent RK relaxation parameter Γ now being given by

$$\Gamma = \frac{\nu}{\left(1 + 2\sqrt{2\omega_e \nu}\, D_{\parallel}\omega_0^{-2}\right)^2}, \tag{8.38}$$

where $\omega_e = |\gamma_e|B$ is the Larmor frequency of the electron spin in the field. When the field is high enough for the second term in the denominator of eqn 8.38 to dominate, Γ becomes independent of ν:

$$\Gamma = \frac{\omega_0^4}{8\omega_e D_\parallel^2},\qquad(8.39)$$

which shows the $1/B$ high field behaviour for Γ that is a characteristic feature of the RK model. Now one only needs to know the hyperfine frequency ω_0 in order to determine $D_\parallel$ from the measured RK relaxation. The field-dependent relaxation is illustrated[23] in Fig. 8.14 . Interchain motion is an extension of the RK model and this is expected to produce

[23]The behaviour of the RK model is explored over a wider range of parameters in F. L. Pratt, J. Phys.: Condens. Matter **16**, S4779 (2004).

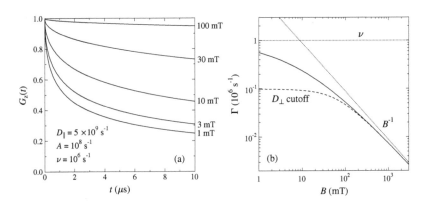

Fig. 8.14 (a) Illustration of the LF dependence of the RK relaxation for the model parameters shown. (b) The corresponding RK relaxation parameter Γ versus field showing $1/B$ dependence at high field. A default low field cutoff is controlled by ν. If $D_\perp$ is sufficiently large, it will shift the cutoff to higher field, allowing both $D_\perp$ and $D_\parallel$ to be determined from the field dependence of Γ.

a low-field cutoff to the $1/B$ dependence of Γ when $D_\perp$ is not too small (Fig. 8.14). Thus, under favourable parameter conditions, both $D_\parallel$ and $D_\perp$ can be estimated from the field dependence of Γ.

Polyaniline

As well as providing a good description of the soliton motion in polyacetylene, the RK relaxation is also found to work well for describing mobile spins in other conducting polymers. Taking as one example the emeraldine base form of polyaniline (PANI:EB), the intra-chain and inter-chain diffusion rates derived from the μSR are shown in Fig. 8.15. This polymer supports mobile electronic excitations in the form of polarons, in which spin and charge are attached to carriers, just as in a metal. The relaxation behaviour in this polymer follows well the RK form (Fig. 8.16), with a $1/B$ region for Γ that allows an estimation of $D_\parallel$ to be made and a T dependent low field cut off that is used to estimate $D_\perp$ (or at least its upper limit). The on-chain diffusion $D_\parallel$ shows a weak metallic T dependence which is controlled by librational modes

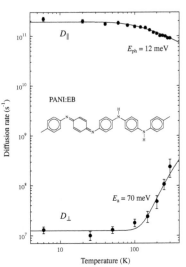

Fig. 8.15 Intrachain and interchain diffusion rates in PANI:EB derived from LF μSR. [Taken from F. L. Pratt *et al.*, Synth. Met. **101**, 323 (1999).]

of the polymer rings, which have energies of order 10 meV. In contrast, the T dependence of the inter-chain diffusion $D_\perp$ has semiconducting character above 100 K, with an activation energy of 70 meV (Fig. 8.15). These measurements are broadly consistent with those of ESR, NMR, and bulk conductivity. This particular polymer does not support charge-neutral solitons, so a muonium electron moving away from the addition site takes the form of negative polaron and it leaves behind a positive charge at the muon site. The muon-induced polaron is therefore expected to be bound to the original site, in contrast to the case of the muon-induced soliton in *trans*-polyacetylene. The coupling of the muon to the mobile polarons present in this conducting polymer is therefore believed to be of the indirect type shown in Fig. 8.8(b).

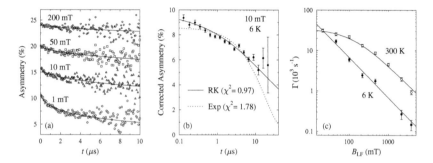

Fig. 8.16 (a) LF relaxation data for PANI:EB at 6 K fitted with the RK relaxation function. (b) Comparison of the RK fit with an exponential fit. (c) Field dependence of the RK relaxation parameter Γ. [Taken from F. L. Pratt *et al.*, Phys. Rev. Lett. **79**, 2855 (1997).]

Limitations of NMR theory for μSR

The RK model provides a good demonstration of the limitations of the perturbative NMR approach to muon-spin relaxation. Although the perturbation theory is a good starting point, it does not always provide a precise description of muon experimental data. This is primarily because the NMR theory is based on assumptions about frequencies and timescales that are not always appropriate for describing μSR experimental conditions. Despite the exponential relaxation model of the perturbation theory failing to accurately describe the precise form of the relaxation, it is notable that taking data described by the RK model and fitting it (poorly) to an exponential function will, in practice, provide a relaxation rate that scales with $B^{-1/2}$. This is, therefore, still qualitatively consistent with the simpler version of the 1D diffusion model.

Besides this RK example, another area where there is a substantial difference between NMR and μSR is in the low-field Kubo-Toyabe relaxation, described in Chapter 5, which has no direct counterpart in traditional magnetic resonance. It is likely that additional relaxation models will be added to this list of special cases in the future, as further

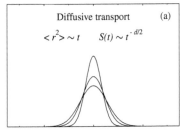

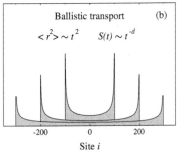

Fig. 8.17 Comparison between diffusive and ballistic transport mechanisms. Probability distributions versus site number are shown for evenly spaced intervals in time t after the initial localization of a particle at the origin. The two mechanisms are distinguished by their different t dependences for the mean square displacement $\langle r^2 \rangle$. (a) Diffusive transport, where the distribution is a Gaussian that becomes progressively broadened with t. (b) Ballistic transport modelled via the quantum mechanics of a fermion chain. The wavepacket forms two sharp wavefronts that move apart linearly with t.

[24] For further information see D. E. Katsanos *et al.*, Phys. Rev. B **51**, 895 (1995). Note that fast oscillations in the wavefunction have been smoothed out in Fig. 8.17(b) to show the slow background evolution of $S(t)$.

[25] See J. X. Zhong and R. Mosseri, J. Phys.: Condens. Matter **7**, 8383 (1995).

μSR-specific theories become needed to describe data more accurately in other fields of study.

Ballistic versus diffusive transport

Besides diffusion, another possibility for the propagation of excitations is a ballistic transport mechanism. The characteristics of these two transport processes are compared in Fig. 8.17. Diffusion is an incoherent stochastic process that is modelled well by the classical random-walk theory. On the other hand, quantum-scale ballistic transport is best modelled by determining the evolution of the particle wavefunction using the time-dependent Schrödinger equation. The specific 1D example of ballistic transport shown in Fig. 8.17(b) is for fermions hopping between sites on a chain with a simple tight-binding Hamiltonian[24].

A key distinguishing feature between the diffusive and ballistic transport processes is the mean squared displacement $\langle r^2 \rangle$. In diffusive transport $\langle r^2 \rangle$ is linear in time t, whereas in a ballistic process $\langle r^2 \rangle$ is quadratic in t. For observation of the transport dynamics via muon spin relaxation, the most important property is the autocorrelation function $S(t)$, which corresponds to the evolution of the probability density at the origin of the plots in Fig. 8.17. In the case of diffusion, the autocorrelation function follows $S(t) \sim t^{-d/2}$ at longer t, where d is the dimensionality, as discussed in the previous two sections. In contrast, for ballistic transport, the autocorrelation function at longer t falls off at a much faster rate, following $S(t) \sim t^{-d}$. For the case of 1D ballistic transport, the corresponding spectral density is expected to show a much weaker frequency dependence than that of diffusion, since the Fourier transform (FT) of t^{-1} is a constant. However, the t^{-1} divergence must actually become cut off at early times, since $S(0) = 1$. The overall form of the ballistic $S(t)$ is given by[25]

$$S(t) = [J_0(2Dt)]^{2d}, \tag{8.40}$$

where $J_0(x)$ is the zeroth order Bessel function and D is the hopping rate. The spectral density obtained from the FT of eqn 8.40 when $d = 1$ is found to be approximated well by the logarithmic expression

$$J(\omega) = \frac{0.318}{D} \ln\left(\frac{16D}{\omega}\right). \tag{8.41}$$

For a $J(\omega)$ of this form, a plot of the relaxation rate versus $\log(B)$ will yield a straight line, which provides a characteristic signature of the presence of ballistic dynamics in a 1D system.

A key area of application for this, where both diffusive and ballistic processes may be present, is in the study of spinons in spin-1/2 chain systems that are close to the Heisenberg AF limit. This is actually an active area of ongoing study, with no overall theoretical agreement to date. There are however some suggestions from theory that both ballistic and diffusive propagation of spin excitations are possible, dependent on certain parameters, such as the degree of Ising character in the exchange interactions. Muon studies in these systems give the opportunity to measure the spectral density versus applied longitudinal field and thus experimentally determine the dominant type of transport in specific spin chains.

One example of a highly ideal spin 1/2 Heisenberg chain system is given by the molecular radical salt DEOCC-TCNQF$_4$. This shows no magnetic ordering down to 20 mK, a temperature that is 5,000 times smaller than the 110 K exchange interaction. The LF relaxation at 1 K is shown in Fig. 8.18. This shows clear evidence for diffusive rather than ballistic transport. The power law is found to be slightly smaller than for the random walk model. Various numerical simulations have indicated that such a reduced 'anomalous diffusion' power law is expected for a spin 1/2 AF chain system at the Heisenberg point.

A contrasting example is provided by the Cu chain system pym-Cu(NO$_3$)$_2$(H$_2$O)$_2$. This has an exchange interaction of 36 K but the magnetic ordering temperature is much lower at 0.22 K. The LF dependent relaxation at 1.6 K is shown in Fig. 8.19. In this case a ballistic dependence is found for the LF relaxation, rather than the diffusive dependence that was found for the DEOCC-TCNQF$_4$.

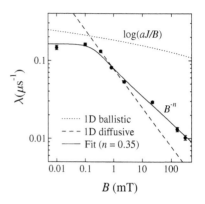

Fig. 8.18 Field dependent relaxation from diffusively propagating spinons at 1 K in the 1D $S = 1/2$ Heisenberg AF DEOCC-TCNQF$_4$. [Taken from F. L. Pratt *et al.*, Phys. Rev. Lett. **96**, 247203 (2006).]

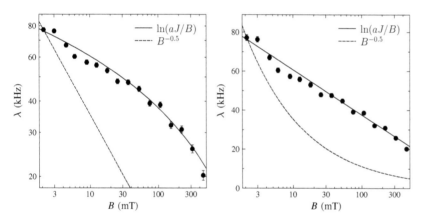

Fig. 8.19 Field dependent relaxation from ballistically propagating spinons at 1.6 K in in the 1D $S = 1/2$ Heisenberg AF pym-Cu(NO$_3$)$_2$(H$_2$O)$_2$. A log-log plot of the data is shown on the left and a log-linear plot on the right. [Data taken from B. M. Huddart *et al.*, Phys. Rev. B **103**, L060405 (2021).]

8.5 Muon diffusion

In some cases, we are not dealing with a stationary muon following the dynamics of its environment but, instead, it is the muon itself that moves. The dynamics of light atoms such as hydrogen and muonium, or particles such as the proton and muon, are interesting in themselves as a subject of study because they can be used to constrain theories concerning the quantum motion of defects and interstitials. The smaller mass of the muon leads to larger tunnelling matrix elements to neighbouring sites and thus enhances the quantum mechanical nature of the motion. Furthermore, because muon-muon (or muonium-muonium) interactions can be neglected, the intrinsic nature of the dynamics can be followed without the complications that can be found in studying the corresponding proton or hydrogen case.

Example 8.8

A good example of this is found in copper, an elemental metal with a face-centred cubic lattice. The muon sits at an octahedral interstitial site (see Fig. 8.20), but is relatively weakly bound to its environment and so it can be easily induced to hop from one site to the next. Copper is not magnetically ordered, but the muon experiences a dipolar field from nearby copper nuclei ($I = \frac{3}{2}$). There is also the possibility of a quadrupole splitting because the nearby copper nuclei experience an electric field gradient from the muon. However, the copper nuclear spins are not coupled with each other, so their arrangement is random. This means that the effective magnetic field at the muon site takes a value determined by some probability distribution. However, when the muon hops to a new site, it is surrounded by a new environment of nuclear spins and so the new magnetic field is assumed to be uncorrelated with the old one. Thus the effect of the muon diffusion (the muon hopping from site to site) can be described within the strong collision model, in which the local field changes discontinuously at the moment of the hop and is uncorrelated with the field at the previous site. This model can be used to estimate the muon hop rate from the spin relaxation rate.

Experimentally, it was found that the muon hop rate decreased as the copper sample was cooled and the spin relaxation rate became constant below around 80 K. This was interpreted as the muon becoming frozen at this temperature, at least on the timescale of several muon lifetimes. However, there was a surprise as the temperature was lowered below about 20 K where the hop rate started to increase again. This was an indication of the possibility of quantum diffusion, as will be discussed below.

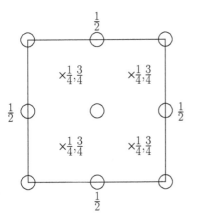

Fig. 8.20 The octahedral interstitial sites (indicated by crosses) in the face-centred cubic lattice of copper.

[26]C. P. Flynn and A. M. Stoneham, Phys. Rev. B **1**, 3966 (1972).

Hopping of the muon between interstitial sites is assisted by phonons and hence rises with temperature, following an approximately activated behaviour [according to a theory by Flynn and Stoneham,[26] the hop rates are proportional to $T^{-1/2} \exp(-E_a/k_B T)$ where E_a is an activation energy]. This occurs because the muon is initially **self-trapped** by its own local distortion of the lattice (Fig. 8.21) and a tunnelling transition is only possible if, by the thermal fluctuations of the lattice which occur because of phonons, two neighbouring energy levels coincide [the coincidence configuration, see Fig. 8.21(b)]. The muon can then tunnel through the barrier and becomes self-trapped in the next site.

As the sample is cooled, the number of phonons decreases and hence so does the hop rate. However, the example of copper shows that something else seems to be going on when the very low temperature regime is reached when the phonons appear to be hindering hopping rather than helping it. The reason is that at low temperatures coherent tunnelling is possible, meaning that the muon starts to delocalize into a band-like state. Phonons now are responsible for inelastic scattering which destroys the coherence of this delocalized state. This coherent effect is known as **quantum diffusion**.

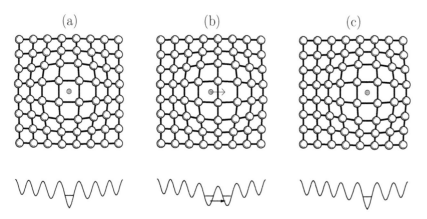

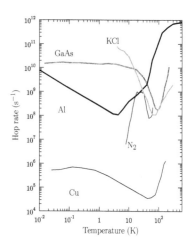

Fig. 8.21 Diffusion process of a muon by phonon-assisted tunnelling [for more details, see e.g. S. F. J. Cox, J. Phys. C **20**, 3187 (1987)]. (a) The muon is stable in an interstitial site and the local distortion leads to self-trapping so that its zero-point energy level lies a little lower than the neighbouring site. (b) Thermal fluctuations provide the opportunity for a coincidence configuration whereby tunnelling is allowed, leading to (c) a new stable configuration.

Example 8.9

Experimental data for various materials are shown in Fig. 8.22 and although there are large differences in the size of the hop rate and the detailed form of the temperature dependence, all show a hop rate that increases with temperature at high temperature, consistent with activated behaviour, whereas the hop rate decreases with increasing temperature at low temperature, consistent with quantum diffusion. The quantum diffusion regime at low temperature can be fitted with an inverse power law $T^{-\alpha}$, and theory[27] predicts that the exponent α should be large, typically ≈ 9. Experiments on Cu and Al (see Fig. 8.22) show a more modest behaviour with $\alpha \approx 0.6$–0.7. This is due to the conduction electrons in these metallic materials; these screen the muon and cannot react fast enough to the diffusing particle and follow it adiabatically. This produces a net drag which reduces the particle hop rate and weakens the temperature dependence.

This dominant role of the electrons has been demonstrated by ingenious experiments on aluminium.[28] In its superconducting state, the presence of the gap in the electronic spectrum effectively decouples the electron bath from the muon. At low temperature the superconductivity can be removed by applying a sufficiently large magnetic field to the sample. This dramatically reduces the muon diffusion rate because the closing of the gap reconnects the muon diffusion process to the electron bath, introducing drag.

In insulators there are no conduction electrons to worry about and in this case the hopping particle is a neutral muonium atom, not a charged muon. The coherent muon hop rate for KCl (Fig. 8.22) rises much more rapidly with decreasing temperature than for metals, and fits to an exponent $\alpha = 3.3$. This still does not quite fit with the earlier theory (Kagan and Klinger, 1974) but agrees with more sophisticated treatments[29] which take into account the phonon spectrum in KCl measured using neutron scattering.

Fig. 8.22 Muon hop rates as a function of temperature for various materials. Copper [G. M. Luke *et al.* Phys. Rev. B **43**, 3284 (1991)], aluminium [O. Hartmann *et al.* Phys. Rev. B **37**, 4425 (1988)], gallium arsenide [R. Kadono *et al.* Hyp. Int. **64**, 635 (1990)], potassium chloride [R. F. Kiefl *et al.* Phys. Rev. Lett. **53**, 90 (1989)], and solid nitrogen [V. G. Storchak *et al.* Phys. Lett. A **193**, 199 (1994)].

[27]Yu. Kagan and M.I. Klinger, J. Phys. C **7**, 2791 (1974). For more details on the theoretical aspects, see the article by Storchak and Prokof'ev in the Further reading.

[28]E. B. Karlsson *et al.*, Phys. Rev. B **52**, 6417 (1995).

[29]Again, see the article by Storchak and Prokof'ev in the Further reading.

In semiconductors muonium is also formed (see Chapter 10) and very similar temperature dependence is found (Fig. 8.22) with $\alpha \approx 3$ below 100 K but the hop rate saturates below ≈ 10 K due to the presence of disorder. (Below this temperature the coherent tunnelling is dominated by the disorder, rather than the phonons, and is therefore temperature independent.) Data are also shown for solid nitrogen in Fig. 8.22. This material shows a very sharp increase in hop-rate with decreasing temperature ($\alpha \approx 7$); as the temperature is lowered further it saturates and thereafter begins to decrease.[30]

[30]This low temperature effect is thought to be due to orientational ordering of the N_2 molecules. For further information, see V. G. Storchak, Low Temperature Physics **29**, 746 (2003).

Note that in many materials the muon forms a chemical bond to the structure (e.g. in oxides) and so the type of quantum diffusion discussed above will not occur. At high temperatures however, it may be possible for the muon to be thermally excited out of this bonded site and to hop to a nearby one. In addition, besides long-range quantum diffusion, another more localized form of quantum diffusion is also possible when the energy barrier between a close pair of sites is not too large. The resultant quantum tunnelling between local sites is analogous to the quantum tunnelling proton dynamics and hydrogen bonding that are found in many materials. Once again, the low mass of the muon enhances the tunnelling rate and the effective range of the bonding compared to those of a proton in a similar configuration. This quantum tunnelling between sets of local potential minima can also complicate the process of estimating the muon stopping sites (Chapter 16).

Chapter summary

- μSR can be used to study the critical dynamics of both magnetic insulators and magnetic metals.
- BPP relaxation can be observed in some situations, but often more complex dynamic models are required.
- A static muon provides a sensitive probe of mobile excitations in low-dimensional systems.
- As a light particle, the muon can diffuse through certain lattices, resulting in relaxation that can give information on the hop rate. At high temperature diffusion is activated, while at low temperature quantum diffusion can occur.

Further reading

- S. J. Blundell, *Magnetism in Condensed Matter*, OUP (2001).
- P. M. Chaikin and T. C. Lubensky, *Principles of Condensed Matter Physics*, CUP (1995).
- P. C. Hohenberg and B. I. Halperin, Rev. Mod. Phys. **49**, 435 (1977).
- A. Abragam, *Principles of Nuclear Magnetism*, OUP (1961).

- V. G. Storchak and N. V. Prokof'ev, Rev. Mod. Phys. **70**, 929 (1998).

Exercises

(8.1) Reproduce the plot of Fig. 8.6 (left), with eqn 8.25 and the following data: $T =$

[50, 65,80, 95, 110, 117, 125, 140, 155, 170, 185, 200, 215, 230, 245, 260, 275, 290],

$\lambda =$

[0.30, 0.28, 0.33, 0.35, 0.36, 0.45, 0.57, 0.96, 0.71, 0.46, 0.34, 0.37, 0.32, 0.29, 0.28, 0.30, 0.28, 0.30],

$e_\lambda =$

[.05, .05, .05, .06, .05, .05, .06, .14, .11, .08, .05, .05, .05, .04, .04, .05, .05, .05].

(*Suggestion:* consider constraints for the maximum values $\tau(T_{max}) = \omega^{-1}$ and a guess for λ_{max}.)

(8.2) Reproduce the plot of Fig. 8.6 (right).

9 Superconductors

Superconductivity is a fascinating *macroscopic* state of quantum matter. It is macroscopic because conduction electrons from the entire superconducting crystal are in a well-defined quantum state, described by a single, coherent wave function ψ. This is an attribute found in very few other physical phenomena (another notable example is superfluidity). The properties of the superconducting state, which exists below a critical temperature, T_c, have effects that are measurable in the muon experiments that we examine in this chapter.

9.1 The discovery

The first piece of evidence for this special phase of matter was the vanishing electrical resistivity ρ of a superconductor, originally demonstrated in mercury by Heike Kamerlingh Onnes in 1911 (Fig. 9.1). Zero electrical resistivity already implies that the superconducting state has macroscopic quantum coherence. This can be understood using the simplest description of a normal metal, the non-interacting Fermi gas, where individual electrons occupy plane-wave states, each characterized by a crystal momentum $\hbar\boldsymbol{k}$. The Drude electrical resistivity of the Fermi gas follows from electrons scattering from a wavevector $\boldsymbol{k}$ to another $\boldsymbol{k}'$ in a mean *relaxation time* τ. In typical metals τ is on the order of 10^{-14} s, making the duration of coherent quantum dynamics of electrons extremely short. In modern parlance we say that electrons *decohere* extremely quickly in metals. Electrical resistance vanishes when τ is infinite so that, for superconductors, decoherence is virtually absent. Superconductors and metals therefore represent the two extremes of the quantum world in this respect. Incidentally, we mentioned the relaxation time in Section 8.3 and it plays a very similar role in the quantum description of the muon spin. From the point of view of their relaxation times, muons and nuclear spin systems sit midway between the two extremes, as shown in Fig. 9.2.

The extremal behaviour of the quantum decoherence of metals and superconductors is therefore directly related to the huge difference in electrical resistivity between the two states (called ρ_n and ρ_s, respectively). Although this might suggest we can describe superconductors within the simple Drude picture, the physics of the vanishing of ρ_s that is found in real superconductors[1] cannot be properly accounted for in this model, and so we are forced to look at the physics in more detail.

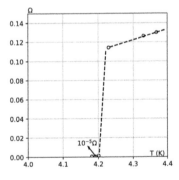

Fig. 9.1 Vanishing resistance of Hg (from H. Kamerlingh Onnes, 1911).

[1] To quantify how close to zero ρ_s is, consider the observation that persistent-mode currents in high-stability superconducting magnets circulate in closed loops with typical decay times of a hundred thousand years. This decay time is set by the extrinsic physics of the so-called pinning defects.

Historically, the second piece of evidence for the special nature of the superconducting state was the **Meissner-Ochsenfeld effect**, discovered in 1935. It consists of a magnetic behaviour not expected from a simple metal with vanishing resistivity (a hypothetical material that we call a **perfect conductor**). We see how in the following example.

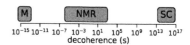

Fig. 9.2 Decoherence time in macroscopic systems: superconductors (SC), metals (M) and nuclear spins (NMR). The nuclear decoherence times are longer in liquids and shorter when coupled to slowly fluctuating electron spins.

Example 9.1

Due to Faraday's law, eddy currents are induced on the surface of any conducting material if we suddenly switch on a magnetic B-field. These eddy currents rapidly decay in normal conductors, but they will circulate forever in both perfect conductors and superconductors, where they screen the B-field from the bulk of the sample.

This screening effect occurs in any superconductor when we cool it below T_c in zero applied magnetic field, and then set the external field to a finite value. The entire protocol is called **zero-field cooling** (ZFC). In a simple cylindrical geometry the vanishing of B in the superconductor implies that $M = -H$ inside the sample, hence that the susceptibility $\chi \equiv \frac{M}{H} = -1$. However if we cool a perfect conductor in a static magnetic field, the so-called **field-cooling** (FC) protocol, Faraday's law does not apply and the field can penetrate the sample.

Surprisingly, the superconductor is found experimentally to *expel* the magnetic flux from its volume, so that even in the FC experiment we find $B = 0$ inside the material. This is the Meissner-Ochsenfeld effect, evident in Fig. 9.3. The important fact shown is that susceptibility also drops in FC and hence χ is a true property of the superconducting phase, different from perfect diamagnetism.

Notice that both the vanishing of the electrical resistivity $\rho = 0$ and the Meissner effect occur only up to a critical applied field B_c, above which normal metallic behaviour is restored. The present simplified discussion will be corrected in Section 9.4 since it is only strictly true for some superconductors, called *type-I*. Others, *type-II* superconductors, behave similarly below a lower critical field B_{c1}, but their negative susceptibility completely disappears only above an upper critical field B_{c2}.

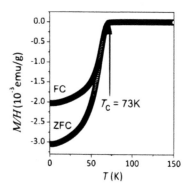

Fig. 9.3 Experimental susceptibility in the Ba_2CuO_{4-y} superconductor, showing a large negative value also in FC (the difference between the two curves is addressed in Section 9.5).

9.2 London penetration depth

The first scientists to postulate that superconductivity implies a coherent wave function for the electrons were Fritz London and his brother Heinz in 1930. They proposed that the supercurrent density $\boldsymbol{J}_s$ must obey the **London equation**

$$\boldsymbol{J}_s = -\frac{ne^2}{m}\boldsymbol{A}, \tag{9.1}$$

where $\boldsymbol{A}$ is the magnetic vector potential and m, e, and n are the electron mass, charge, and density, respectively.

[2]This reflects the fact that a superconductor is the result of a symmetry-breaking phase transition. The symmetry that is broken is the phase invariance of the superconducting wavefunction, the properties of which depend on gauge invariance. See the book by Annett in the Further reading for a discussion.

Example 9.2

Equation 9.1 poses a problem: the current density is a measurable quantity (an observable) but the vector potential A is not. Electromagnetic potentials, both scalar and vector ones, are not unique. This means the same measurable electromagnetic fields can be described with different potential functions, a property known as *gauge invariance*. Using a specific choice of potentials is known as *fixing the gauge*. For the stationary case, current conservation implies that $\nabla \cdot J_s = 0$, so that the London equation requires $\nabla \cdot A = 0$. This forces a specific form of the gauge potential on us: the one that obeys this equation, known as the Coulomb gauge. Hence the London equation is manifestly not gauge invariant.[2]

Equation 9.1 forces the magnetic field to obey a self-consistent equation whose solutions actually describe the Meissner effect. Example 9.3 demonstrates that substituting eqn 9.1 into the curl of the Ampère's law one obtains

$$\nabla^2 B = \frac{\mu_0 e^2 n}{m} B, \tag{9.2}$$

(and the identical eqn 9.5 for A). By dimensional analysis the right-hand side of eqn 9.2 is the product of the B-field with the inverse square of a length, λ^{-2}, which allows us to define

$$\lambda = \left(\frac{m^*}{\mu_0 e^2 n} \right)^{\frac{1}{2}}, \tag{9.3}$$

[3]This is sometimes written λ.

known as the **London penetration depth**,[3] a fundamental quantity that characterizes the superconductor. This quantity depends on two independent properties of the material, the density of carriers n and their **effective mass** m^*. In the earlier London interpretation these were the normal electron density and mass. We now know that the carriers of the current in a superconductor (the supercurrent) are actually Cooper pairs of electrons (see Section 9.6), with **superconducting carrier density** $n_s = n/2$ and superconducting carrier mass $m_s = 2m^*$, twice the electron effective mass.[4]

[4]The resulting factor of 1/4 in eqn 9.3 is cancelled by the factor resulting from the square of the Cooper pair's effective charge 2e.

Equation 9.2 shows indeed that a supercurrent causes the magnetic field to vanish deep inside a sample, justifying the Meissner effect. This is straightforward to prove in the simple geometry of Fig. 9.4, that is close to the situation found in a real μSR experiment.

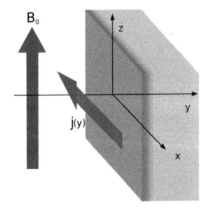

Fig. 9.4 A thick, flat sample placed in a uniform external B-field. Superconducting screening currents are induced parallel to the surface of the sample and perpendicular to the magnetic field.

Example 9.3

The derivation of eqn 9.2 is based on applying the identity

$$\nabla \times (\nabla \times F) = \nabla(\nabla \cdot F) - \nabla^2 F, \tag{9.4}$$

to Ampère's law, $\nabla \times B = \nabla \times (\nabla \times A) = \mu_0 J_s$, where the London equation may be substituted on the right-hand side. We imagine the superconductor of Fig. 9.4 immersed in a uniform field $B_0 \hat{z}$, as that produced e.g. by an infinite source current sheet on the far left. We can set the boundary condition $B(y_0) = B_0 \hat{z}$ at some distance $y_0 < 0$ from the slab and ignore the source current. Using the Coulomb gauge ($\nabla \cdot A = 0$) the first term on the right-hand side of eqn 9.4 vanishes, and we obtain

$$-\nabla^2 A = -\frac{\mu_0 n e^2}{m^*} A, \tag{9.5}$$

i.e. the equivalent of eqn 9.2 for the vector potential. Equation 9.2 itself is obtained directly by taking the curl of Ampère's law and applying similar arguments (see Exercise 9.1).

Now consider a sample placed in a uniform field $\boldsymbol{B} = B_0 \hat{\boldsymbol{z}}$. We can approximate a thick, flat sample in this field as a semi-infinite space (Fig. 9.4) with its surface perpendicular to the y-axis at $y = 0$. When the sample is below T_c it is easy to check by substitution that the solution of eqn 9.2 is

$$\boldsymbol{B}(x, y, z) = B_0 \hat{\boldsymbol{z}} \, \mathrm{e}^{-y/\lambda} \qquad (9.6)$$

This solution is the curl of a vector potential that may be written $\boldsymbol{A}(x, y, z) = B_0 \lambda \mathrm{e}^{-y/\lambda} \hat{\boldsymbol{x}}$ for $y > 0$ (see Exercise 9.2). The meaning of this equation is evident, since inside the sample $\boldsymbol{A}$ is proportional to the supercurrent (eqn 9.1), which is progressively shielding the external field over a thin surface layer of thickness λ, and eventually becomes zero in the bulk. A more realistic case is the not-so-thick slab, where the slab thickness d cannot be neglected and the field for $y > 0$ is given by

$$\boldsymbol{B}(x, y, z) = B_0 \frac{\cosh(y/\lambda)}{\cosh(d/\lambda)} \hat{\boldsymbol{z}}. \qquad (9.7)$$

Low-energy muon (LEM) experiments can closely approximate the geometry of Fig. 9.4, when muons, with spins polarized along the beam in the $\hat{y}$-direction, impinge perpendicularly on a superconducting slab in the x-z plane. The field $\boldsymbol{B}$ is along $\hat{z}$, as in the figure. We recognize this as a TF-geometry experiment where the implanted muon spin precesses around the field direction.

As explained in Chapter 18, low-energy muons can be implanted at a chosen average depth $y(E)$ from the surface by tuning their kinetic energy E. Their depth-distribution width Δy is smaller than the maximum achievable depth $y(E_{\max})$, typically slightly exceeding 100 nm, i.e. the same order of magnitude as λ for high T_c superconductors. Thus, distinct measurements for each value of $y(E)$ reveal spin precession at an average Larmor frequency $\omega(y) = \gamma_\mu B(y)$, where the field $B(y)$ varies according to eqn 9.7.

Three representative asymmetries at different depths are displayed in Fig. 9.5 for a LEM experiment on a superconductor. Figure 9.6 shows the exponential decrease with depth of the local field, demonstrating the Meissner effect. Actually, two sets of data are shown, due to the fact that the single crystal sample of orthorhombic $\mathrm{YBa_2Cu_3O_{6.92}}$, is an **anisotropic superconductor**. This finer point is explained in the next example.

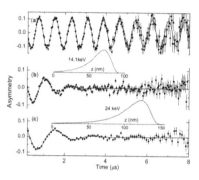

Fig. 9.5 LEM asymmetries in a superconducting slab at different average depths y from the surface, in the same transverse applied field $B_0 = 9.46$ mT. The precession frequency, averaged over the depth distributions shown as insets, decreases with increasing y, due to the Meissner effect. [Taken from R. Kiefl *et al.*, Phys. Rev. B **81**, 180502 (2010).]

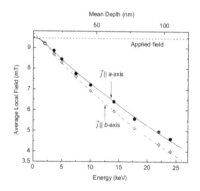

Fig. 9.6 Depth-dependent average field in a single crystal, in an applied field $B_0 = 9.46$ mT parallel to the a- (open symbols) and b-axis (filled symbols). [Taken from R. Kiefl *et al.*, Phys. Rev. B **81**, 180502 (2010).]

Example 9.4

For anisotropic metals, such as the **cuprate superconductors**, the effective mass is a tensor. Therefore the inverse-square penetration depth $1/\lambda^2$ depends on the effective-mass tensor components in the plane where the superconducting currents circulate. Rotating a single crystal slab in Fig. 9.4 with the c-axis parallel to the beam, either a or b can be aligned to the field. Thus, two different values of the penetration depth can be measured, λ_a and λ_b, with field parallel to a or to b (currents parallel to b or a) respectively. Their ratio is given by

$$\frac{\lambda_a}{\lambda_b} = \left(\frac{m^*_{bb}}{m^*_{aa}} \right)^{\frac{1}{2}}. \qquad (9.8)$$

For this reason the experiments performed with two different orientations of the same thin single crystal produce the two distinct depth-dependent exponential decays of the local field, in Fig. 9.5. In the following we consider isotropic superconductors for simplicity, unless otherwise specified.

9.3 Ginzburg-Landau model

In this section we introduce the **Ginzburg-Landau model**. This model makes it clear that, in addition to the penetration depth λ, superconductors are also characterized by second length-scale: the **coherence length** ξ. This quantity has several equivalent definitions. One of them comes from considering the Landau free energy density of the superconductor,[5] which is reduced below T_c by the onset of a complex order parameter ψ, which we identify with the macroscopic wave function of the superconducting state. The free energy density is then written

$$f_s(B) = f_n(B) + a(T - T_c)|\psi|^2 + \frac{b}{2}|\psi|^4 + \frac{1}{2\mu_0}B^2, \qquad (9.9)$$

where f_n is the contribution of the normal state and, in the spirit of Landau theory, the magnitude of the order parameter $|\psi|$ is assumed small. The condition $a, b > 0$ guarantees that in zero magnetic field the order parameter is non-vanishing (i.e. $|\psi|^2 > 0$) for $T \leq T_c$, where the free energy density gain is $f_s(0) - f_n(0) < 0$. This negative difference corresponds to a very small condensation energy density, favouring the superconducting state.

Example 9.5

We mentioned previously that a finite field B_c suppresses the superconducting state. We can see that this occurs when the gain in the zero-field energy density, $f(0)$, equals the maximum magnetic energy density $B_c^2/2\mu_0$ term. This term corresponds to the energy of the maximum field that shielding supercurrents are able to produce (i.e. to expel, thanks to the negative sign of the susceptibility). For larger values of the field the free energy of the superconductor is larger that that of the normal metal and we have[6]

$$f_n(0) - f_s(0) = \frac{1}{2\mu_0}B_c^2. \qquad (9.10)$$

Note that the condensation energy for a cell ranges from μeV to meV, much smaller than the Fermi energy ε_F, which is the typical energy scale for a metal. In fact, the condensation energy corresponds to the opening of a **superconducting gap** *at* the Fermi energy.

[6]We have ignored the very small Pauli susceptibility of the normal metal with respect to that of the superconductor.

Variations in the order parameter (such as that occurring at the surface) can be straightforwardly treated in Landau theory by adding a gradient term, $c|\nabla\psi|^2$ to the free energy, which imposes an energy penalty for rapid changes in the order parameter. Vitaly Ginzburg introduced a gradient term $\hbar^2|\nabla\psi|^2/2m_s$ into eqn 9.9 to treat the theory for superconductors, and the resulting equations are known as the Ginzburg-Landau (GL) theory.

Example 9.6

The free energy in eqn 9.9, with the addition of the gradient term, must include supercurrents as sources of magnetic field. If we assume that the order parameter can be written $\psi = \psi_0 e^{i\theta}$ and represents a quantum wave function for the electrons, the magnetic field due to their current density is included using the minimal coupling prescription,[7] substituting the moment $-i\hbar\boldsymbol{\nabla}\psi$ with $(-i\hbar\boldsymbol{\nabla} - q\boldsymbol{A})\psi$. Now the free energy density can be minimized self-consistently in a variational approach, and, since there are two independent functions to vary (ψ and $\boldsymbol{A}$), the procedure produces two GL equations of motion that are valid below T_c.

The first equation is for the order parameter, and is written

$$\left[\frac{1}{2m_s}(-i\hbar\boldsymbol{\nabla} - q\boldsymbol{A})^2 + a(T - T_c) + b|\psi|^2\right]\psi = 0. \tag{9.11}$$

Close to T_c the square modulus of the order parameter, $|\psi|^2$, is small and eqn 9.11 may be linearized, neglecting the b term. Equation 9.11 then becomes equivalent to a Schrödinger equation[8] for ψ, in the presence of the magnetic field $\boldsymbol{B} = \boldsymbol{\nabla} \times \boldsymbol{A}$.

The second equation links the supercurrent $\boldsymbol{J}_s$ to the vector potential

$$\boldsymbol{J}_s = -\frac{iq\hbar}{2m_s}(\psi^*\boldsymbol{\nabla}\psi - \psi\boldsymbol{\nabla}\psi^*) - \frac{q^2}{m_s}|\psi|^2\boldsymbol{A}. \tag{9.12}$$

Exercise 9.3 shows that for uniform order parameter $\psi = \psi_0 e^{i\theta}$ where ψ_0 and the phase θ do not vary spatially, the gradient terms vanish and this second equation reduces to the London eqn 9.1. In this way we can identify the square modulus of the order parameter with the density of superconducting carriers, $n_s = |\psi|^2$, that appears in the London penetration depth (eqn 9.3).

In the general case the two GL equations from the last example are very complicated to solve. We shall see that in simple geometries like that of Fig. 9.4, and in many other interesting cases, the solutions provide a full description for the complex order parameter and for the magnetic field, including the contribution of the supercurrents.

Example 9.7

The current density of eqn 9.12 contains a gradient term. Let's assume that only the phase $\theta(\boldsymbol{r})$ has a spatial dependence, while the amplitude $\psi_0 = \sqrt{n_s}$ does not. Then, on substituting $\psi = \psi_0 e^{i\theta(\boldsymbol{r})}$, the current density becomes

$$\boldsymbol{J}_s = \frac{qn_s}{m_s}(\hbar\boldsymbol{\nabla}\theta - q\boldsymbol{A}). \tag{9.13}$$

Now stationary currents $\boldsymbol{\nabla} \cdot \boldsymbol{J} = 0$ imply $\boldsymbol{\nabla} \cdot (\hbar\boldsymbol{\nabla}\theta - q\boldsymbol{A}) = 0$, and also imply that the gradient term of the free energy becomes

$$\frac{|(-i\hbar\boldsymbol{\nabla} - q\boldsymbol{A})\psi|^2}{2m_s} = \frac{\hbar^2 n_s}{2m_s}\left(\frac{\boldsymbol{\nabla}\theta}{2\pi} - \frac{\boldsymbol{A}}{\Phi_0}\right)^2, \tag{9.14}$$

where[9]

$$\Phi_0 = \frac{h}{2e} = 2 \times 10^{-15} \text{ Wb} \tag{9.15}$$

is the **flux quantum**, a quantity that will play a very important role below. Equation 9.14 shows that there is an energy cost in having a non-uniform phase for the order parameter, proportional to the so-called **superfluid stiffness** of the superconductor, $\hbar^2 n_s/2m_s$. We recover again the London case in eqn 9.13 whenever the phase θ of the order parameter is constant. But spatially varying solutions are possible and eqn 9.14 indicates that a tiny loop along which the phase changes by 2π encircles a flux quantum Φ_0 at no (local) energy cost.[10] This connection between a 2π periodicity and the flux quantum corresponds to a local vortex of supercurrent.

We can now clarify the meaning of the coherence length ξ in the context of Landau theory by considering the interface between a superconductor and the vacuum. Due to the gradient term, the order parameter cannot vanish abruptly at the surface of the superconductor. This gives rise to a normal-to-superconducting (n-s) transition layer, as shown in Fig. 9.7 and as we examine in the next example.

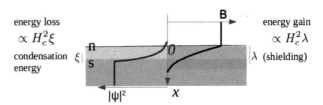

Fig. 9.7 Interface between a superconductor (lower region) and vacuum (upper region). The overlaid curves show how the magnetic field and the order parameter vanish in the superconductor bulk and at the surface, respectively. The suppression of $|\psi|^2$, over a length ξ, and the expulsion of $\boldsymbol{B}$, over the length λ, must balance their respective energy density loss and gain.

Example 9.8

Dimensional analysis shows that the gradient term in the free energy implicitly defines a length scale, via the square root of the ratio between c and the term $a(T - T_\mathrm{c})$, of

$$\xi_{\mathrm{GL}}(T) = \frac{\hbar}{\sqrt{2m_\mathrm{s}a(T_\mathrm{c} - T)}}. \tag{9.16}$$

The value of $\xi_{\mathrm{GL}}(0) = \hbar/\sqrt{2m_\mathrm{s}aT_\mathrm{c}}$ at $T = 0$ tells us the distance over which the superconducting state vanishes.[11] If $\xi_{\mathrm{GL}}(0)$ is somewhat less than the penetration depth λ, the layer through which the magnetic field penetrates is larger than the region where the order parameter is suppressed. In this case it is energetically favourable for the system to build the n-s interface region.

[11] When the interface is between a superconductor and a normal metal the same approach dictates that the gap actually vanishes inside the metal, so that its surface layer actually becomes weakly superconducting. This is called the *proximity* effect.

9.4 Type-II superconductors

The ratio $\kappa = \lambda/\xi$ is known as the Ginzburg-Landau κ parameter. Its value classifies superconducting materials, discriminating between homogeneous **type-I superconductors** ($\kappa < 1/\sqrt{2}$), where interfaces are limited to a minimum possible volume at the surface, and inhomogeneous **type-II superconductors** ($\kappa > 1/\sqrt{2}$), where interfaces become energetically favourable in magnetic field.

From here onwards we shall be mainly dealing with type-II superconductors, for which magnetic fields vary in space with scale λ, a longer scale than their microscopic sources, that change over the shorter scale ξ.[12] Fortunately, all type-II materials are *local* superconductors, where thermodynamic fields (obtained as averages over the source inhomogeneities, governed by ξ) coincide with the microscopic fields and we

[12] We also drop the subscript GL under the coherence length.

avoid the complications of non-local electrodynamics, that are inherent in type-I superconductors.

In these type-II superconductors a state of alternating normal and superconducting regions, called the **mixed state**, exists in moderate fields. A marked difference in magnetization versus applied field characterises their distinct behaviour. Type-I superconductivity is destroyed in $B > B_c$, whereas the coherent state exists in type-II materials up to a much higher field $B_{c2} > B_c$ (see Fig. 9.8), where the energy gain at the n-s interfaces generates an inhomogeneous field configuration.

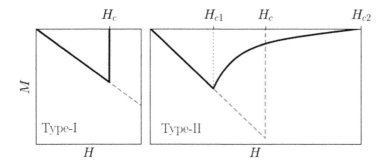

Fig. 9.8 Magnetization vs $H = B/\mu_0$ in type-I and type-II superconductors

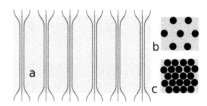

Fig. 9.9 (a) Sketch of the flux lattice in a section containing the field direction (the grey area is the superconductor). Cross section perpendicular to the field direction (b) at intermediate fields, and (c) at $B \approx B_{c2}$.

The flux lattice

We will need to discuss the type-II superconducting state in some detail since it is the key to understanding a standard type of muon experiment on a superconductor. For fields above a **lower critical field** B_{c1} the magnetic field can penetrate in type-II superconductors as localized *normal metal* flux-line defects, each corresponding to a **fluxon**, a tube of quantized flux of value Φ_0. This connects to the conclusion of Example 9.7 where we inferred the possibility of vortices of super-current. The defects organize themselves in a regular two-dimensional lattice of flux tubes, known as the **flux-line lattice, flux lattice** or **vortex lattice**, whose presence reduces the absolute value of the magnetic susceptibility.[13] The flux lattice is most often found to be triangular like those sketched in Fig. 9.9, although square lattices are observed under some circumstances.

A flux tube of cross section $\pi\xi^2$ first nucleates at $B = B_{c1}$ inside the superconductor. Upon increasing B, other fluxons are formed in a regular two-dimensional lattice, progressively decreasing the spacing between them, up to the maximum field $B_{c2} = \frac{\Phi_0}{2\pi\xi^2}$. For this field value the fluxon cross sections cover the area orthogonal to $\boldsymbol{B}$ in a close-packed arrangement, as shown in Fig. 9.9(c). Substitution of eqn 9.16 into the full expression of the free energy yields a very useful expression for the critical field, which is given by

$$B_c = \frac{\Phi_0}{\sqrt{2}(2\pi)\xi\lambda}.$$

[13]The solution of the GL model for Type II superconductors that justifies the susceptibility in Fig. 9.8 was discovered by Alexei Abrikosov in 1950, and earned him the 2003 Nobel Prize for physics.

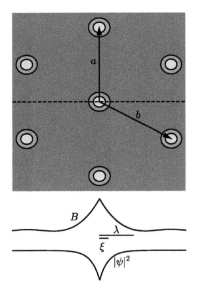

Fig. 9.10 Top: Abrikosov triangular flux line lattice in a plane orthogonal to the magnetic field B. Bottom: the order parameter and the magnetic field across a normal core, along the dashed line in the top panel.

The second critical field and the *thermodynamic* critical field are in the ratio

$$\frac{B_{c2}}{B_c} = \sqrt{2}\kappa. \tag{9.17}$$

The expression for the lower critical field is $B_{c1} = \ln\kappa B_c/\sqrt{2}\kappa$, although it is only strictly valid in the large-κ limit.

Between the lower and upper critical fields ($B_{c1} \le B \le B_{c2}$) the periodic flux lattice is the solution of the two coupled GL equations. Normal cores of radius ξ decorate the lattice sites and the flux lattice parameter, $a = b = (4/3)^{\frac{1}{4}}(\Phi_0/B)^{\frac{1}{2}}$ for the triangular case (see Exercise 9.4), provides an additional length scale which is tunable by field. Figure 9.10 shows a flux lattice and its inhomogeneous field distribution. Since the flux lattice parameters are generally much larger than, and incommensurate with, the crystal lattice parameters, implanted muon sites are uncorrelated with the core positions and uniformly sample the magnetic field distribution. The bottom part of the figure shows how the square modulus of the order parameter $|\psi|^2$ and the magnetic field vary around a fluxon, governed by the two independent length scales ξ and λ. Notice that the field does not vanish in between cores because of the overlap of contributions from many fluxons.

Local magnetic field in a flux lattice

The flux lattice is obtained as a solution of the GL equations. In the London approximation they reproduce the London field equation, eqn 9.2, with an extra term

$$\boldsymbol{B}(\boldsymbol{r}) - \lambda^2\nabla^2\boldsymbol{B}(\boldsymbol{r}) = \Phi_0\hat{\boldsymbol{z}}\sum_{\boldsymbol{R}}\delta^{(2)}(\boldsymbol{r} - \boldsymbol{R}), \tag{9.18}$$

containing a two-dimensional delta function $\delta^{(2)}(\boldsymbol{r} - \boldsymbol{R})$ for each of the flux lattice sites $\boldsymbol{R} = n\boldsymbol{a} + m\boldsymbol{b}$. Equation 9.18 dictates that the exponential decay of the field, taking place at the sample surface in the Meissner case, also occurs around each fluxon. The total field is the sum of many such terms. In a typical flat μSR sample, the field lines are parallel to the applied field $\boldsymbol{B} = B(\boldsymbol{r})\hat{\boldsymbol{z}}$ and a periodic solution is obtained by making use of a Fourier transform (see Exercise 9.5)

$$B(\boldsymbol{r}) = \frac{\Phi_0}{|\boldsymbol{a} \times \boldsymbol{b}|}\sum_{\boldsymbol{Q}}\frac{1}{1 + \lambda^2 Q^2}\, \mathrm{e}^{\mathrm{i}\boldsymbol{Q}\cdot\boldsymbol{r}}, \tag{9.19}$$

with the reciprocal lattice vectors $\boldsymbol{Q} = (2\pi/a)(n\hat{\boldsymbol{a}}^* + m\hat{\boldsymbol{b}}^*)$ expressed in terms of their unit vectors $\hat{\boldsymbol{a}}^*$ and $\hat{\boldsymbol{b}}^*$. Each Fourier component $B(\boldsymbol{Q})$ is a simple Lorentzian function that reflects the exponential field decay from each fluxon core.

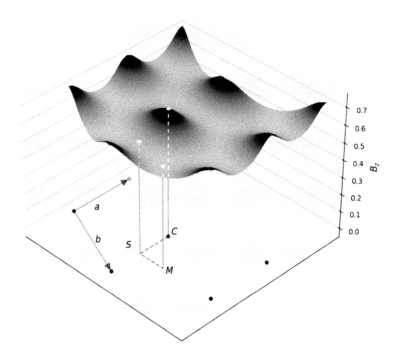

Fig. 9.11 Surface plot of $B_z(\boldsymbol{r})$ for $B = 0.64$ T, $\lambda = 70$ nm, $\xi = 9$ nm, and a Gaussian cut-off (see Example 9.9). In the London limit the surface would have a cusp in the white dot at core sites C. The flux lattice cell is shown in the bottom plane.

Example 9.9

The exponential decay from each core in eqn 9.19 gives an unphysical cusp at the core center, an artefact of the London approximation that ignores the finite size of the core, ξ. In a better approximation $B_z(Q)$ is multiplied by a cut-off function $C(Q) = \exp(-Q^2 \xi^2 / 2)$, to cure the London singularity. For $\kappa > 10$ the best GL approximation is provided by the cut-off

$$C(u) = (1 - b^4) u K_1(u), \tag{9.20}$$

where $u = 2\xi Q(1 + b^4)(1 - 2b + 4b^2)$ is a function of the reduced field, $b = B/B_{c2}$, and K_1 is the modified Bessel function. [See for instance A. Yaouanc *et al.*, Phys. Rev. B **55**, 11107 (1997).]

It is straightforward to obtain an accurate numerical estimate of eqn 9.19 by the fast Fourier transform algorithm, as shown in Fig. 9.11, with any choice of cut-off. The bottom plane shows the flux lattice cell in the sample cross section orthogonal to the field. The dashed line indicates a path from the core centre (C) to the saddle point (S) between two adjacent cores, to the minimum field site (M) at the cell centre.[14]

[14]Exercise 9.8 suggests how to reproduce the plot.

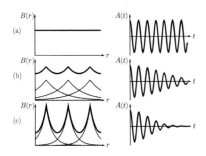

Fig. 9.12 The field distribution in a superconductor as a function of position, along with the corresponding muon-spin relaxation function for three cases: (a) the normal state, (b) the superconducting state, (c) the superconducting state but with a shorter penetration depth. [Figure from S. J. Blundell, Contemp. Phys. **40**, 175 (1999).]

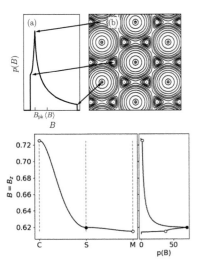

Fig. 9.13 (a)-(b): the relationship between the field distribution $p(B)$ and the contours of the vortex lattice field surface 9.11. [Figure from S. J. Blundell, Contemp. Phys. **40**, 175 (1999).] (c): $B(\boldsymbol{r})$ along the C-S-M path in Fig. 9.11; (d): the field distribution $p(B)$, for $B = 0.64$ T, $\xi = 9$ nm, $\lambda = 70$ nm, $\Delta B = 0.1$ mT. Note that (d) shares the B-axis with (c), so is a mirror image of (a), rotated by 90°.

9.5   Measuring the penetration depth

Important quantities can be extracted from a TF-μSR measurement on a superconductor when we apply fields $B > B_{c1}$ and realise the flux lattice, with its characteristic field distribution given by eqn 9.19. Here we examine the most widely used method to obtain the penetration depth by μSR.

Example 9.10

The principle of the TF measurement is delightfully simple. In the normal state ($T > T_c$) with a transverse field B, all muons precess with frequency $\omega = \gamma_\mu B$ [Fig. 9.12(a)]. In the superconducting state, muons implanted close to the vortex cores experience a larger magnetic field than those implanted between vortices. Consequently there is a distribution in precession frequency, resulting in relaxation of the observed precession signal [Fig. 9.12(b) and (c)]. The larger the penetration depth, the smaller the magnetic field variation and the less pronounced the relaxation [Fig. 9.12 (b)].

We know from Chapter 6 that the muon-spin polarization is given by $P_z(t) = \int_0^\infty p(B) \cos\gamma_\mu Bt$ (the case $\theta = \pi/2$ in eqn 6.28) and so the field distribution $p(B)$ of the flux line lattice is fully accessible via a transform of the asymmetry spectra into the frequency domain (see Section 15.5). It is easy to obtain $p(B)$ from eqn 9.19 numerically, to compare experiment with predictions. The same quantity could be calculated as $p(B) = |dB(\boldsymbol{r})/dS|^{-1}$, where dS is the infinitesimal area centred at $\boldsymbol{r}$ (the derivation of this expression is common to any lattice density of states, but we do not pursue it since the result is an implicit function of B, and is not readily invertible).

From the last example, we see that the field distribution in the flux lattice is directly obtained from the experiment via the muon asymmetry lineshape, i.e. the Fourier amplitude spectrum of the asymmetry $A(\nu) = p(B/2\pi\gamma_\mu)$. The standard geometry for this experiment at a continuous source employs a thin, flat sample with normal parallel to the applied field and to the muon beam direction, $\hat{\boldsymbol{z}}$, to allow the application of a high magnetic field without beam deflection. This TF-μSR geometry requires a large component of the muon spin orthogonal to $\hat{\boldsymbol{z}}$, which is then obtained with a spin rotator (see Chapter 13). With pulsed muon beams the reduced frequency passband requires that we use low magnetic fields, and in that case TF-μSR may be carried out with $\boldsymbol{B} \parallel \hat{\boldsymbol{x}}$ and no spin rotation. In both cases the experiments are preferentially performed after field-cooling (FC, see Example 9.1), to avoid the disorder induced by switching on the field in the superconducting state (ZFC), which typically nucleates domains with frustrated superconducting phases and results in a broadened glassy lineshape.

The resulting field distribution $p(B)$ is shown in Fig. 9.13(a) with its features attributed to the different parts of the flux line lattice. The full range of the function $B(\boldsymbol{r})$ is also shown in Fig. 9.13(d) (with $p(B)$ plotted vertically along the shared B-axis). The distribution is highly asymmetric and displays three singularities: two cut-offs at the maximum and minimum field, in C and M, respectively [Fig. 9.13(c), open symbols], and a singularity at the saddle point S (filled symbols), where the field derivative vanishes. Looking back at Fig. 9.11 we recognize that the cusp and cut-offs of $p(B)$ correspond to the extrema of $B(\boldsymbol{r})$. To reflect sources of disorder, the distribution is often convoluted with a Gaussian smearing function of standard deviation ΔB, replacing the divergence with a peak. Exercise 9.9 discusses a numerical derivation of $p(B)$.

Example 9.11

Figure 9.14 (left) shows the Fourier transform of the muon asymmetry for a single crystal of Nb. Notice that here both the data and the simulated $p(B)$ (center and right, for varying parameter values) are plotted against a horizontal B-axis. They both display the skewed appearance typical of the flux-lattice distribution, still evident despite the Gaussian smearing, much larger that that of Fig. 9.13 (a) and (d).

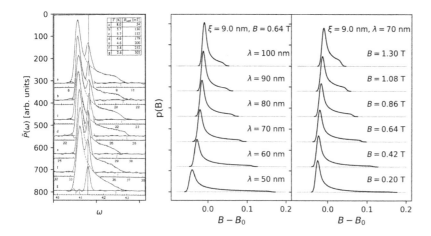

Fig. 9.14 Single crystal data (left) [taken from D. Herlach *et al.*, Hyp. Int. **63**, 41 (1991)]; $p(B)$ from numerical simulations vs κ (centre) and B (right), with $\Delta B = 4$ mT.

Best fit parameters for a single distribution are highly correlated and the model often requires a global fit of extensive data sets for proper convergence. Furthermore, polycrystalline samples, together with anisotropic metallic properties, tend to result in large smearing of the distribution that wipes out the sharper features and, in extreme cases, even its skewed shape. In these cases it may be possible to measure only the second moment of the distribution (e.g. through a Gaussian fit of the muon asymmetry). We discuss these considerations in the next example.

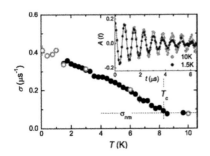

Fig. 9.15 Full TF-μSR σ in superconducting FeSe$_{0.85}$ with residual σ_n above $T_c \approx 8$ K; inset: Muon asymmetry $A(t)$ above and well below T_c. [Taken from R. Khasanov *et al.*, Phys. Rev. B **78**, 220510(R) (2008).]

Example 9.12

Often the skewed lineshape is still experimentally apparent when singularities are totally smeared out. In this case a multi-Gaussian fit of the distribution

$$A(t) = a_0 \sum_i \frac{f_i}{\sqrt{2\pi}\sigma_i} e^{-\sigma_i^2 t^2/2} \cos\gamma_\mu B_i t, \tag{9.21}$$

where $\sum_i f_i = 1$, might be the best option. The first and second moments of this distribution are readily computed (Exercise 9.7) to be

$$\langle B \rangle = \sum_i f_i B_i,$$

$$\langle \Delta B^2 \rangle = \sum_i \left[f_i \left(\frac{\sigma_i}{\gamma_\mu} \right)^2 + \frac{1}{2} \sum_{j \neq i} f_i f_j (B_i - B_j)^2 \right] = \left(\frac{\sigma_\mu}{\gamma_\mu} \right)^2. \tag{9.22}$$

In additional to the contribution to the Gaussian linewidth from the flux lattice, random magnetic fields from nuclear dipoles also contribute to the broadening, yielding a nearly Gaussian relaxation of the spin precession with rate σ_n above T_c (Fig. 9.15). This independent contribution can be determined at high temperature and subtracted in quadrature from the fitted value of σ_μ. As a result, the Gaussian width due to the presence of the flux lattice parameter is

$$\sigma_{\mathrm{FL}}^2 = \sigma_\mu^2 - \sigma_n^2 = \gamma_\mu{}^2 \langle \Delta B^2 \rangle - \sigma_n^2, \tag{9.23}$$

which is a commonly used method for determining σ_{FL}.

The second moment of the field distribution of eqn 9.19, including a simple cut-off function $C(Q\xi)$, is not too difficult to calculate (see Exercise 9.6 and eqn 9.39), and we obtain

$$\langle \Delta B^2 \rangle = \left\langle [B(\boldsymbol{r}) - \langle B(\boldsymbol{r})\rangle]^2 \right\rangle = \frac{\Phi_0^2}{|\boldsymbol{a} \times \boldsymbol{b}|^2} \sum_{\boldsymbol{Q} \neq 0} \left[\frac{C(Q\xi)}{1 + \lambda^2 Q^2} \right]^2. \tag{9.24}$$

For extreme type-II superconductors ($\kappa \gg 1$) we have $\lambda \gg \sqrt{\Phi_0/B}$, and we can neglect $1 (\ll \lambda^2 Q^2)$ in the denominator of eqn 9.24, so that the muon Gaussian parameter σ_μ is proportional to λ^{-2}. A useful expression for the second moment is obtained by performing the numerical sum, e.g. in a triangular lattice in the London limit [$C(Q\xi)=1$]. The result is

$$\langle \Delta B^2 \rangle = 0.00371 \frac{\Phi_0^2}{\lambda^4}, \tag{9.25}$$

which is valid both in SI and in c.g.s. units.[15] This corresponds to the following simple relation for λ in units of nm, where σ_{FL} is in units of μs^{-1}:

$$\lambda = 327.5/\sqrt{\sigma_{\mathrm{FL}}}. \tag{9.26}$$

This direct relationship between the measured relaxation rate for the flux lattice and the penetration depth makes μSR a powerful technique

---

[15]Other useful relations in this limit are

$$B_{\mathrm{saddle}} = \langle B \rangle - 0.0368 \frac{\Phi_0}{\lambda^2},$$

$$B_{\min} = \langle B \rangle - 0.0436 \frac{\Phi_0}{\lambda^2},$$

for the saddle point and minimum field respectively. For the square lattice we have

$$\langle \Delta B^2 \rangle = 0.00387 \, \Phi_0^2/\lambda^4.$$

in the general study of superconductors. Its effectiveness is also due to the fact that μSR provides a *bulk* determination of the penetration depth, insensitive to both the surface state of the superconductor (unlike alternative surface impedance techniques) and to the muon site of implantation (thanks to the incommensuration between flux and crystal lattices). This aspect has gained μSR a prominent place in the search for new superconductors.

If the mean internal field is calculated correctly from the data and the sample approximates the infinite slab geometry, we expect a vanishing shift ($\langle B \rangle - B$) of the flux lattice precession, since $\Phi_0/|a \times b|$ equals the applied field B. A shift is, however, often observed and is taken as a signature of the onset of superconductivity. It often has one of two causes: (i) the first is the finite sample demagnetization (or that of polycrystalline superconducting domains); (ii) the second is the bias of multi-Gaussian fits, which neglect the longer tail of the magnetic field distribution that arises from the vortex core. The peak at the saddle point of the distribution dominates these fits, with its negative shift given in Note 15.

Example 9.13

We learned from Fig. 9.6 that the penetration depth in anisotropic superconductors depends on the field orientation, and the Meissner-effect experiment reveals a small anisotropy in the ab-plane of orthorhombic $YBa_2Cu_3O_{6.92}$. This cuprate superconductor actually displays a much larger ac-anisotropy, characterized by the ratio

$$\gamma_{ac} = \sqrt{\frac{m_{cc}^*}{m_{aa}^*}} = \frac{\lambda_a}{\lambda_c}. \tag{9.27}$$

Another single-crystal measurement of the flux lattice may be performed by rotating the field in the ac-plane, as shown in Fig. 9.16. The solid line shows that in this case the square root of the second moment (after subtraction of the nuclear dipolar term) is

$$\sqrt{\langle \Delta B(\theta)^2 \rangle} = \sqrt{\langle \Delta B(0)^2 \rangle}(\cos^2 \theta + \gamma_{ac}^2 \sin^2 \theta). \tag{9.28}$$

Frequently, newly discovered superconductors are available only as powders. In this case the measured second moment is the average over the angle θ. In the London approximation, the ratio of this average to the largest component $[\langle \Delta B(0)^2 \rangle$ in Fig. 9.16] is a fixed factor that depends on γ_{ac}, but saturates at $\overline{\langle \Delta^2 B(\theta) \rangle} \approx 0.44 \langle \Delta B(0)^2 \rangle$ for $\gamma_{ac} > 3$.

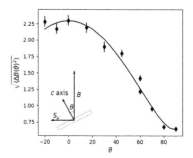

Fig. 9.16 Square root second moment of the flux lattice field distribution in a $YBa_2Cu_3O_{7-\delta}$ single crystal, with the geometry shown in the inset. [Taken from C. Ager *et al.*, Phys. Rev. B **62**, 3528 (2000).]

Equation 9.25, or its variant from anisotropic superconductors, yield the $T = 0$ value of the penetration depth. From eqn 9.3 we see that this quantity is proportional to the superconducting carrier density n_s, itself equal to the square modulus of the order parameter $n_s = |\psi|^2$. This also provides the normalized temperature-dependent superfluid density $\rho = n_s(T)/n_s(0)$, which is usually plotted against the reduced temperature $t = T/T_c$. A frequently used empirical expression is provided by a **two-fluid model**, where two distinct fluids represent the normal electron gas and the superconducting fluid. They are assumed to have densities $\propto t^4$ and to $\propto 1 - t^4$, respectively,[16] so that

The normalized superfluid density ρ introduced here should not be confused with resistivity, with which it shares a symbol.

[16]The large exponent catches empirically the rapid disappearance of the normal density for $t \to 0$.

$$\rho(t) = \frac{\sigma_{\mathrm{FL}}(T)}{\sigma_{\mathrm{FL}}(0)} = 1 - t^4, \qquad (9.29)$$

as shown in the example of Fig. 9.17, where σ is already corrected for nuclear dipoles by eqn 9.23.

The methods discussed in this section represent the most common use of muon spectroscopy in investigating superconductors. Further progress requires that we briefly make contact with the microscopic models of superconductivity.

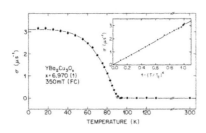

Fig. 9.17 Early TF-μSR data from polycrystalline YBa$_2$Cu$_3$O$_{7-\delta}$: σ is σ_{FL} and the solid line is eqn 9.29; inset: σ_{FL} vs. t^4 demonstrates a good fit by the two fluid model. [Taken from B. Pümpin *et al.*, Phys. Rev. B **42**, 8019 (1990).]

[17]This is equivalent to the Born-Oppenheimer approximation, based on the fact that nuclear dynamics is much slower than electron dynamics, by virtue of their mass ratio.

9.6   The microscopic model

John Bardeen, Leon Cooper, and Robert Schrieffer first proposed the microscopic model of superconductivity that captures the quantum essence of the phenomenon, sharing the Nobel Prize in Physics 1972 for this fundamental achievement. Two simple universal ingredients account for the Bardeen, Cooper, and Schrieffer (**BCS**) model.

(1) An effective *attractive* interaction between electrons, mediated by a boson (a weak phonon coupling, in the original case). The electron-boson susceptibility is of the general form $\chi \propto (\omega^2 - \omega_{\mathrm{b}}^2)^{-1}$ (typically with the Debye frequency $\Omega_{\mathrm{D}} \equiv \omega_{\mathrm{b}}$ for the electron-phonon interaction). The adiabatic approximation[17] requires that the boson frequency ω_{b} is intermediate between the thermal frequency of the condensate $\omega \approx k_{\mathrm{B}}T/\hbar$ and the normal-electron Fermi scale $\varepsilon_{\mathrm{F}}/\hbar$. As a consequence, the boson-mediated Coulomb interaction $\overline{V} = \chi V$ is of opposite sign to the repulsive bare Coulomb potential V.

(2) The presence of the Fermi sea, which screens the direct e-e repulsion V, so that the effective potential, instead of being the expected $\varepsilon V = (1 + \chi)V = V + \overline{V}$, reduces almost to $\overline{V}$. This resulting attractive interaction creates an instability in the Fermi sea, leading to the formation of **Cooper pairs** of electrons for any value of $\overline{V}$, however weak.

With these two ingredients, the Fermi surface is unstable against the opening of an energy gap around ε_{F}, which, in the BCS weak limit at $T = 0$, is of magnitude $\Delta(0) = 1.764 \, k_{\mathrm{B}}T_{\mathrm{c}}$.

Superconductivity results from Cooper pairs condensing into a ground state wave function, corresponding to the order parameter ψ. Normal electrons correspond to quasiparticle excitations across the gap, having energy $E_{\boldsymbol{k}} = \sqrt{(\varepsilon_{\boldsymbol{k}} - \varepsilon_{\mathrm{F}})^2 + \Delta^2}$. The celebrated weak-coupling prediction is that the critical temperature is given by

[18]Incidentally, the BCS model implies a different definition of the coherence length from the GL expression, eqn 9.16, $\xi = \hbar v_{\mathrm{F}}/\pi\Delta$. The value of $\xi_{\mathrm{GL}}(0)$ however roughly coincides with the BCS value.

$$k_{\mathrm{B}}T_{\mathrm{c}} = \frac{2}{1.763}\hbar\omega_{\mathrm{b}}e^{-\frac{1}{N(0)\overline{V}}}. \qquad (9.30)$$

This expression involves the metal's density of states at the Fermi energy $N(0)$ within a pseudopotential $\mu = N(0)\overline{V}$.[18]

Example 9.14

In the original phonon-mediated BCS theory the gap is constant (isotropic, or s-wave) in k-space. However, depending on the nature of the pairing interaction, the gap inherits the symmetry of the Cooper pair wavefunction. This may be expressed as $\Delta_{\mathbf{k}} = \Delta(0)g(\hat{\mathbf{k}})$ at $T = 0$, where the polar coordinates of $\hat{\mathbf{k}}$, θ, ϕ are the explicit variables of the gap symmetry function $g(\hat{\mathbf{k}})$. Cuprate superconductors, for instance, (Example 9.15) feature a $d_{x^2-y^2}$-wave gap described by $g(\hat{\mathbf{k}}) = \cos 2\phi$, a function that vanishes (having nodes) for $\phi = \pi(2n+1)/4$. Figure 9.18 compares it with the conventional s-wave case.

The thermodynamic properties of a BCS superconductor are governed by the Fermi-Dirac distribution $f(E_{\mathbf{k}}) = 1/(\exp(E_{\mathbf{k}}/k_BT)+1)$ and the presence of the gap in the density of states for the quasiparticle excitations. A very useful expression for the gap itself is given by

$$\Delta(T) = \Delta(0)\tanh\left(\frac{\pi k_B T_c}{\Delta(0)}\sqrt{a\frac{1-t}{t}}\right), \tag{9.31}$$

where $a = 1, 4/3$ and $\Delta(0)/k_B T_c = 1.763, 2.14$ for s and d-wave, respectively[19]. The temperature dependence of the normalized superfluid density is obtained from the normalized muon second moment, $\rho(T) = \frac{\lambda(0)^2}{\lambda(T)^2} = \frac{\sigma_{FL}(T)}{\sigma_{FL}(0)}$. For an anisotropic conductor ($i,j = a,b,c$) the BCS model predicts

$$\rho_{ij}(T) = \lambda_{ij}^2(0)\oint_{FS} dS_k \frac{v_{Fi}v_{Fj}}{v_F}\left(1 + 2\int_{\Delta_k}^{\infty}\frac{\partial f}{\partial E_{\mathbf{k}}}\frac{E_{\mathbf{k}}}{\sqrt{E_{\mathbf{k}}^2 - \Delta_{\mathbf{k}}^2}}dE_{\mathbf{k}}\right), \tag{9.32}$$

where v_F is the Fermi velocity. Thus μSR can determine the gap and its symmetry through the temperature dependence of the penetration depth. For $T = T_c$ the right-most integral in eqn 9.32 yields $-1/2$ (since $\Delta = 0$), and the superfluid density vanishes (Exercise 9.10). By inspection, the gap determines the temperature dependence of $\lambda(T)$, but $\lambda(0)$ is independent of Δ. We can write a useful simpler expressions for the isotropic s-wave gap of

$$\rho(t) = 1 - \frac{1}{2t}\int_0^{\infty}\frac{dx}{\cosh^2\left[\frac{\sqrt{x^2+\delta^2(t)}}{2t}\right]}, \tag{9.33}$$

with $\delta(T) = \Delta(T)/\Delta(0)$, and for the in-plane component of a d-wave gap:

$$\rho(t) = 1 - \frac{1}{2\pi t}\int_0^{2\pi}d\phi\int_0^{\infty}dx\frac{\cos^2\phi}{\cosh^2\left[\frac{\sqrt{x^2+\delta^2(t)\cos^2 2\phi}}{2t}\right]}. \tag{9.34}$$

These two functions, compared in Fig. 9.19, can be calculated numerically making use of eqn 9.31 (Exercise 9.11). Their low-temperature behaviour is qualitatively very different: the flat isotropic gap curve indicates excitations vanishing exponentially as $T \to 0$, whereas the presence of gap nodes allows excitations down to the lowest T, producing a linear slope in $\rho(T)$.

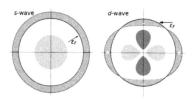

Fig. 9.18 Cooper-pair functions s-wave, left, d-wave, right (center) and corresponding gaps (shaded external areas).

[19] F. Groß-Altag *et al.*, Zeitschrift für Physik B: Condensed Matter **82**, 243 (1991), R. Prozorov *et al.*, Superconductor Science and Technology **19**, R41 (2006).

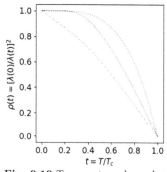

Fig. 9.19 Temperature dependence of the supercarrier density, corresponding to $\sigma_{FL}(T)/\sigma_{FL}(0)$, for s-wave (solid), d-wave (dashed) gaps, and two-fluid model (dotted).

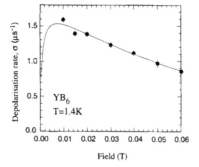

Fig. 9.20 Low temperature square-root second moment σ_{FL}/γ_{μ} vs applied field and its best fit by eqn 9.24, with a modified London model ($C(Q\xi) = \exp\left[-Q^2\xi^2(1-b)\right]$, $Q \to Q/\sqrt{1-b}$, $b = B/B_{c2}$), yielding $\xi = 33$ nm. [From A. D. Hillier and R. Cywinski, Appl. Magn. Reson. **13**, 95 (1997).]

The coherence length ξ (our second important length scale) may be determined by μSR by measuring the second moment $\sigma_{FL}^2/\gamma_{\mu}^2$ as a function of applied magnetic field. An example is shown in Fig. 9.20 where the reduction of σ can be extrapolated to zero to infer B_{c2} (and hence ξ, by eqn 9.17). The precise functional form is dictated by the expression derived in Exercise 9.6 (i.e. by the choice of cut-off function), but independently of these details σ_{FL} must vanish as B approaches B_{c2}.

9.7 Example materials

The history of superconductivity can be seen as the quest to achieve the highest critical temperature, which was extended in sudden jumps by the discovery of new classes of materials, as shown in Fig. 9.21. The timeline shows that, despite much effort, superconductivity remained a rather low-temperature phenomenon until the Nobel-prize-winning discovery of high T_c cuprates by Bednorz and Müller. The most recent record-breaking values of T_c have at last taken superconductivity up to room temperature, as shown in Fig. 9.21. However, this can only be achieved using extremely high pressures.

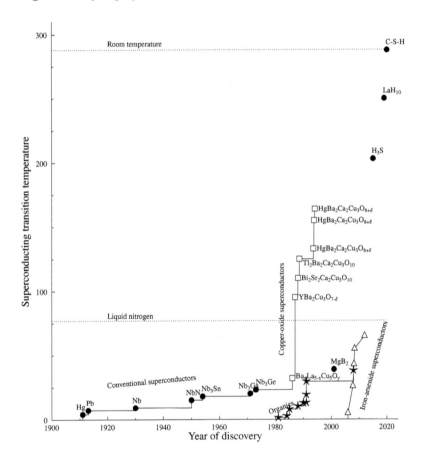

Fig. 9.21 Timeline of superconductivity from 1900 to 2020, with different symbols for different classes of materials. Notice the sudden increase following the discovery of high-T_c compounds in 1986.

Some very important features of the copper oxides, the materials that marked the transition towards high T_c, are very briefly summarized in Example 9.15.

Example 9.15

The original high-T_c cuprate oxide is $La_{2-x}Sr_xCuO_4$, the layered perovskite shown in Fig. 9.22. This material displays nearly two-dimensional metallic bands, with very small overlap along the c-axis. The figure shows that cation substitution (swapping Sr^{2+} for La^{3+}) controls a large variation in the electronic properties of the compound. This is due to the different valence of the two cations, that produces hole doping, like that of a p-type semiconductor (with $p = x$).

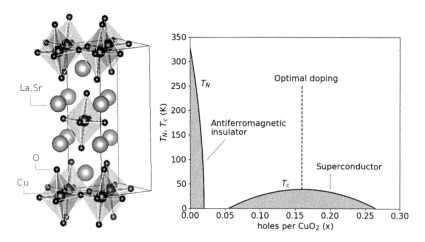

Fig. 9.22 Left: $La_{2-x}Sr_xCuO_4$ crystal structure; right: its phase diagram (ordering temperatures T_N and T_c) versus charge doping.

Low doping rapidly suppresses the Nèel temperature of the insulating *parent* antiferromagnet $LaCuO_4$, a charge-transfer insulator.[20] The superconducting phase appears for higher x, under a characteristic *dome*, with a maximum T_c at an optimal doping $x_o = 0.16$. The maximum distinguishes an *underdoped* and an *overdoped* superconducting regime, for $x < x_o$ and $x > x_o$, respectively.

The presence of a magnetic *parent* compound is just the first hint on the role of spin fluctuations in the pairing mechanism of cuprates, supported also by the d-wave symmetry, discussed in Example 9.14.

This is accompanied by a very rich variety of other phenomena, showing up in different regions of the phase diagram of Fig. 9.22, either in the pure material or under small non-doping substitutions of the chemical elements. Outstanding among them are charge and spin density wave instabilities, both independently and in the structurally linked form of stripes of alternating charge and spin order, first detected by neutron scattering [J. M. Tranquada *et al.*, Nature **375**, 561 (1995)]. Their observation signals the presence of many low-lying ordered states, with comparable total energies, competing to be the true ground state of the system, so that the balance may be easily tipped by a small perturbation. Several of them have been considered as candidates for the superconducting pairing mechanism.

A connected phenomenon is the nanoscopic **phase coexistence** of antiferromagnetism and superconductivity. This was originally observed[21] by ZF-μSR as a local spontaneous magnetic field B_μ of electronic origin in a fully superconducting nearly-optimal-doping sample of $La_{1.85}Sr_{0.15}CuO_4$. It is demonstrated in more detail for $YBa_2Cu_3O_{6+x}$ in Fig. 9.23.

[20]The prototypical metal-insulator transition is the Mott transition, induced by large on-site repulsion that prevents double occupation and opens a gap at half band filling. The cuprate variant is due to the fact that the ligand oxygen band partially overlaps with this gap.

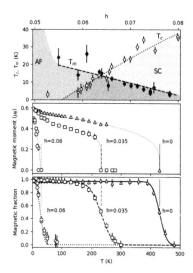

Fig. 9.23 Many $YBa_2Cu_3O_{6+x}$ compositions display both bulk superconducting and bulk magnetic transitions (top, T_c, T_m). All muons (a unit volume fraction, bottom) detect a local magnetic field of electronic origin, rescaled to a magnetic moment in the center panel. [Taken from F. Coneri *et al.*, Phys. Rev. B **81**, 104507 (2010).]

[21]A. Weidinger *et al.*, Phys. Rev. Lett. **62**, 102 (1989).

[22]This distance represents a typical scale for the domains of the two phases and is compatible with the extremely small coherence length ξ of the cuprates.

[23]There is no universally accepted definition of an 'unconventional superconductor'. The terms implies a different pairing mechanism to phonons, and/or a non s-wave gap. A simple theoretical definition could be $\sum_k \Delta_k = 0$ (see Fig. 9.18). For an experimental definition see Section 9.9.

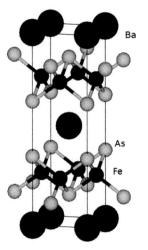

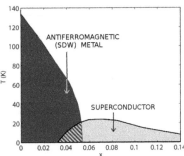

Fig. 9.24 Top: $BaFe_{2-x}Co_xAs_2$ crystal structure; bottom: its phase diagram (ordering temperatures T_N and T_c) versus Co content x.

Example 9.16

The electronic origin of B_μ is demonstrated by the value of the magnetic moment deduced from it, close to 0.6 μ_B, which is the $T = 0$ moment for Cu $3d^9$, $S = \frac{1}{2}$ holes in two dimensions (Fig. 9.23 central panel). The coexistence is nanoscopic when the fraction of muons experiencing B_μ, also called the *magnetic volume fraction* (bottom panel in the figure), approaches unity.

How can muons experience spontaneous internal fields, a signature of magnetic order, everywhere inside a bulk superconductor? There are two possibilities: either the two order parameters coexist at the atomic scale in a very exotic state, or else the sample is inhomogeneously separated into two phases. In both cases antiferromagnetism means that there is no net magnetic flux through the sample. However, even in the second case, the two phases must always be within a distance of order 1 nm from each other,[22] since, for larger distances than this, an increasing fraction of muons would experience a vanishingly small local field. The very few features we have described, including the last, are common to many cuprate families, among which $La_{2-x}Sr_xCuO_4$ ($T_c = 39$ K) and $YBa_2Cu_3O_{6+x}$ ($T_c = 92$ K) are the most thoroughly investigated.

Some other important breakthrough discoveries of new superconductors are summarized in Fig. 9.21. Different families of compounds mark the paradigm shift from type-I and type-II phonon-mediated superconductors to **unconventional superconductors**.[23] Among them, besides cuprates, are actinide-based heavy fermion systems, molecular and organic superconductors and **iron-based superconducting** pnictides and chalcogenides. The latter are discussed in the next example.

Example 9.17

Iron pnictides such as $SmFeAsO_{0.89}F_{0.11}$, $BaFe_{1.93}Co_{0.07}As_2$ (reaching $T_c = 55$ and 20 K, respectively), and chalcogenides such as FeSe (reaching $T_c = 30$-40 K under either high pressure or intercalation) are collectively known as iron-based superconductors. They represent another broad family of layered materials in which a transition metal element, Fe in this case, is tipped by either chemical (substitution) or physical means (pressure) from an antiferromagnetic state to a rather high T_c superconductivity. This is an important feature that iron-based superconductors share with cuprates, together with relatively large critical fields, hence relatively small coherence lengths. The similarity is seen in phase diagrams, such as that shown in Fig. 9.24. Nanoscopic coexistence of magnetism and superconductivity is detected in iron-based superconductors as well, both by muons and by NMR.

However, there are several notable differences between the iron-based superconductors and cuprates. The magnetic parent compounds of iron-based superconductors are metals rather than charge-transfer insulators. As a result, the antiferromagnetic state is a commensurate spin-density wave (Fig. 6.14). There are as many as five Fe bands crossing the Fermi surface, instead of just one Cu band in the cuprates. The electronic properties of iron-based superconductors are drastically modified by relatively moderate pressures. Finally, the symmetry of the gap is neither the plain BCS s-wave, nor the d-wave version, characteristic of the cuprates. Several of these features are directly or indirectly accessible via μSR experiments. Consequently, muons represent a very important experimental tool to address these unconventional superconductors.

Another conceptual breakthrough is represented by MgB_2, a simple compound, whose superconducting properties, with a relatively high

$T_c = 39$ K, were discovered in 2002 by Akimitsu and coworkers. This material demonstrated that the phonon-based BCS mechanism is not intrinsically limited to the low T_c values obtained before Müller and Bednorz's discovery of the cuprates.[24] A further reason for interest in MgB_2 is that it represents a textbook example of a multi-gap superconductor. Thanks to its simple nature and good sample quality, the presence of a two-component gap stands out in many experiments. However, before we describe this aspect, we need a brief introduction covering the extension of the BCS theory to a strong coupling regime, originally proposed by Eliashberg. This is necessary to understand the high T_c of MgB_2, which is not well represented by eqn 9.30.

[24]The record onset temperature for superconductivity, at the time of writing, is $T_c \approx 288$ K at $P \approx 268$ GPa for a C-H-S compound, via a phonon-coupled BCS mechanism (see Fig. 9.21).

[25]The parameters μ and λ introduced here should not be confused with the muon probe and the penetration depth.

Example 9.18

The adiabatic Eliashberg theory is an extension of BCS theory that goes beyond weak-coupling in the electron pseudopotential μ (see Example 9.14). In this extension we include details of the boson dispersion, represented by a realistic average, λ, of its q-dependent coupling.[25] The former is renormalized as $\mu^* = \mu/[1 + \mu \ln(\varepsilon_F/\hbar\omega_b)]$ to better treat the repulsive Coulomb interaction. The Eliashberg prediction replaces eqn 9.30 with

$$k_B T_c = \frac{\hbar\omega_b}{1.45} \exp\left[-\frac{1.04(1+\lambda)}{\lambda - \mu^*(1+0.62\lambda)}\right]. \quad (9.35)$$

This model may be extended to multi-band metals by means of multiple λ and μ^* parameters. MgB_2 is a layered hexagonal material whose cohesion is due to a network of three-dimensional π-bonds and two-dimensional σ-bonds. The two orbitals give rise to two distinct energy bands, both crossing the Fermi energy. The eight parameters λ_{ij} and μ^*_{ij}, with $i, j = \sigma, \pi$ are arranged into diagonal, intraband, and off-diagonal, interband couplings and pseudopotentials. The limiting case of vanishing interband couplings would result in independent superconductivity, with a distinct T_c for each of the two bands. Increasing interband couplings first yields a unique T_c and eventually, for large enough values, a single common gap. MgB_2 is an intermediate case: two energy gaps were observed, first of all by tunnelling spectroscopy. These gaps were found to vanish at a common T_c of 39 K.

Figure 9.25 shows how μSR data allow the identification of two gaps from a small kink in the low temperature dependence of $\sigma_{FL}(T)$. The feature is not very evident because the π-band density is quickly suppressed by field (see below, a fact not yet recognized at the time of the experiment, that was thus performed in non-optimal condition). The extracted values are $\Delta_\sigma = 6.0(1)$ meV and $\Delta_\pi = 2.6(2)$ meV.

The sharp drop of $\sigma_{FL}(B)$ with field is shown in the experimental data of Fig. 9.26. The disagreement with one-gap fits, like that of Fig. 9.20, is quite evident. The solid line is a fit to a phenomenological $\langle \Delta B^2 \rangle$ function, consisting of the sum of two weighted contributions, each one according to eqn 9.24, with two distinct values of $\xi_\pi = 23(1)$ nm and $\xi_\sigma = 5.1(2)$ nm. Incidentally, this implies that a temperature scan at lower fields ($B = 0.2 - 0.3$ T), where the π-component weight is larger, would produce a much more evident kink than that of Fig. 9.25 in the temperature dependence of $\sigma(T)$. Together, these two measurements show the potential of μSR for detecting multi-gap superconductivity. We

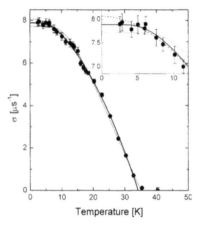

Fig. 9.25 Temperature dependence of σ_{FL} at $B = 0.6$ T in MgB_2, and best fit to one (dotted), two (solid) isotropic gaps, or two-fluid model (dashed). [Taken from Ch. Niedermayer *et al.*, Phys. Rev. B **65**, 094512 (2002).]

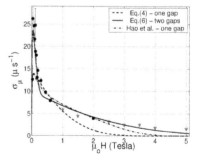

Fig. 9.26 Field dependence of σ_{FL} at $T/T_c = 0.05$ in MgB_2 with best fits to one-gap functions (dashed and dash-dotted lines, for two different cut-off function) compared to a two-gap function (solid line) [Taken from S. Serventi *et al.*, Phys. Rev. Lett. **93**, 217003 (2004).]

remark, however, that these features will be blurred for a superconductor featuring strong interband scattering.

9.8   Clean versus dirty

In the presence of impurities, a third length scale becomes important in superconductors, in addition to λ and ξ. This is the **mean-free path** ℓ. As in the case of normal metals, ℓ can greatly influence the transport and magnetic properties of superconductors. This issue was first investigated by Brian Pippard, who was originally dealing with type-I superconductors. In that case the fields acting on an electron at position r must be averaged over a volume, centred at r, of linear dimension ξ, which often turns out to be the largest of the three lengths for these type-I materials. As we remarked at the beginning of Section 9.4, this makes the treatment of electromagnetic fields in the superconductor (both the London model and the GL equations) very difficult to tackle analytically because of non-local electromagnetism. Pippard's treatment, however, becomes simple for type-II superconductors, for which the electromagnetic fields are local thermodynamic functions. In the following we restrict ourselves to this case.

Example 9.19

Pippard defined a *clean* coherence length $\xi_0 = a\,\hbar v_{\mathrm{F}}/k_{\mathrm{B}}T_{\mathrm{c}}$, with a of order unity, to be determined by comparison with experiment in pure materials. In the presence of electron scattering the resulting *dirty* coherence length is defined by

$$\frac{1}{\xi} = \frac{1}{\xi_0} + \frac{1}{\ell}. \tag{9.36}$$

This idea was introduced before both the GL and the BCS models were developed, and the expression suggested by Pippard has a striking similarity with that of the latter (see Note 18). Now we can identify ξ_0 with the $T=0$ BCS value, and assume the same relation as Pippard's for the mean-free path correction. Equation 9.36 may be straightforwardly inserted whenever the coherence length appears in an expression for the field distribution and its second moment (from eqn 9.19, with the cut-off functions described in Example 9.9), producing the equivalent **dirty superconductor** curves.

Material	B_{c2} (T)	ξ_0 (nm)
$YBa_2Cu_3O_{6.92}$	≈ 200	≈ 1
$LaFeAsO_{0.9}F_{0.1}$	≈ 100	≈ 1.8
K_3C_{60}	50	2.8
MgB_2	40	2.9

Table 9.1 Clean limit coherence length in a number of extreme type-II unconventional superconductors.

For type-II materials the deviation between ξ and ξ_0 may become significant for modest values of the GL paramater κ. A corresponding correction should be applied to the penetration depth, and therefore on the muon σ_{FL} parameter

$$\sigma_{\mathrm{FL}} \propto \frac{1}{\lambda^2} = \frac{\mu_0 e^2 n_{\mathrm{s}}}{m_{\mathrm{s}}}\left(1 + \frac{\xi_0}{l}\right)^{-1}. \tag{9.37}$$

Unconventional superconductors are generally extreme type-II superconductors, with $\kappa \gg 1/\sqrt{2}$. Indeed, the coherence length for these materials, as shown in Table 9.1, is as short as a few nanometers. A huge impurity density n_{imp} would be required for ξ to start deviating from ξ_0, according to eqn 9.36, since $\ell \propto n_{\text{imp}}^{-3}$. Hence these materials may be safely assumed to be in the clean limit.

Another important consequence of defects in superconductors may be explored by μSR, namely fluxon pinning. Fluxon-creeping motion produces residual resistivity and dissipation. Therefore pinning is essential for applications. Figure 9.27 shows the effect of pinning in the Fourier spectrum of the muon asymmetry in a superconductor, field-cooled (FC) in a field B_1 (top panel) and then measured again (bottom panel) after lowering the field to B_0. The sharp peak is due to muons implanted outside the sample, and its Larmor frequency is a measure of the applied field $\gamma_\mu B$ shifting from B_1 to B_0 in the two panels. The asymmetric fluxon spectrum remains centred at $\gamma_\mu B_1$ in both panels, demonstrating the persistence of the field-cooled magnetic flux, thanks to strong pinning in this $YBa_2Cu_3O_{6.95}$ sample.

9.9 The Uemura plot

Superconductivity, as the BCS theory shows in the weak-coupling limit, is a very general low-temperature instability of the metallic state. We have mentioned that different excitations may act as the coupling boson. These can range from phonons in conventional superconductors (including more recently discovered materials like fullerides, MgB_2, and high pressure H_3S phases) to spin fluctuations (as it is most likely the case in heavy fermions, cuprates, and pnictides) and possibly to other exotic boson excitations. In view of the many differences among these mechanisms, it is perhaps too ambitious to assume that a general phase diagram for superconductivity might be drawn. However, common trends have been sought incessantly, and μSR has provided one of the earliest successful examples of an experimental diagram, known as the **Uemura plot**. This diagram offers an insight into all superconducting materials. Figure 9.28 shows its original version, limited to 2D superconductors. Different compounds are represented by their T_c versus the value of the TF-μSR σ_{FL}. Materials with the smallest value of σ_{FL} line up along a striking linear relation: $T_c \propto n_s/m_s$.

The original idea of Uemura and coworkers is simple: the Fermi energy of a two-dimensional Fermi gas is precisely $\varepsilon_F \propto n_s/m_s$. The critical temperature in weak-coupling BCS theory is governed by eqn 9.30, where $T_c \propto \omega_b$, whereas for the high T_c materials plotted in Fig. 9.28, it appears to line-up along a line proportional to the Fermi energy scale ε_F of the metal.

Many versions of the Uemura plot exist, with additions by many authors. Several of them try to reinforce the original statement and a few attempt to refute it. Figure 9.29 shows the addition of more cuprate high T_c materials, indicating that the *underdoped* ones lie on the line, whereas the *overdoped* ones deviate downwards. More importantly, the plot includes a number of 3D metals, as opposed to the 2D cuprates.

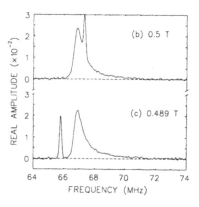

Fig. 9.27 Fourier amplitude in $YBa_2Cu_3O_{6.95}$, FC in $B_1 = 0.5$ T (top), then lowered to $B_0 = 0.489$ T (bottom). [Taken from J. Sonier et al., Phys. Rev. Lett. **72**, 744 (1994).]

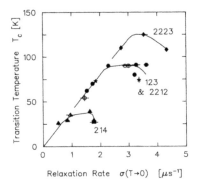

Fig. 9.28 The original Uemura plot: critical temperature T_c vs $\sigma_{\text{FL}}(0) \propto n_s/m_s$. [Taken from Y. J. Uemura et al., Phys. Rev. Lett. **62**, 2317 (1989).]

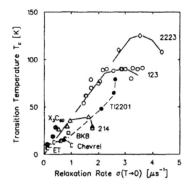

Fig. 9.29 A more populated Uemura plot. [Taken from Y. J. Uemura et al., Phys. Rev. Lett. **68**, 2665 (1992).]

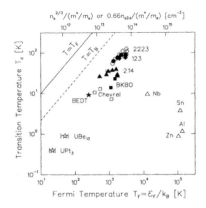

Fig. 9.30 A renewed version of the Uemura plot explicitly comparing T_c with the Fermi temperature of the normal metal. [Taken from Y. J. Uemura *et al.*, Phys. Rev. Lett. **68**, 2665 (1992).]

[26]Cooper pairs break up at T_c, whereas the hypothetical boson gas is also made of bosons above T_c (so-called preformed pairs).

They are plotted against their experimental σ_{FL} parameter, which is still proportional to n/m^*, but no longer to T_{F}.

For this reason the authors replotted the data in Fig. 9.30 against T_{F}. Since it is easier to obtain the Sommerfeld constant γ than the Fermi energy itself, the new plot is produced by extracting T_{F} from the muon relaxation σ_{FL}, as

$$T_{\mathrm{F}} = \frac{\hbar^2}{2m^*} \frac{(3\pi^2 n)^{\frac{2}{3}}}{k_{\mathrm{B}}} \propto \left(\frac{\sigma_{\mathrm{FL}}^3}{\gamma} \right)^{\frac{1}{4}}. \qquad (9.38)$$

The new figure still shows that the data lie below a hypothetical diagonal limit, which is parallel, and well below the diagonal line $T = T_{\mathrm{F}}$ (the superconducting gap cannot be larger than the Fermi energy itself). There is a second parallel line above the data, $T = T_{\mathrm{B}}$, where the temperature $T_{\mathrm{B}} \approx \hbar^2(n_{\mathrm{s}}/2)^{2/3}/k_{\mathrm{B}}(m_{\mathrm{s}}/2)$ represents the condensation temperature of an ideal three-dimensional boson gas with the same density and effective mass.[26] This plot therefore classifies superconductors according to how close they are to the limiting diagonal, which could be taken as an experimental definition of unconventional superconductivity. It shows further that all known superconductors are well separated from the Bose condensation regime, although some get close. The Uemura plot poses a challenge to our understanding of the boundaries of a universal superconducting phase diagram. It has established μSR as the technique of choice for this purpose.

Extension of the Uemura plot

Some examples are given next of work that extends and builds on the ideas introduced by the Uemura plot. First we look at the scaling found for iron-based superconductors and compare with the distinct scaling forms found for hole-doped and electron-doped cuprates. Next we take a look at the particular behaviour of molecular superconductors. Finally we consider a modified version of the Uemura scaling plot that brings the normal state conductivity into the mix.

Example 9.20

When placed on an Uemura plot, the iron-based superconductors are mostly found to lie close to the standard scaling line that is characteristic of the hole-doped cuprates (Fig. 9.31). One clear exception is LiFeAs, whose carrier density and T_c can be tuned by replacing some of the Fe with Ni or Co. These LiFeAs superconductors all have a stronger superfluid response than would be expected from their T_c values with the standard scaling relationship. Their properties are actually found to match well to a scaling line with a reduced slope, that was originally used to describe the electron-doped cuprates.

Both electron-doped cuprates and iron-based superconductors have a multiband character that may lie behind these differences in behaviour. At the time of writing the exact origin of these different scaling properties is still an open question and an ongoing area of research.

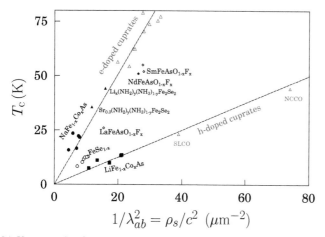

Fig. 9.31 Uemura plot for some iron based superconductors. [Adapted from M. J. Pitcher *et al.*, J. Am. Chem. Soc. **132**, 10467 (2010) and F. R. Foronda *et al.* Phys. Rev. B **92**, 134517 (2015).]

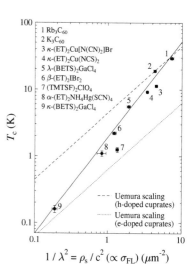

Fig. 9.32 Uemura plot for μSR data on molecular superconductors. [See F. L. Pratt and S. J. Blundell, Phys. Rev. Lett. **94**, 097006 (2005); T. Lancaster *et al.*, Phys. Rev. B **83**, 024504 (2011); F. L. Pratt *et al.*, Phys. Rev. Lett. **110**, 107005 (2013).]

Example 9.21

Organic and molecular superconductors are a group of materials that span a very wide range of T_c, whilst all having very similar carrier concentrations. Unlike high T_c cuprates, they do not rely on substitutional or vacancy doping to vary T_c, hence they are very clean materials that provide a good test bed for ideas about the intrinsic scaling of superconducting properties.

Collected μSR results are shown in Fig. 9.32 for nine examples of molecular superconductor. While the fullerides that are the highest T_c examples are seen to be consistent with the linear Uemura scaling law of the hole-doped cuprates, the lower T_c examples depart significantly from this linear scaling and follow a steeper scaling law that is close to $T_c \propto \sigma_{FL}^{3/2}$. This behaviour reflects the point that changes in T_c for the molecular superconductors are due to differences in electron interactions rather than changes in carrier concentration and T_F. It is notable that, as with the iron-based superconductors, all of the measurements lie within the boundaries set by the scaling lines for h-doped and e-doped cuprates. The ability of μSR to simultaneously measure T_c and σ_{FL} spanning a range from tens of mK to tens of K or more is a great benefit for this type of investigation.

Example 9.22

By considering the normal state conductivity in addition to T_c and λ^{-2}, Homes *et al.* were able to demonstrate a type of scaling that works, not just for underdoped samples, but also for optimally doped and overdoped samples. In this case the superfluid density is plotted on the vertical axis against the product of $\sigma_{DC}T_c$, where σ_{DC} is the normal state conductivity just above the transition An example plot is shown in Fig. 9.33.

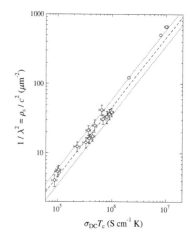

Fig. 9.33 Homes plot of the superfluid density versus the product of the normal state conductivity and the transition temperature. [Taken from C. C. Homes *et al.*, Nature **430**, 539 (2004).]

The basic idea behind the Homes scaling form is that the superfluid response at zero frequency is determined by the area of the normal state frequency-dependent conductivity spectrum below the gap. Assuming a Drude form for this conductivity and a scattering rate that is smaller than the gap and linear in T leads to the observed scaling. High-T_c cuprates are well known to have a T-linear scattering rate, ensuring that the Homes scaling is obeyed. On the other hand, molecular superconductors have scattering rate power laws in T that are typically between 1.5 and 2; the Homes scaling is not obeyed for this group of superconductors.[27]

[27]F. L. Pratt and S. J. Blundell, Phys. Rev. Lett. **94**, 097006 (2005).

9.10   Spontaneous fields

The sensitivity of the muon probe to small local fields has made it a valuable tool in the search for such fields, which are often predicted in unconventional superconductors. For some unconventional superconductors, it is possible for time-reversal symmetry (TRS) breaking to occur. This implies that there will be a spontaneous field associated with the superconductor. The conventional s-wave singlet BCS pairing conserves TRS, but if the pairing is triplet or has finite orbital angular momentum, then TRS breaking can occur for some particular types of pairing.

There is currently no accepted theory which predicts how large the spontaneous field should be, to what extent these spontaneous fields should be screened by supercurrents, whether these fields are particularly associated with defects, interfaces, and domain boundaries, or indeed whether the presence of the muon itself might play the role of a detect. Nevertheless, in the absence of time-reversal symmetry breaking a spontaneous field would not be expected, but if present then any spontaneous field which it produces can in principle be detected by a sensitive local probe such as ZF-μSR. A prime example of this is provided by Sr_2RuO_4.

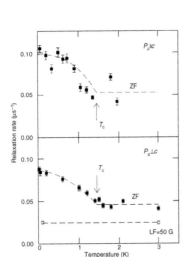

Fig. 9.34 ZF μSR in Sr_2RuO_4. [Taken from G. M. Luke *et al.*, Nature **394**, 558 (1998)].

[28]The T_c of Sr_2RuO_4 is highly sensitive to small concentrations of non-magnetic impurities, which was interpreted as being consistent with p-wave superconductivity.

Example 9.23

When Sr_2RuO_4 is cooled in zero field through its superconducting transition at $T_c = 1.4$ K, a significant increase in the relaxation rate is found (Fig. 9.34), which suggests the onset of a spontaneous local field at the transition. The effect is quenched in a small LF. Further μSR measurements showed that the effect tracks the differing T_c in different samples,[28] a fact that has subsequently been reproduced by many other groups.

Since the original work on Sr_2RuO_4, spontaneous fields have been observed with ZF-μSR in many other examples of unconventional superconductor. In particular, materials with non-centrosymmetric structures often provide the necessary conditions to enable these fields to be observed.

Chapter summary

- Muons measure the London penetration depth λ, the distance over which the superconductor screens an external magnetic field. This distance is proportional to a very important superconducting quantity, the density n_s of superconducting carriers.

- Low energy muons in TF geometry directly measure λ by detecting the decrease of the Larmor precession frequency at increasing depth from the superconductor surface.

- In type-II superconductors that support the Abrikosov flux lattice in the mixed state, the square root of the second moment of the precession frequency distribution at the muon site produced by the flux lattice, σ_{FL}, is proportional to λ^{-2}.

- Model fitting of the temperature dependence of $\sigma_{FL}(T)$ gives access to the number of superconducting gaps and their symmetry.

- The field dependence of the same parameter, $\sigma_{FL}(B)$, provides access to the coherence length, ξ.

- TF-μSR determines the position of every superconductor in the Uemura plot, a comprehensive attempt to classify superconductors and to identify the limit for the ratio between T_c and T_F.

- ZF-μSR allows spontaneous fields to be observed in the superconducting state. These can be produced by TRS-breaking superconducting pairing.

Further reading

- A compact introduction to superconductivity is given in J. F. Annett *Superfluidity, Superconductivity and Condensates*, OUP (2004).

- A classic reference on the subject is M. Tinkham *Introduction to Superconductivity*, McGraw-Hill (1996).

Exercises

(9.1) Obtain eqn 9.2 by using the curl of the curl identity (eqn 9.4) on the curl of Ampère's law.

(9.2) Obtain a vector potential $\boldsymbol{A}$ for the $\boldsymbol{B}$ field of the infinite slab (Example 9.3).

(9.3) Show that eqn 9.12 becomes London equation 9.1 for a homogenous, real order parameter $\psi = \psi_0$.

(9.4) Calculate the lattice parameter of the triangular and square flux line lattice. (*Hint:* calculate the magnetic flux.)

(9.5) Show, by direct substitution, that eqn 9.19 is the solution of eqn 9.18. (*Hint:* remember that $\sum_{\boldsymbol{R}} \delta^{(2)}(\boldsymbol{r} - \boldsymbol{R}) = \frac{1}{v_c} \sum_{\boldsymbol{Q}} e^{i\boldsymbol{Q}\cdot\boldsymbol{r}}$ and $v_c = |\boldsymbol{a} \times \boldsymbol{b}|$.)

(9.6) Show that the second moment of the flux lattice field distribution

$$B(r) = \frac{\Phi_0}{|a \times b|} \sum_Q \frac{C(Q\xi)}{1 + \lambda^2 Q^2} e^{iQ \cdot r} \qquad (9.39)$$

with the simple cut-off function $C(Q\xi)$, is that of eqn 9.24, recalling that

$$\int_{\text{cell}} dr\, e^{i(Q+Q') \cdot r} = v_c\, \delta_{Q+Q',0}\,, \qquad (9.40)$$

where $v_c = |a \times b|$ is the area of the unit cell of the flux lattice.

(9.7) Demonstrate that the first and second moment of the multi-Gaussian Fourier spectrum of eqn 9.21 corresponds to eqn 9.22.

(9.8) Write a routine that computes the space variation of the magnetic field $B_z(r)$ of a square lattice and plot it, as it is done in Fig. 9.11 for the triangular lattice. (*Hint:* choose a cut-off function, generate a reciprocal lattice of $N \times N$ sites and the corresponding direct cell, use a Fast Fourier Transform algorithm.)

(9.9) Write a routine that computes the field distribution $p(B)$ for the square lattice and plot it, as it is done in Fig. 9.13 for the triangular lattice. (*Hint:* use the result of the previous exercise, convolute the result with a Gaussian smearing function and histogram it.)

(9.10) Knowing that $\Delta(T_c) = 0$, demonstrate that $\rho(T_c) = 0$, using eqn 9.32.

(9.11) Write a routine that computes $\rho(t)$ for the *s*-wave BCS superconductor, using eqns 9.33 and 9.31. (*Hint:* calculate the integral with Simpson's quadrature rule, e.g. `scipy.integrate.quad` or matlab **quad** routines.)

Semiconductors and dielectrics

10

Semiconductors are among the materials with the biggest impact on everyday life, since they are the basis of most of modern technology. Examples of applications range from transistors in electronic devices through LEDs in optical displays to photovoltaics in solar panels. From the perspective of the contemporary physicist, the physics of semiconductors is perhaps one of the best advertisements for the discipline of physics, connecting fundamental contributions (both theoretical and experimental) with applied physics and the transfer of knowledge to technology.[1] The designation *semiconductors* arises from electrical transport properties, with electrical conductivities intermediate between those of metals and those of insulators. We recall that when submitting a wire of material with length L and cross-section A to a voltage V, an electrical current is produced, which is found to be proportional to the voltage V (Ohm's law)

$$V = RI, \tag{10.1}$$

where R is the electrical resistance. Assuming a constant electric field E along the wire ($V = EL$) and a constant current density J across the transverse section ($I = JA$), we may express Ohm's law in its local form

$$J = \sigma E, \tag{10.2}$$

where $\sigma = L/RA$ is the conductivity of the material (its inverse $\rho = 1/\sigma$ being the resistivity). Typical values of conductivities for metals are around $10^8 \, \Omega^{-1} \mathrm{m}^{-1}$, whereas the conductivity for an insulator such as SiO_2 can be 26 orders of magnitude lower at around $10^{-18} \, \Omega^{-1} \mathrm{m}^{-1}$. In order to account for these extreme variations, we recall that the current density J can be easily related to the (drift) velocity v of the electric charges and the respective concentration n by

$$J = -env, \tag{10.3}$$

where we are assuming, as occurs typically with metals, that the charge carriers are negatively charged electrons with charge $-e$. Ohm's law then clearly expresses the fact that the charge carriers acquire a constant (average) velocity which is simply proportional to the electrical force $-eE$ applied to them, implying the existence of a drag-like force counterbalancing the electrical force. The combination of eqns 10.2 and 10.3 allows the relation between the drift velocity and the applied electric

[1]This can be expressed by summarizing the contributions distinguished with the Nobel prize of physics that have occurred since the birth of this sub-discipline in the second half of the 20th century: transistor (1956), quantum electronics (1964), super-lattices (1973), disordered systems (1977), quantum Hall effect (1985), fractional quantum Hall effect (1998), integrated circuit and hetero-structures (2000), CCD camera (2009), graphene (2010), blue LEDs (2014).

field to be expressed as

$$v = -\frac{\sigma}{ev}E = -\mu E, \tag{10.4}$$

where μ is the mobility of the charge carriers.[2] The combination of eqns 10.2 and 10.4 allows us to express the conductivity as a function of the concentration of charge carriers and of the mobility by

$$\sigma = -en\mu. \tag{10.5}$$

The values of n and μ do not vary greatly between different metals, implying that the conductivities of metals are all very similar.[3] This is emphatically not the case for semiconductors and insulators, where conductivities are observed to vary by several orders of magnitude, not only between different materials, but even for different samples of the same material.[4] The understanding of this surprising behaviour only became possible after the development of the quantum theory of solids. Within quantum band theory, the energy levels of electrons in solids are grouped into quasi-continuous energy bands, separated by energy gaps. Metallic behaviour is characterized by partially filled bands that allow electrons to move upon application of an electric field. Semiconducting and insulating behaviour is characterized by fully occupied bands, so electrons need an energy of the order of the bandgap energy E_g (around 1 eV for silicon) to be promoted from the lower (full) valence band to the upper (empty) conduction band so that they can then contribute to the conductivity. This can only be achieved with stimulation by light with energy larger than the bandgap (photoconductivity) or by thermal excitation at high temperatures.[5] The significant room-temperature conductivities found in semiconductors reflect additional levels introduced into the bandgap by impurities and defects. These levels only require a much lower energy than the bandgap for their thermal excitation, either to the upper conduction band, creating **n-type conductivity**, or from the lower valence band, leaving an empty level in the valence band (known as a **hole**). In the latter case, holes can be modelled as quasi-particles with positive charge and are responsible for **p-type conductivity**.

[3]Silver and copper are very good conductors with $\sigma \approx 6 \times 10^7\,\Omega^{-1}\,\mathrm{m}^{-1}$ at room temperature. Lead is a poor conductor with $\sigma \approx 5 \times 10^6\,\Omega^{-1}\,\mathrm{m}^{-1}$, but the variation is not so great.

[4]This was the cause of great distress in the early days of the discipline, even leading Wolfgang Pauli to write to Rudolph Peierls that '*Über Halbleiter soll man nicht arbeiten, das ist eine Schweinerei; wer weiss, ob es uberhaupt Halbleiter gibt.*' ('*One shouldn't work on semiconductors, that is a filthy mess; who knows whether any semiconductors exist.*'), Letter to Peierls, September 29, 1931.

[5]To thermally excite across the energy gap, one needs $T \approx E_\mathrm{g}/k_\mathrm{B}$, where k_B is the Boltzmann constant.

Impurities and defects therefore play a major role in the physics of semiconductors. Various types of behaviour can be found, leading to multiple classifications. One of the most important parameters is the position of the defect level in the bandgap, leading to a distinction between **shallow** and **deep impurities**. In the case of shallow levels, the defect level is very close to the bottom of the conduction band (**shallow donor**) or to the top of the valence band (**shallow acceptor**), making them very effective for n-type or p-type doping, respectively. The electronic wave-function of the shallow defect can be adequately described by a superposition of conduction band or valence band states (for shallow donors or shallow acceptors, respectively) leading to hydrogen-like levels. The deep levels found closer to mid-gap are usually effective recombination centres (sometimes known as traps). Some impurities can present multiple configurations in the semiconductor: if both donor and acceptor configurations are possible, the impurity is designated *amphoteric* (for example the Si impurity in GaAs).

Usually, more than one impurity/defect may be present in the material. Two important cases must be distinguished. In **compensation**, impurities remain isolated from each other, but have opposite effects in the electrical conductivity (donor versus acceptor), whereas in **passivation** electrically active impurities react to form a complex without a defect level in the bandgap, therefore removing the electrical levels of the isolated impurities.[6]

[6]The reverse **activation** process can also happen, with the reaction of electrically inactive impurities to form an electrically active complex.

10.1 Ubiquitous hydrogen impurities

Hydrogen plays a particularly important role among all impurities. Protons can be incorporated in semiconductor materials and semiconductor devices in many of the fabrication steps, from initial growth conditions to ageing of fully developed devices. Hydrogen is thus omnipresent, sometimes achieving concentrations as high as 1%, which can strongly affect the properties of the material. A classic example is the passivation by hydrogen of dangling bonds in amorphous silicon, thus removing the corresponding trap levels from the bandgap and leading to more efficient performance of the corresponding photovoltaic devices. It is also notable that the identification of the role of hydrogen in the passivation of Mg acceptors in GaN was a crucial step in the development of blue LEDs.[7]

[7]This was the subject of the Nobel Prize in Physics in 2014, awarded to I. Akasaki, H. Amano, and S. Nakamura.

Understanding hydrogen as an impurity is therefore a particularly important topic within the physics of semiconductors. From the fundamental point of view, an essential step for the understanding of hydrogen in semiconductors is the microscopic description both of the isolated impurity and of its possible interaction with other impurities and defects. This is the arena where muon spin spectroscopy has provided its most successful contributions to the physics of semiconductors. Hydrogen is an experimentally elusive atom and high concentrations are typically required for detection; studying muonium as a hydrogen analogue has turned out to be a very fruitful approach.

Apart from μSR, the use of experimental techniques sensitive to isolated hydrogen itself is limited to a couple of systems where hydrogen is usually present in high concentrations (notably ZnO and TiO_2). Microscopic information about isolated hydrogen configurations, electronic structure, and electronic levels is even more difficult to obtain. Most of the existing information about isolated hydrogen in semiconductors and dielectric insulators has been therefore been derived from μSR, using muonium as a light pseudo-isotope of hydrogen.[8]

[8]See Chapter 4.

A comparison of the atomic properties of muonium and of hydrogen can be found in Table 3.1. As explained in Chapter 3, since $m_\mu \approx 207 m_e$, the reduced mass of muonium is basically the same as that of hydrogen, leading to nearly identical electronic properties. The nuclear magnetism is however very different and this leads to the hyperfine interaction of muonium being more than three times larger than that of hydrogen.

Although the main electronic properties of atomic muonium are expected to be nearly the same as those of atomic hydrogen, there are isotopic effects for the dynamical properties that are related to the difference in the total mass of these atoms (muonium being approximately nine times lighter than hydrogen).[9] The contribution of the mass difference to the static electronic properties, if any, is limited to zero-point effects.[10] Of course, the electronic state of atomic muonium (or hydrogen) inside a solid is not expected to be the same as in the vacuum, and so we can also expect the isotropic hyperfine constant to change.

10.2 Muonium

The formation of muonium in solids was first proposed on the basis of experiments on silicon with a magnetic field applied parallel to the initial muon spin polarization (longitudinal geometry). Whereas in zero-field the initial state $| \uparrow^\mu \uparrow^e \rangle$ is an eigenstate of the Hamiltonian of isotropic muonium,[11] the other possible initial state $| \uparrow^\mu \downarrow^e \rangle$ is not and leads to a component of the spin polarization varying at a frequency corresponding to the hyperfine constant, which is usually too high to be observable. Therefore, only 50% of the initial spin polarization is observed. However, the application of a magnetic field parallel to the muon spin polarization leads, in the high-field limit, to the complete decoupling of the hyperfine interaction and recovery of the full spin polarization.[12] Although spectroscopic measurements provide the most accurate measurement of the hyperfine coupling, this repolarization method remains a very valuable technique, particularly when no direct spectroscopic measurements are possible. In general, the muonium state is most directly characterized in transverse-field spectroscopy measurements. Figure 10.1 shows the observable transition frequencies of isotropic muonium in vacuum as a function of applied transverse field, and the corresponding precession amplitudes. Three different regimes can be distinguished:

- For **low fields** (and for hyperfine interactions close to the value in vacuum) the transition frequencies ν_{14} and ν_{34} (amounting to 50% of the total muonium fraction) are usually unobservable and the muonium signal is limited to an oscillation frequency of about 13.9 GHz/T (i.e. 1.39 MHz/G) corresponding to the ν_{12} and ν_{23} transitions (see Example 4.11.)

- At **intermediate fields**, ν_{12} becomes the most easily observable signature of muonium (together with ν_{34}). Around the 'magic' field of order 1 T, both these frequencies become approximately field-independent.

- In the **high-field regime** (which for vacuum-like hyperfine interactions would only be attained at fields of order 100 T) both ν_{12} and ν_{34} converge towards the Larmor frequency $\gamma_\mu B$ of the muon (see Example 4.11) and the frequency difference between them simply corresponds to the hyperfine interaction (see eqn 4.65).[13] As we shall see below, this high-field regime is particularly important

---

[9]See Section 8.5.

[10]When in an equilibrium position, an atom has the zero-point energy $E_0 = \hbar\omega/2$ characteristic of the quantum oscillator, where the frequency decreases with the mass as $\omega \propto 1/\sqrt{m}$. This zero-point energy is different for muonium and hydrogen, and may affect the overall electronic wave function (and consequently its energy levels and hyperfine interaction).

[11]The Hamiltonian for isotropic Mu was presented in eqn 4.45.

[12]The corresponding curve of the polarization as a function of the applied longitudinal field is usually known as a repolarization curve (see eqn 4.62 and Fig. 4.10) and was the experimental signature leading to the first findings of muonium in solids.

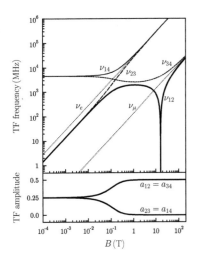

Fig. 10.1 Transition frequencies of isotropic muonium in vacuum as a function of applied transverse field B. The lower panel shows the precession amplitudes.

[13]Whether one needs to take the difference or the sum depends on whether one is below or above the level crossing between levels 1 and 2 because ν_{12} changes sign (see Example 4.10.)

for the identification of shallow-donor states, where the hyperfine interaction is four orders of magnitude lower than in vacuum, so that the high-field regime is already reached at the extremely low field of 10 mT.

10.3   Silicon: the foundations

The formation of muonium in silicon was first proposed in longitudinal-field repolarization experiments,[14] which revealed the presence of a muonium state with a hyperfine interaction about $0.45 \times A_{\text{vac}}$. Bearing this in mind, the first spectroscopic measurements[15] were very puzzling. These historical results are shown in Fig. 10.2, where the frequency spectra are shown for quartz at room temperature and for a p-type silicon sample at $T = 77$ K, both for an applied magnetic field $B = 10$ mT. The interpretation of the quartz spectrum is straightforward, the lower frequency at around 1.36 MHz corresponding to the Larmor frequency and the pair of frequencies centred around 139 MHz corresponding to the Mu frequencies ν_{12} and ν_{23}. For silicon, however, an additional pair of frequencies is prominent around 44 MHz and was then assigned to an 'anomalous' muonium state, initially given the label Mu*.

Further investigations helped to clarify the nature of this anomalous muonium in silicon. An important finding was that this muonium state corresponded to a distorted electronic cloud around the muon, leading to an anisotropic hyperfine interaction, which is axially symmetric around the [111] axis.[16] The corresponding spin Hamiltonian now becomes

$$\hat{\mathcal{H}} = h\boldsymbol{S}_\mu \cdot \underline{\boldsymbol{A}} \cdot \boldsymbol{S}_\text{e} - \boldsymbol{\mu}_\mu \cdot \boldsymbol{B} - \boldsymbol{\mu}_\text{e} \cdot \boldsymbol{B}, \qquad (10.6)$$

where $\boldsymbol{\mu}_\mu$ and $\boldsymbol{\mu}_\text{e}$ represent the magnetic moments of the muon and of the electron, respectively,[17] and the hyperfine interaction is now described by a second rank tensor which can be written as

$$\underline{\boldsymbol{A}} = \begin{pmatrix} A_\perp & 0 & 0 \\ 0 & A_\perp & 0 \\ 0 & 0 & A_\parallel \end{pmatrix}, \qquad (10.7)$$

as described in Section 4.4.[18] Although it is possible to express analytically the energy eigenvalues and eigenvectors of the Hamiltonian in eqn 10.6 for some important particular cases, the general solution can only be obtained numerically. The precession amplitudes and frequencies are now dependent not only on the applied magnetic field $\boldsymbol{B}$, but also on the angle θ between $\boldsymbol{B}$ and the symmetry axis of the hyperfine centre, leading to an effective angle-dependent hyperfine interaction

$$A(\theta) = A_{\text{iso}} + \frac{D}{2}\left(3\cos^2\theta - 1\right). \qquad (10.8)$$

The interpretation of the anomalous muonium state represented a big challenge. S. F. J. Cox and M. C. R. Symons finally proposed that it corresponded to hydrogen sitting in the middle of a silicon-silicon bond

[14]D. G. Andrianov *et al.*, Soviet Physics JETP **31**, 1019 (1970).

[15]J. H. Brewer *et al.*, Phys. Rev. Lett. **31**, 143 (1973).

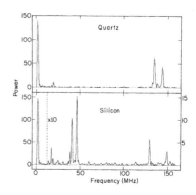

Fig. 10.2 TF-μSR frequency spectra for quartz at room temperature and that for a p-type silicon sample at $T = 77$ K, both for an applied magnetic field $B = 10$ mT. [Taken from J. H. Brewer *et al.*, Phys. Rev. Lett. **31**, 143 (1973).]

[16]B. D. Patterson *et al.*, Phys. Rev. Lett. **40**, 1347 (1978).

[17]See Section 4.4.

[18]Recall from Section 4.4 that $A_\perp$ perpendicular and $A_\parallel$ can be expressed in terms of an isotropic component A_{iso} and a dipolar component D as $A_\perp = A_{\text{iso}} - D/2$ and $A_\parallel = A_{\text{iso}} + 2D$.

[19]S. F. J. Cox and M. C. R. Symons, Chem. Phys. Lett. **126**, 516 (1986).

[20]R. F. Kiefl *et al.*, Phys. Rev. Lett. **60**, 224 (1988).

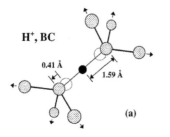

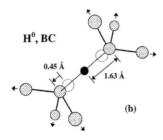

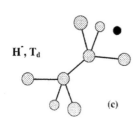

Fig. 10.3 Schematic illustration of the minimum-energy atomic configurations for hydrogen interstitials in silicon, as derived from first-principles calculations: (a) the positive charge state, (b) neutral, and (c) the negative charge state. [Taken from C. Herring *et al.*, Phys. Rev. B **64**, 125209 (2001).]

[21]See Chapter 3.

[22]See Chapter 3.

[23]Moreover, it has also been suggested that some fast-relaxing diamagnetic lines may correspond to transient fast-fluctuating paramagnetic states. The discussion of this is however outside the scope of the present introductory chapter. See R. C. Vilão *et al.*, Phys. Rev. B **96**, 195205 (2017), for more details.

(a bond-centre configuration),[19] which was confirmed by R. F. Kiefl *et al.*[20] via μSR measurements of the hyperfine structure of the isotope ^{29}Si. This represented a major breakthrough in the physics of hydrogen in semiconductors, since the bond-centre configuration had not been anticipated by any other theoretical or experimental results.

Basic configurations and metastability

Coming back to Fig. 10.2, we immediately recognize that it informs us of the presence, at $T = 77$ K, of at least three distinct muonium/hydrogen configurations in Si:

- **Mu_T^0**, which is a neutral (paramagnetic) configuration with a vacuum-like hyperfine interaction corresponding to atom-like 'normal' isotropic muonium; this atom-like configuration (as well as its negatively charged version) is stable in the highly symmetrical tetrahedral site in the silicon crystallographic structure, as depicted in its negatively charged form in Fig. 10.3(c); it thus corresponds to a (deep) acceptor configuration.

- **Mu_{BC}^0**, which is a neutral (paramagnetic) configuration with a much smaller hyperfine interaction, which is highly anisotropic, being distorted by bonding. This state corresponds to the anomalous bond-centre muonium represented in Fig. 10.3(b); it thus corresponds to a (deep) donor configuration.

- **Mu_{BC}^+**, which is a charged (diamagnetic) configuration corresponding to the ionized version of the bound bond-centre muonium, as represented in Fig. 10.3(a).

The observation of multiple configurations (including two different neutral configurations corresponding to the neutral donor and to the neutral acceptor) is a common signature of the μSR method. Since muons are typically implanted with 4 MeV energy, in the final stage of the rapid thermalization process[21] they can probe several configurations, including metastable ones with energy higher than the ground state. The ability of the μSR method to access metastable configurations gives it a unique advantage over other techniques, providing a wide overview of the most important configurations that may play a role in the dynamics of hydrogen inside the material.

The negatively charged diamagnetic component Mu$^-$

A more severe problem is that of the interpretation of the diamagnetic component.[22] Both μ^+ (or better, Mu$^+$) and Mu$^-$ appear as a diamagnetic signal and are difficult (i.e. usually impossible) to distinguish directly.[23] The usual argument is that the prompt formation of Mu$^-$ is highly unlikely in the thermalization process, since it requires the capture of two electrons by the muon. It is thought, however, that the formation of Mu$^-$ can be forced in highly doped n-type materials (at temperatures above the ionization of the donors). It was this line of thought, together with the accumulation of two decades of experimental data pointing to the presence of an additional diamagnetic fraction, that finally led to the proposal of Mu^- in heavily doped n-type Si samples.[24]

[24]B. Hitti *et al.*, Phys. Rev. B **59**, 4918 (1999).

Donor and acceptor levels

The presence of several interconverting configurations required the use of radio-frequency μSR techniques[25] to determine accurately the fraction of muons forming each configuration, from which it was possible to extract, by examining the variation of the fractions with temperature, the activation energies associated with the transitions and establish the position of the muonium donor and acceptor levels.[26]

[25]See Chapter 19.

The proposed donor level $E_D = E(0/+)$ has been determined at 0.21 eV below the bottom of the conduction band and the proposed acceptor level $E_A = E(-/0)$ has been determined at 0.56 eV above the top of the valence band.

[26]A detailed model relating the assigned configurations and the respective dynamics has also been proposed, which exceeds the scope of this chapter (see B. Hitti *et al.*, *op. cit.*).

Insights from first-principle calculations

In order to rationalize these level positions, it is useful to consider the way these levels are calculated using first-principle methods. These methods usually proceed to a computation of the formation energy $E_{form}(H^q)$ of a given charged state q of hydrogen inside a supercell of the material by taking into account that this formation energy is dependent on the abundance of the additional species (the hydrogen atom H^0 and the additional electron contributing to the charge q). This abundance is dependent, in thermodynamical equilibrium, on the chemical potential for each of these species (μ_H for the hydrogen atom and E_F for the electron).[27] The formation energy $E_{form}(H^q)$ of H^q is then simply computed as the difference between the total energy $E_{tot}(H^q)$ of the supercell with H^q and the total energy $E_{tot}(bulk)$ of the neutral supercell without H, corrected by the chemical potentials μ_H and E_F, so that

[27]In statistical thermodynamics, the chemical potential represents the average change in free energy of the system per particle, for a given species. The chemical potential of electrons is usually designated the Fermi level E_F.

$$E_{form}(H^q) = E_{tot}(H^q) - E_{tot}(bulk) - \mu_H + qE_F. \quad (10.9)$$

The resulting calculations for the different charge states in Si are shown in Fig. 10.4, where the neutral configuration H^0 shown is the ground-state bond-centred configuration. The formation energy of H^0 is computed after subtracting the free energy associated with the introduction of a hydrogen atom in the system (the chemical potential of H, which is typically taken as half the energy of the H_2 molecule at $T = 0$ K). The formation energy of H^0 thus obtained is not expected to depend on the chemical potential of the electrons (the Fermi level), since the additional energy associated with the electron brought in by the H atom is taken into account by μ_H.

If an additional electron is introduced in the supercell, making it negatively charged, the formation energy of the negatively charged H^- configuration is computed similarly, but we now have to take into account that an additional electron has been placed in the system and subtract

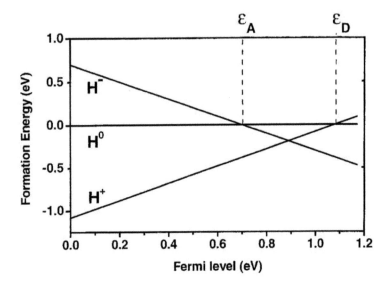

Fig. 10.4 First-principles results for relative formation energies for the different charge states of hydrogen interstitials in Si shown in Fig. 10.3, as a function of Fermi level E_F, with $E_F = 0$ corresponding to the top of the valence band. From C. Herring *et al., op. cit.*

the corresponding energy as well from the total energy of the negatively charged supercell, in order to obtain the correct formation energy. The free energy of the additional electron is the chemical potential and is thus larger for n-type samples than for p-type samples. The corresponding formation energy of H^- is thus much smaller for n-type samples than for p-type samples (the line with negative slope in the diagram). Conversely, the positively charged configuration implies the removal of an electron, implying that the free energy of the removed electron must be added to the total formation energy. This leads to a formation energy of the H^+ configuration that is larger in n-type samples than in p-type samples (the line with positive slope in the diagram). Several important aspects of the diagram shown in Fig. 10.4 can be noted:

- The value of E_F for which the formation energies H^+ and H^0 are the same defines the $E(+/0)$ transition point corresponding to the donor level E_D.

- Similarly, the value of E_F for which the formation energies of H^- and of H^0 are the same defines the $E(0/-)$ transition point corresponding to the acceptor level E_A.

- The position of the donor level E_D is *higher* in energy than E_A, contradicting a naïve expectation from pure atomic physics that the second electron should be less bound than the first one due to electron repulsion; this line of reasoning forgets that the corresponding configurations are very different, implying a substantial contribution to the effective electron-electron interaction U from the lattice itself. The lowering of the acceptor level with respect to the donor level thus implies that it pays energetically (from the lattice point of view) to accommodate the negatively charged configuration, the lattice contribution to the total formation energy being strongly negative (the result is called a negative-U state). In this situation, the neutral configuration does not correspond to a thermodynamically stable state for any value of E_{F}.

- The important **pinning level** $E_{\mathrm{P}} = E(+/-)$ is defined by the value of E_{F} for which the formation energies H^+ and H^- are the same. This level has been proposed to have a universal value, on an electrochemical scale, for a wide class of materials.[28]

[28]C. G. Van de Walle and J. Neugebauer, Nature **423**, 626 (2003).

- The position of the pinning-level in the middle of the bandgap implies that the positively charged configuration is the stable one (with lower formation energy) for p-type materials and that the negatively charged configuration is the stable one for n-type materials. Isolated hydrogen behaves as a donor in p-type materials and as an acceptor in n-type materials. It thus presents amphoteric behaviour in Si. It also behaves as a compensating impurity, acting to reduce the dominant contribution to the conductivity.

10.4   Shallow donor states

The basic behaviour described above for silicon has also been found for the most important elemental semiconductors (Ge and diamond), as well as for GaAs, one of the most prominent III-V compounds. The discovery in CdS (a II-VI compound) of a muonium state with a hyperfine interaction amounting only to approximately 10^{-4} of the vacuum value represented a major breakthrough. In Fig. 10.5 the Fourier transforms of these spectra are shown, revealing the presence of the ν_{12} and ν_{34} lines of a muonium centre around the diamagnetic line, in a dramatic demonstration of the Paschen-Back effect (i.e. the high field regime of Fig. 10.1) being found at the low field of 10 mT. The shallow state ionizes around $T = 20$ K, with an ionization energy of just 26 meV.

This discovery was followed by the theoretical prediction that the hydrogen was a negative-U centre in ZnO, with a pinning level E_{p} lying above the bottom of the conduction band, meaning that the donor configuration should be the stable one in all cases. This implied that hydrogen could act as a possible source of n-type conductivity in ZnO (instead of the usual compensating behaviour). It also implied shallow-donor behaviour, which was quickly confirmed by μSR measurements in ZnO, giving results similar to CdS. The above-mentioned theoretical prediction that E_{p} is universal on an electrochemical scale allowed a clear separation between materials where hydrogen is expected to behave as a compensating impurity (E_{p} is located inside the gap) from those where it is expected to behave as a shallow donor (E_{p} lies above the bottom of the conduction band) or as a shallow acceptor (E_{p} lies below the top

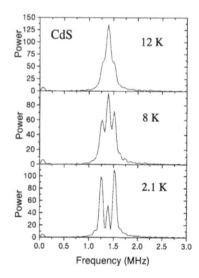

Fig. 10.5 Fourier transforms of μSR spectra for undoped CdS at $B = 10$ mT and three different temperatures. [Taken from J. M. Gil *et al.*, Phys. Rev. Lett. **83**, 5294 (1999).]

of the valence band). This prompted a review of existing results in order to check alignment of E_p as determined from experimental results, as well as a large experimental program in order to find the predicted shallow-donor behaviour.

10.5   Related techniques

Isolated hydrogen is an elusive species that is very hard to investigate experimentally, taking into account that most methods require relatively high concentrations for detection. The most significant examples are the donor and energy levels in silicon, which have been the object of particularly careful determination using macroscopic methods, and the hyperfine interaction in Si, ZnO, and TiO$_2$, which have been determined for hydrogen using Electron Spin Resonance (ESR).[29] Some examples of these complementary experimental studies are given here, starting with the donor and acceptor levels in silicon:

[29] ESR is also known as Electron Paramagnetic Resonance (or EPR).

Example 10.1

The energy levels of isolated hydrogen donor and acceptor levels in silicon have been measured by time-resolved capacitance-transient measurements in Schottky diodes under changes of bias, both in experiments with implanted protons[30] and in experiments based on the dissociation of hydrogen–containing complexes.[31] The positions of the (bond-centre) donor E_D and (tetrahedral) acceptor E_A energy levels below the bottom of the conduction band, for isolated hydrogen in silicon, are summarized in Table 10.1, together with the results from first-principle calculations mentioned above. The slight differences with respect to μSR are expected from differences in the proton and muon zero-point motion.

[30] K. Bonde Nielsen *et al.*, Phys. Rev. B **65**, 075205 (2002).

[31] C. Herring *et al.*, *op. cit.*

Method	E_D	E_A
Muon implantation	0.21 eV	< 0.56 eV
Proton implantation	0.16 eV	0.65 eV
Dissociation of P-H complexes	0.16 eV	0.67 eV
First-principles calculations	0.2 eV	0.6 eV

Table 10.1 Position of the (isolated) hydrogen donor E_D and acceptor E_A energy conversion level below the bottom of the conduction band, as determined by several experimental methods and by first-principles calculations.

Complementary studies have also been made using magnetic resonance techniques such as ESR and ENDOR to measure hyperfine parameters:

Example 10.2

The electron spin resonance technique probes the electronic spin transitions in the presence of an applied magnetic field and is therefore able to perform a microscopic evaluation of paramagnetic defects in solids, including the measurement of the hyperfine interaction. A complementary technique is Electron-Nucleus DOuble Resonance (ENDOR), where the spin transitions of the interacting nucleus are directly probed by appropriate excitation with electromagnetic radiation. These measurements require that the investigated defect is relatively abundant, so the characterization of hydrogen defects is limited to a few relevant cases, namely Si (the ground-state BC configuration)[32], ZnO (the shallow donor configuration), and TiO_2 (a peculiar example where hydrogen binds to an oxygen, but the corresponding electron is trapped by a Ti^{4+} ion, reducing it to Ti^{3+} and forming a polaron configuration).[33] Both for Si and for TiO_2 the measurements reveal that the hyperfine interaction obtained for muonium with μSR and that obtained for hydrogen using ESR techniques are remarkably similar, after taking into account the ratio of the magnetic moment of the muon μ_μ and of the proton μ_N ($\mu_\mu/\mu_N = 3.183$). For ZnO, however, these values differ by about one order of magnitude, suggesting that μSR and ESR/ENDOR are not probing the same configuration.

[32] Yu. V. Gorelkinskii and N. N. Nevinnyi, Physica B **170**, 155 (1991).

[33] R. C. Vilão *et al.*, Phys. Rev. B **92**, 081202(R) (2015).

Although the μSR studies of isolated hydrogen impurities are very important, the contribution of μSR to the physics of semiconductors is not just restricted to these isolated hydrogen impurities. We briefly present three important examples of μSR studying other aspects of semiconductors, starting with the factors affecting the passivation of impurities and defects:

Example 10.3

The motional properties of the several charge states has also been investigated together with the characterization of the electronic configurations. These results[34] allow an extension of isotopic studies in hydrogen diffusion research to the light muonium pseudo-isotope. Moreover, the onset of diffusion allows the muon to probe a large volume of the solid, thus interacting with defects and impurities (usually present in concentrations low-enough to make little probable interaction during the implantation stage). Reaction of the muon with these impurities and defects usually corresponds to the respective passivation processes (for example, the passivation of the Zn impurity in GaAs or of the Cu vacancy in $CuInSe_2$).

[34] See Chapter 8.

The technique of photo-excited μSR (Chapter 19) is a method that can be used for measuring carrier recombination in semiconductors:

Example 10.4

The presence of the energy bandgap in semiconductors is responsible for their important optical properties, with the possibility of absorbing or emitting light at the corresponding wavelength to excite across the gap (in the visible region in many cases). The investigation of the recombination mechanisms of the photo-excited carriers has always played a major role in semiconductor research. The possibility of measuring the μSR signal in photo-excited samples has been pursued for many years and a dedicated laser facility is now available at ISIS, where the possibility of measuring excess carrier lifetimes in semiconductors has been demonstrated.[35]

[35]See Chapter 19.

The final example given here is the use of the muon as a probe of interface layers in semiconductors:

Example 10.5

The stopping range of the usual 4 MeV muons is several hundreds of micrometers, preventing the realization of spatially resolved investigation in the nanometer range, necessary for addressing all-important phenomena occurring in semiconductor interfaces (which are the basis of most semiconductor applications). The development of slow-muon beams with depth-resolution around tens of nanometers has therefore opened up the possibility of using muons as a microscopic probe of these important interfaces (for which microscopic information is usually lacking and/or not very informative). A dedicated low energy muon user facility is available at PSI.[36] The use of slow muons in semiconductor interfaces is still in its infancy and the nature of the obtained information is the matter of much debate, taking into account the multiple charged configurations accessible in principle to the muon, as well as the effects of the implantation stage. However, it has already been possible to demonstrate the interaction of muons with defect regions in the interfaces and obtain profound physical insight not accessible by any other technique.[37]

[36]See Chapter 18.

[37]H. V. Alberto *et al.*, Phys. Rev. Materials **2**, 025402 (2018).

Chapter summary

- Hydrogen is an important impurity in semiconductors and muonium can be used as a light pseudo-isotope of hydrogen. The spectroscopy of this state can give information about the hyperfine coupling and dynamics of this impurity.

- In silicon there are three types of muon state: a neutral, paramagnetic, isotropic state at the tetrahedral site (Mu^0_T), a neutral, paramagnetic, anisotropic state at the bond-centre site (Mu^0_{BC}), and a charged, diamagnetic, ionized bond-centre state (Mu^+_{BC}).

- Shallow donor states have been found in various materials such as CdS in which the hyperfine interaction is only around 10^{-4} of the value for muonium in vacuum.

Further reading

- B. D. Patterson, Rev. Mod. Phys. **60**, 69 (1988).
- K. H. Chow, B. Hitti, and R. F. Kiefl, in *Identification of Defects in Semiconductors, Semiconductors and Semimetals* Vol. 51A, edited by M. Stavola, Academic Press (1998), pp. 137-207, treatise edited by R. K. Willardson and E. R. Weber.
- S. F. J. Cox, Rep. Prog. Phys. **72**, 116501 (2009).
- S. F. J. Cox, R. L. Lichti, J. S. Lord, E. A. Davis, R. C. Vilão, J. M. Gil, T. D. Veal, and Y. G. Celebi, Physica Scripta **88**, 068503 (2013).
- P. W. Mengyan, in *Characterisation and control of defects in semiconductors*, edited by F. Tuomisto, IET (2019).

Exercises

(10.1) Using the data in Fig. 10.5, estimate the hyperfine coupling in CdS. Identify all the lines in the frequency spectrum.

(10.2) Using the data in Fig. 10.2:
i) estimate the hyperfine coupling for the isotropic muonium state in quartz and in silicon;
ii) estimate an upper bound for the isotropic hyperfine coupling for the anomalous muonium state in silicon;
iii) identify all of the lines in the frequency spectra.

11

Ionic motion

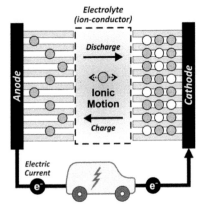

Fig. 11.1 Schematic view for the structure and general operating principle of an electrochemical device (e.g. rechargeable battery or fuel cell) that efficiently converts chemical energy into electrical energy.

An important scientific problem for modern society to solve is how to harvest and store clean energy. Here it is important to remember the first law of thermodynamics, i.e. that *'energy can be transformed from one form to another, but can be neither created nor destroyed'*. Hence, for all types of energy device and on the atomic scale inside an energy material, we are generally dealing with energy *conversion* that occurs through different types of *dynamic* processes. Two well-known energy-conversion devices are rechargeable batteries and fuel cells, which are both electrochemical devices that directly and efficiently transform chemical energy into electrical energy. The general structure and functionality of both devices are very similar. As shown in Fig. 11.1, they consist of a cathode/electrolyte/anode stack where the energy conversion occurs via a motion of charged ions through the electrolyte and a corresponding flow of electrons in the outer circuit. In order to improve performance, lifetime, safety, and sustainability of these devices, a new generation of energy materials needs to be developed. Here one of the main obstacles is the understanding, control, and enhancement of the ion diffusion both inside and between the individual components/materials. A key to unlocking a paradigm shift in this field is to develop and utilize state-of-the-art experimental techniques to study ion dynamics inside the materials on the atomic length scale.

11.1 Why use muons?

During recent decades, energy devices and materials have gone through substantial developments and improvements in performance. For non-specialists, this is most visible within the area of rechargeable batteries that we use in our daily life (computers, smart phones, electric cars, etc.). Up until recently, the majority of the material developments within this field has been conducted using electrochemical methods, e.g. Potentiostatic/Galvanostatic Intermittent Titration Techniques (PITT/GITT) or Electrochemical Impedance Spectroscopy (EIS). One cannot deny the success of such techniques, however, from a pure materials science point of view, it is important to understand that they do not directly yield intrinsic material properties. The reason is that such techniques require the construction of a working device, either a full battery cell or half of one. Consequently, the extracted data will depend not only on the material properties themselves, but also on how the test cell has been put together. This includes, for example, the composition of the electrolyte and the surface area of the electrodes, etc. As a result, the extracted and published values of the important ion diffusion coefficient (D_{ion}) are unreliable, with obtained values spanning several orders of magnitude for the same material.

Another in-house method that is frequently used for extracting D_{ion} is Nuclear Magnetic Resonance (NMR). Such a technique is ideal for studying (for example) Li-ion diffusion, since the NMR signal is element selective, even within rather complex compounds. For normal cases, the dynamic contribution (and D_{ion}) is extracted from the spin-lattice relaxation ($1/T_1$). However, for many energy compounds, such as battery cathode materials, it has been shown that the extracted values of D_{ion} are several orders of magnitude lower than expected. This is because many battery materials contain magnetic elements, such as the transition metal ions in transition metal oxides (TMO), which are included in order to keep charge-neutrality during the charge/discharge cycles. Such magnetic ions tend to cause new relaxation pathways and therefore lead to unreliable values for D_{ion} derived from $1/T_1$. Instead, the absolute values of D_{ion} have to be extracted from the more complex spin-spin relaxation ($1/T_2$), which then yields less robust results.

To take the next step within this field, new characterization methods are needed. Synchrotron X-ray techniques are the most established methods and supply both high intensity and high resolution. However, specifically for energy materials, that often involve light elements (H, Li, ...), X-rays have very low scattering lengths, making investigations challenging. Here the closely-related neutron techniques are more suitable, since the neutron scattering length is optimal for the same light elements. For studies of ion dynamics the Quasi-Elastic Neutron Scattering (QENS) technique is very powerful, but requires a large sample volume and substantial measurement time.

To improve the possibilities for studying ion dynamics, μSR has been developed as an effective probe. When positive muons are implanted into a TMO compound (for example), the muon will generally stop in the vicinity of the oxygen ions. If the material is in its paramagnetic state, the muon will mainly experience the field distribution (with width Δ/γ_μ) from the random nuclear magnetic dipole field. If we suppose a completely random and static field distribution at the muon stopping site we can describe the muon spin depolarization function in zero magnetic field (ZF) using the static Gaussian Kubo-Toyabe function that was introduced in Chapter 5

$$G_{KT}^{Stat}(\Delta, t) = \left[\frac{1}{3} + \frac{2}{3} \left(1 - \Delta^2 t^2 \right) \exp\left(-\frac{1}{2}\Delta^2 t^2 \right) \right]. \qquad (11.1)$$

For a battery or hydrogen-storage material this would correspond to the situation where Li or H ions are immobile and the ZF μ^+SR time spectrum will ideally match the bottom curve ($\nu/\Delta = 0$) shown in Fig. 11.2(b). For a situation where the ions instead become mobile (e.g. at higher temperature), $G_{KT}^{Stat}(\Delta, t)$ is replaced by the dynamic

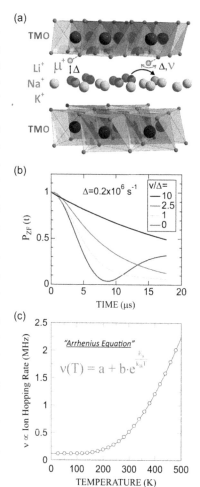

Fig. 11.2 (a) Muons implanted into a paramagnetic battery material at room temperature will mainly experience nuclear magnetic dipole moments. (b) For static ions the μ^+SR time spectrum is described by a static Kubo-Toyabe function and the field distribution width (Δ). For dynamic ions the spectrum is described by a dynamic Kubo-Toyabe function with the additional field fluctuation rate (ν). (c) For the ideal case ν translates directly into the ion hopping rate and the temperature dependence $\nu(T)$ is fitted to an Arrhenius equation.

Kubo-Toyabe function $G_{\mathrm{KT}}^{\mathrm{Dyn}}(\Delta, \nu, t)$. This function also includes the field fluctuation rate represented by the additional parameter ν. For the ideal case where only the ions of interest (Li, H, ...) become mobile, ν directly translates into the **ion hopping rate**. As a result, the ZF μ^+SR time spectrum will change its shape significantly with increasing ν as schematically shown in Fig. 11.2(b). A standard measurement is carried out as a function of temperature from which $\nu(T)$ can be extracted. For a thermally activated process such as ionic diffusion, above its activation energy (E_a), the ion hopping rate will initially increase exponentially and is typically fitted by the Arrhenius equation

$$\nu(T) = a + b \exp\left(\frac{E_a}{k_B T}\right). \tag{11.2}$$

Here a is a constant related to the background contribution from fluctuating electronic spins that usually is virtually temperature independent. Further, b is a constant related to the detailed characteristics of the diffusion process (e.g. dimensionality) and is known as the pre-exponential factor, which for most reactions has only a very weak temperature dependence. Finally the activation energy (E_a) is an important parameter that can be extracted on the atomic level from μSR experiments and compared to values obtained from other experiments such as the electrochemical or NMR measurements mentioned above.

To finally extract the ion self-diffusion coefficient (D_{ion}) using ν as the ion hopping rate, we need additional information about the possible hopping paths in the crystallographic lattice. Such information can be obtained from logical assumptions or modelling, but a better approach is to conduct complementary investigations using neutron diffraction, for example. With such additional input we are able to calculate D_{ion} using the expression[1]

$$D_{\mathrm{ion}} = \sum_{i=1}^{n} \frac{1}{N_i} Z_{\mathrm{v},i} s_i^2 \nu, \tag{11.3}$$

where N_i is the number of possible ion sites in the i-th path, $Z_{\mathrm{v},i}$ is the vacancy fraction, and s_i is the jump distance for such a path.

Within the field of energy materials and devices, it is becoming clear that not only the intrinsic material properties, but also surface and interface effects are the limiting factors for improving performance, lifetime, safety, and sustainability. However, experimental investigations of such effects are complicated, since most cases require direct access to buried interfaces. μSR has recently been shown to be a suitable technique for investigating these aspects in two different ways. In the first method, μSR is able to separate the signal into different volume fractions. By synthesizing the material of interest in the form of nano-particles of different size and surface coatings, it is possible to tune the ratio of μSR signals from surface versus bulk (smaller particles equals larger surface contribution). Hence, through systematic studies using bulk μSR techniques, it is possible to discern between intrinsic materials properties

[1] See e.g. R. J. Borg and G. J. Dienes, *An Introduction to Solid State Diffusion* (Academic Press, San Diego) 1988.

(a) **Low-energy μ⁺SR (LEM)**

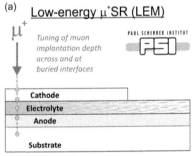

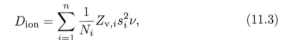

(b)

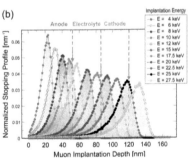

Fig. 11.3 (a) Schematic view of the low energy μSR (LEM) technique, which enables ion-diffusion studies in thin film and multi-layer samples or devices. (b) Depth-resolved studies are possible by tuning the muon momentum, i.e. changing the muon implantation depth into the sample.

and surface contributions. A concrete example of such an application is shown in Example 3 in the next section.

A second more direct method is to use low energy μSR (LEM), which allows the tuning of the muon momentum and, therefore, the muon implantation depth ($d_{\mathrm{imp}} \approx 10 - 350$ nm) in thin samples (see Fig. 11.3). This enables the study of surface effects, and also the growth of multilayers of working energy devices (e.g. a thin film solid state battery) from which we can extract depth and temperature-dependent ion diffusion both at and across the interfaces. Such non-destructive characterization is not possible using other experimental techniques, making LEM a unique tool for scientists working in this field.

11.2 Science examples

Selected examples where μSR has been utilized to study ionic motion are discussed below.

Example 11.1

Battery Materials. As mentioned above, μSR has been frequently utilized to study alkali-ion diffusion in battery related materials. The majority of such studies have been performed on Li-ion cathode materials including[2] $Li_x CoO_2$, $LiFePO_4$, and $LiNiO_2$. In the latter case, μSR was an important technique to show that, on the microscopic scale, a small number of Ni-Li anti-site defects strongly affect the Li-ion self diffusion. Using LEM the first measurements of Li-ion diffusion in $Li_4 Ti_5 O_{12}$ and $LiTi_2 O_4$ solid-state thin-film anode materials were conducted, with results indicating a difference between bulk and surface/interface effects. This was also the first step towards conducting an *in-operando* study of ion diffusion in a working thin-film battery device. It is important to emphasize that for anode and electrolyte materials we require a material or a state where the alkali ions are inserted, and thereby can be sensed and studied by μSR. Note that for electrolytes, anode and fully de-intercalated cathode materials (i.e. no alkali ions present), μSR is not usable and β-NMR is a better approach.

Beyond the Li-ion based battery materials it is also possible to study the dynamics of other ions, as long as they contain an isotope that possesses a nuclear magnetic moment. One example is the Na-ion cathode-type material $Na_x CoO_2$. Such studies reveal a strong dependence on the Na-ion content for both E_a and D_{Na} (see Fig. 11.4). Further, from complementary neutron diffraction measurements[3] it is found that subtle structural transitions are responsible for the opening of the ion-diffusion channels, thereby controlling the Na-ion motion.

Example 11.2

Hydrogen storage materials. In addition to rechargeable batteries, solid-state hydrogen-storage materials are one of the most important areas in energy storage. This is especially true for applications within fuel-cell vehicles where both energy density and safety are key issues. For these materials μSR has contributed information on the thermodynamic stability of the borohydride family of materials: $M(BH_4)_2$ (where M = Li, Na, Mg, K, Ca).[4] For such studies, both the muon's capability to form an HμH type bond (see Chapter 4 for a description of the analogous F$-\mu-$F signal) and its ability to detect H-ion dynamics were utilized.

[2]See J. Sugiyama *et al.*, Phys. Rev. Lett. **103**, 147601 (2009); J. Sugiyama *et al.*, Phys. Rev. B **84**, 054430 (2011); J. Sugiyama *et al.*, Phys. Rev. B 85, 054111 (2012); J. Sugiyama *et al.*, Phys. Rev. B **82**, 224412 (2010); I. Umegaki *et al.*, Phys. Chem. Chem. Phys. **19**, 19058 (2017); J. Sugiyama *et al.*, Phys. Rev. B **92**, 014417 (2015); H. Nozaki *et al.*, JPS Conf. Proc. **2**, 010303 (2014); M. Månsson and J. Sugiyama, Physica Scripta **88**, 068509 (2013).

[3]See M. Medarde *et al.*, Phys. Rev. Lett. **110**, 266401 (2013).

[4]See J. Sugiyama *et al.*, Phys. Rev. B, **81**, 092103 (2010).

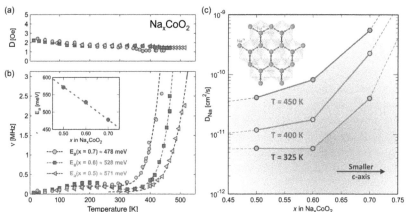

[5]See J. Sugiyama *et al.*, Sustainable Energy & Fuels **3**, 956 (2019) for further details.

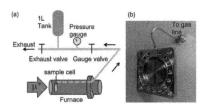

Fig. 11.5 (a) *In-situ/in-operando* setup and (b) sample cell for studying solid-state H-storage materials using μSR at ISIS and J-PARC. [Taken from J. Sugiyama *et al.*, Sustainable Energy & Fuels **3**, 956 (2019).]

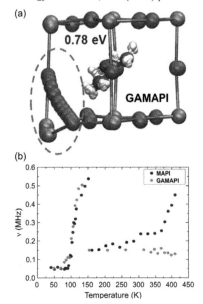

Fig. 11.6 (a) *Ab-initio* simulations of GAMAPI revealing local lattice relaxations near the iodine diffusion paths (lighter grey atoms are Pb and darker grey are I). (b) Field fluctuation rate $\nu(T)$ for MAPI and GAMAPI. [Taken from D. W. Ferdani *et al.*, Energy & Environmental Science **12**, 2264 (2019).]

Fig. 11.4 (a-b) Temperature dependences of $\Delta(T)$ and $\nu(T)$, respectively, for the Na-ion battery cathode type materials Na_xCoO_2 with different Na-content ($x = 0.5$, 0.6, and 0.7) as acquired from μSR measurements. The inset shows how E_a is inversely proportional to x. (c) Calculated Na-ion self-diffusion constant (D_{Na}) as a function of x and T. Inset shows the Na-ion jumping paths within the Na-layers as extracted from complementary neutron powder diffraction measurements. [Taken from M. Månsson and J. Sugiyama, Physica Scripta **88**, 068509 (2013).]

In a further study of the related magnesium hydride material MgH_2, an *in-situ/in-operando* setup was developed[5], including a double-sealed sample cell connected to a gas handling system with pressure gauge (see Fig. 11.5). Such a setup allows measurements of H-ion dynamics inside the sample using μSR while ramping up the temperature in the furnace. It is then possible to simultaneously and directly investigate microscopic H-ion dynamics as well as the desorption of hydrogen gas via the pressure gauge inside the gas-tank. Among other things, such *in-operando* studies reveal that microscopic hydrogen diffusion sets in well below the desorption temperature (T_d) and that optimizing such process will be important for efficient H-storage materials within fuel-cell applications.

Example 11.3

Photovoltaic materials. For energy-harvesting devices, solar cells and photovoltaic (PV) materials represent key areas. The operational mechanism of such devices are not directly linked to ion diffusion. However, for many PV materials, ionic motion related to phase separation is a limiting factor for device lifetime and long-term performance. Here one example is the so-called halide perovskite PV materials, such as methylammonium lead iodide, $MAPbI_3$ (MAPI), where iodine transport is considered a main concern. A study of A-site substitution was made by inserting guanidinium (GA) to create the $MA_{1-x}GA_xPbI_3$ (GAMAPI) compound. For this material there have been indications that iodine diffusion is suppressed by such substitution, thereby stabilizing the material's long-term operational stability. By combining structural studies, *ab-initio* modelling and μSR of I-ion diffusion, information on the potential stabilization mechanism was extracted. As shown in Fig. 11.6(a), modelling indicates that inserting GA creates a local lattice relaxation near the Iodine diffusion paths. From comparable μSR studies of MAPI and GAMAPI compounds [Fig. 11.6(b)] there are indications that iodine diffusion at high temperatures is strongly suppressed, even by a very moderate (5%) GA substitution.

Example 11.4

Nanostructured materials. For many energy materials and devices it is becoming increasingly clear that not only intrinsic material properties but also surface and interface effects are important for the final performance, lifetime, and safety. In recent publications, nano-particles of the one-dimensional (1D) LiFePO$_4$ battery cathode material were studied through a combination of μSR and neutron scattering.[6] Nanocrystals were synthesized in different sizes with controllable crystallographic orientation, which therefore determines the length of the quasi-1D diffusion channels along the crystallographic [010] direction (see Fig. 11.7). In addition, the amount of surface carbon coating is also tunable. By a separation of surface and bulk contributions (via μSR volume fractions) it was shown that Li-ion diffusion at the surface of the particles is key for the operation of devices made from this material. Complementary inelastic neutron scattering (INS) studies show that new phonon modes appear at the surface, which are of importance for the ion diffusion in this material. Such phonons are found by *ab-initio* calculations and molecular dynamics simulations to be related to a subtle structural change in the surface of the particles. These results emphasize how atomic surface engineering can be used to enhance ion transport of energy materials, as well as ultimately, the device performance.

[6] See P. Benedek *et al.*, Sustainable Energy & Fuels **3**, 508-513 (2019); P. Benedek *et al.*, ACS Applied Materials & Interfaces **12**, 14, 16243 (2020).

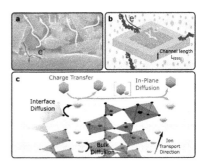

Fig. 11.7 (a) Schematic of lithium ion and charge dynamics in a Li-ion battery electrode. (b) Same mechanism within a single coated LiFePO$_4$ nano-crystal. Here the charges (electrons) move through the carbon-black network, while lithium ions are transferred by the surrounding electrolyte (hexagons). (c) Mechanisms of ionic motion within a system containing 1D ion diffusion channels. Four different processes are discernable: bulk diffusion, charge transfer, in-plane diffusion, and interface diffusion. [Taken from P. Benedek *et al.*, ACS Applied Materials & Interfaces **12**, 14, 16243 (2020).]

11.3 Limitations

Investigations of ionic motion using μSR have some limitations, like any experimental technique. Firstly, muons will probe only the motion of ions that have a nuclear magnetic moment, and their success in probing the motion will depend on the abundance of that magnetic nucleus, and to a lesser extent its moment size. Secondly, it is important to remember that we *implant* muons into the material, and our goal is that they stay immobile while detecting the diffusion of ions. However, without any other input or support it is impossible to distinguish whether an increase in hopping rate (ν) comes from diffusion of ions or of muons (or both). Therefore it is important to obtain further experimental and theoretical input to complement the μSR measurements. In previous studies, such supporting information has included for instance:

- Electrochemical measurements showing that ions are mobile in the relevant temperature range.

- Temperature-dependent NMR data showing that the onset temperature of diffusion for the specific ion matches the onset temperature for an increase in $\nu(T)$. Similarly, QENS measurements can also be used.

- Computer modelling demonstrating that the implanted muon is energetically more stable inside the lattice than the ions in question.

- μSR measurements on other members of the same family of compounds, where the ions are known to be immobile in the entire temperature range. If no increase of $\nu(T)$ is found, it is clear that the muons are stable in this type of lattice.

- Positive μSR studies can also be complemented by negative μ^-SR measurements, where atomic capture guarantees that the muon is immobile (see Chapter 22.)

Chapter summary

- Muons provide a means of probing the diffusion of ions in a range of technologically important materials.
- The muon spectra are frequently modelled using a dynamic Kubo-Toyabe function to determine diffusion rates.
- Applications of the technique include battery-cell and hydrogen-storage materials, as well as photovoltaics.
- Particular advantages of muons in this field are the sensitivity of muons to volume fraction effects and the ability of low-energy muons to probe surfaces and interfaces.

Further reading

- J. Sugiyama, J. Phys. Soc. Jpn. **82**, SA023 (2013).
- I. McClelland *et al.*, Ann. Rev. Mat. Res. **50**, 371 (2020).

Exercises

(11.1) The width of the nuclear field distribution Δ can be calculated using eqn G.42. A simple material for which Δ can be calculated is Cu. To get a simple estimate we can consider the muon site at the centre of an octahedron of ^{63}Cu with a nuclear spin $I = 3/2$ and gyromagnetic ratio $\gamma = 7.11 \times 10^7$ rad T^{-1}s^{-1}. The nearest neighbour atoms are at the centres of the faces of a cube of edge $a = 0.36$ nm. What value of Δ is obtained from the contributions of these six nearest neighbours? What is the value of Δ if the contributions of the eight Cu nuclei on the corners of the cube surrounding the muon are also included? For reference, the experimental value is $\Delta = 0.39$ μs^{-1}. The expression G.42 does not take account of the orientation of quadrupolar nuclear spins by the electric field gradient associated with the muon. To account for this, the terms in eqn G.42 must be multiplied by an additional factor. The factor is given in eqn G.45 and for half integer I, as we have here, the factor can be expressed as $2/3 + (2I + 1)/(8I(I + 1))$. The Δ values can now be reevaluated using these corrections and checked for closer agreement with the experimental value.

(11.2) Muon measurements provide the rate at which ions move past the muon. To compare this with bulk measurements it is useful to calculate D_{ion} as given in eqn 11.3. A material with a simple primary diffusion pathway is LiFePO$_4$, where Li$^+$ ions move $b/2 = 0.3$ nm in each jump. Estimating the vacancy fraction is more challenging because it varies over orders of magnitude as a battery is charged. For this example, we can choose 0.1 as a representative value of the vacancy fraction. The hopping rate at room temperature is roughly 1×10^6 s^{-1}. From these values estimate D_{Li}.

Chemistry

12

This chapter is concerned with the structure, properties, dynamics, and reactions of chemical species containing the positive muon and the ways that this information can be obtained using μSR.

12.1 Chemical environments

Muons can be implanted into solid, liquid, or gaseous samples. The implanted muons initially have very high energy but are slowed down to thermal energies on the nanosecond timescale with the polarization being mostly conserved. The final chemical environment of the muon depends on the chemical properties of the material in which it has been implanted. A fraction of the implanted muons will end up in diamagnetic chemical environments, as solvated muons,[1] or substituted for the proton of a diamagnetic molecule (i.e. a molecule without an unpaired electron such as MuOH or $C_6H_{11}Mu$). The short lifetime of the muon limits the spectral resolution to ≈ 70 kHz so it is not possible to resolve chemical shifts or nuclear couplings and distinguish between muons in different diamagnetic environments. Another fraction of muons can pick up an electron during the slowing down process and form muonium (Mu = μ^+e^-), a one-electron atom with the positive muon as the nucleus.[2] The fraction of muons forming Mu depends strongly on the material and its physical state, ranging from 0 in liquid CCl_4 to 1 in gaseous Kr. The Mu fraction depends strongly on the phase, temperature, and pressure; as an example, the Mu fraction in water is 0.2 for liquid water at ambient temperature and pressure (298 K and 1 bar), and increases to 0.8 at 673 K and 245 bar.

Mu is a paramagnetic atom. As discussed in Chapter 3 there is a hyperfine interaction between the unpaired electron and nuclear spins with the strength of this interaction, the isotropic hyperfine coupling constant, proportional to the unpaired spin density at the nucleus, $|\psi(0)|^2$, and the nuclear gyromagnetic ratio. This is much larger than muon-nuclear interactions. The isotropic hyperfine coupling constant (in frequency units) is given by eqn 3.4 which we will rewrite here as

$$A_X = \frac{2\mu_0}{3h} g_e \mu_B g_X \mu_N |\psi(0)|^2, \tag{12.1}$$

where g_e is the free electron g factor, μ_B is the Bohr magneton, g_X is nuclear g factor, and μ_N is the nuclear magneton.[3] In the condensed phase, the hyperfine coupling constant of Mu and the other hydrogen

---

[1] In this chapter, the symbol Mu^+ is used rather than μ^+ to refer to thermalized muons in a material. Free Mu^+ doesn't exist in condensed matter for the same reason as H^+ is not found; it is too reactive to occur in many liquids, even though it is sometimes visualized to do so by students of chemistry. Mu^+ reacts with a molecule of the liquid to form a more complicated cation.

[2] See Chapter 3.

[3] In the chemistry literature, it is common to write the Bohr magneton as β_e and the nuclear magneton as β_X.

isotopes are altered by the interaction between the atom and the neighbouring molecules. Comparisons between different isotopes are made by considering the reduced hyperfine coupling constant, A'_X, which is given by multiplying the hyperfine coupling constant by the ratio of the nuclear and proton magnetic moments ($A'_X = \gamma_X A_X / \gamma_p$). There is also a dipolar hyperfine interaction that is important in the solid state.

The main chemical properties of muonium, protium (H), deuterium (D), and tritium (T) are listed in Table 3.1 in Chapter 3. The chemical properties of Mu are very similar to those of the hydrogen isotopes, so it is common to refer to it as a light isotope of hydrogen, even though the conventional definition of isotopes are atoms that have nuclei with the same number of protons but a different number of neutrons. The mass of Mu is $\approx 1/9$ the mass of H, but the chemical properties of Mu are similar to those of the other hydrogen isotopes because the positive muon is much heavier than the electron and the chemical properties depend on the reduced mass.

abstraction	$\mathbf{Mu} + H_2 \rightarrow MuH + \mathbf{H}$
addition to unsaturated molecules	$\mathbf{Mu} + CH_2{=}CH_2 \rightarrow \mathbf{MuCH_2\text{-}\dot{C}H_2}$
oxidation-reduction	$\mathbf{Mu} + Ag^+ \rightarrow \mathbf{Ag^0} + Mu^+$
acid-base	$\mathbf{Mu} + OH^- \rightarrow MuOH + \mathbf{e^-}$
electron spin exchange	$\mathbf{Mu(\uparrow)} + \mathbf{NO(\downarrow)} \rightarrow \mathbf{Mu(\downarrow)} + \mathbf{NO(\uparrow)}$
combination	$\mathbf{Mu} + \mathbf{O_2} \rightarrow \mathbf{MuO_2}$

Table 12.1 Chemical reactions of muonium. The bold species are paramagnetic.

Muonium is extremely reactive due to the unpaired electron, which makes it a free radical. The reactions of Mu are listed in Table 12.1. The chemistry of Mu is studied for two main reasons:

- It is studied in place of H, as it is possible to observe Mu using μSR under conditions where it is not possible to observe H.[4] Muonium studies should be viewed as complementary to other methods of investigating hydrogen atom chemistry. Reaction mechanisms are often deduced or tested by following the fate of labelled molecules.

- It is studied to see how it reacts differently than H, D, or T due to its light mass. The ratio of rate constants caused by such an isotopic exchange is termed the kinetic isotope effect (KIE). Mu and H should have similar chemical reactions, although the rates may be affected by kinetic isotope effects and tunnelling.

[4]Hydrogen atoms are produced by photolysis or radiolysis, and this results in many other reactive species. In water radiolysis produces the H atom along with the hydroxyl radical (OH), the solvated electron (e^-_{aq}), the hydroperoxyl radical (HO_2), and hydrogen peroxide (H_2O_2), see Fig. 12.1. Scavengers can be added to remove unwanted reactants, but this leads to complex samples and possible unwanted side reactions. Mu is thermalized far from the reactive species in the track and these other species can be ignored.

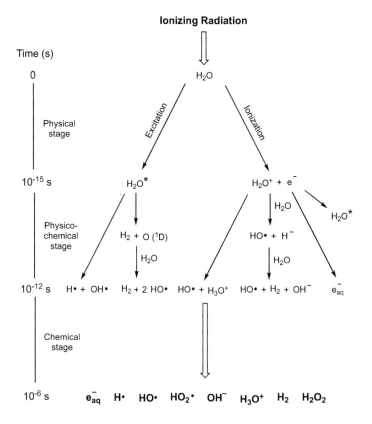

Ionizing Radiation

Time (s)

0 —— H_2O

Physical stage

Excitation   *Ionization*

10^{-15} s H_2O^* $H_2O^+ + e^-$

H_2O

H_2O^*

Physico-chemical stage $H_2 + O\,(^1D)$ $HO\bullet + H^-$

H_2O H_2O

10^{-12} s $H\bullet + OH\bullet$ $H_2 + 2\,HO\bullet$ $HO\bullet + H_3O^+$ $HO\bullet + H_2 + OH^-$ e^-_{aq}

Chemical stage

10^{-6} s e^-_{aq} $H\bullet$ $HO\bullet$ $HO_2\bullet$ OH^- H_3O^+ H_2 H_2O_2

Fig. 12.1 Reactions involved in the radiolysis of water and the resulting radiolytic products.

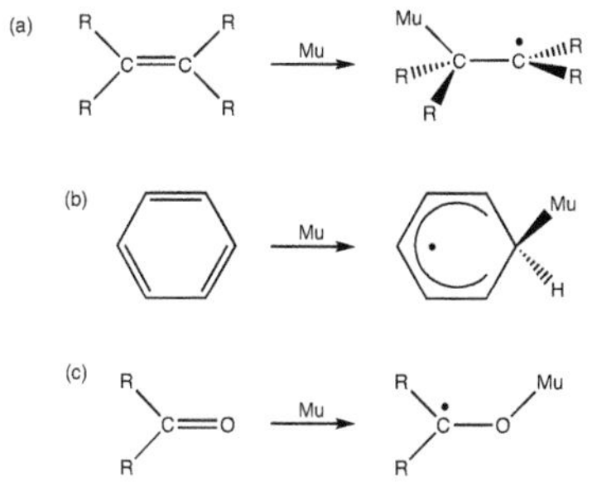

(a)
(b)
(c)

Fig. 12.2 Formation of muoniated radicals; (a) Mu addition to an alkene to form a muoniated alkyl radical; (b) Mu addition to benzene to form the muoniated cyclohexadienyl radical; (c) Mu addition to a ketone to form a muonoxy-alkyl radical.

Muoniated radicals are formed by the addition of Mu to an unsaturated bond (Fig. 12.2).[5] They have the same structure as the corresponding non-muoniated radical except that one proton has been replaced by a positive muon. In muoniated radicals there is a hyperfine interaction of the unpaired radical electron with the muon, but also hyperfine interactions with all of the nuclei having spin $I > 0$, such as ^1H, ^2H, ^{13}C, and ^{14}N. By measuring the muon and nuclear hyperfine coupling constants one can map out the distribution of the unpaired electron and determine the structure of the radical and by measuring the temperature dependence of the hyperfine coupling constants one can learn about the dynamics and three dimensional structure of the radical.

Muoniated radicals are studied for similar reasons to Mu:

- The radicals can be studied under conditions where it would be difficult or impossible to observe the corresponding non-muoniated radical with conventional spectroscopic techniques such as EPR.
- One is interested in studying the effect of the light mass of the muon on the conformation[6] and dynamics of a radical.

[5]Older literature uses the term muonated radical. This has been discontinued as 'muonation' is now defined to be the equivalent of protonation. 'Muoniation' is used to denote replacement of hydrogen by muonium.

[6]Conformation is any of the spatial arrangements which the atoms in a molecule may adopt and freely convert between, especially by rotation about individual single bonds.

Energy eigenstate	Frequency
$\lvert 1 \rangle = \lvert \uparrow^{\mu} \uparrow^{e} \rangle$	$E^{\lvert 1 \rangle}/h = \frac{1}{4}A_{\mu} + \frac{1}{2}\left(\nu_{e} - \nu_{\mu}\right)$
$\lvert 2 \rangle = s\lvert \uparrow^{\mu} \downarrow^{e} \rangle + c\lvert \downarrow^{\mu} \uparrow^{e} \rangle$	$E^{\lvert 2 \rangle}/h = \frac{1}{4}A_{\mu} + \frac{1}{2}\left(\sqrt{A_{\mu}^2 + (\nu_{e} + \nu_{\mu})^2} - A_{\mu}\right)$
$\lvert 3 \rangle = \lvert \downarrow^{\mu} \downarrow^{e} \rangle$	$E^{\lvert 3 \rangle}/h = \frac{1}{4}A_{\mu} - \frac{1}{2}\left(\nu_{e} - \nu_{\mu}\right)$
$\lvert 4 \rangle = c\lvert \uparrow^{\mu} \downarrow^{e} \rangle - s\lvert \downarrow^{\mu} \uparrow^{e} \rangle$	$E^{\lvert 4 \rangle}/h = -\frac{3}{4}A_{\mu} - \frac{1}{2}\left(\sqrt{A_{\mu}^2 + (\nu_{e} + \nu_{\mu})^2} - A_{\mu}\right)$

$$c^2 = \frac{1}{2} + \frac{1}{2}\frac{\nu_{e}+\nu_{\mu}}{\sqrt{A_{\mu}^2+(\nu_{e}+\nu_{\mu})^2}}; \; c = \frac{1}{\sqrt{2}} \text{ at } B = 0 \text{ and } c = 1 \text{ when } \nu_{e} + \nu_{\mu} \gg A_{\mu}$$

$$s^2 = \frac{1}{2} - \frac{1}{2}\frac{\nu_{e}+\nu_{\mu}}{\sqrt{A_{\mu}^2+(\nu_{e}+\nu_{\mu})^2}}; \; s = \frac{1}{\sqrt{2}} \text{ at } B = 0 \text{ and } s = 0 \text{ when } \nu_{e} + \nu_{\mu} \gg A_{\mu}$$

Table 12.2 Eigenvectors and energies for muonium (see also Table 4.2).

12.2 Muonium spectroscopy

Transverse field muon spin rotation of muonium

In TF-μSR one observes transitions between spin states. There are four spin states for Mu that are linear combinations of the product Zeeman states and whose energies depend on the applied magnetic field, as shown in Table 4.2 [the same information is given in Table 12.2 where the expressions are given directly in terms of ν_{e} and ν_{mu}; see also Fig. 12.3(a)]. As discussed in Section 4.4, in zero field three of the states are degenerate and are referred to as the triplet states [$\lvert \uparrow^{\mu} \uparrow^{e} \rangle$, $\lvert \downarrow^{\mu} \downarrow^{e} \rangle$, and $\frac{1}{\sqrt{2}}(\lvert \uparrow^{\mu} \downarrow^{e} \rangle + \lvert \downarrow^{\mu} \uparrow^{e} \rangle)$]. The remaining state is lower in energy and called the singlet state [$\frac{1}{\sqrt{2}}(\lvert \uparrow^{\mu} \downarrow^{e} \rangle - \lvert \downarrow^{\mu} \uparrow^{e} \rangle)$]. The degeneracy of the triplet states is broken by applying a magnetic field. The dependence of the spin eigenstates with applied field is shown in a Breit-Rabi diagram.[7]

There are four precession frequencies (Table 12.3) whose amplitudes are proportional to the coefficients in front of the cosine terms, c^2 and s^2. The formulae for c and s are given in Table 12.2. All four frequencies have equal amplitude at zero field. The amplitudes of the ν_{23} and ν_{14} frequencies decrease with increasing field and disappear when when $\nu_{e} + \nu_{\mu} \gg A_{\mu}$. Transitions between the $\lvert 1 \rangle$ and $\lvert 3 \rangle$ states, and the $\lvert 2 \rangle$ and $\lvert 4 \rangle$ states are forbidden due to the selection rule that $\Delta M = \pm 1$, where M is the total spin quantum number.

In zero field the intra-triplet transitions (ν_{12} and ν_{23}) are 0, while the ν_{14} and ν_{43} transitions are degenerate and equal A_{μ}, which is 4463 MHz for Mu in vacuum. Detecting such high frequencies requires special effort but is the most direct and precise way to measure A_{μ}. The hyperfine constant A_{μ} can also be determined from the splitting of ν_{12} and ν_{23} in fields of a few tens of mT.

Measurements of Mu are typically done in very low fields of a few tenths of a mT [Fig. 12.3(b)] where the ν_{12} and ν_{23} precession frequencies are degenerate and have a frequency[8] of $\frac{1}{2}(\nu_{e} - \nu_{\mu})$, which gives an effective gyromagnetic ratio (γ_{Mu}) of ≈ 1.39 MHz/G. The low TF-μSR

[7]The Breit-Rabi diagram was introduced in Fig. 4.7. It is named after Gregory Breit and Isidor Isaac Rabi (Nobel Prize in Physics 1944). Rabi developed the magnetic resonance method on beams of atomic hydrogen and deuterium and measured the magnetic moments of protons, deuterons, and, by inference, neutrons. We recall again here that he famously said "Who ordered that?" when informed about the discovery of the muon.

[8]See Example 4.11.

Frequency	Amplitude
$\nu_{12} = \frac{1}{2}\left(\nu_e - \nu_\mu\right) - \frac{1}{2}\left[\sqrt{A_\mu^2 + \left(\nu_e + \nu_\mu\right)^2} - A_\mu\right]$	$\propto c^2$
$\nu_{23} = \frac{1}{2}\left(\nu_e - \nu_\mu\right) + \frac{1}{2}\left[\sqrt{A_\mu^2 + \left(\nu_e + \nu_\mu\right)^2} - A_\mu\right]$	$\propto s^2$
$\nu_{14} = \frac{1}{2}\left(\nu_e - \nu_\mu\right) + \frac{1}{2}\left[\sqrt{A_\mu^2 + \left(\nu_e + \nu_\mu\right)^2} - A_\mu\right] + A_\mu$	$\propto s^2$
$\nu_{43} = \frac{1}{2}\left(\nu_e - \nu_\mu\right) - \frac{1}{2}\left[\sqrt{A_\mu^2 + \left(\nu_e + \nu_\mu\right)^2} - A_\mu\right] - A_\mu$	$\propto c^2$

Table 12.3 Transition frequencies for muonium

spectra can be fitted to the following equation:

$$A(t) = A_{\mathrm{Mu}}\cos\left(2\pi\nu_{\mathrm{Mu}}t + \phi_{\mathrm{Mu}}\right)e^{-\lambda_{\mathrm{Mu}}t} + A_{\mathrm{D}}\cos\left(2\pi\nu_{\mathrm{D}}t + \phi_{\mathrm{D}}\right), \quad (12.2)$$

where A_{Mu} and A_{D} are the Mu and diamagnetic asymmetries,[9] respectively, $\nu_{\mathrm{Mu}} = \gamma_{\mathrm{Mu}}B$ is the Mu precession frequency in low field, ν_{D} is the muon Larmor frequency, ϕ_{Mu} and ϕ_{D} are the Mu and diamagnetic muon phase shifts, respectively, and λ_{Mu} is the damping of the Mu precession signal. The relaxation rate of the diamagnetic signal is negligible (on the μs timescale imposed by the muon lifetime), except for the case of concentrated solutions of paramagnetic ions.

In magnetic fields of several tens of mT, the ν_{12} and ν_{23} precession frequencies are no longer degenerate [Fig. 12.3(c)]. The ν_{14} and ν_{43} frequencies in this range have values of GHz and can only be measured with specialized equipment. The muon hyperfine coupling constant of Mu can be measured from the splitting of the ν_{12} and ν_{23} frequencies using eqn 4.64.

[9] These can be converted to fractions of muon polarization, P_{Mu} and P_{D}, by calibration against the diamagnetic signal of a standard such as CCl_4 run under identical conditions.

Muonium in a longitudinal field

The initial muon spin polarization (i.e. at $t = 0$) in a longitudinal magnetic field is given by

$$P_z^{\mathrm{Mu}}\left(t = 0\right) = P_{\mathrm{Mu}}\left\{\frac{1}{2} + \frac{1}{2}\left[\frac{(B/B_0)^2}{1 + (B/B_0)^2}\right]\right\}, \quad (12.3)$$

where $B_0 = A_\mu/(\gamma_\mu + \gamma_e)$. The polarization $P_z^{\mathrm{Mu}}\left(t = 0\right)$ is $P_{\mathrm{Mu}}/2$ at zero magnetic field and P_{Mu} at high fields when $\nu_e + \nu_\mu \gg A_\mu$.

In the repolarization technique, the initial Mu polarization is measured as a function of the applied longitudinal magnetic field.[10] As the strength of the magnetic field increases, the muon and electron spins are progressively decoupled from the hyperfine field as the Zeeman energy becomes dominant. For isotropic Mu, half of the initial polarization of implanted muons is lost because of the hyperfine coupling, but this is recovered in a sufficiently large applied field, from which one can estimate the strength of A_μ. The repolarization method is the least accurate method to measure A_μ but it is popular at pulsed facilities

[10] The polarization P_{D} is determined from a 100 G transverse field spectrum. The spin relaxation of diamagnetic muons is negligible on the μs timescale and can be fixed at zero in fits.

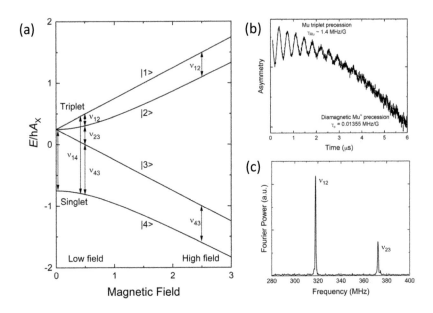

Fig. 12.3 (a) Breit-Rabi diagram showing the magnetic field dependence of the spin states (Table 12.2) and the allowed transitions. (b) TF-μSR time-domain spectrum of Mu in degassed H_2O at 2 G. The fast damped precession is due to Mu (degenerate ν_{12} and ν_{23}) and the slow precession is due to diamagnetic muons. (c) TF-μSR Fourier-transform spectrum of Mu in degassed H_2O at 250 G. The ν_{12} and ν_{23} frequencies are no longer degenerate.

where one cannot measure high frequencies (>10 MHz) and so is often the only method for quantifying Mu in a sample. It remains an important technique for characterizing unknown systems, and particularly for those with state-to-state transitions or dynamical properties and has been useful as well in characterizing anisotropic muon environments in the solid state. The inflection point, $B/B_0 = 1$ ($P_{Mu} \approx 0.75$), is seen at ≈ 1585 G for vacuum Mu and at somewhat smaller values for Mu in condensed matter where there is a small amount of spin density transferred to its surroundings. This hyperfine coupling to other nuclei also results in a low-field polarization that is observed to be considerably lower than 50%. The LF repolarization technique is discussed in more detail in Chapter 20.

Example 12.1

Repolarization curves for Mu in N_2 gas and liquid water are shown in Fig. 12.4. In the gas phase there is no hyperfine coupling to any nucleus apart from the muon and this results in the Mu polarization being approximately 50% at low fields. Mu is a hydrophobic solute and produces a transient clathrate-like cage or bubble around itself. There is a small amount of spin density on the water molecules in the primary solvation sphere. This results in the low-field Mu polarization being reduced to $\approx 30\%$ due to hyperfine interaction with the H nuclei in the water.

Fig. 12.4 Repolarization curves of Mu in nitrogen gas and liquid water at 298 K. [N_2 data from D. G. Fleming *et al.*, Phys. Chem. Chem. Phys. **14**, 10953 (2012).]

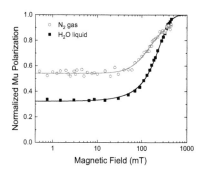

The shape of the measured curve is critically dependent on the accuracy with which the diamagnetic component can be determined over the full longitudinal magnetic field range of the spectrometer (a non-trivial exercise) and the concomitant effect of the field on both the incident μ^+ and decay e^+ trajectories, affecting both the parameter α and the intrinsic asymmetry of the spectrometer.

12.3 Reactions of muonium

Measuring muonium reaction rate constants

The bimolecular reaction of Mu with the solute S follows second-order kinetics, i.e. rate $= k_{\text{Mu}}[\text{Mu}][\text{S}]$. Here, k_{Mu} is the second-order rate constant for the reaction between Mu and S and has units of $\text{M}^{-1}\,\text{s}^{-1}$. The concentration of Mu decreases exponentially with time with a decay rate λ_{Mu}, while the solute concentration, [S], which is present in large excess, does not change significantly.[11] The rate constant for the reaction of Mu with a solute S can be determined by measuring how λ_{Mu} changes with [S]. This is given by

$$\lambda_{\text{Mu}} = \lambda_0 + k_{\text{Mu}}\,[\text{S}]\,, \tag{12.4}$$

where λ_0 is the Mu decay rate in the absence of S, which can be due to magnetic field inhomogeneities and reaction with the solvent.

[11] One can ask what does [Mu] mean when we only have one Mu atom at a time? According to the ergodic principle it doesn't matter if the atoms are present at the same time or spread over an interval. The average of a parameter over time and the average over the statistical ensemble are the same.

Example 12.2

The rate constant of Mu + H_2O_2 at 295 K and pH = 7 was determined by measuring λ_{Mu} versus [H_2O_2] (Fig. 12.5). The slope gives k_{Mu} of $1.8(1) \times 10^6\ \text{M}^{-1}\ \text{s}^{-1}$.

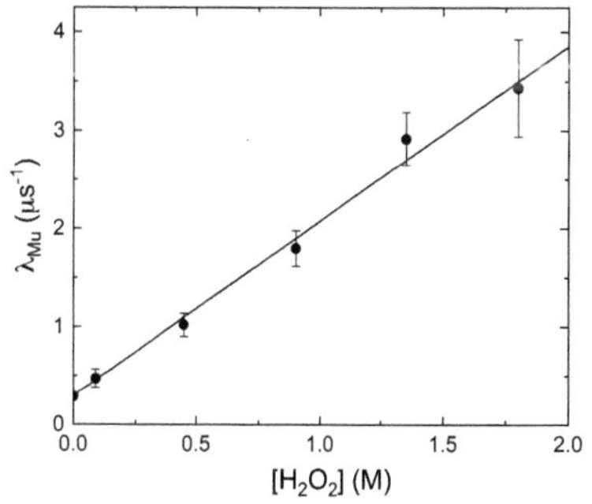

Fig. 12.5 Muonium damping rate in low transverse field versus hydrogen peroxide concentration.

Addition reactions

Mu adds to multiple bonds, which produces a muoniated radical where the muon takes the position of a proton attached to an atom next to the radical centre (the β-position) (Fig. 12.2).

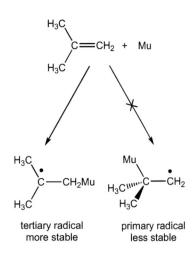

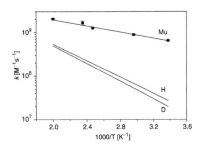

tertiary radical
more stable

primary radical
less stable

Fig. 12.6 Muonium adds exclusively to the less substituted carbon of 2-methylpropene.

Fig. 12.7 Arrhenius plots for the addition reaction of hydrogen isotopes to benzene in the gas phase. [Taken from E. Roduner, Chapter 15 'Muonium - An Ultra-Light Isotope of Hydrogen' in *Isotope Effects In Chemistry and Biology, 1st Edition*, Eds. A. Kohen and H.-H. Limbach, Taylor and Francis (2005).]

The rate constants for Mu addition to a single C=C bond in solution are on the order 10^9–10^{10} M^{-1}s^{-1}, which is close to the diffusion-controlled limit. The addition rate depends on the substituents attached to the C=C bond. Electron donating substituents decrease the Mu addition rate constant (e.g. Mu + styrene; $k_{Mu} = (1.1 \pm 0.1) \times 10^9$ M^{-1} s^{-1}) while electron withdrawing substituents increase the Mu addition rate constant (e.g. Mu + acrylamide; $k_{Mu} = (1.90 \pm 0.13) \times 10^{10}$ M^{-1} s^{-1}).

Mu adds to the side of the double bond that generates the more stable muoniated radical. This is typically the less substituted side of the double bond and is referred to as anti-Markovnikov addition. For example, Mu adds exclusively to the primary carbon of 2-methylpropene (isobutene) to give the muoniated-*tert*-butyl radical $MuH_2CC(CH_3)_2$, as shown in Fig. 12.6. No addition is observed at the tertiary carbon of 2-methylpropene, as this would produce the primary radical $H_2\dot{C}C(CH_3)_2Mu$. This pattern of reactivity is observed in more complicated alkenes. In the case of dienes and allenes, one must also consider the stabilization afforded by delocalization of spin density.

Mu adds to benzene to produce the muoniated cyclohexadienyl radical (C_6H_6Mu). The rate constant for Mu addition to benzene is $(3.7 \pm 0.8) \times 10^9$ M^{-1} s^{-1} in *n*-hexane and $(8.9 \pm 0.6) \times 10^9$ M^{-1} s^{-1} in benzene. The rate of addition of Mu to an isolated double bond is comparable to that of addition to a benzene ring. Mu adds preferentially to the least substituted carbon of a substituted aromatic molecule.

Mu addition to C=O bonds is considerably slower than addition to C=C bonds; the rate constant for Mu addition to acetone is 8.7×10^7 M^{-1} s^{-1}. In aliphatic carbonyl compounds Mu addition to the oxygen atom of the carbonyl group is thermodynamically and kinetically favoured. Mu addition to an isolated phenyl ring is much faster than Mu addition to an isolated C=O bond, but Mu addition is observed to carbonyl groups conjugated with double bonds or aromatic groups.

The addition of Mu to benzene is about 20 times faster than its heavier isotopes at room temperature (Fig. 12.7). The activation energy of the Mu reaction is only 37% of that of H, and the pre-exponential factor is also significantly lower. These are clear indicators for a significant contribution of tunnelling, which offsets any zero-point energy effects in the transition state. It is well known from calculations on model barriers that tunnelling is promoted by a low mass, and a narrow and not too high barrier. Addition reactions are often considerably exothermic, so that their transition state is reactant-like, and the barrier is narrow and relatively low. Therefore, tunnelling plays a dominant role, rendering the reaction of the light isotope faster than that of the others.

Abstraction reactions

Mu can react with organic compounds by abstracting an atom such as H, which generates a radical and the MuH molecule, which is indistinguishable from other diamagnetic muon states. H abstraction reactions are less exothermic than addition reactions, so their reaction barrier is normally higher, and often much wider, than the barrier to addition. The size of the barrier impedes tunnelling, so that the KIE in abstraction reactions is dominated by zero-point energy of the transition state. According to classical transition-state theory the ratio of frequency factors is expected to correspond to the inverse square root of the reduced mass ratio. Thus, the barrier increases for the light isotope and makes its reaction slower. This is seen for the reactions of Mu, H, and D with H_2 and D_2, where the reaction of Mu with H_2 is slower than that of H or D by more than three orders of magnitude at room temperature (Fig. 12.8).

Abstraction is faster when the resulting radical is stabilized by alkyl substituents. This means there is a preference for abstraction of an H attached to a carbon with the most number of alkyl substituents. Consider the reaction of Mu with ethanol; abstraction occurs from the secondary carbon. Abstraction involving Mu is much slower than most addition reactions, and considerably slower than the corresponding H or D reactions both in the gas phase and in solution (Fig. 12.9). The rate of abstraction reactions can be slowed down by using deuterated solvents.

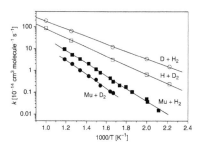

Fig. 12.8 Arrhenius plot for the reaction of Mu with H_2 and with D_2, in comparison with D with H_2 and H with D_2. [Taken from E. Roduner, Chapter 15 'Muonium - An Ultra-Light Isotope of Hydrogen' in *Isotope Effects In Chemistry and Biology, 1st Edition*, Eds. A. Kohen and H.-H. Limbach, Taylor and Francis (2005).]

Competing reaction pathways

In many materials there are multiple Mu addition and abstraction pathways. The amount of each type of radical that *could* form is the result of competition kinetics. The disappearance of Mu depends on the rate constants for all of the reaction pathways via

$$\lambda_{Mu} = \lambda_0 + \sum_i k_{Mu}^i [i], \tag{12.5}$$

where k_{Mu}^i is the second order rate constant for the reaction of Mu with reaction partner i, which has a concentration $[i]$. The relative yield of a product R_i is given by

$$P_{R_i} = \frac{k_{Mu}^i [i]}{\Sigma_j k_{Mu}^j [j]}, \tag{12.6}$$

where the summation is over all reaction partners. This can be different molecules or different sites on the same molecule.

Often we are interested in producing a certain radical to study, which means that one would want to eliminate reaction pathways that don't generate the desired species. This can be accomplished by carefully choosing the chemicals present in the system, particularly the solvents. Unsaturated solvents would compete for Mu and thus are not suitable solvents. One should use solvents that react slowly with Mu, such as THF. The reactivity of the solvent can be further minimized by deuteration.

$$H_3C\!-\!OH \;+\; X \longrightarrow H_2\overset{\bullet}{C}\!-\!OH \;+\; XH$$

X	$k_{abstraction}$ (M^{-1} s^{-1})
Mu	3×10^4
H	3.1×10^6
D	4.4×10^6

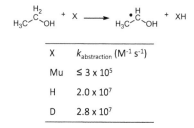

X	$k_{abstraction}$ (M^{-1} s^{-1})
Mu	$\leq 3 \times 10^5$
H	2.0×10^7
D	2.8×10^7

Fig. 12.9 Rate constants at 298 K for Mu/H/D abstraction reactions with methanol and ethanol.

Predicting the products of Mu reactions

In general, the reaction rate is fastest for the reaction that generates the most stable products. According to the Hammond postulate, exothermic reactions generally result in early transition states where the transition state is closer in energy to the reactants than to the products and the transition state geometry will be more similar to the reactants than to the products. Abstraction and addition reactions involving Mu fall into this category; it is energetically favourable to delocalize the unpaired electron. According to the Bell-Evans-Polanyi hypothesis the activation energy (E_a) is linearly related to the reaction enthalpy (ΔH) for a series of related single-step reactions by $E_a = E_0 + \alpha \Delta H$, where α is a measure of 'lateness' of the transition state ($0 < \alpha < 1$). This indicates that the activation energy is lower for more exothermic reactions (i.e. more negative ΔH). The corollary to this is that the reaction is fastest for the pathways that generate the most stable free radical. The stability order of free radicals is tertiary > secondary > primary; this can be accounted for by hyperconjugation.

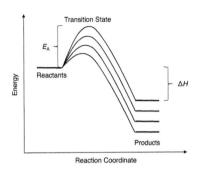

Fig. 12.10 Variation of activation energy according to the Bell-Evans-Polanyi hypothesis for exothermic reactions.

12.4 Muoniated radicals

Spectroscopy of muoniated radicals

μSR offers several advantages over conventional spectroscopic methods:

- Muons can be injected into any sample.
- The muons arrive with almost 100% spin polarization, so very low quantities of muoniated species can be detected (typically 10^7 spins for μSR, compared with 10^{12} for EPR or 10^{18} for NMR).
- Very short-lived muoniated radicals can be detected.
- Unlike conventional EPR or NMR, μSR does not require an external electromagnetic field to stimulate spin-level transitions.
- The lifetime of the muon is comparable to the chemical lifetime of many organic radicals.

Perhaps the most significant advantage of μSR is its high selectivity of detection, which minimizes complications from radiation damage and secondary radical species.

There have been a large number of studies over the last 30 years of the structure and dynamics of muoniated organic alkyl radicals. The structure of a muoniated radical can be determined by measuring the muon hyperfine coupling constant and as many nuclear hyperfine coupling constants as possible, which maps out the distribution of the unpaired electron. This is compared with hyperfine coupling constants obtained from *ab initio* calculations on the possible radical structures, as described in Chapter 16. μSR is a valuable tool because it is possible to determine not just the magnitude of the nuclear hyperfine coupling constants, but their signs as well, relative to the sign of the muon hyperfine coupling constant, which is very difficult to achieve using EPR.

Transverse field muon spin rotation of muoniated radicals

The radical precession frequencies are due to transitions between spin states and are frequently labelled according to the spin states involved. The magnetic moment of the unpaired electron couples to the magnetic moments of the muon and any other nuclei, resulting in a multitude of spin states. In low magnetic fields, the muon polarization is distributed over many transition frequencies, which makes detection of muoniated radicals difficult or impossible (Fig. 12.11). In high magnetic fields, where the Zeeman energy is much larger than the hyperfine interactions, the frequency spectrum is considerably simplified, and two radical frequencies are observed at the same frequencies as observed in electron-nuclear double resonance spectroscopy. The radical frequencies are given by

$$\nu_{12} = \nu_{\text{mid}} - \frac{1}{2}A_\mu \qquad \text{and} \qquad \nu_{43} = \nu_{\text{mid}} + \frac{1}{2}A_\mu, \qquad (12.7)$$

where

$$\nu_{\text{mid}} = \frac{1}{2}\left[\sqrt{A_\mu^2 + (\nu_e + \nu_D)^2} - \nu_e + \nu_D\right], \qquad (12.8)$$

and ν_e and ν_D are the electron and muon Larmor frequencies. In high fields, where $(\nu_e + \nu_D)^2 \gg A_\mu^2$, $\nu_{\text{mid}} \approx \nu_D$, the precession frequency of muons in diamagnetic environments ($\nu_D = 13.55$ kHz G^{-1}). The hyperfine constant A_μ (in units of frequency) is calculated from the difference in the two precession frequencies and the isotropic muon hyperfine coupling constant of an organic radical is typically between a few megahertz and approximately 700 MHz. In some cases the higher frequency radical precession frequency is not observed due to the time resolution of the μSR spectrometer, and A_μ must be determined from the lower frequency radical precession frequency and ν_{mid}. In situations where ν_{mid} is less than $A_\mu/2$ one of the radical frequencies will be negative (i.e. precess in the opposite direction to diamagnetic muons), which can be distinguished from positive frequencies by using two sets of orthogonal positron detectors. It is good practice to perform TF-μSR measurements at more than one magnetic field and confirm that the radical frequencies shift appropriately.

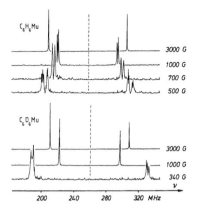

Fig. 12.11 TF-μSR spectra of C_6H_6Mu and C_6D_6Mu. Smaller magnetic fields are needed for C_6D_6Mu to be in the high-field limit due to the smaller magnetic moment of D compared with H. [Taken from E. Roduner, *The Positive Muon as a Probe in Free Radical Chemistry: Potential and Limitations of the μSR Techniques*, Springer-Verlag (1988).]

Example 12.3

The TF-μSR spectrum of benzene at 280 K and 22 kG is shown in Fig. 12.12. This spectrum is due to the muoniated cyclohexadienyl radical (C_6H_6Mu), whose structure is shown in Fig. 12.2b. There are three precession frequencies at 43.3, 300.1, and 559.1 MHz. The peak at 300.1 MHz is the diamagnetic muon precession frequency. This can be checked by multiplying γ_μ by the applied magnetic field. The peaks at 43.3 and 559.1 MHz are the ν_{12} and ν_{43} frequencies. A_μ is determined from the difference between these frequencies and is 515.8 MHz. The lower amplitude of ν_{43} compared with ν_{12} is due to a fall-off in frequency response of the spectrometer due to limited time resolution.

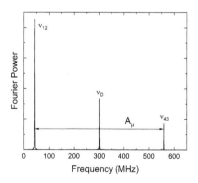

Fig. 12.12 TF-μSR spectrum of the muoniated cyclohexadienyl radical in benzene at 280 K and 22 kG.

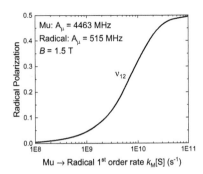

Fig. 12.13 Amplitude of ν_{12} line as a function of the 1st-order conversion rate, which is $k_{\text{Mu}}[S]$, using the parameters for the muoniated cyclohexadienyl radical ($A_\mu = 515$ MHz). If $k_{\text{Mu}} = 3.7 \times 10^9$ M^{-1} s^{-1} and the concentration of benzene in *n*-hexane is 0.27 M, the amplitude of the ν_{12} line would be approximately 0.05, which is $\approx$10% of the value in neat benzene.

The muoniated radical must be formed promptly in order for the muon spin polarization to be transferred from the Mu precursor to the radical. A slower formation reduces polarization because of dephasing of the precessing muons in the transverse field. The fraction of spin polarization transferred to a radical (P_R) may be calculated for the general case. In the high-field limit, the corresponding expression simplifies to

$$P_R \approx \frac{\lambda_{\text{Mu}}^2}{\lambda_{\text{Mu}}^2 + \delta\omega^2}, \tag{12.9}$$

where λ_{Mu} is the Mu reaction rate and $\delta\omega$ is the difference between the muon precession frequencies in Mu and in the radical. The minimum concentration at which a radical can be observed in a TF-μSR spectrum depends on the muonium reaction rate (k_{Mu}), but is typically 0.1–0.5 M for k_{Mu} on the order of 3.7×10^9 M^{-1} s^{-1}, which is the rate of addition of Mu to benzene in *n*-hexane.

The reaction rate of a muoniated radical with some reacting species, S, can be determined from the damping of the radical precession signal, which is related to the linewidth of the TF-μSR line. The damping rate (λ_R) is related to the second-order rate constant, k_R, by

$$\lambda_R = \lambda_0 + k_R \,[S], \tag{12.10}$$

where λ_0 is the damping in the absence of chemical reaction and is frequently due to magnetic field inhomogeneity, and [S] is the concentration of the reacting species.

Muoniated free radicals studied by RF resonance

RF-μSR is the type of μSR that most closely resembles conventional magnetic resonance, since it relies on resonance between an exciting RF field and one of the energy transitions discussed above. This technique is more time-consuming than TF-μSR and is only used in situations where TF-μSR or ALC-μSR cannot be applied, such as in the measurement of A_μ for radicals in dilute solution. There is no loss of polarization going from Mu to the muoniated radical, as the muon polarization is initially aligned along the applied magnetic field. A linearly polarized radio frequency field perpendicular to the external field induces transitions between spin states with opposite muon spins when the frequency of the RF (ν_{RF}) matches the transition frequency. This is seen as a reduction of the time-integrated muon spin polarization at a resonance field.

In the high-field limit (i.e. where the transition frequencies are given by Equation 12.8) the resonance field is given by

$$B_{\text{res}} = \frac{\nu_{\text{RF}} \pm \frac{1}{2}A_\mu}{\gamma_\mu}. \tag{12.11}$$

In this simplified picture the center position of these two resonances is determined by the muon hyperfine coupling constant (RF transitions in a hypothetical muonium system neglecting the proton hyperfine coupling constant) and the splitting by the proton hyperfine coupling constant. Difficulties arise when one of the resonances occurs at a low magnetic field when one must include hyperfine couplings to other nuclei in the radical. RF is used more frequently at the pulsed facilities as continuous use can heat a sample. RF methods are covered in more detail in Chapter 19.

Example 12.4

Fullerenes are not sufficiently soluble in any suitable solvent for TF-μSR measurements. Mu can add to the outside surface of C_{60} to give the muonium-like C_{60}Mu radical in which there is only hyperfine coupling to the muon. The RF-μSR spectrum of the C_{60}Mu radical is shown in the inset of Fig. 12.14. There are resonances at 26.8 and 250 mT where $\nu_{43} = \nu_{RF}$ The expression for ν_{43} is given in Table 12.3. In this measurement the other allowed transition frequencies, ν_{12}, ν_{23}, and ν_{14}, did not match ν_{RF} in the accessible magnetic field range.

Muoniated free radicals studied by longitudinal field repolarization

In cases, such as dilute systems, where the reaction rate for Mu addition, k_{Mu}, is particularly low, the formation of the muoniated radical can be relatively slow, though still on a fast ($\approx$ ns) time scale at the concentrations of interest here. In these cases a two component repolarization curve is measured, with a shape indicative of the product of the polarizations for the prompt initial (Mu) and final state (radical) species. An example of this technique is given in Fig. 12.15, which shows the magnetic field dependence of the initial polarization (the repolarization curve) for pure benzene and a dilute solution of benzene in cyclohexane, together with simulations carried out with the QUANTUM program by J. S. Lord (see Chapter 17 for further examples of simulations using this program).

Avoided level crossing muon spin resonance

Avoided level crossings were introduced in Chapter 4. The ALC-μSR technique involves measuring the time-integrated asymmetry of the muon decay as a function of a magnetic field applied parallel to the initial direction of the muon spin. In contrast to TF-μSR, there is no restriction on the number of muons in the sample at one time so it is possible to run at a much higher incident muon rate, and the dephasing problem of TF-μSR does not apply to longitudinal fields, so it is possible to observe species using ALC-μSR that are formed within $\approx$ 1 μs of muon implantation. This makes it possible to study samples with a low concentration of the precursor molecule (on the order of millimolar for molecules with aromatic functional groups).

In high magnetic fields, the eigenstates of the radical can be approximated by as the product of Zeeman states of the muon, electron, and

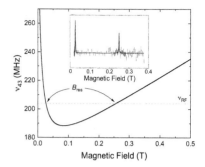

Fig. 12.14 The magnetic field dependence of the ν_{43} transition of C_{60}Mu ($A_\mu = 330.89$ MHz). The applied RF frequency ($\nu_{RF} = 204$ MHz) is denoted by the dotted line. The inset shows the RF-μSR spectrum of the C_{60}Mu radical in dilute decalin solution at 293 K. [Spectrum from P. W. Percival *et al.*, Appl. Magn. Reson. **11**, 315 (1996).]

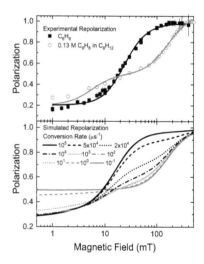

Fig. 12.15 (Top) Repolarization curves of pure benzene and a dilute solution of benzene in cyclohexane. (Bottom) Simulations for different formation rates for Mu addition to benzene. The muon hyperfine coupling constant in the radical and in Mu were fixed at their known gas-phase values, 514.3 and 4463 MHz, respectively. Proton hyperfine coupling constants for the ipso and para protons were included. [Taken from D. G. Fleming *et al.*, Phys. Chem. Chem. Phys. **14**, 10953 (2012).]

nuclei, so there is no evolution of the muon's spin with time and the asymmetry is independent of the magnetic field. At specific values of the applied magnetic field, nearly degenerate pairs of spin states can be mixed through the isotropic and anisotropic components of the hyperfine interaction. The muon polarization oscillates between the two mixing states and this leads to a loss of time-integrated asymmetry. There are three types of resonances, which are characterized by the selection rule $\Delta M = 0, \pm 1$, and ± 2, where M is the sum of the m_z quantum numbers of the muon, electron, and proton spins. The resonances are referred to as Δ_0, Δ_1, and Δ_2 resonances, respectively. The Δ_2 resonance is extremely weak and is rarely observed. There can be only one Δ_1 resonance for each type of radical, but there can be as many Δ_0 resonances as there are nuclei with $I > 0$ in the radical, although in practice, the number of observable Δ_0 resonances is less than the maximum, as many of these nuclei can have small hyperfine coupling constants, which results in very small resonance amplitudes.

The Δ_0 resonance is due to mixing between spin states that have the same electron spin but opposite muon and proton spins, and is observed for muoniated radicals in the solid, liquid, or gas phases. The Δ_0 resonance field depends on both the muon hyperfine coupling constant and the nuclear hyperfine coupling constant, A_X, and is given by

$$B_{\text{res}}^{\Delta_0} = \frac{1}{2}\left[\frac{A_\mu - A_X}{\gamma_\mu - \gamma_X} - \frac{A_\mu + A_X}{\gamma_e}\right], \qquad (12.12)$$

where γ_X is the nuclear gyromagnetic ratio.

The Δ_1 resonance field is given by:

$$B_{\text{res}}^{\Delta_1} = \frac{1}{2}\left[\frac{A_\mu}{\gamma_\mu} - \frac{A_\mu}{\gamma_e}\right], \qquad (12.13)$$

where A_μ is the muon hyperfine coupling constant, γ_μ is the muon gyromagnetic ratio, and γ_e is the electron gyromagnetic ratio. The Δ_1 resonance arises from mixing between spin states with the same electron and nuclear spins but different muon spin directions. These spin states are only mixed in the presence of anisotropy, so the presence of a Δ_1 resonance can be considered to be diagnostic of a frozen state or of anisotropic motion. The resonances are averaged out by isotropic motion on a critical time scale given by the inverse of the hyperfine anisotropy (typically ≈ 50 ns for a cyclohexadienyl-type radical).

Example 12.5

ALC-μSR spectra of the muoniated cyclohexadienyl radical are shown in Fig. 12.16. There are four resonances in the spectrum of C_6H_6Mu in neat benzene. These are proton Δ_0 resonances of the four types of protons in the radical. Analysis of the ALC-μSR data gives the ^1H hyperfine coupling constants in C_6H_6Mu at 298 K listed in the following table.

Nucleus	A_μ (MHz)	$B_{res}^{\Delta_0}$ (T)	A_p (MHz)
H(6)	514.47	2.0774	126.11
H(2,4)	514.47	2.7176	7.47
H(1,5)	514.47	2.8936	−25.14
H(3)	514.47	2.9532	−36.19

Four additional resonances are observed in the ALC-μSR spectra of the $^{13}C_6H_6Mu$ radical. These are ^{13}C Δ_0 resonances of the four types of ^{13}C nuclei in the radical, giving rise to the ^{13}C hyperfine coupling constants in $^{13}C_6H_6Mu$ (again at 298 K) listed in the following table.

Nucleus	A_μ (MHz)	$B_{res}^{\Delta_0}$ (T)	A_C (MHz)
C(3)	512.28	1.8261	53.81
C(1,5)	512.28	1.8840	39.42
C(6)	512.28	2.1803	−34.05
C(2,4)	512.28	2.1849	−35.19

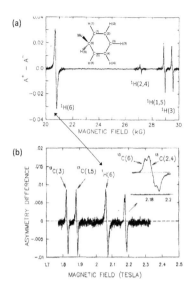

Fig. 12.16 (a) ALC-μSR spectrum of the muoniated cyclohexadienyl radical (C_6H_6Mu) in neat benzene at 300 K. [Taken from P. W. Percival *et al.*, Chem. Phys. Lett. **133**, 465 (1987).] (b) ALC-μSR spectrum of the muoniated ^{13}C-labelled cyclohexadienyl radical ($^{13}C_6H_6Mu$) in cyclohexane at 300 K. [Taken from R. F. Kiefl *et al.*, Chem. Phys. Lett. **143**, 613 (1988).]

12.5 Structure and dynamics

The magnitude and temperature dependence of the hyperfine coupling constants provides information about the structure and dynamics of radicals.

Hyperfine coupling constants of β nuclei

The majority of muoniated radicals to have been studied have the muon in the β position with respect to the unpaired electron, i.e. attached to an atom one removed from radical centre, which is the α position. The reduced hyperfine coupling constant of a nucleus X in the β-position obeys the empirical McConnell relation,

$$A'_X = \left(L + M \left\langle \cos^2 \theta \right\rangle\right) \rho^\pi, \qquad (12.14)$$

where L and M are constants (generally $|L| \ll |M|$), θ is the angle between the $C - X$ bond and the axis of the singly occupied molecular orbital (SOMO), and ρ^π is the spin density in the π orbital on the

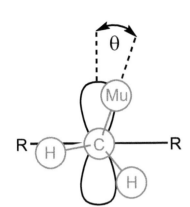

Fig. 12.17 Newman projection of a pseudo-methyl group showing overlap between β-nuclei and SOMO.

α-carbon (Fig. 12.17). Here L accounts for the effect of spin polarization and M arises from hyperconjugation (or direct overlap) between the p_z-orbital containing the unpaired electron and the $C-X$ bond. The hyperfine coupling constant is modulated with the dihedral angle between the p_z-orbital and the $C-X$ bond direction. In the high temperature limit, where the system has an energy far above the potential barriers that hinder internal rotation, all angles θ contribute equally to A'_X, so that $\langle A'_X \rangle$ approaches $L + M/2$ (assuming that $\rho^\pi = 1$). This is one way that the spin density on the α-carbon can be determined. As the temperature goes to zero the $C-X$ bond will adopt the minimum energy geometry, and the hyperfine coupling constant will provide information about θ. As an example, if the $C-X$ aligns along the axis of the SOMO ($\theta = 0°$) then the low temperature limit of the hyperfine coupling constant is $L + M$, whereas if the $C-X$ is aligned at $\theta = 60°$ then the low temperature limit of the hyperfine coupling constant is $L + M/4$. For a CH_3 group, the minima all have equal energy, so due to symmetry, the value of $\langle A'_p \rangle$ is $L + M/2$, independent of temperature. The temperature dependence of the hyperfine coupling constants is given by a Boltzmann average over ground and excited internal rotational states, and the torsional potential can be determined from the temperature dependence of the hyperfine coupling constants.

Hyperfine coupling constants of α-nuclei

The hyperfine coupling constant of an α muon or proton also depends on spin density in the π orbital on the α-carbon. If the radical centre is planar then the α-nucleus would reside in the nodal plane of the SOMO and, as a first approximation, there would be no spin density at the α-nucleus leading to $A_X = 0$. This is not the case due to spin polarization of the $C-X$ bond by the unpaired electron, which results in *negative* spin density at the α-nucleus and a *negative* A_X given by

$$A'_X = -Q\rho^\pi, \tag{12.15}$$

where Q is a proportionality constant that is ≈ 70 MHz. The dominant vibrational mode at the α-carbon is the out-of-plane bend, which moves the substituents out of the nodal plane and into direct overlap with the SOMO, which results in a contribution of positive spin density and leads to the magnitude of A_X to decrease. If the radical centre is non-planar, which will be the case for everything except the methyl radical, then the hyperfine coupling constant of the α-substituent will depend on the spin density at the radical centre as well as the out-of-plane angle, and it will be a delicate balance between spin polarization and hyperconjugation. The hyperfine coupling constants can change substantially with temperature due to the thermal population of vibrational states.

Isotope effects on muoniated radicals

Isotopic substitution has some effects on the vibrationally averaged structure of a radical. The simplest isotope effect to consider is that on the $C - X$ bond. The zero-point energy (ZPE) of the $C - X$ ($X = Mu, H, D$) bond increases as the mass of hydrogen isotope decreases, and the larger ZPE leads to a longer vibrationally averaged bond length due to the asymmetric bond stretching potential i.e. $\langle R_{C-Mu} \rangle_\nu > \langle R_{C-H} \rangle_\nu > \langle R_{C-D} \rangle_\nu$. The vibrationally averaged $C - Mu$ bond length is estimated to be approximately 4.9% longer than the corresponding $C - H$ bond length (Fig. 12.18). Longer bonds tend to result in larger hyperfine coupling constants, either because of better overlap with the singly occupied molecular orbital or because of increased spin polarization of the bond by the unpaired electron. In the muoniated cyclohexadienyl radical A'_μ is 28 % larger than the methylene proton A_p value primarily due to the longer $C - Mu$ bond.

Isotopic substitution of a methyl proton destroys the symmetry of the methyl group by distinguishing one of the particles and Mu substitution has a particularly large effect. The torsional potential is the sum of the electronic potential, which is isotopically invariant, and the zero-point energy, which depends on the isotope. In the case of CH_3, the wells of the torsional potential have the same depth and the barriers have the same height, while the torsional potential of CH_2Mu has one well, which corresponds to the $C - Mu$ bond aligned along the SOMO, that is deeper than the other two and the barrier heights are no longer equal. This change in the torsional potential results in the hyperfine coupling constants changing dramatically as a function of temperature. An instructive example is the temperature dependence of the muon and β-proton hyperfine coupling constants in the muoniated *tert*-butyl radical (Fig. 12.19). The muon hyperfine coupling constant decreases with increasing temperature while the proton hyperfine coupling constant of the protons in the CH_2Mu group increase with temperature. This shows that the $C - Mu$ bond of the CH_2Mu group is aligned along the axis of the orbital containing the unpaired electron at low temperature.

Reorientational dynamics of muoniated radicals

The Δ_1 resonance is suitable for studying anisotropic reorientational motion of radicals in single crystals as well as in polycrystalline or amorphous states. The lineshape of the Δ_1 resonance is extremely sensitive to and characteristic of the type of reorientational motion that the muoniated radical is undergoing. The dipolar hyperfine is described by a tensor with three principal values and which is traceless (i.e. the three components add to equal zero). Theoretical lineshapes have been simulated for axial hyperfine tensors where once component $(D_\mu^\parallel)$ has the opposite sign and twice the magnitude of the other two components $(D_\mu^\perp)$. The resonance has a characteristic asymmetry depending on the sign of the dipolar coupling constant when λ is small. The resonance is asymmetric with the cusp falling on the low-field side when both A_μ^{iso} and $D_\mu^\parallel$ are positive and on the high-field side when A_μ^{iso} is positive and $D_\mu^\parallel$ is negative. Motion of the radical can average the dipolar hyperfine coupling constants and change their value and sign. The result is that

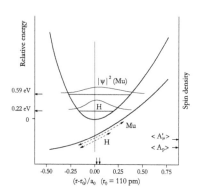

Fig. 12.18 Zero-point energies for H (0.22 eV) and Mu (0.59 eV) are indicated and zero-point distributions for the nuclear positions sketched for the stretch mode, these latter skewed towards the softer side of the potential. The anharmonic contribution comes from the different average bond lengths and the harmonic contribution from the different RMS excursions, via the first and second derivatives of the spin-density curve. [Taken from S. F. J. Cox, Rep. Prog. Phys. **72**, 116501 (2009).]

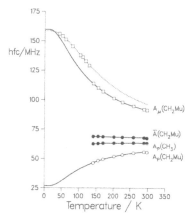

Fig. 12.19 Temperature dependence of hyperfine coupling constants in the muoniated *tert*-butyl radical (MuCH$_2$-$\dot{C}$-(CH$_3$)$_2$ in 2-methylpropene). [Taken from P. W. Percival *et al.*, Chem. Phys. **127**, 137 (1988).] There is a discontinuity in the value of A_μ at the melting point of 2-methylpropene.

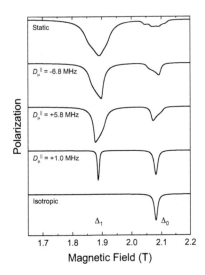

Fig. 12.20 Simulated Δ_1 and methylene proton Δ_0 resonances of the C_6H_6Mu radical with different dipolar couplings. The hyperfine parameters used in the simulation were the values of the C_6H_6Mu radical in high-silica ZSM-5: $A_\mu = 513.8$ MHz, $A_p = 124.9$ MHz, and the muon dipolar tensor of $+9.4$, $+1.0$, and -10.4 MHz.

the shape of resonances in ALC-μSR is sensitive to the motion of the radical and can be used to determine whether the motion of the radical is anisotropic and, in some cases, what is the preferred reorientation axis, as motion modulates the dipolar coupling constant. An axial hyperfine tensor arises if the radical is undergoing rapid uniaxial rotation.

Figure 12.20 shows the simulated spectra of the muoniated cyclohexadienyl radical around the Δ_1 and the Δ_0 methylene proton resonances for different types of reorientational motion. Fast uniaxial rotation about an axis perpendicular to the molecular plane results in a negative $D_\mu^\parallel$ while fast uniaxial rotation about the long axis of the radical results in a positive $D_\mu^\parallel$. The different types of motion will reverse the high- and low-field sides and change the width of the Δ_1 resonance, so it is possible to determine the preferred rotational axis from the line shape of the powder spectra. An axial lineshape with a further reduced width would indicate more extensive averaging by tumbling or wobbling motion superimposed on the rapid uniaxial rotation. The effect of more complex processes on the ALC lineshapes must be simulated numerically.

Muoniated radicals as probes in condensed matter

The structure and dynamics of a number of muoniated radicals, particularly alkyl and cyclohexadienyl radicals, are very well understood and these radicals are now used as probes in more complex systems. Spin labelling with stable free radicals is a well established technique. It is advantageous to have the probe stand out dramatically from the background, which is certainly the case for radicals in a sea of diamagnetic molecules. The benefit of using muoniated radicals is that they are formed *in-situ* and have structures that are very similar to their diamagnetic parents.

Moreover, the hyperfine coupling constants of muoniated radicals are sensitive to the local environment, so measuring the hyperfine coupling constants can provide information about the location of a muoniated spin probe in a system containing hydrophobic and hydrophilic regions. The spin probes also provide information about the local viscosity. In highly viscous environments, where the reorientation rate is comparable to the inverse of the muon dipolar hyperfine coupling constant, a Δ_1 resonance is observed. The shape of the Δ_1 resonance provides information about the type of motion the probe is undergoing.

Example 12.6

ALC-μSR was used to study the reorientational dynamics of polycrystalline norbornene in its plastic phase. Mu can add to the double bond of norbornene. The Δ_1 resonance of the exo adduct is shown in Fig. 12.21 at several temperatures in the plastic phase. The asymmetric shape of the resonance indicates that $D_\mu^{\parallel}$ is negative and the magnitude decreases with increasing temperature due to motional averaging.

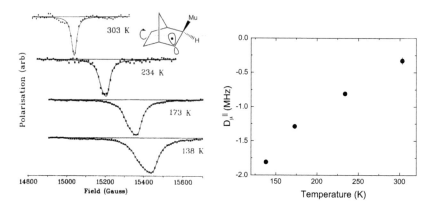

Fig. 12.21 (Left) The Δ_1 resonance of the exo Mu adduct of norbornene at several temperatures in the plastic phase of polycrystalline norbornene. The preferred reorientation axis of the radical is denoted as a dotted grey line. (Right) Temperature dependence of the dipolar muon hyperfine coupling constant of the exo Mu adduct of norbornene in the plastic phase of polycrystalline norbornene. [Taken from E. Roduner *et al.*, Ber. Bunsenges. Phys. Chem. **93**, 1194 (1989).]

Example 12.7

Surfactants are molecules with functional groups that have widely differing affinities for water and that can self-assemble in aqueous solution to form a number of remarkable structures including *lamellar* liquid crystalline phases where the surfactant molecules form arrays of regularly separated bilayer sheets, separated by solvent and micelles. Subtle changes in the microstructure of the bilayer can have dramatic effects on macroscopic behaviour, which can be brought about by changes in surfactant and electrolyte concentrations, or by the addition of a cosurfactant. The cosurfactant may have additional roles, such as being a fragrance or a dye. They are typically present only in very low concentrations, which makes them difficult to study using traditional spectroscopic techniques as the signals from these molecules are swamped by signals from the more numerous solvent molecules.

ALC-μSR has been used to study the partitioning of 2-phenylethanol (PEA) in bilayers of the dichain cationic surfactants 2,3-diheptadecyl ester ethoxypropyl- 1,1,1-trimethylammonium chloride (DHTAC). PEA is a common ingredient in perfumes as it smells like roses. Mu adds to the phenyl ring of PEA to give three isomers shown in Fig. 12.22. DHTAC forms bilayers that undergo a structural change between 330 K; below this temperature, the system is in the L_β phase where the motion of the alkyl chains is limited, and above this temperature, it is in the L_α phase, and the alkyl chains have considerable mobility.

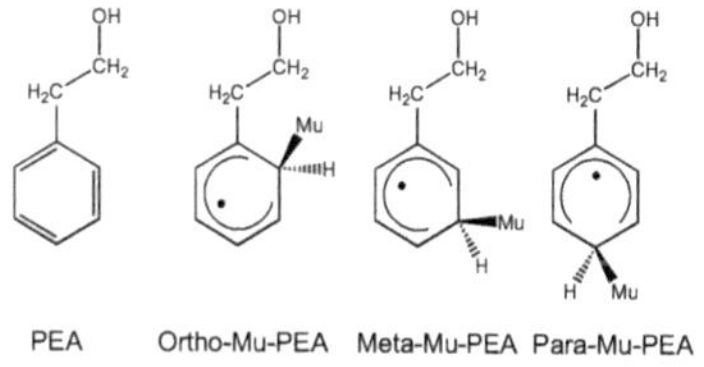

PEA   Ortho-Mu-PEA   Meta-Mu-PEA   Para-Mu-PEA

Fig. 12.22 Structure of PEA and the three Mu adducts of the phenyl ring.

The hyperfine coupling constants of the muoniated spin probe provide information about the local environment, as they are sensitive to interactions of the radical with the neighbouring molecules. The methylene proton hyperfine coupling constants of the three Mu adducts of 2-phenylethan-1-ol (PEA) were measured in a variety of solvents and were observed to shift in a systematic way with the polarity of the solvent (Fig. 12.23). The '% aqueous character' of the radical in a solvent X was defined in terms on the relative position of the Δ_0 resonance in the solvent compared with the positions of the Δ_0 resonances in H_2O and $C_{18}H_{38}$.

$$\% \text{ aq. char.} = \frac{B_{res}^{X} - B_{res}^{C_{18}H_{38}}}{B_{res}^{H_2O} - B_{res}^{C_{18}H_{38}}}. \tag{12.16}$$

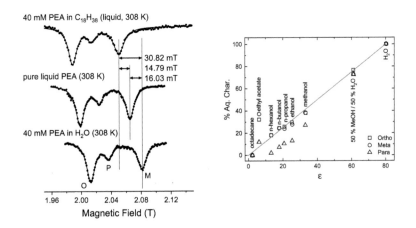

Fig. 12.23 (Left) Δ_0 resonances of the three isomers of the muoniated cyclohexadienyl radical in a dilute solution of PEA in n-octadecane (top), bulk liquid PEA (middle), and in a dilute solution of PEA in water (bottom) at 308 K. (Right) Polarity diagram for 40 mM PEA in various solvents, obtained from the relative resonance positions of PEA-Mu compared to H_2O and $C_{18}H_{38}$. [Taken from R. Scheuermann *et al.*, Langmuir **20**, 2652 (2004).]

The % aqueous character of the ortho and meta isomers of the Mu adduct of PEA were shown to depend linearly on the dielectric constant. This means that one can determine the polarity of the local environment of the cosurfactant by measuring the methylene proton Δ_0 resonance field.

The ALC-μSR spectrum of PEA in DHTAC is shown in Fig. 12.24. In the low temperature L_β phase it has three Δ_0 methylene proton resonances but no Δ_1 resonances, indicating that the motion of the radical is isotropic.

The position of the Δ_0 resonances indicate that the radicals are in an aqueous environment. The PEA-Mu radicals (and by inference the PEA molecules) have difficulty penetrating into the bilayer in the L_β phase because of the tight packing of the alkyl chains. Δ_1 resonances are present in the L_α phase, which indicates that the motion of the radical is anisotropic, likely around the long axis of the molecule, and the Δ_0 resonances have shifted to a lower field, which indicates that the radicals are in a less polar environment. The meta-PEA-Mu isomer appears to be in a more non-polar environment than the ortho isomer and this could be evidence of a water gradient within the bilayer. The low polarity sensed by the para isomer is likely overestimated due to the reasons discussed above. Near the $L_\alpha - L_\beta$ transition there is coexistence between radicals in the bilayer and in the aqueous region.

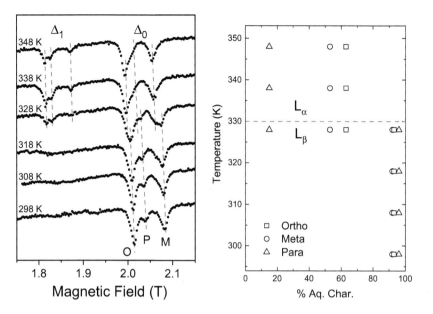

Fig. 12.24 (Left) ALC-μSR resonance spectra of PEA in DHTAC at various temperatures. (Right) Local polarity determined from the Δ_0 resonance positions as a function of temperature. [Taken from A. Martyniak *et al.*, Phys. Chem. Chem. Phys. **8**, 4723 (2006).]

Chapter summary

- Muoniated radicals are formed by addition of Mu to an unsaturated bond. Muonium can also generate radicals by abstracting an atom such as H, forming a MuH molecule.
- Spectroscopic experiments used in chemical applications of μSR include transverse field studies, longitudinal field repolarization, RF resonance, and ALC. Information extracted from these experiments includes reaction rate constants, hyperfine coupling constants, and information about reorientational dynamics.

Further reading

- E. Roduner, *The Positive Muon as a Probe in Free Radical Chemistry: Potential and Limitations of the μSR Techniques*, Springer-Verlag (1988).
- D. C. Walker, *Muon and Muonium Chemistry*, CUP (1983).
- I. McKenzie and E. Roduner, Naturwissenschaften **96**, 873 (2009).
- I. McKenzie, Annual Reports Section C (Physical Chemistry) **109**, 65 (2013).

Exercises

(12.1) Positronium (Ps $= [e^+e^-]$) is another one-electron atom. Why is it not considered to be an isotope of H like Mu? How do the reduced mass, ionization potential, and Bohr radius of Ps compare with Mu, H, D, and T?

(12.2) Why is O_2 removed from samples for μSR measurements? Consider water where the solubility of O_2 at 1 bar and 298 K is 2.6×10^{-4} M and the pseudo-first-order rate constant is $(1.8\pm0.1) \times 10^{10}$ M^{-1} s^{-1}.

(12.3) A TF-μSR spectrum of the muoniated methyl radical (CH_2Mu) obtained at 5 kG has three lines with absolute frequencies of 32.93, 67.77, and 168.46 MHz. Assign the three lines and calculate A_μ of the radical. Predict what the spectrum would look like at 1 kG.

(12.4) Given the proton hyperfine coupling constants of the C_6H_6Mu radical listed in the first table in Example 12.5, how is the unpaired spin density distributed around the cyclohexadienyl ring? What are the values of ρ^π for the carbon atoms and what do they sum to?

Part III

Practicalities of muon spectroscopy

In this part we discuss the practical details of how to do muon spectroscopy: how to make the muons in the first place, what instrumentation is required to do a muon experiment, and what you actually need to do to perform an experiment.

- Before an experiment can even begin, you need to make your muons! The physics and technology behind this is described in Chapter 13.

- For the experiment to work, a chain of instrumentation needs to be in place. Chapter 14 describes how spectrometers are designed for pulsed and continuous sources, as well as what is needed for studying small samples.

- Chapter 15 describes various practicalities that are important in performing muon experiments and doing the analysis.

Making muons

[1]Accelerator, target, and beamline design are all well-established fields in their own right, and are discussed in the specialist literature. Some of the simulation techniques that are used for designing and optimizing muon beamlines are briefly described in Chapter 17.

While all muons are equal, some are more useful for particular experiments than others. The experiment to be performed therefore decides what source of muons is needed and how they can be transported to the sample. Muons are produced by natural and man-made sources. These both rely on the same physical processes – protons colliding with light nuclei – but the muons produced naturally as secondary cosmic rays allow little manipulation after they are produced. Producing muons in a particle accelerator gives far greater control and the possibility of far higher fluxes. The control comes largely from the design of the beamline transporting the muons from a production target to the experimental sample.[1]

Obvious experimental requirements for a beam of muons are the muon charge, spin polarization, and beam-spot size at the sample position, but more subtle details like beam contamination with other particles and when muons arrive in the sample relative to each other, also make a huge practical difference. Cosmic ray muons give little selectivity in any of these respects, so we just mention them briefly before moving on to concentrate on the muons produced by particle accelerators. Such muons have carefully controlled parameters suitable for the types of μSR experiments described in this book.

13.1 Muon production

While particle physics offers many processes that produce muons, the process most suitable for making intense muon beams is to collide energetic protons with the protons and neutrons in light nuclei within a production target to produce charged pions. These charged pions then almost exclusively decay into muons. Depending on the energy of the protons, processes either produce single pions, e.g.

$$p^+ + p^+ \rightarrow p^+ + n + \pi^+, \tag{13.1}$$

$$p^+ + n \rightarrow p^+ + p^+ + \pi^-, \tag{13.2}$$

or at higher energies, two pions can be produced, e.g.

$$p^+ + p^+ \rightarrow p^+ + p^+ + \pi^+ + \pi^-. \tag{13.3}$$

$$p^+ + n \rightarrow p^+ + n + \pi^+ + \pi^-. \tag{13.4}$$

In the laboratory frame of reference, the threshold for the proton energy E_p to produce single pions is $E_p > 280$ MeV and production peaks between 500 and 1000 MeV. The threshold for double pion production is $E_p > 600$ MeV and production increases up to $E_p \approx 1.5$ GeV, beyond which it saturates.[2]

Pions have spin quantum number $S = 0$ and decay with a time constant of 26.03 ns, via the two-body reactions

$$\pi^+ \rightarrow \mu^+ + \nu_\mu \tag{13.5}$$

and

$$\pi^- \rightarrow \mu^- + \bar{\nu}_\mu. \tag{13.6}$$

These pion decays are caused by the **weak interaction**, outlined in Appendix E. In the rest frame of the pion, the muon and neutrino emerge with equal and opposite linear momentum $\boldsymbol{p}$ so that linear momentum is conserved [Fig. 13.1]. Neutrinos have spin $S = 1/2$ and necessarily possess negative helicity (i.e. their spin is always directed antiparallel to their momentum). In order that angular momentum is conserved, the positive muon produced in eqn 13.5 is then also forced to have negative helicity in the pion rest frame, so *its* spin is directed opposite to that of the neutrino. This means that the positive muons produced via pion decay at rest are 100% spin polarized, with their spin directed antiparallel to their momentum.[3] The polarization loss in a practical beamline due to finite angle is discussed in the next section.

Example 13.1

Muons produced by primary cosmic rays striking the upper atmosphere, at altitudes around 15 km, have a typical energy of 4 GeV and a flux at the Earth's surface of around 1 cm^{-2}min^{-1}. They follow an intensity distribution $\propto \cos^2 \theta$, where θ is the zenith angle, and have a typical stopping range of about 20 m in water. The energy spectrum however extends to $E_\mu \gg 1$ TeV and it is possible to use these high energy muons to measure several kilometres through the Earth, with enough patience. This useful property has led to the primary contemporary use of cosmic ray muons being tomography, i.e. examining what is inside very large objects that are otherwise inaccessible. Some examples are probing magma chambers inside active volcanoes, finding unknown chambers inside pyramids, or scanning cargo containers.

The standard approach to producing muons at a proton accelerator is to use a thin graphite target at a focal point in the proton beam. This provides a source of protons and neutrons to allow pions to be produced. Most muon facilities have other targets further down the proton beamline in order to make more muons, or other particles such as neutrons. This configuration favours a target that allows most of the protons to pass through (> 95 %) with minimal scattering and ideally minimal production of neutrons that would activate the target assembly. Graphite is an excellent material for this, as it has a low atomic number, is easy to handle, and has a high melting point, compatible with the challenge of cooling the target in vacuum. With proton accelerators operating at powers significantly below 1 MW, it is possible to use fixed muon targets with edge cooling, but for MW-class accelerators, rotating

[2]Among contemporary muon sources, only J-PARC ($E_p = 3$ GeV) has significant double pion production, with ISIS ($E_p = 800$ MeV), PSI ($E_p = 590$ MeV), TRIUMF ($E_p = 520$ MeV), and MuSIC ($E_p = 392$ MeV) all primarily using single pion production.

[3]The antineutrino produced in eqn 13.6 has positive helicity, but again this results in the negative muons being spin polarized.

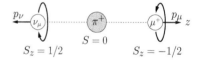

Fig. 13.1 The decay of a positive pion. Conservation of linear and angular momentum forces the spins of both the neutrino and the muon to be antiparallel to their respective linear momenta.

wheel targets are used to reduce the heating of individual points on the target.

13.2   Surface and decay muons

One could classify the types of particle accelerators used to produce muons by the technicalities of the acceleration process, but the main practical division is between pulsed accelerators that produce bunches of muons with significant time separation and (quasi-)continuous accelerators that deliver a steady beam of muons. The time structures of these two types of source are illustrated in Fig. 13.2. Each type of muon source has particular characteristics (Table 13.1) that provide advantages and disadvantages for different types of experiments.

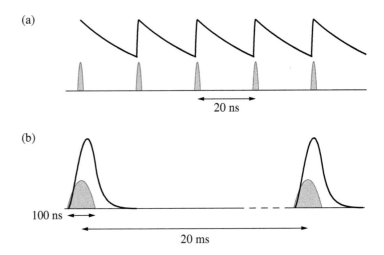

Fig. 13.2 Proton and muon time structures of (a) a typical quasi-continuous source (PSI) and (b) a typical pulsed source (ISIS). Proton pulses are shown in grey and the solid line shows the corresponding time structure for the rate of muons in the beam. Since the 26 ns pion lifetime is comparable to the gap between the pulses in the quasi-continuous accelerator, the time structure of the muons produced is substantially smoothed out, however the residual sawtooth modulation leads to spurious signals at the cyclotron operating frequency (50 MHz in this case) and its harmonics, which can be a problem for some types of experiment. For the pulsed accelerator the pion lifetime is comparable to the proton pulse width and acts to broaden the muon pulse relative to the proton pulse. Note that for clarity in each case, the muon intensity has been scaled up against the proton intensity by an arbitrary factor.

There are two principal types of muon beamline: **surface beamlines** that transport muons produced by pions decaying at rest close to the surface of the target and **decay beamlines** that initially collect and transport pions. These pions then decay in flight to produce muons, which are transported by the beamline to the sample.

Surface beamlines can only work for positive muons (since negative pions would be captured in the target before decay) but offer a high

	Continuous	Pulsed
Accelerator type	Cyclotron	Synchrotron
Beam structure	Effectively DC	Periodic pulse
Count system (TD)	Individual muon trigger	Muon pulse trigger
Rate limit (TD)	Single muon in sample ≈ 6 kHz	Detector deadtime ≈ 400 Hz per detector
Rate limit (TI)	> 1 MHz	≈ 400 Hz per detector
Background	Undetected second muon	Only cosmic rays
Time range	≈ 0.1 ns–10 μs	≈ 0.1–30 μs
Time resolution	Detectors/electronics ≥ 80 ps	Muon pulse width ≈ 100 ns

Table 13.1 Properties of continuous and pulsed muon sources. Muon measurements either record the muon spin polarization as a function of time after implantation (time-differential, TD) or else measure a time averaged muon spin polarization (time integral, TI). The numbers given here are typical limits.

[4]For details, see Appendix E.

[5]Estimation of the penetration range for muons is covered in Chapter 17.

flux and spin polarization at a modest momentum. They have a peak flux at around[4] $p_\mu = 29$ MeV/c and $E_\mu = 4$ MeV. As discussed above, because the pions decay at rest and the helicity of the neutrino forces the positive muon spin to be antiparallel to its momentum, the muons collected in one direction are perfectly polarized with respect to that direction. Since a real beamline collects muons from a range of angles it cannot preserve the polarization perfectly with respect to the average beam direction, but the spin polarization at the sample is normally > 95 %. The muons transported can penetrate up to 200 mg/cm^2, dependent on material, which leads to a typical penetration depth of ≈ 500 μm, again dependent on material.[5] This limits the thickness of beamline and instrument components, and sample environment windows between the production target and the sample.

In decay beamlines, the muons that are selected are those either emitted along the direction of pion momentum ('forward') or opposite to it ('backward'). These are the directions with maximum spin polarization. The minimum forward muon momentum is therefore ≈ 29 MeV/c, as for surface muons, but backward muons can be obtained down to almost zero momentum with decreasing intensity. The maximum momentum is constrained by the accelerator energy or by the magnetic fields provided by the beamline components. Forward muon beams are contaminated by π and e, but backward beams are relatively clean, with both able

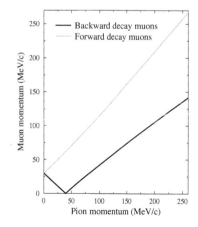

Fig. 13.3 Two types of decay muons that can be produced in a beam line for a given pion momentum. The muon momentum is either forward or backward with respect to the pion reference frame. Backward muons originating from pions with momenta below 39 MeV/c are travelling backwards in the laboratory frame as well as in the pion frame.

to provide up to ≈ 80 % spin polarization. The polarization of a backward decay beam of μ^+ is parallel to the momentum, in contrast to the antiparallel property of the μ^+ in forward decay and surface beams. Penetration into materials can typically be several g/cm^2. Decay beamlines select a particular momentum of pions to transport while they decay and must also select the corresponding muon momentum (forward or backward) to be transported to the sample (Fig. 13.3). They are generally more expensive than surface muon beamlines since a long solenoid is the component of choice for pion transport after momentum selection. This can be justified by their flexibility in providing both positive and negative muons, and enabling the use of momenta that can penetrate thick walled sample containers, such as pressure cells[6] or glassware. A specific example of a decay beamline is described in more detail in Chapter 21.

[6]High pressure experiments are covered in Chapter 21.

13.3 Beamline components

A trivially simple muon beamline would be an evacuated pipe connecting the target producing the muons to the sample. While this would result in at least some of the muons being implanted in the sample, allowing a basic type of experiment to be performed, significant improvements in the flux of muons and the rejection of the background from other particles can be achieved by adding other components between the production target and the sample. An example of a beamline is shown in Fig. 13.4, from which it can be seen that various elements are incorporated along the path of the muons. We will now look in turn at each of the main elements in a beamline.

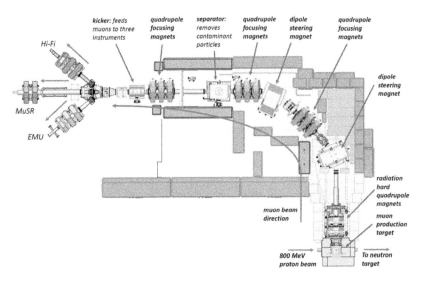

MuSR is the name of a spectrometer at the ISIS facility. Curiously the word 'musr' is the word for 'Egypt' in Egyptian Arabic. The origin of the word is unknown, but is thought to predate the Arabic language.

Fig. 13.4 An example of a muon beamline at ISIS. The muons are produced at the production target at the bottom right and are transported to the HiFi, MuSR, and EMU instruments at the top left.

Focusing magnets: quadrupoles and solenoids

Focusing the muon beam from the production target to capture muons emitted over a larger solid angle and focusing onto the sample to give a smaller beam size both increase the muon flux on the sample. Intermediate focal points allow for longer beamlines to include the components described below. A beamline that only focuses along a straight line will still transmit many other charged and neutral particles, and the charged particles will be poorly focused except for a specific momentum.

The two conventional choices of focusing components are quadrupole magnets and solenoids. Quadrupole magnets individually focus along one axis and defocus along the other axis perpendicular to the beam direction (z), as shown in Fig. 13.5. They are therefore used in pairs (doublets) that give different magnification along x and y or in triplets that preserve the image shape. Solenoids focus both axes simultaneously and can be built around curves to act as bending elements. Quadrupoles are generally cheaper than solenoids.

A third option is the *Dai-Omega* (large solid-angle) beamline, which uses superconducting coils on axis and at carefully chosen positions close to the production target to increase the solid-angle capture roughly twenty-fold compared to using a quadrupole magnet close to the target. This considerable increase in beam intensity trades off against the complexity of construction, the lower muon polarization, and the higher cost.

Bending magnets: dipoles and curved solenoids

Introducing bends into the beamline both allows flexibility in placing the instrument relative to the target and also prevents uncharged particles and those with the wrong sign of charge from reaching the sample. Because the magnetic Lorentz force is $\boldsymbol{F} = q\boldsymbol{v} \times \boldsymbol{B}$, where $\boldsymbol{v}$ is velocity, the radius of curvature R for a particle with charge q and mass m in a magnetic field B is $R = qB/mv$ (see Exercise 13.4). For this reason, a curved beamline with a magnetic field applied selects particles with a particular sign of charge and containing particles with a particular momentum. Such a beamline would therefore transport particles (potentially a mixture of muons, electrons, and pions) with different velocities (because of their different masses).

Placing a collimating slit at a focal point between two dipole magnets in a bend allows control of the range of particle momenta transmitted to the sample, commonly known as the **momentum bite**. Alternatively, combining two opposite dipoles and quadrupoles without an intermediate slit can give an **achromatic beam** with unrestricted momentum. Curved solenoids can do a similar job. In either case, further rejection of unwanted particles will be required using the velocity filtering properties of a separator.

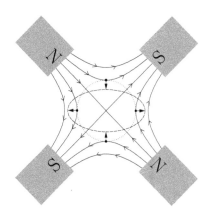

Fig. 13.5 A schematic diagram, showing the magnetic field pattern inside a quadrupole magnet. A beam of charged particles (with circular cross-section, as shown in the dotted circle) travelling along the axis of the magnet will experience a Lorentz force which focuses the beam along one axis and defocuses along the other axis (resulting in a beam of elliptical cross-section, as shown by the dashed curve).

Separators and spin rotators

The magnetic beamline components described above are sufficient to produce a well-focused beam of particles with the desired momentum and charge, but cannot select the particle type. To remove the remaining contamination, largely positrons, a device called a **Wien filter** or **separator** is used. This uses the Lorentz force $\boldsymbol{F} = q(\boldsymbol{E} + \boldsymbol{v} \times \boldsymbol{B})$ acting on the charged particles to act as a velocity selector with the $\boldsymbol{E}$ and $\boldsymbol{B}$ fields perpendicular to each other and to the particle velocity. The values are chosen to exactly balance for the muons, whereas the positrons are deflected.[7] Typical separators are around 1 m long, with an electric field of ≈ 1 MV/m and magnetic field of ≈ 5 mT (see Example 13.5).

If we ignore relativistic corrections and hence neglect the tiny value of $g - 2$, the cyclotron frequency $\omega_{\rm c} = eB/m$ (measuring the frequency a particle with charge e orbits in space) and the Larmor frequency $\omega_{\rm L} = ge/2m$ are equal. This implies that muons moving in transverse magnetic fields essentially keep their spins aligned with their momentum.[8] Electrostatic deflection leaves the spin direction unchanged but changes the momentum. The combined effect within the separator therefore acts to rotate the muon spin in flight. Devices dedicated to separating positrons from muons in the beam typically rotate the muon spin by a few degrees, but by increasing their length and the applied fields, this angle can be significantly increased towards 90°, producing a **spin rotator**. This higher amount of spin rotation is used for experiments that need the ability to apply significant magnetic fields to the sample region that are perpendicular to the muon spin without bending the beam (if the magnetic field is applied perpendicular to the muon momentum, the beam is deflected and in fields as small as ≈ 0.1 T this is significant compared to the typical size of samples). The signal size for a rotation angle of θ grows as $\sin^2 \theta$, whereas practical considerations make higher rotation angles difficult, so often $60° < \theta < 70°$ is used.[9]

Example 13.2

Charged particles with velocity $\boldsymbol{v}$ will pass undeflected through a Wien filter of length L if the force due to the uniform $\boldsymbol{E}$ field is perfectly balanced by the force due to the $\boldsymbol{B}$ field. Taking $\boldsymbol{v} = v\hat{\boldsymbol{z}}$, $\boldsymbol{E} = E\hat{\boldsymbol{x}}$, and $\boldsymbol{B} = B\hat{\boldsymbol{y}}$, then balance occurs if $q\boldsymbol{E} + q\boldsymbol{v} \times \boldsymbol{B} = 0$ and hence $v = E/B$, and so will only work with a particular velocity of particle (irrespective of the value of q). A muon will take a time L/v to pass through a Wien filter (assuming $v \ll c$) and hence its spin will rotate by an angle $(\gamma_\mu B)L/v$. Thus, in a spin rotator, the ratio E/B is held fixed (so that only muons with speed v pass through), but B (and hence $E = vB$) is varied to adjust the degree of spin rotation.

Kickers

Beamline components can also control the time structure of the muon beam. This is particularly significant at pulsed muon sources driven by synchrotrons, which generate trains of multiple proton pulses, typically closely-separated pairs that correspond to diametrically opposite particle bunches in the synchrotron. Proton-pulse pairs produce corresponding pairs of muon pulses separated by less than the muon lifetime. The consequence of this is that the spectra from each pulse that enters the sample overlap and any oscillations present in the decay asymmetry partially or completely cancel, apart from at specific fields related to the pulse separation. The simplest way to overcome this issue is to throw one pulse of muons away.[10] More modern beamlines make use of both pulses by sending each to specific beamlines using electrostatic or electromagnetic **kickers**. These apply a large electric or magnetic field perpendicular to the beam momentum while one pulse passes through and then they change this field over to another value in the gap between the pulses.

Similar devices can be used in continuous muon beamlines to prevent a second muon from entering the sample while waiting for another muon to decay. This offers a significant reduction in the background counts (see Fig. 15.9), increasing the effective width of the time window available for the measurement. However, this is achieved at the expense of some spectral distortion at early times. At pulsed sources, kickers can also be used to slice off the ends of the pulses, giving a smaller uncertainty in when muons arrive in the sample and therefore a better time resolution and an extended frequency range.[11]

[10]This was the method deployed with the so-called UPPSET device at ISIS before 1992.

[11]See Fig. 14.5.

Collimation

A small size of the muon beam spot incident on the sample is the key to minimize the fraction of muons stopping outside the sample. Reducing this size is a vital part of controlling the background in the muon signal. It is also commonly used to control the rate of muons entering the sample to match the maximum rates of the detectors (see Table 13.1). There are two ways to control the size of the beam spot. Local collimation uses lead rings shaped to match the sample placed at the end of the beamline just before the muons reach the sample. Relatively thick lead (> 2 cm) is needed to minimize the contamination of the detector signal by decay positrons originating from muons stopping in the collimator. Remote collimation generally uses metal slits at a focal position earlier in the beamline, which has the advantage that decay positrons from muons stopping in the slits cannot reach the detectors. In analogy with optics, alternate focal points are momentum and real space foci, respectively. Collimation at a momentum focus reduces the momentum bite, while collimation at a real space focus reduces the beam waist; either can be reduced linearly to zero.

Shutters

It is also important to stop the muon beam for tasks like sample changes and maintenance. Switching off the accelerator producing the muons at the whim of individual instrument users is rarely practical when many instruments share the same accelerator. Fortunately, muon beams are relatively easy to stop. For surface muons, 5 cm lead plates are sufficient, compared to ≈ 1 m of steel required to stop beams of thermal neutrons. High-energy decay-muon beamlines generally switch off a bending magnet as well as lowering a shutter to stop the beam. Switching off bending magnets is also used in surface beamlines to provide a quicker stop if safety interlocks are triggered.

We now have a picture of how muons can be produced by an accelerator and transported to our μSR experiment via a beamline. In the next chapter we discuss the instrumentation that allows the measurement to be carried out.

Chapter summary

- Cosmic ray muons have typical energies of several GeV and at the highest energies can penetrate kilometres through the Earth.
- The principal means of producing muons is the collision of high energy protons with light nuclei such as carbon.
- Accelerator-based muon sources produce controlled beams of muons suitable for particular experiments.
- Surface beamlines transport muons produced when pions decay at rest.
- Decay beamlines transport muons produced from pions decaying in flight.
- Beamline components allow control of the muon beam charge, momentum, spin polarization, and direction, as well as the beam size and time structure.

Further reading

- J. J. Beatty, J. Matthews, and S. P. Wakely *Chapter 30. Cosmic rays* in C. Patrignani *et al.* (Particle Data Group), Chin. Phys. C **40**, 100001 (2016) and 2017 update.
- A wealth of information on particle accelerators and beamline design is available in the proceedings of the CERN accelerator schools, such as CERN Report: 94-01 and CERN Report 2005-004. More recent lecture slides are available from https://cas.web.cern.ch.

Exercises

(13.1) What order of magnitude is the total rate of cosmic ray muon production over the visible sky from a typical point on the Earth? The following pieces of information may be helpful in making an estimate: the peak in muon flux production occurs at a height above the Earth's surface of $h = 15$ km and is approximately 100 muons/m^2/s/sr, and the area of sky visible above the Earth's surface at a height h is given by $A = 2\pi Rh$, where $R = 6400$ km is the radius of the Earth.

(13.2) Instruments at a pulsed muon source count around 10^8 muon decays per hour. For how long would such an instrument have to count to record a muon decay more than 30 μs after implantation? A significant source of background on such instruments comes from cosmic ray muons hitting the detector array. Typical detector arrays on such instruments present an area of 600 cm^2 to the vertical. What rate of cosmic ray muon events would be expected on such an instrument? How long after muon implantation would this exceed the count rate from muons decaying in the sample? You might want think about the consequences of the detectors only being active for ≈ 30 μs every 20 ms at a typical pulsed source.

(13.3) How small a field can you measure using muons at a pulsed source? How big? Compare with frequencies easy to resolve at a continuous source.

Assuming that a useful number of muon decays can be measured 25 μs after implantation, what is the smallest magnetic field that can be measured? At a pulsed source the total width of the muon pulse can be ≈ 100 ns, introducing an uncertainty in when the muons enter the sample. What upper limit does this place on the magnetic field that could measured? At a continuous muon source detector timing is the principal limit on observing high precession frequencies. Supposing a detector system had a time-resolution of 200 ps, what would be the largest observable field?

(13.4) Show that the radius of curvature R of a beam of particles with charge q and momentum p in the presence of a magnetic field B is given by $R = qB/p$.

(13.5) The separator in a muon beamline removes parasitic particles and also rotates the muon spin. How fast do surface muons ($E_\mu = 4$ MeV) travel along the beamline compared to the speed of light? How fast do positrons with the same momentum travel? Supposing a particular separator has length $L = 2$ m and magnetic field strength $B = 40$ mT, what electric field is needed to permit surface muons to travel through? Calculate the spin rotation angle for surface muons travelling through this separator.

Instrumentation

In this chapter we will briefly discuss the instrumentation that is used in a typical muon experiment and which forms the signal measurement chain for a muon spectrometer.

14.1 Spectrometer elements

The basic components of a muon spectrometer were introduced in Chapter 2. Figure 14.1 shows an example of a spectrometer, opened up for insertion of a sample.

The key concept for the operation of a spectrometer is that when the muon decays, the positron is emitted preferentially along the direction of the muon spin at the time of the decay. The positron from a muon decay is detected and its individual lifetime is measured by a Time-To-Digital converter (TDC). A simplified flow diagram of this process is shown in Fig. 14.2. The individual muon lifetime is the length of time between the arrival of the muon and the outgoing positron signal. It is digitized into an integral number of units of the TDC time resolution as j. The set of times j is turned into the familiar data that the experimenter can analyse by incrementing the j-th element of a positron **histogram**[1] that is stored within the Data Acquisition (DAQ) system. We now describe each piece of the logical chain in turn.

[1] A histogram is an array representing frequency of events (here a count rate, in counts/bin, where the bin is the time interval between the centres of two successive elements). The TDC resolution defines the minimum bin width that can be used in subsequent analysis, but the counts in adjacent sets of bins can added together to produce a lower resolution histogram with a greater number of counts in each enlarged bin, this process is known as **bunching** or **rebinning**.

(1) **Scintillators** - the first stage in the detector chain. These are usually pieces of a plastic material that emit a pulse of light when hit by a positron (or a muon). Scintillator materials are chosen to have a very high efficiency and a fast nanosecond response with a rapid recovery. The scintillators are typically wrapped in a reflective coating to increase the intensity of the light pulse and they are usually attached to an optical wave guide so the light can be transported to a photomultiplier. The amplitude of the light pulse depends on the track length of the positrons within the scintillator material and so typically these are of the order of 10 mm in thickness. Thin scintillators, typically 0.1 to 0.2 mm in thickness, are used to detect incoming muons for spectrometers at continuous muon sources.

(2) **Photomultipliers** (PMs) convert a pulse of light into a voltage pulse. This voltage pulse can then be turned into a digital signal further down the chain. The traditional PM is a vacuum tube, with

Fig. 14.1 The MuSR instrument at ISIS. It is composed of two opposite rings of detectors, one of which is visible. Here each detector is composed of a scintillator and a long light guide ending in a photomultiplier tube. Muons enter the spectrometer from the left along the indicated path. The scintillators cover the internal cylindrical space of the Helmholtz coil pair that provides the magnetic field at the sample position. These coils are hidden behind the light guides with position indicated by the dashed circles. The cryostat (here a closed cycle refrigerator, its radiation shield and vacuum tail removed) is rotated into place along the dashed arrow, after the sample has been mounted and the shield and tail have been replaced.

a photocathode converting the light pulse into electrons. These are accelerated in N successive stages by an electric field. Each electron impinging on an intermediate electrode (dynode) produces m secondary electrons, resulting in a very large total gain, $\approx N^m$. This device requires high voltages (typically in the kV region) and overall electron paths of tens of cm, inevitably affected by the magnetic field of the spectrometer. Thus PMs are coupled to scintillators by long plastic light guides to remove them from high magnetic field regions. Recently, all-solid-state silicon photomultipliers (SiPM) have become available, which are based on avalanche photodiode technology. They have been widely incorporated into recent spectrometer designs, since they are much more

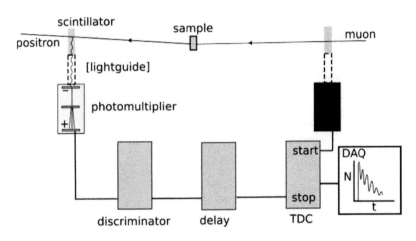

Fig. 14.2 The TDC receives a start trigger and a stop signal from logically similar detector chains. The stop chain is shown in detail and starts with a scintillator converting the charge passage into a light pulse and a photomultiplier then converts the light pulse into an analogue voltage pulse. This is followed by a discriminator that is used to convert the analogue pulse into a standard pulse for the TDC. A delay can also be inserted at this point for fine tuning the time alignment of different detectors. Note that the TDC used at a pulsed source must be able to handle multiple stop signals originating from the many muons that arrive within the same pulse.

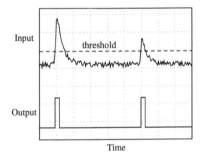

Fig. 14.3 Illustration of the discrimination of two pulses from a PM. The input is the raw output voltage from the PM with a degree of noise. The output after discrimination shows well-defined logic pulses, that are triggered when the leading edge of the incoming pulse passes the preset voltage threshold.

[2] At a pulsed source, it would typically be a Cherenkov detector situated upstream from the spectrometer. At a continuous source, it would typically be a thin scintillator detector placed in the beam just before the sample position. The relative timing of the signals reaching the electronics depends on the cable lengths. In both cases, the cable length of the start detector is shorter than that of the positron detectors, and time zero must be calibrated (Section 15.3).

[3] At ISIS the time bins are usually 16 ns, whereas at a continuous source, sub ns bins can be used.

compact and far less sensitive to the magnetic field. It may not be necessary to use light guides with SiPMs.

(3) **Discriminators** - a discriminator is used to turn the analogue voltage signal generated by the PM into a well-defined logic pulse. The leading edge of the input voltage passes through a preset threshold, set to maximize the detection of the chosen particles and minimize the detection of other particles or spurious triggering by analogue electronic noise. The output is a logic pulse of preset width which can then be counted as one **event**. Figure 14.3 shows this process.

(4) **Time-to-Digital Converter** (TDC) - as noted earlier, this is the part of the measurement chain which times the muon decay events. The TDC must be triggered by a start pulse, which is correlated with the **time zero** reference. This usually comes from a detector in the beam line before the sample position.[2] The decay events are then grouped into evenly spaced time bins[3] allowing the DAQ system to store a histogram of counts versus time for each detector.

(5) **Data reduction and analysis** - after the detector histograms have been produced by the TDC and associated data acquisition software, data analysis software can be used to group detector signals and convert the signal to asymmetry, ready for the process of data fitting and analysis.

The type of muon source has a significant effect on the spectrometer characteristics, leading to inherent strengths and limitations, which we describe next for the two cases of pulsed and continuous sources.

14.2 Pulsed sources

At a pulsed source, protons are accelerated using a synchrotron, and extracted and transported towards a muon production target, which is usually graphite. The synchronous nature of the acceleration produces a bunched time structure. In the case of the ISIS synchrotron, the protons have a double-pulse time structure, where each pulse has a FWHM of 80 ns. This pulse width limits the useful time resolution, leading to a maximum measurable frequency of approximately 10 MHz.

Finding a way to deal with the high instantaneous event rates is essential at a pulsed source, as there are many muons per pulse, but relatively few pulses per second. As a result of the pulse structure, the rate of decay positrons is highest just after the muon pulse arrives. This means that the detectors take the largest 'hit' at short times, where there are a large number of positrons to count. The minimum time between pulses in order for the detector system to register them as distinct events is called the **dead time**. If a second positron arrives after a first positron within this time window, it will not be counted and so a spectral distortion will occur due to this counting loss.

In order to work around this, one needs to reduce the probability of two positrons arriving within the dead time for a particular detector by building an array of detectors with a large amount of segmentation, so that the peak positron rates for individual detectors are reduced to manageable levels. For example, the EMU spectrometer at ISIS has a detector system with 96 detector segments, where count rates of over 120×10^6 events per hour can be achieved with minimal distortion.

Another issue to consider is data timing. The TDC is triggered by a pulse that is typically provided by a Cherenkov detector further upstream in the beamline and close to the muon production target. This trigger then starts the collection of the positron histogram. The time zero for data analysis purposes is the centre of the muon pulse, however due to the finite pulse width, the end of the pulse is a little later. Analysis of the muon relaxation signal should start from the time at which all the muons have finished arriving, this time is called the first good data time. Figure 14.4 summarizes these times.

One of the features of a pulsed source is the relatively limited frequency response compared to a continuous source. This is illustrated in Fig. 14.5. Although it might be expected that making the accelerator proton pulse narrower would improve the frequency response, the pion lifetime is actually a strong limiting factor [Fig. 14.5(a)]. A more effective approach is to improve the time structure of the muon pulse by extracting a short slice from the original muon pulse [Fig. 14.5(b)]. This results in a significant loss of muon flux, so the method requires a high intensity pulsed source in order to maintain an acceptable muon flux after the pulse slicing.

Taking all these factors into account, we can summarize the general strengths and limitations of spectrometers at pulsed sources. The strengths are:

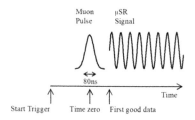

Fig. 14.4 A schematic showing the difference between the start trigger, time zero, and first good data.

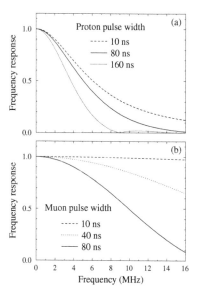

Fig. 14.5 (a) The frequency response for a pulsed source versus proton pulse width. Note that the pion lifetime limits the degree of improvement of the frequency response by making the proton pulse narrower. (b) Effect of direct muon pulse slicing. The frequency response can be significantly improved in this way (at the expense of muon flux).

- High data rates are possible with no intrinsic count rate limit.
- Low intrinsic background and the ability to measure asymmetry out to long times.
- Separate pulses can easily be directed to different instrument spectrometers.
- Straightforward synchronization to pulsed excitation sources such as lasers and RF.
- Much higher repetition rates than those currently used are possible, in principle, for a future dedicated muon source. This could lead to greatly increased data rates compared to those available at present.

On the other hand, the general limitations are:

- Relatively broad pulse structure that limits the time resolution and frequency response.
- The large number of muon decays at short times can produce spectral distortion via count loss due to detector dead time.
- Since all the muons arrive within one narrow pulse, it is not possible to use active collimation with veto detectors to reject individual events and define a small beam spot.
- All existing pulsed sources have been designed for a primary purpose other than producing muons, with the result that pulse structure and repetition rate are not best optimized for muon production and spectroscopy.

14.3   Continuous sources

The characteristics of a continuous source, such as the SμS at the Paul Scherrer Institut, Switzerland, are in many ways complementary to those of a pulsed source. Particular strengths of spectrometers at continuous sources are:

- The quasi-continuous time structure of the source allows high frequencies to be measured, limited only by the timing resolution of the detector system.
- There is no need to use large numbers of detectors to avoid problems with dead time.
- Continuous sources were historically purpose-built for muon production (although not primarily for μSR), resulting in narrower muon beams compared to those at pulsed sources.
- It is possible to define a more accurate logical event using gating of several physical detectors for coincidence (logical *and*) and veto (logical *not and*). Consequently, veto detectors can be used for active collimation and for dealing with small samples.

- Having a small spot size together with simpler and smaller detectors means that they can be placed very close to the sample position, which is particularly beneficial for high transverse field measurements, since one can reduce both the phase errors from positron bending and the timing errors from variable light paths in long light guides.

The limitations one has to consider are:

- Because there is no distinct accelerator time structure to trigger the measurement, one needs another way to trigger the start of the data collection. This requires an additional incoming muon detector to be inserted in the beam before the sample.

- Muon **pile-up** should be avoided, this is where a second muon arrives before the first muon has decayed and makes paternity of the detected positron uncertain. This means that the incoming muon rate has to be reduced to match the measurement time window.

- The requirement for only one muon at a time in the sample provides an intrinsic count rate limit.

- In line with muon pile-up, uncorrelated detector hits lead to a non-negligible steady background count, which limits the maximum time available for measurement.

The detector requirements are different from those at a pulsed source, in particular they do not need a fast recovery, as they do not have to cope with large instantaneous rates, but they require a very fast signal rise time, to reduce their timing jitter and increase the frequency passband. The significant background counts at a continuous source mean that studying slow frequencies or relaxation rates is relatively hard at continuous sources, imposing a practical time limit of around 8 μs in the normal mode of operation, although special fast kicker techniques can also be used to remove the background and extend the usable time range (see Chapter 15).

14.4 Small samples

It is often necessary to study samples that are smaller then ideal, due to challenging synthetic procedures or because one needs to study a single crystal with limited dimensions. Two different approaches have been developed at ISIS and PSI to allow such measurement to be made. At ISIS the **fly-past** technique can be used. This involves having a beam spot larger than the sample, in which case some of muons will pass by, or fly-past, the sample without stopping near the sample. At PSI, a different approach is employed, which involves being able to exclude any muons that pass around or through the sample by placing a **veto detector** behind the sample. This means that very small samples (>1 mm^2) can be measured. More details on these fly-past and veto techniques are provided in Chapter 15.

Chapter summary

- Muon spectrometers incorporate a chain of elements to detect muon decay positron events and turn these events into the raw data of the experiment.
- Spectrometers at pulsed and continuous sources each have specific characteristics that can suit them to particular types of experiment.

Further reading

- G. H. Eaton *et al.*, Nucl. Instr. and Meth. A **269**, 483 (1988).
- T. Matsuzaki *et al.*, Nucl. Instr. and Meth. A **465**, 365 (2001).
- G. H. Eaton *et al.*, Nucl. Instr. and Meth. A **319**, 342 (2004).
- D. Tomono *et al.*, Nucl. Instr. and Meth. A **600**, 44 (2009).
- S. R. Giblin *et al.*, Nucl. Instr. and Meth. A **751**, 70 (2014).
- A. Amato *et al.*, Rev. Sci. Instr. **88**, 093301 (2017).
- D. J. Arseneau *et al.*, Hyp. Int. **106**, 277 (1997).

Doing the experiment

15

In the previous two chapters we have met the components of a muon beamline and spectrometer. We are now ready to discuss how to do the experiment itself! This chapter covers some of the practicalities associated with setting up a muon spectroscopy measurement and analysing the data obtained from the study.

15.1 Experimental setup

Some of the important factors to consider when setting up the experiment are set out below.

Matching sample to beam

The sample area and areal mass density are important factors for a muon spectroscopy experiment. These determine how many muons are implanted in the sample compared to those outside, when the sample is placed in a muon beam with a particular spot size and beam momentum. The muons stopping outside the sample will produce an additional signal, which should ideally be minimized. Maximizing the match between the muon beam spot and the sample can be done either by controlling the spot size or by optimizing sample geometry, or both. The beam spot size is typically controlled by using a collimator close to the sample position or by using slits placed at an intermediate focal point in the beamline. For controlling sample geometry, the requirement is to maximize the area presented to the beam, subject to maintaining enough thickness to stop all of the incident muons in the sample. At pulsed sources the beam spot is relatively large and is frequently larger than the sample. In this case a silver mask can be used to catch muons stopping outside the sample (see Fig. 15.1). The muons implanted in the silver mask[1] relax very slowly due to the small nuclear moment and it is relatively straightforward to allow for this background component in the data analysis.

Surface muons emerge from the production target at one quarter of the speed of light and are slowed by interactions with all of the material between the production target and the sample. These materials are split between unavoidable materials and additional materials that are placed in the beam to act as degraders. Typical fixed and unavoidable materials would be the beamline windows, and the cryostat windows and radiation shields. The muon stopping range after these fixed materials would

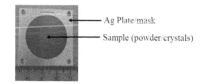

Fig. 15.1 Standard silver sample holder and mask for MuSR at the ISIS muon facility.

[1] Any non-magnetic material with negligible nuclear moments will do as long as there is negligible muonium/radical formation. Titanium is used for furnaces.

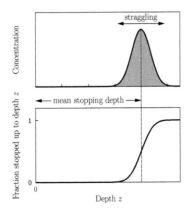

Fig. 15.2 A schematic diagram showing the stopping profile for muons as a function of depth, z, measured from the surface of the sample. Notice that it is often helpful to think of depth in terms of *areal density* (measured in mg/cm^2) since the stopping of muons is largely controlled by the mass density of material, although it is also a weaker function of the type of material.

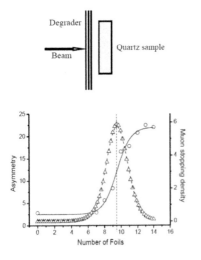

Fig. 15.3 A variable thickness stack of 25 μm Ti foils acting as a degrader to tune the muon stopping in a quartz sample held in a CCR cryostat at the ISIS muon facility. The simulated stopping profile (triangles) and measured asymmetry (circles) are also shown.

[2]The limit is the saturation of the pile-up veto system, when too many muons hit the veto counter.

typically be 100 to 200 mg/cm^2 (note that the stopping power is material dependent). Muons will stop at a mean depth inside the sample, but they do not all stop at precisely the same depth due to the phenomenon of **straggling**; this means that there is a *distribution* of stopping depths, as shown in Fig. 15.2, which shows a schematic stopping profile. If large amounts of sample are available, then placing 200 mg/cm^2 across the area of the beam will result in a large signal. For thinner samples, where the areal density is not sufficient to completely stop the beam, degraders must be used to optimize the stopping profile of the muons against the sample thickness. The thickness of the degraders is chosen so that the centre of the sample is located at the mean stopping depth, with the objective that all the straggling occurs within the sample.

Figure 15.3 shows an example exploring the effect of using Ti foil as a degrader in front of a quartz sample. The plotted asymmetry is obtained in successive TF experiments, where Ti contributes with a full diamagnetic Mu$^+$ asymmetry signal and quartz with negligible Mu$^+$ asymmetry. From this figure the optimum degrader setting corresponds to seven foils, the point at which the muons are just able to pass through the degrader. The muons will then be completely stopped in the sample if it has a stopping power equivalent to five Ti foils.

A similar result may be obtained by using kapton foils in front of a metal plate. Since kapton in TF experiments yields a negligible diamagnetic Mu$^+$ signal, the experimental plot is reversed (full asymmetry with no foils, nearly zero asymmetry with many foils). Further information about muon stopping profiles and their modelling is given in Chapter 17.

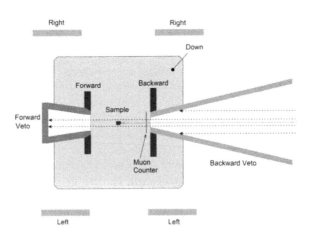

Fig. 15.4 Studying smaller sized samples at a continuous source. Veto detectors are used to exclude events from muons that do not stop in the sample.

When the sample is very small (down to a beam area of a few mm^2), a special experimental setup is required to improve the signal to noise ratio. In the case of a continuous beam, very small samples can be measured without significant loss of count rate by actively vetoing the muons not stopping in the sample using veto detectors (Fig. 15.4).[2] For a pulsed

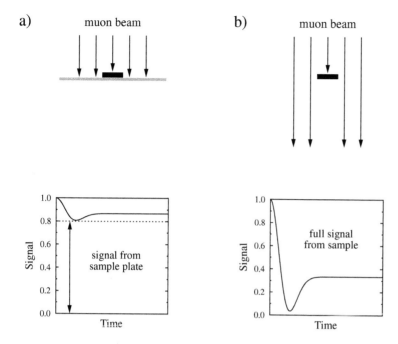

a) muon beam

b) muon beam

Fig. 15.5 Studying smaller sized samples at a pulsed source. (a) The regular configuration using a sample plate results in a significantly reduced signal fraction from the sample. (b) Full signal fraction from the sample can be recovered using the fly-past configuration.

source, small sample measurements are possible by suspending the sample in the beam, which allows muons not implanting in the sample to fly-past, and implant and decay at a position that is well downstream from the spectrometer, where they do not contribute significantly to the signal (Fig. 15.5).[3]

[3]To ensure that the flown-past muons stop downstream from the detectors, a long and broad vacuum-chamber extension is employed.

Thermal environment

In the case of cold-finger cryostats, the joint to the finger and thermal shielding are crucial and powder samples can be difficult to manage due to poor thermal contact. In this case thermal grease or GE varnish diluted with alcohol can be used to make good thermal contact between the sample grains and the sample holder. Single crystal samples can be glued to the sample holder using a small amount of GE varnish. N-grease (or H-grease depending on the temperature range of study) is used to make a good thermal contact between the sample holder and cold-finger/blade of the cryostat. For exchange gas cryostats, it is not necessary to glue the samples tightly to the sample plate. In this case, a powder sample can be packed tightly in silver or aluminium foils and attached to the sample plate with thin adhesive tape that is suitable for the temperature range. For single crystal samples, adhesive tape can also be used to hold them tightly to the sample plate. Knowing the properties of the materials used for the mounting configuration is very important, as any superconductivity or magnetism not originating from the sample is likely to be a serious problem for the measurement.

Time resolution and measurement window

At a continuous source it is necessary to choose the width of the measurement time window and the choice has implications on the data rate. For example, measurements can be done with a 5 or 10 μs time window at PSI and the data rate is nearly double for a 5 μs window compared to a 10 μs window. In situations where the relaxation signal is fast and the relaxation is completed at early times, a smaller time window and a finer time resolution will usually be used to optimize the data rate.

In contrast, at a pulsed source there is no link between the measurement time window and the data rate. The window is then usually fixed to 32 μs, which is around 14 times the muon lifetime. The time resolution in this case is generally limited by the muon pulse, however techniques such as RF spin rotation or laser excitation can produce data with much finer intrinsic time resolution (see Chapter 19).

15.2   Calibrations

Balancing factor and full asymmetry

When working in longitudinal geometry, a small transverse field is usually applied to the sample to calibrate the relative efficiency of the forward and backward detectors, as discussed Chapter 2. A calibration can also be done with a metallic reference sample (typically a silver plate) to determine the full asymmetry a_0 of the instrument configuration. The calibration parameter[4] α is dependent on sample position and detector efficiencies. It needs to be determined for each sample above any magnetic ordering temperature and it is adjusted such that the signal oscillates symmetrically about the time axis.

[4]Recall from eqn 2.8 that we have

$$A(t) = \frac{N_F(t) - \alpha N_B(t)}{N_F(t) + \alpha N_B(t)}.$$

Background asymmetry

[5]Hematite is an antiferromagnet with a Néel temperature of 948 K and a large internal field at the muon site, so a pure sample does not produce a diamagnetic muon signal in an applied TF. Moisture can affect the magnetism and lead to a small diamagnetic asymmetry, so hematite powder is baked before use to remove any absorbed moisture.

For quantitative studies of asymmetry fractions, it is desirable to make an independent measurement of the signal fractions originating from muons stopping in the sample and the background fraction from muons stopping in other places, such as the sample holder. Calibrations can be made using weak transverse field for materials in the sample holder with different, but well-known levels of diamagnetic asymmetry, e.g. silver, which shows full asymmetry, compared with hematite (Fe_2O_3), which has negligible diamagnetic asymmetry.[5]

Zero field

For ZF studies where the relaxation rate is expected to be weak, it is important to avoid any small residual transverse magnetic fields, as this would lead to a slow precession component in the data that would be indistinguishable from a slow Gaussian-like relaxation. Spectrometers often have compensation coils that allow the X, Y, and Z components of the residual field, as measured by a field probe at the sample position, to be reduced to the level of μT. In pulsed source spectrometers, such as those at ISIS, this adjustment is done continuously for ZF measurements using a permanently installed triple-axis field sensor driving the compensation coils via three independent feedback systems.

15.3   Data characteristics

Data analysis is the crucial step for turning a muon spectroscopy measurement into a potentially meaningful scientific result. The data analysis for muon spectroscopy can be carried out in either the time domain or the frequency domain. Time domain analysis is much more common in muon spectroscopy. Analysis in the time domain is also recommended before a frequency domain analysis, in order to ensure that a correct setup is being used for the frequency domain analysis. This section describes some of the general features of muon data that need to be taken into account when doing the data analysis. The subsequent two sections then describe some specific aspects of time domain and frequency domain analysis.

Raw detector arrays

The muon decay positrons are random decay events and thus the distribution of counts expected within each detector array time bin follows the Poisson probability distribution. This distribution is completely defined by the count rate r, which is the expectation value of the number of counts n within the time bin, i.e. $r = \langle n \rangle$. The Poisson probability distribution for n takes the form

$$P(n) = \frac{r^n e^{-r}}{n!}, \tag{15.1}$$

which is illustrated for several different average count rates in Fig. 15.6.

The variance of the Poisson distribution is equal to its average r, hence an estimate of the measurement error can be provided by the square root of the measured counts. This is a good estimate for a large number of counts, but becomes less reliable as the count rate reduces at later times in the histogram. In evaluating measurement errors in muon experiments, a second approximation is usually made that the errors are normally distributed, i.e. the Poisson distribution is approximated by a

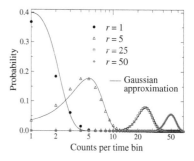

Fig. 15.6 The Poisson distribution eqn 15.1 for the counts per bin for several different average rates r. The solid lines show the Gaussian approximation to the Poisson distribution given by eqn 15.2. This approximation works well at high count rates, but becomes less accurate at lower count rates.

corresponding Gaussian distribution

$$P_{\mathrm{G}}(n) = \frac{1}{\sqrt{2\pi r}} \exp\left[-\frac{(n - r)^2}{2r} \right]. \qquad (15.2)$$

This approximation is very convenient, as it allows straightforward propagation of errors in the data analysis. Again, the approximation is good at larger count rates, but becomes less reliable at lower count rates (see Fig. 15.6). As the count rate decays with time, the signal to noise ratio also becomes smaller, following $\sqrt{n}$, which has an exponential decay rate that is half that of the raw muon signal. This results in the signal to noise ratio roughly halving every 3 μs [see Fig. 15.7(a)]. It is a particular characteristic of muon spectroscopy that measuring the muon relaxation signal out to long times requires the accumulation of a large number of decay positron events to compensate for this reduction in signal to noise ratio.

It is often useful to bunch together the data from successive raw array bins to give an array with wider bins and more counts per enlarged bin [Fig. 15.7(b)], providing a better signal to noise ratio at the expense of time resolution. The new bin width is typically an integer multiple of the raw bin width and this ratio is called the bunch, or rebinning factor. In this fixed binning mode, the bunch factor should be chosen carefully so as not to unintentionally lose any high frequency component in the data. On the other hand, a large bunch factor can be useful to intentionally suppress high frequencies when focusing the analysis on low frequency components.

Another way of binning is to use a variable bin width that starts narrow at early times, but increases steadily in width as time increases to provide some compensation for the falling count rate [Fig. 15.7(c)]. This variable rebinning can be used when the evolution of polarization at long times is relatively slow. Variable rebinning can also be used with a bin width that is tuned to maintain a constant error versus time [Fig. 15.7(d)], however the differences in bin width become very large in this case.

Rebinning is often used for graphical purposes to make a clearer presentation of data which would otherwise be hard for the eye to follow at the raw resolution level.

Pulsed source data

The high instantaneous muon rate associated with the pulsed time structure poses a challenge for the detector system, as the large gain of a photomultiplier implies a finite time for restoring its high voltage level after one detection. This corresponds to a **dead time** τ_{d} after each event during which counts are missed. This counting loss leads to distortion in the asymmetry signal at early times for pulsed source data and puts a limitation on the use of high counting rates. Figure 15.8 illustrates the distortion of the detector signals at early times as the event rate is increased. Without correction this leads to early time distortion of

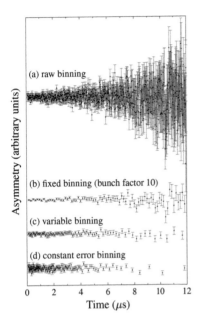

Fig. 15.7 (a) For the raw array bin width (16 ns in this case) the errors become very large at long times. (b) Adding the content of m raw bins together improves the signal to noise ratio at the expense of time resolution; here the bunch factor m is 10. (c) Using a steadily increasing m factor provides a variable rebinning that controls the signal to noise ratio at longer times. (d) The pattern of increasing bin width can be chosen to maintain a constant error.

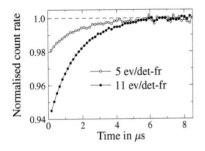

Fig. 15.8 An example of the uncorrected count rate at early times normalized to the undistorted count rate for a high intensity pulsed source. The early time distortion increases with overall event rate specified as events per detector per frame (a frame is one cycle of the proton accelerator, which will usually contain one muon pulse per instrument, but may contain more than one pulse if no kicker has been used).

the asymmetry. This effect however can be modelled, calibrated, and corrected for each detector. Data analysis programs designed for pulsed source data are able to carry out this correction automatically.

Continuous source data

The time difference between a pair of signals in the muon detector and in a positron detector must be recorded as a good muon decay event only when the 'muon → positron' correlation is unambiguous. A few ambiguous conditions can be encoded and rejected by appropriate gating of the logic pulses from different detectors. For instance, if two or more positron detectors fire within the chosen time window after a proper muon TDC start, or else if a second muon arrives within the time window before any detected positron (the pile-up event already mentioned in Chapter 14), then the muon-positron correlation cannot be assigned uniquely and the whole sequence is rejected by the instrument electronics.

In addition to this detected pile-up condition, positron detectors may be hit by particles not correlated with the decay of a muon stopping in the sample. These events, for example those due to decay positrons from muons stopping upstream in the beamline, escape the coded rejection conditions and lead to a steady **time-independent background signal**. In a continuous source this constant uncorrelated background count rate is particularly significant at long times where it outweighs the signal from muon correlated events.

In contrast, a pulsed spectrometer, typically recording events for 32 μs every 20 ms, is exposed to an effective background flux that is much lower, being reduced by the duty factor $0.032/20 = 1.6 \times 10^{-3}$. The uncorrelated background is suppressed in comparison by the same factor and is virtually unmeasurable. Figure 15.9 compares the background signals between a continuous and a pulsed source. When analysing continuous source data out to longer times, it is essential to take account of this constant background signal.

Since the background originates mainly from particles coming down the beamline, it is possible to use a fast kicker system to isolate the spectrometer from the beamline once a muon has been detected in the spectrometer. This gives the possibility of making low background measurements at a continuous source, although at reduced count rate compared to that available at a pulsed source (Fig. 15.9, dotted line).

Time zero and detector phases

The **time zero** must be determined by examining each detector array at early times (never rely on information stored in the data file, if you have not recorded it yourself!). For a pulsed source, time zero corresponds to the mid point of the rising edge of the histogram counts. For a continuous source the implantation time zero is generally indicated by a sharp **prompt peak** in the histogram.[6] However, the time zero relevant for the analysis is the beginning of the spin dynamics. An apparent

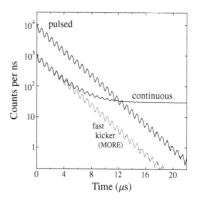

Fig. 15.9 Comparison of typical detector signals at continuous and pulsed sources. The continuous source has a significant background that is absent at the pulsed source. The continuous background can be removed however by a fast kicker system, such as the 'Muons On Request' (MORE) system provided at the PSI GPS instrument (result shown as the dotted line).

[6] The prompt peak is due to a single particle, e.g. a positron travelling along the beam, detected by both the muon and the positron counters. It gives time zero within few tenths of a ns, the difference being the time of flight of a positron travelling at the speed of light over spectrometer distances.

difference between the two arises in transverse field geometry, where the stray magnetic field along the incoming muon path induces a spin precession before muons are implanted. The effective timing shift is of a few ns at most and it is hardly relevant for measuring relaxation rates. But detector phases can be determined in the transverse field measurement of a diamagnetic signal and any offset in time zero shows up as an apparent linear shift of the phase ϕ with frequency f of the form $\phi = \phi_0 + qf$. The parameters ϕ_0 and q can be determined by making a series of transverse field calibration measurements at different fields. If q is expressed in degrees per MHz then a time zero correction of $q/360$ in μs should be added to the original implantation time zero value to get the correct spin time zero setting.[7]

[7]Beware that delayed state formation of one muon species alters its apparent time zero with respect to other species.

15.4 Time domain analysis

Muon spectroscopy data are measured in the time domain and analysis in the time domain is thus the most direct and straightforward approach. At continuous sources the first logical step is removing the background. This can be achieved in two ways: either by determining it from the bins before time zero and subtracting from the data,[8] detector by detector, or by including a separate parameter in the fit model of each individual detector. In the second option, formation of the pair asymmetry function eqn 2.8 and its further analysis can only be done after the background signals have been fitted and subtracted from each detector signal. The main steps in the time domain analysis procedure are outlined here.

[8]In error propagation, the analysis software must allow for the fact that the Poisson estimate of the standard deviation $\sqrt{n}$ refers to the total number of counts, n, including the background.

Identifying the model

In the following we describe models for the time-dependent asymmetry $A(t)$, the key quantity both for direct fitting of the experimental data array reduced by eqn 2.8 and for fitting individual detector counts defined by eqn 2.11. Physics dictates the ingredients of the model, which may be very complex.

In general the asymmetry will contain a number of components and each component can be labelled by an index i. For m components the asymmetry can be typically expressed as

$$A(t) = \sum_{i=1}^{m} A_i G_i^{\text{osc}}(t) G_i^{\text{rel}}(t) + A_{\text{bg}} G_{\text{bg}}^{\text{rel}}(t). \tag{15.3}$$

There are many possible origins for these multiple components, e.g. there may be more than one muon site and more than one physical phase contributing components to the recorded data. The evolution of the transverse and longitudinal polarization is in general different, e.g. as seen in eqn 5.20, so it is often useful to treat these two polarization terms as separate components. The transverse response has both an oscillatory behaviour described by a function $G^{\text{osc}}(t)$ and a relaxation $G^{\text{rel}}(t)$, whereas the longitudinal[9] response is purely relaxational, i.e. it

[9]Often longitudinal terms are hard to distinguish between multiple sites or phases, and one longitudinal term will be used to account for many components.

has $G^{\text{osc}}(t) = 1$. Transverse and longitudinal terms have distinct asymmetry fractions that in general depend on specific sites in a crystal, and the orientation of the spin polarization and applied field. However for a polycrystalline sample in ZF they are in the simple ratio 2:1. The term A_{bg} in eqn 15.3 is a separate background component that is often needed to describe muons that do not stop in the sample.[10] The relaxation functions $G_i^{\text{rel}}(t)$ would typically be a Lorentzian, a Gaussian, the Abragam relaxation function, or any other of those discussed in Chapter 5. Note that, since for all relaxation functions $G(t) = 1$ for $t = 0$, asymmetry fractions must obey the sum rule $\sum_i A_i + A_{\text{bg}} = a_0$, where a_0 is calibrated experimentally.

The oscillatory factor G_i^{osc} may be a simple cosine or else a more complex oscillatory function, such as a Bessel function reflecting a spin density wave. We have also encountered functions that encompass their own transverse and longitudinal components, such as the Kubo-Toyabe function reflecting disordered spins, or its dynamicized version (see Chapter 5). Additional uncorrelated relaxation mechanisms may be present, which allows any of these functions to be multiplied by a simple relaxation function.

At this point we give a couple of examples of how eqn 15.3 would be used in practice for typical experimental situations.

[10]This background term in the asymmetry function should not be confused with the uncorrelated background signal found in the count rate histograms of continuous source data, as discussed in Section 15.3.

Example 15.1

One common type of measurement is to study the ZF asymmetry response in an ordered magnet. For a single muon site and a polycrystalline sample eqn 15.3 becomes

$$A(t) = A_{\text{T}} \cos(\gamma_\mu B_0 t) G_{\text{T}}^{\text{rel}}(t) + A_{\text{L}} G_{\text{L}}^{\text{rel}}(t) + A_{\text{bg}} G_{\text{bg}}^{\text{rel}}(t), \qquad (15.4)$$

where G^{osc} has been assigned to a simple cosine function of B_0, the internal field at the muon site, the symbols L and T label the longitudinal and transverse components at the single stopping site, and $A_{\text{L}} = A_{\text{T}}/2$. $G_{\text{L}}^{\text{rel}}(t)$ would typically be a Lorentzian relaxation function reflecting the spin dynamics, whereas a slow Gaussian would typically be used for $G_{\text{bg}}^{\text{rel}}(t)$ to account for the weak nuclear dipolar relaxation in a metallic sample holder. $G_{\text{T}}^{\text{rel}}(t)$ could be more complex, potentially having contributions from any static inhomogeneity in B_0 and also from dynamic fluctuations.

Example 15.2

Another standard type of measurement is to study the TF rotation signal in an applied field B_{app} where the single sample rotation signal with amplitude A_{s} reflects the local field at the muon site, B_0, which is shifted from the applied field by an amount that depends on the effective local susceptibility at the muon site. We consider for simplicity a forward-backward pair asymmetry. In the high field limit there is no longitudinal term, only the transverse oscillation term. In this TF configuration the background term will have, not only the relaxation term, but also an oscillation term corresponding to B_{app} and eqn 15.3 becomes

$$A(t) = A_{\text{s}} \cos(\gamma_\mu B_0 t) G^{\text{rel}}(t) + A_{\text{bg}} \cos(\gamma_\mu B_{\text{app}} t) G_{\text{bg}}^{\text{rel}}(t). \qquad (15.5)$$

Quality of fit

The quality of fits using different relaxation functions can be compared using the residual deviations of the fit model with respect to the data points, quantified by the reduced χ^2 parameter, χ_r^2, which is χ^2 normalized to the number of degrees of freedom in the fit, i.e.

$$\chi_r^2 = \frac{1}{N-\nu} \sum_{i=1}^{N} \left(\frac{y_i - y_i^m}{\sigma_i} \right)^2, \tag{15.6}$$

where N is the number of data[11] points y_i with errors σ_i, y_i^m are the corresponding fitted model values, and ν is the number of variable parameters in the fit. The number of degrees of freedom is $d = N - \nu$. The values of χ_r^2 produced by random data sets that are well described by the model are distributed about a mean value of 1.

When d is large, as is usually the case for muon data, the χ_r^2 probability distribution approaches a normal distribution with standard deviation σ given by $(2/d)^{1/2}$. If χ_r^2 is significantly higher than expected according to this probability distribution, then the model is likely to be a poor description of the data. On the other hand, if χ_r^2 is significantly lower than expected according to the distribution, one can conclude that the model is likely to be too sophisticated for the level of accuracy provided by the data. In this case the model is essentially fitting the noise as well as the data. A good quality fit of the data aims for the 'Goldilocks zone' where χ_r^2 is neither too large nor too small, but within the main peak region of the probability distribution (Fig. 15.10).

Let us close this section with a few general suggestions.

- The construction of the model should always be guided by a guess on the physical origin of each component. It should be kept as simple as possible.

- A very useful tool is the plot of the residuals, the difference between the data and the model versus time. These residuals can be compared to the Poisson statistical noise: a systematic deviation beyond the statistical error bars should be addressed by improving the model.

- In complex models one should keep an eye on parameter standard errors and on the correlation among parameters, contained in the error matrix. Its diagonal terms are the parameter variances, and the off-diagonal terms provide the correlation between parameters, normalized by their standard deviations. Certain parameters are intrinsically correlated, e.g. asymmetry fraction and decay rate (when the decay rate is very slow), but normalized correlations above ≈ 0.7 may indicate that the model is over-parameterized. One should also be aware that the standard errors can be trusted only if you are making independent use of each parameter. When computing a function of several parameters, error propagation should include their cross correlations.

[11]The **data array** y_i may be either the individual histogram, or the asymmetry appropriate for a group of histograms, in which case σ_i is calculated by error propagation.

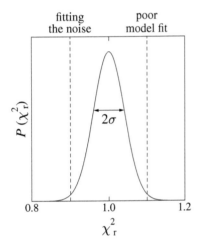

Fig. 15.10 The expected probability distribution of χ_r^2 for $d = 1250$. Obtaining too large a value of χ_r^2 in a fit suggests a poor match between model and data, whereas too small a value suggests that the model is fitting the noise as well as the data.

Fitting individual and multiple groups

The fit of the ZF or the LF asymmetry, obtained by data reduction on the FB group of detectors, is the simplest option. However, in several experimental situations it is preferable to fit multiple groups simultaneously within a single run. This is the case for instance for TF data, where the precession plane shown in Fig. 2.6 can be spanned by multiple sets of detectors, leading to more groups than just F and B.

A multiple-group fit has many data arrays, each with its appropriate version of the model used for calculating the chi-squared function of that data array. In multiple-group fitting there are local parameters that change for each group, such as initial count rate, phase, and background, along with global parameters that apply to all groups, such as frequency and relaxation rate. The global fit optimizes the total chi-square, equal to the sum of all chi-squares for the groups. It reduces the overall number of free parameters, compared to making sequential independent fits of the groups. This comes at the cost of dealing with a larger number of free parameters in the single combined optimization, compared to the more limited number in an individual group optimization.

Sequential and global fitting

Results of an experiment often consist of a sequences of runs scanning a physical quantity, such as the temperature or applied field. The initial approach to analysis is usually to fit the runs in the sequence one after the other to establish which models provide a good description of the data and to examine the trends in the parameters obtained. At this stage in the data analysis it is often found that certain parameters are not varying significantly across the sequence. For instance the total longitudinal asymmetry fraction A_L and the background asymmetry fraction A_bg, typically determined by sample geometry, are not expected to vary across the sequence. These constant parameters can then be fixed at their average values while the data sequence is refitted, to obtain a more stable assessment of the parameters that are varying with the scanned quantity, typically frequencies and relaxation rates.

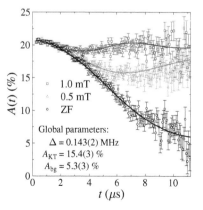

Fig. 15.11 An example of global fitting to a sequence of runs measured in LF. The simultaneous fit is made to a static Kubo-Toyabe function with three global fitting parameters.

Example 15.4

For the case of global fitting of temperature dependent data, consider instead data having a longitudinal relaxation rate in the fast fluctuation regime where it is proportional to a characteristic time $\tau(T)$ that follows an activated behaviour. The global parameters in this case are the activation energy E_a and the pre-exponential factor τ_∞ for the characteristic time entering the relaxation rate, $\tau(T) = \tau_\infty \exp(E_a/k_B T)$, plus A_L and A_{bg}. This activated relaxation model implies using the set of experimental temperature values T of each run as fixed parameters in the model.

Care should be taken with the use of global fitting, since one is imposing a particular model on the data, which may not be fully justified. However, by testing different sequential and global models against the data, and by keeping an eye on the partial chi-squares of all individual runs, it should be possible to gain confidence that the finally chosen global model is a good description of the data. It should also be noted that global fitting can be very computationally demanding when working with a large number of runs, since a very large number of data points and fitting parameters are involved in a single optimization.

Rotating reference frame

When studying slowly relaxing high frequency rotation signals with a relatively narrow frequency spectrum, it can be convenient to transform them into a rotating reference frame (RRF) representation, which allows the slow relaxation envelope of a fast oscillating signal to be followed more easily.[12] The frequency of the rotating reference frame ν_0 is chosen to be close to the main frequency in the measured spectrum, but in a region without any significant spectral intensity.

The simplest configuration is where 4 detectors are placed at $90°$ intervals in the precession plane of the muon, which is a detector geometry often found in μSR spectrometers at continuous sources (Fig. 15.12).

Since this geometry ensures that the detector pair signals 1,3 and 2,4 are in quadrature, a complex asymmetry signal in the laboratory reference frame can be defined as

$$A^{lab}(t) = A_R^{lab}(t) + iA_I^{lab}(t), \tag{15.7}$$

where

$$A_R^{lab}(t) = [A_1(t) - A_3(t)]/2 \tag{15.8}$$

and

$$A_I^{lab}(t) = [A_2(t) - A_4(t)]/2, \tag{15.9}$$

with $A_1(t)$ to $A_4(t)$ being the asymmetry signals in the four detectors. Following the RRF transformation the signals become

$$A_R^{RRF}(t) = A_R^{lab}(t)\cos(2\pi\nu_0 t) + A_I^{lab}(t)\sin(2\pi\nu_0 t),$$
$$A_I^{RRF}(t) = A_I^{lab}(t)\cos(2\pi\nu_0 t) - A_R^{lab}(t)\sin(2\pi\nu_0 t). \tag{15.10}$$

[12]T. M. Riseman and J. H. Brewer, Hyperfine Interactions **65**, 1107 (1991).

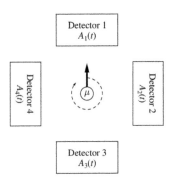

Fig. 15.12 The geometry used for RRF analysis. Four detectors are arranged at $90°$ intervals in the precession plane of the muon (the magnetic field is perpendicular to the page). The initial positron emission asymmetry is aligned with the first detector.

The RRF transformation is a frequency modulation process that produces both sum and difference frequencies in the output. The slow difference frequency is the desired signal and the fast sum frequency must be filtered out using a smoothing or rebinning procedure on the transformed data. Figure 15.13 demonstrates the use of the RRF transform on some high transverse field data.

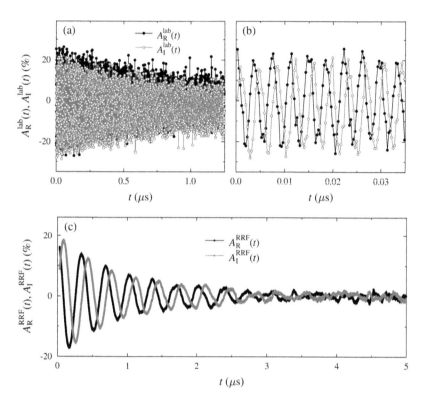

Fig. 15.13 (a,b) Example of high TF data measured in a field of 2 T, shown before the RRF transformation is applied. A wider time region is shown in (a) and an expansion of the early time region is shown in (b). (c) The quadrature RRF signals produced using a value of ν_0 corresponding to a 1.98 T muon Larmor reference frame.

A more sophisticated treatment can take account of phase offsets for the individual detectors and having more than four detectors, which is invariably the case for pulsed source spectrometers. In this method[13] all of the detectors are mapped onto just two signals oscillating in quadrature, so that the full data set can simply be represented again by eqn 15.7 and the RRF transformation eqns 15.10 can be applied.

The accurate conversion of data to RRF requires a good knowledge of the background counts, the geometric phases, and the spin time zero for each detector, which are not always easy to establish unambiguously at high field. The RRF filtering process can also introduce line shape distortion, It is therefore preferable to use the RRF transformation as a visualization tool, rather than using it to preprocess the data before

[13]B. D. Rainford, p. 463 in *Muon Science*, eds. S. L. Lee, S. H. Kilcoyne, and R. Cywinski, CRC Press (1999). This method is currently implemented within the Mantid data analysis package.

[14]U. Locans and A. Suter, JPS Conf. Proc. **21**, 011051 (2018).

fitting. One of the original motivations for the RRF was to reduce high field data sets to a manageable size for fitting, however progress in computer power, such as that provided by GPUs (graphical processing units), makes it feasible to use direct fitting of unprocessed high resolution time domain data from high field measurements.[14]

Another approach to analysing challenging high transverse field data is to switch from the time domain to the frequency domain and this is the next topic we cover.

15.5 Frequency domain

Rather straightforward analysis can be performed on the frequency spectrum of the time-domain data array. The frequency spectrum is obtained either via the Fourier transformation method, using the highly efficient Fast Fourier Transformation (FFT) algorithm, or else using the Maximum Entropy (MaxEnt) method to reconstruct the most probable spectrum that is consistent with the data. Frequency domain analysis is a particularly efficient procedure for analysing high transverse field data, since the fitting usually only needs to be done in a relatively narrow frequency band around the frequency corresponding to the applied field, whereas the corresponding fitting in the time domain would require a very large number of data points to be included, which is very computationally demanding.

Fourier transform

The Fourier transform (FT) is the standard long-established way of moving between the time domain and the frequency domain, which is widely used across many areas of science.

The general definition for the frequency spectrum $f(\nu)$ for a time domain signal $y(t)$ is defined as the infinite integral

$$f(\nu) = \int_{-\infty}^{\infty} y(t) \exp(2\pi i \nu t) dt. \tag{15.11}$$

The function $y(t)$ can be complex, or it can be regarded as the real part of a complex function, which allows for a rotating signal with non-zero phase. For the experimental situation found in muon spectroscopy, the data array y_k is a discrete set of N data points sampled for time intervals Δ_t spanning the finite time range from 0 to a maximum of $t_m = N\Delta_t$ [Fig. 15.14(a)], i.e.

$$y_k \equiv y(k\Delta_t), (k = 0, ..., N-1). \tag{15.12}$$

The corresponding discrete frequency spectrum f_n is

$$f_n \equiv f(n\Delta_f), (n = -N/2, ..., N/2), \tag{15.13}$$

where Δ_f is the intrinsic frequency resolution

$$\Delta_f = \frac{1}{t_m}, \tag{15.14}$$

(a)

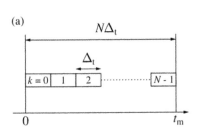

(b)

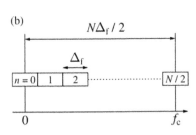

Fig. 15.14 (a) The discrete sampling properties of the time domain data array A_k. (b) The elements of the corresponding spectrum f_n for positive frequencies, that extends up to the Nyquist frequency f_c.

and the range of frequencies spans $\pm f_c$, where f_c is the Nyquist frequency

$$f_c = \frac{1}{2\Delta_t}. \tag{15.15}$$

This is illustrated for positive frequencies in Fig. 15.14(b). The transformation between the data array and its frequency spectrum uses the discrete version of the FT

$$f_n = \sum_{k=0}^{N-1} y_k \exp(2\pi i k n/N). \tag{15.16}$$

Notice that uncorrelated random noise in the time domain, such as the Poissonian noise of the counting statistics (eqn 15.1), is transformed into correlated noise in the frequency domain in the form of a constant random distribution that affects all frequencies equally.

Aliasing

The Nyquist frequency interval places an important constraint on the frequencies that should be present in the array of data. Any frequencies outside this interval will be appear at folded back or **aliased** positions in the frequency spectrum, which would cause a big problem for the analysis, so this effect should be strenuously avoided. At pulsed sources the breadth of the muon pulse suppresses higher frequencies before they are recorded in the array (Fig. 14.5), acting as a passband filter. The bin width of the raw data array is then chosen to be short enough to capture the remaining spectrum of frequencies that could be present. A wise rule is to check for the absence of high frequencies before introducing a bin width that reduces f_c. Continuous sources have an intrinsically much wider frequency response than pulsed sources and so aliasing is potentially very relevant.

Phases and phase correction

If the phase ϕ of the signal is 0 or 180°, then a cosine FT can be used

$$f_n^c = \sum_{k=0}^{N-1} y_k \cos(2\pi k n/N) \quad (n = 0, ..., N/2). \tag{15.17}$$

Likewise for ϕ values of -90 and $90°$, the sine FT can be taken

$$f_n^s = \sum_{k=0}^{N-1} y_k \sin(2\pi k n/N) \quad (n = 0, ..., N/2). \tag{15.18}$$

For intermediate phase values, a phase corrected spectrum f_n^0 is given by a linear combination of the sine and cosine transforms

$$f_n^0 = \cos(2\pi\phi) f_n^c - \sin(2\pi\phi) f_n^s. \tag{15.19}$$

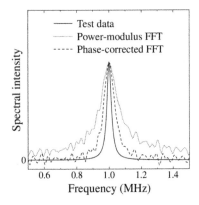

Fig. 15.15 Comparison between taking a power-modulus spectrum and taking a phase-corrected cosine transform for the same muon test data set. Gaussian apodization with a 16 μs time constant has been applied in both cases.

Alternatively, the power spectrum f_n^{p}, or its square root, the power-modulus spectrum, can be derived from f_n^{c} and f_n^{s}, which avoids any need to deal directly with phases

$$(f_n^{\mathrm{p}})^{1/2} = [(f_n^{\mathrm{c}})^2 + (f_n^{\mathrm{s}})^2]^{1/2}. \tag{15.20}$$

This however has the effect of broadening the spectral features and introducing extended tails, which could be a problem when trying to resolve features of unequal intensity.

The two versions of FFT, amplitude and power-modulus, are compared in Fig. 15.15. In general, for the simplest and most usual case where all spectral features share the same spin time zero, t_0, the phase-corrected cosine FFT spectrum gives the best result. If this is not the case, e.g. due to delayed formation of one muon state with respect to others, frequency lines can be individually phase-corrected using eqn 15.19 for a portion of the spectrum. This analysis may be complemented by a power FFT, providing a spectrum that is independent of phase, but broader than the corresponding cosine FFT spectrum.

FFT and zero padding

The simplest direct approach to calculating the discrete FT requires N^2 multiplications. However the very efficient FFT algorithm reduces this to $N \log_2 N$ operations. For just a moderate N value of 1024 this corresponds to a speed factor increase of order 100 and the advantage increases rapidly with N, so the FFT is now used almost universally for solving FT problems numerically. A key feature of the FFT algorithm is its constraint that N must be a power of 2. Therefore data must either be truncated to the next lowest power of 2 or else padded out with zeroes to reach a higher power of 2. Zero padding is often chosen, as it can also be used to increase the number of points in the spectrum for a given intrinsic resolution. This gives a curve smoothing effect, which can be useful for improving the appearance of spectral data without affecting the actual resolution.

Apodization filtering

It is usually necessary for the data to be processed before applying the FT to avoid troublesome artefacts in the frequency spectrum. This involves multiplying the data by a filtering window function $W(t)$ before the transformation, a procedure known as **apodization**, i.e.

$$A^{\mathrm{apod}}(t) = A(t)W(t). \tag{15.21}$$

As a result of the convolution theorem, the FT of $A^{\mathrm{apod}}(t)$ is the convolution of $f(\nu)$ with the FT of $W(t)$, which we call the instrumental function $I(\nu)$. Apodization is necessary, since without it, the effective $W(t)$ is a rectangular top-hat function, whose corresponding $I(\nu)$ follows the function $\mathrm{sinc}(\pi\nu/\Delta_f)$. This leads to undesirable oscillatory artefacts in the final spectrum. To avoid this the apodization function $W(t)$ is chosen to gradually taper the signal down towards zero at the end of the data window. The penalty for this is an additional instrumental broadening that depends on the choice of $W(t)$.

Another reason for using apodization is that it provides the ability to control the balance between the signal amplitude, the instrumental broadening, and the noise in the spectrum. If the signal is strongly damped and only present at early times, then a rapidly tapering apodization function is used to suppress the noise at later times, thus acting as a type of noise filter. On the other hand, if the signal is only weakly damped, then useful signal remains present at longer times and a more slowly tapering apodization function can be used and correspondingly less instrumental broadening will be produced in the spectrum. The apodization function time constant should ideally match the damping of the signal of interest for optimum signal to noise ratio and broadening. Gaussian or Lorenztian apodization functions are commonly used in muon spectroscopy.

Sometimes it is not immediately obvious that apodization is being used, as in one approach to obtaining the FT of individual histograms, where the exponentially decaying average signal is subtracted before applying the transform to the remaining count rate (without first converting this to asymmetry). The signal used here is the asymmetry weighted by the muon decay and so this procedure is equivalent to applying Lorentzian apodization to the asymmetry with a time constant equal to the muon lifetime.

FT summary

In summary, when applied to muon spectroscopy the FT method has a number of advantages compared to other methods, but also a number of disadvantages. The advantages of the FT method for spectral estimation of muon data are:

- The FT is a widely used and well understood linear transform.
- No optimization of coefficients is involved.
- It is well suited to data with broad spectral features.
- The highly efficient FFT algorithm is available, enabling fast data processing.

On the other hand, the disadvantages of the method for muon data are:

- Preprocessing of the data is required, which introduces spectral distortion and degrades resolution.
- The non-uniform and uncorrelated noise in the muon data is redistributed across the whole spectrum and becomes correlated.
- Matched apodization filtering is required to control the noise, suppress undesirable artefacts, and optimize specific features in the spectrum.

Maximum entropy methods

Informational entropy is an important concept in probability theory that originates from information theory. The informational entropy S for a discrete random variable X with K states in terms of its probability distribution $p(X)$ can be defined as follows

$$S = -\sum_{k=1}^{K} p(X_k) \ln \frac{p(X_k)}{b}, \tag{15.22}$$

where b is a constant. A change in $p(X)$ that increases S corresponds to reducing the structure in the spectrum, which reduces the information content. The probability P of the overall distribution $p(X)$ is given by

$$P[p(X)] \propto \exp(\alpha S), \tag{15.23}$$

where α is a constant, so that increasing S increases the probability of the distribution at the same time as decreasing the structure in the spectrum. Maximum entropy methods look for the most probable distribution for $p(X)$, given the data, by finding the maximum value of S that is consistent with the constraint provided by the data. The data constraint is quantified by χ_r^2, which is defined here as

$$\chi_r^2 = \frac{1}{N} \sum_{i=1}^{N} \frac{(y_i - m_i)^2}{\sigma_i^2}, \tag{15.24}$$

where there are N data points y_i with errors σ_i and the predicted model values are m_i.

Two different maximum entropy methods have been used for estimating frequency spectra in muon spectroscopy. The first method, called autoregression (AR), is actually a type of transform that is complementary in some ways to the FT, but with a reduced number of coefficients compared to the FT. These coefficients are thus not completely defined by the data, but are chosen on the basis of maximum entropy principles. The second method goes by various names, here we call it MaxEnt. This is a more straightforward implementation of the maximum entropy method, using a constrained optimization algorithm to arrive at the spectrum and performing no numerical operations on the raw data set, other than comparing it with the model prediction.

Method 1: Autoregression

This method makes use of the autocorrelation properties of the signal, i.e. the correlation of the signal with a delayed version of itself. The autocorrelation function ϕ_j is the average correlation across the data for a particular delay j, i.e.

$$\phi_j = \phi_{-j} = \frac{1}{N-j+1} \sum_{i=0}^{N-j} y_i y_{i+j} \quad (j = 0, .., N). \tag{15.25}$$

The power spectrum is then obtained as the square of the FT of ϕ_j via the Wiener-Khinchin theorem. A convenient z-transform representation of the FT maps the real Nyquist frequency range onto a complex variable z that forms a unit circle in the complex plane, with

$$z \equiv e^{2\pi i \nu \Delta_t}. \tag{15.26}$$

The FT power spectrum derived from the autocorrelation then becomes a polynomial in z

$$P(\nu) = \left| \sum_{j=-N}^{N} \phi_j z^j \right|^2. \tag{15.27}$$

The data used for the estimation eqn 15.27 are from a limited measurement window, whereas the true spectrum corresponding to the full time dependent signal can be represented by the infinite series

$$P(\nu) = \left| \sum_{j=-\infty}^{\infty} c_j z^j \right|^2. \tag{15.28}$$

Equation 15.27 can be regarded as a truncated version of eqn 15.28. However eqn 15.28 is not the only way to represent the spectrum and the alternative form

$$P(\nu) = \frac{a_0}{\left| 1 + \sum_{k=1}^{M} a_k z^k \right|^2}, \tag{15.29}$$

is a function having M poles in the z-plane. Such poles are well suited to approximating spectra with sharp features using relatively few terms in the expansion, i.e. M could be very much smaller than N. This is why it is sometimes called the **all-poles method** to contrast it with the **all-zeroes method** of the standard FT. The AR method for order M involves estimating the coefficients a_0, a_k in eqn 15.29 in terms of the ϕ_j in eqn 15.27 by matching the lower order coefficients in the z series expansions. An efficient algorithm for solving this was developed by Burg,[15] which is based on maximum entropy principles. The only choice to make with this method is the number of poles. A property known as the final predictive error (FPE) can be calculated for each order M and the FPE goes through a minimum versus M. The M value at the minimum indicates the optimum level of approximation to use for the spectrum.

Figure 15.16 shows the effect of applying the all-poles AR method to the same data as was used for Fig. 15.15. The performance of the AR method can be seen to be comparable to that of the phase-corrected FT (Fig. 15.15), but without the need to deal with phase. For the AR spectrum shown in Fig. 15.16, a Gaussian apodization with a 16 μs time constant has been applied and the optimum number of poles is $M = 38$, whereas the data array has $N = 2048$ points.

[15] J. P. Burg, Geophysics **37**, 375 (1972).

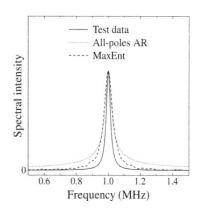

Fig. 15.16 Application of the maximum entropy-based methods AR and MaxEnt to the same data as was used in Fig. 15.15.

AR method summary

We can summarize the advantages of the AR method over the FFT method as:

- Better intrinsic frequency resolution, making the method well suited to data with sharp spectral features.

- No phase correction needed.

- Works well with short data sets.

On the other hand, the disadvantages of the method are:

- Sometimes it produces spurious splittings of strong features and spurious peaks in the baseline.

- It can produce small offsets to the positions of spectral features.

- As with the FT method, the time dependent errors of muon data are not taken into account.

Recent implementations of AR algorithms should suffer less from these spectral distortions, but these have not yet been applied to muon data. One reason for this is that the MaxEnt method of spectral estimation based on maximum entropy principles has now become dominant.

Method 2: MaxEnt

The most widely used method for applying maximum entropy principles to muon spectroscopy was pioneered by Rainford and Daniell in 1994.[16] This method is based on the Maximum Entropy algorithm of Skilling and Bryan,[17] which was originally developed in the context of image reconstruction in astronomy. The muon implementation has been extensively discussed in a series of papers by Riseman and Forgan.[18]

The core algorithm of MaxEnt proceeds iteratively to maximize the informational entropy of the frequency spectrum S, subject to χ_r^2 being reduced to a target value, which is usually set to be slightly larger than 1. Iterations proceed until convergence is reached, which is determined by the gradient vectors of S and χ_r^2 with respect to the elements of the spectrum becoming parallel. The muon specific aspects of the data set, such as detector phases, amplitudes, and backgrounds, are optimized in an outer loop and the final spectrum is obtained when both loops have converged. The optimization procedure is illustrated in Fig. 15.17 for the simplest case, where the spectrum and the data set both have just two points.

The result of applying the muon MaxEnt method to our test data is shown in Fig. 15.16. Note that this is the only method that does not have to use apodization. It can be seen that, of the four methods illustrated here, MaxEnt provides the least broadening with respect to the actual width of the test spectrum.

The key advantages of the MaxEnt method for spectral estimation of muon spectroscopy data can be summarized as follows:

[16]B. D. Rainford and G. J. Daniell, Hyp. Int. **87**, 1129 (1994).

[17]J. Skilling and R. K. Bryan, Mon. Not. Roy. Astr. Soc. **211**, 111 (1984).

[18]T. M. Riseman and E. M. Forgan, Physica B **289-290**, 718 (2000); **326**, 226 (2003); **326**, 230 (2003); **326**, 234 (2003).

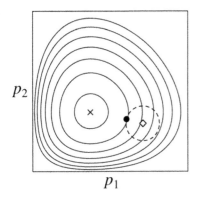

Fig. 15.17 Illustration of the MaxEnt algorithm for a simple spectrum with just two points p_1 and p_2. Contours are shown for the entropy S, with the unconstrained maximum indicated by the cross. With a two point data set χ_r^2 is zero at the point shown by the diamond symbol. The dashed circle indicates the data constraint given by the target value $\chi_r^2 = 1$ and the maximum S subject to this constraint is indicated by the solid circle, which corresponds to the required MaxEnt spectrum (p_1, p_2).

- Structure is placed in the spectrum only where there is statistically significant evidence for it in the data.

- The raw data are never processed and degraded, the problems associated with apodization and zero padding are completely avoided.

- The particularly challenging noise profile of muon data is taken fully into account.

- Corrections for detector phases, time zero, muon pulse shape, frequency response, dead times, and background can all be included.

- Deconvolution is possible for 'nuisance' effects, such as nuclear dipolar broadening.

On the other hand, the main limitations of the method in its current implementation are:

- One phase parameter is required for each detector, and it must be constant across the whole spectrum.

- The method is currently optimized better for TF rotation rather than for ZF precession.

- The errors associated with the reconstructed spectrum are difficult to extract and interpret, being highly correlated.[19]

[19]T. M. Riseman and E. M. Forgan, Physica B **326**, 230 (2003).

Spectral fitting

Once a frequency spectrum has been obtained, whether by Fourier or maximum entropy methods, it is often desirable to take the analysis further by fitting the spectral features to parameterize their behaviour. This is typically done using Lorentzian or Gaussian peak functions or Bessel functions in the case of incommensurate magnetism (Chapter 6). For studies of superconductors, more complex asymmetric functions can be used, representing the field distribution within a flux line lattice (Chapter 9). These spectral features can also be characterized by moments, which can easily be calculated from the obtained spectra.

Finally, it should be noted that time domain fitting is generally more accurate and straightforward than the two stage process of first obtaining the frequency domain spectrum and then fitting it. However the time domain analysis needs to know the number and type of spectral components present. Frequency domain analysis provides this key information about the spectral components that can be fed back to the time domain analysis. Taken together, the combination of frequency domain and time domain methods can thus provide a very robust approach to muon data analysis.

Chapter summary

- Practical considerations for setting up and carrying out muon measurements include matching the sample to the beam spot size, choosing the sample environment, and setting the required time resolution and measurement window.
- Muon spectroscopy data are often analysed in the time domain, which might involve fitting individual or multiple detector groups in a sequence or simultaneously.
- Data analysis in the frequency domain can be carried out using Fourier and maximum entropy methods.

Further reading

- Further practical information about muon spectroscopy at various muon facilities can be found via links on the book website: `https://musr.org/muon-spectroscopy-book`.
- Current data analysis software for muon spectroscopy can be accessed via links at the book website.
- W. H. Press, S. A. Teukolsky, W. T. Vetterling, and B. P. Flannery, *Numerical Recipes: The Art of Scientific Computing, 3rd Edition*, CUP (2007).
- E. T. Jaynes, *Probability Theory: the Logic of Science*, CUP (2003).
- B. Buck and V. A. Macaulay (editors), *Maximum Entropy in Action*, OUP (1990).
- D. S. Sivia, *Data Analysis: a Bayesian Tutorial*, OUP (1996).

Exercises

(15.1) Assume that events arrive at a detector at an average rate f (counts/second), but their arrival is Poisson distributed so that the probability $P(t) \, dt$ that there are no events from time 0 up to just before time t, but there is an event between time t and $t + dt$, is given by $P(t) \, dt = f \, e^{-ft} \, dt$. Check that $P(t)$ is normalized and show that the expected time between counts is $\langle t \rangle = f^{-1}$. In a real detector, an event is undetectable if it arrives at a time t after the previous detection where $t < \tau$, where τ is the detector dead time. Show that the expected time between detected events is now $f^{-1} + \tau$ and hence deduce that if $N_{\exp}$ counts are measured during a time interval of length T, the corrected number of counts N is given by $N_{\exp}/(1 - N_{\exp}\tau/T)$.

Part IV

Further topics in muon spectroscopy

Further topics that extend and enhance the μSR technique are presented in this part of the book.

- In order to make μSR experiments more quantitative, it is important to be able to understand the site at which the muon is implanted and any effect it may have. The ideas behind muon site estimation are described in Chapter 16.

- Chapter 17 describes other numerical modelling techniques that are important in muon spectroscopy.

- Muons are usually relatively high energy probes, but an important advance is the development of technology to slow muons down to very low energy in order to study surfaces and interfaces. Chapter 18 describes low energy μSR.

- At pulsed sources it is possible to stimulate the muon and/or the sample in various different ways. The ideas behind these techniques are described in Chapter 19.

- A major effort in μSR is extending the sample environment to include extreme conditions, one of which is very high magnetic field, as presented in Chapter 20.

- Another extreme sample environment, described in Chapter 21 is high pressure.

- Everything in this book so far has been about positive muons. However, it is also possible to work with negative muons, even though the experiments are more difficult. The science behind negative muons is described in Chapter 22, together with some applications in diverse areas.

16 Calculating muon sites

Ever since μSR was first deployed in solid state research, much attention has been paid to the two perceived drawbacks in the method: (i) lack of knowledge of the muon's stopping site in materials raises the question of which fields are actually being probed; (ii) the unknown effect that the positively charged muon has on its local environment raises the question of whether intrinsic behaviour is being measured, or whether instead we measure something caused by the presence of the implanted muon. In many cases these worries can be argued to be unfounded, but this collection of concerns, often called the **muon site problem**, is something in which users of muon spectroscopy techniques should certainly take an interest.

If the final muon stopping state involves relatively little contact hyperfine coupling (i.e. a diamagnetic site), the stopped muon resembles something like a bare particle in the role of an interstitial defect. Although such states are usually assumed to be passive, the bare muon can be 'dressed' by interactions with the electronic system. This might involve the muon acquiring a screening cloud of electronic charge in a metal, or acquiring a strain field in an insulator as it deforms atoms in its vicinity. Knowledge of the muon's stopping site in such cases allows us an insight into the local fields and interactions at a known position. If the muon instead forms a bound paramagnetic state such as muonium, or induces a large spin density at its position, knowledge of the muon site allows us to identify the electronic state that gives rise to the measured hyperfine coupling constants, providing a direct insight into the local electronic structure.

For many years a range of approaches to locating muon sites were commonly employed:

- Angle-dependent Knight-shift measurements in single crystals reduce the number of possible muon sites by constraining them by symmetry, often leading to the exact muon site being deduced. The same applies to single crystal measurements in applied field, taking due care of the demagnetization.

- Nuclear dipolar fields can be calculated at candidate sites and compared with measurements.

- In single crystal systems with quadrupolar nuclei the angular dependence of the LF decoupling can provide site information.

- In magnetically ordered materials where the magnetic structure is known, dipole field calculations allow the magnetic fields at candidate muon sites to be approximately computed. The hyperfine contribution is often unknown here and has to be either assumed or neglected.

- The electrostatic potential can be computed in a target material. A positive muon might then be expected to adopt positions in potential wells where the potential is most strongly negative.[1]

In each of these methods, the presence of the charged muon is assumed not to make any difference to the material, so that the magnetic structure or the potential is that for the material in the absence of the muon impurity. However, the use of these techniques, along with some chemical intuition, led to a number of ad hoc rules for guessing muon sites. These include the rules of thumb that:

- Positive muons favour sites close to negatively charged anions.
- Muons often sit in positions of high symmetry in metals.
- A muon is usually 1 Å from an O^{2-} ion in an oxide.
- Muons form $F-\mu-F$ bonds in fluorides.

These rules are often useful and lead to the identification of highly plausible muon sites, albeit in a limited range of materials. More sophisticated attempts to address the muon site problem have involved the development of numerical calculations targeted specifically to determine and model the muon stopping state. We look at these in this chapter.

16.1 The site problem

The location of the muon site is determined by non-relativistic quantum mechanics. The system comprising the stopped muon and its host material can therefore be described by the Schrödinger equation, and so a computation of the wavefunction of the system should, in principle, be possible. In addition to the muon, the Hamiltonian of the system should also describe the large number of electrons and nuclei in the material. This makes the quantum mechanical description of the muon in a material an example of a *many-body problem*. Of course, it is the electrons that determine most properties of a material (e.g. electrons bond atoms together and determine mechanical qualities, they conduct electricity and heat, and they give rise to magnetic fields), and the behaviour of the electrons also ultimately determines the muon site. Therefore, it is really the electronic part of the wavefunction that we need to determine.

Unfortunately, for any systems of interest, computing this wavefunction is an impossible task. For a typical many-body system, the electronic wavefunction is a function of the positions of all of the electrons and atomic nuclei in the system, and there are no analytical solutions for the electronic wavefunctions for systems of several interacting electrons. In fact, even numerical solutions are often impossible to calculate, as illustrated in the following example.

[1] A simple approach for ionic materials uses Coulomb's law to calculate the potential, in this case it is necessary to include some form of nuclear repulsion effect to prevent a positive muon sitting exactly on top of a negative ion. However, the inclusion of this repulsive effect often dominates the potential such that the position of the muon is very sensitively dependent on the form of the repulsive interaction assumed. More sophisticated calculation methods can also be used for deriving the potential. All of these methods are classified as unperturbed electrostatic potential (UEP) techniques (see Example 16.6).

Example 16.1

To work with differential equations such as the Schrödinger equation, computer programs often rely on defining a grid on which the function is calculated. In three spatial dimensions, the wavefunction for a system containing N indistinguishable particles is a $3N$-dimensional function, $\psi(x_1, y_1, z_2, x_2, y_2, z_2, ...x_N, y_N, z_N)$. If each dimension is divided into L intervals, the grid becomes a $3N$-dimensional hypercube with L^{3N} points. For a large value of N, this quickly becomes unmanageable. Even the memory required to store so many numbers would be beyond the capability of the most powerful computers currently in existence (see Exercise 16.1).

Fortunately, approximations exist that allow us to compute the electronic wavefunction of a many-body system with sufficient accuracy to be useful for many problems.

In the general case of interest to us, the quantum mechanical description of a system of atomic nuclei and electrons, such as a crystal or a molecule, is given by the wavefunction that solves a Hamiltonian that includes all electrostatic interactions between electrons and electrons, nuclei and nuclei, and electrons and nuclei:

$$\hat{H} = \hat{T}_{\mathrm{e}} + \hat{T}_{\mathrm{n}} + \hat{V}_{\mathrm{ee}} + \hat{V}_{\mathrm{en}} + \hat{V}_{\mathrm{nn}}, \tag{16.1}$$

where T_{e} is the kinetic energy of the electrons, T_{n} is the kinetic energy of the nuclei, V_{ee} is the electron-electron interaction potential, V_{en} is the electron-nuclear interaction potential, and V_{nn} is the nucleus-nucleus potential.

One thing we can immediately do to simplify the problem is to make the **Born-Oppenheimer** approximation. This makes use of the fact that atomic nuclei are much heavier than electrons and therefore that the energy scale for the electronic part of the wavefunction is significantly larger than that of the nuclear part of the wavefunction, allowing these to be separated.[2] The Born-Oppenheimer approximation allows us to treat the atomic positions as parameters of an electron-only Hamiltonian that obeys a Schrödinger equation

$$\hat{H}(\boldsymbol{R})\psi(\boldsymbol{X}) = E(\boldsymbol{R})\psi(\boldsymbol{X}), \tag{16.2}$$

where here $\boldsymbol{X} = (\boldsymbol{r}_1, \boldsymbol{r}_2, \boldsymbol{r}_3, ...)$ is a vector made up of the electronic position vectors[3] $\boldsymbol{r}_i$ and $\boldsymbol{R} = (\boldsymbol{R}_1, \boldsymbol{R}_2...)$ is a vector formed from the nuclear coordinates.[4] The Born-Oppenheimer approximation basically clamps the nuclei at given fixed positions $\boldsymbol{R}$ and the electron wavefunction $\psi(\boldsymbol{X})$ is the one that's appropriate for the clamped nuclear configuration we have chosen. With fixed nuclear positions we have $\hat{T}_{\mathrm{n}} = 0$ and $\hat{V}_{\mathrm{nn}}$ constant, so the electronic Hamiltonian is reduced to three terms

$$\hat{H}(\boldsymbol{R}) = \hat{T}_{\mathrm{e}}(\boldsymbol{R}) + \hat{V}_{\mathrm{ee}}(\boldsymbol{R}) + \hat{V}_{\mathrm{en}}(\boldsymbol{R}). \tag{16.3}$$

Despite the vast simplification that the Born-Oppenheimer approximation brings, producing a solution to eqn 16.2 is still an extremely complex problem, even for relatively simple systems. There are many possible approaches, with different methods available for different situations,

[2]High energy is equivalent to short times, so one could say that the electrons solve their Schrödinger equation before the nuclei have time to solve theirs!

[3]The vector $\boldsymbol{X}$ therefore has $3N$ components in order to describe N electrons.

[4]The vector $\boldsymbol{R}$ has $3M$ components to describe M nuclei.

usually offering some degree of trade-off between precision of the answer and length of the calculation. One of the most widely used of these is **Density Functional Theory (DFT).**

DFT has proven to be one of the most successful methods of addressing the N-particle quantum problem in both condensed matter and molecular systems. It is both an exact theory for interacting electrons and a practical prescription for calculating exact or approximate values of the properties of the system. The point of DFT is to make no attempt to compute the $3N$-particle electronic wavefunction $\psi(\boldsymbol{X})$ itself, but instead to aim to calculate the *electron density* $n(\boldsymbol{r})$, which represents the probability of finding any electron at a particular point in space.

16.2 What is DFT?

Let's examine density functional theory. Firstly, what is a functional? A function is a rule that inputs a number (or set of numbers) and outputs a number. An example is $y = f(x) = x^2$, where we input a number $x = 2$ and output a number $y = 4$. A functional is a rule that inputs a function [like $f(x)$] and outputs a number.

Example 16.2

An example of a functional F is a definite integral

$$F[f(x)] = \int_0^1 \mathrm{d}x\, f(x) = \int_0^1 \mathrm{d}x\, x^2 = \frac{1}{3}. \tag{16.4}$$

We provide the functional F with our choice of the function $f(x)$, in this case $f(x) = x^2$, and the functional F outputs a number, in this case $1/3$.

In DFT, we deal with the functional $E[n(\boldsymbol{r})]$ whose output is the ground state energy of the N-electron system and whose input is the electron density function $n(\boldsymbol{r})$. The electron density is not a function of all of the coordinates of all of the electrons; it is simply a function of the three position coordinates $\boldsymbol{r} = (x, y, z)$. It is therefore a quantity averaged over $3N - 3$ of the degrees of freedom of the N-particle wavefunction $\psi(\boldsymbol{X})$ (leaving only the three represented by $\boldsymbol{r}$ as variables). It seems surprising that we can make such a severe simplification and still compute the energy of a system. The reason that this miracle occurs is based on the two **Hohenberg-Kohn (HK) theorems.** We briefly examine these in the next section.

The Hohenberg-Kohn theorems

The first HK theorem establishes a one-to-one correspondence between an external potential $V(r)$ that defines an electronic system, and the ground state electronic density $n(r)$. The theorem comes in two parts:
1. Two potentials differing by more than a constant lead to different ground state wavefunctions.
2. Different ground state wavefunctions originating from different potentials lead to different ground state densities.

The corollary to this two-part theorem is that the ground state density $n(r)$ uniquely *determines* the potential. This implies that it uniquely determines *all* of the properties of the system and therefore also the many-particle wavefunction. In short, if we have $n(r)$ we have access to the important physics of the system.

The first HK theorem allows us to write the properties of an interacting many-electron system, such as the ground state energy E, as a unique functional of the electron density $n(r)$ via an expression of the form

$$E[n(r)] = T[n(r)] + E_{ee}[n(r)] + E_{en}[n(r)]$$
$$= F[n(r)] + \int d^3r \, n(r) V_{en}(r), \qquad (16.5)$$

where, in the last expression, we have separated out the electron-nuclear part of the energy E_{en} and have written the rest as $F[n(r)] = T[n(r)] + E_{ee}[n(r)]$, which is a functional encoding all of the purely electronic contributions to the energy.

If we have access to $E[n(r)]$ then we might be able to solve the problem of finding the electronic configuration corresponding to it by minimizing E with respect to the electron density n. That is, we continually adjust the function $n(r)$ until we find a form that gives us the lowest-energy electron density E. Is this a safe strategy? The second HK theorem tells us that it is.

The second HK theorem states that $E[n(r)]$ gives the lowest energy if, and only if, the input density $n(r)$ is the true ground state electron density. If we do not use the true density, then we obtain an upper bound on the ground state energy.

We therefore have a strategy: minimize the function $E[n(r)]$ with trial functions $n(r)$. If we find the true minimum then we will have the electron density. There is, however, a problem, in that we do not actually know the exact form of the electron functional $F[n(r)]$. One of our tasks is then to find an approximate form for $F[n(r)]$.

The Kohn-Sham methodology

A major simplification of the calculation of the electron density is made if we use the **Kohn-Sham formulation**, the idea of which is to reformulate the problem by replacing the *interacting* system of many electrons by a *non-interacting* system of many electrons constrained to have the same electron density.

In the Kohn-Sham (KS) formulation, we approximate the interacting kinetic energy $T[n(\boldsymbol{r})]$ of the N interacting particles by the kinetic energy $T_s[n(\boldsymbol{r})]$ of N non-interacting particles with the same density $n(\boldsymbol{r})$. We also extract from $F[n(\boldsymbol{r})]$ the **Hartree component** $U[n(\boldsymbol{r})]$. The Hartree component is simply the classical Coulomb interaction between regions of charge density, and can be easily written[5] in terms of $n(\boldsymbol{r})$. We then write

$$F[n(\boldsymbol{r})] = T_s[n(\boldsymbol{r})] + U[n(\boldsymbol{r})] + E_{xc}. \tag{16.6}$$

We have therefore bundled up our remaining ignorance into the functional E_{xc}, known as the **exchange-correlation** energy.[6]

To summarize, the HK theorems tell us we can find the ground state energy by minimizing a functional $E[n(\boldsymbol{r})] = F[n(\boldsymbol{r})] + \int \mathrm{d}^3 r\, V_{en}(\boldsymbol{r})n(\boldsymbol{r})$ with respect to the density $n(\boldsymbol{r})$. The KS formulation, with its replacement of the interacting system with an equivalent non-interacting one, allows us to say that this minimization will be equivalent to a simpler procedure where we minimize the total energy of a non-interacting system subject to an **effective potential** V_s, whose form is

$$V_s(\boldsymbol{r}) = V_{en}(\boldsymbol{r}) + V_H(\boldsymbol{r}) + V_{xc}(\boldsymbol{r}). \tag{16.8}$$

The first term in the effective potential represents the electron-nuclear potential, the second the Hartree interaction, and the third the exchange-correlation potential. The end result of the KS approach is that we can solve the ground state density for the interacting system by first solving the Schrödinger equation for a single particle in an effective potential, using the set of **KS equations** defined as:

$$\left[\frac{\hat{p}^2}{2m} + V_s(\boldsymbol{r}) \right] \phi_i(\boldsymbol{r}) = \varepsilon_i \phi_i(\boldsymbol{r}). \tag{16.9}$$

The solutions of these KS equations give a set of KS wavefunctions and energy levels labelled by i. These states are filled up by the set of available electrons and the density can be constructed from these non-interacting wavefunctions as $n(\boldsymbol{r}) = \sum |\phi_i(\boldsymbol{r})|^2$, where the sum runs over the occupied states.

Although we now have a set of single-particle problems to solve, we don't actually know the potential. Our ignorance has two aspects: the first is the functional form of V_{xc}, which we postpone to Section 16.5, the second is that the Hartree potential V_H and the exchange correlation potential V_{xc} both depend on $n(\boldsymbol{r})$, the solution to the problem, which in turn is constructed from $\phi_i(\boldsymbol{r})$. We therefore solve the equations iteratively: (i) an initial guess is first made of $n(\boldsymbol{r})$ and the potential V_s is computed; (ii) the single-particle wavefunctions are computed; (iii) the set of occupied ϕ_i is used to compute $n(\boldsymbol{r})$, and then we start again by returning to step (i) and recalculating the potential. This procedure is repeated until some suitable convergence criterion is reached.

[5]The Hartree component is written (in atomic units) as

$$U[n(\boldsymbol{r})] = \frac{1}{2} \int \mathrm{d}^3 r \mathrm{d}^3 r' \frac{n(\boldsymbol{r})n(\boldsymbol{r}')}{|\boldsymbol{r} - \boldsymbol{r}'|},$$

or $U[n(\boldsymbol{r})] = \frac{1}{2} \int \mathrm{d}^3 r V_H(\boldsymbol{r})n(\boldsymbol{r})$.

[6]Rearranging, we have

$$\begin{aligned} E_{xc}[n(\boldsymbol{r})] = &(T[n(\boldsymbol{r})] - T_s[n(\boldsymbol{r})]) \\ &+ \{E_{ee}[n(\boldsymbol{r})] - U[n(\boldsymbol{r})]\}. \end{aligned} \tag{16.7}$$

This equation expresses our ignorance of two parts of the problem: the deviation of the true kinetic energy from the KS kinetic energy, plus the difference between the full electron-electron potential and the classical Hartree energy.

Optimizing the geometry

Recall that we started by using the Born-Oppenheimer approximation to write eqn 16.2, before formulating the DFT approach to finding its ground state energy. However, in order to write eqn 16.2 we needed to specify the positions of the nuclei. Nothing tells us whether the configuration of nuclei we have chosen is an equilibrium one. Since one of our main purposes is finding the details of the nuclear structure that accommodates the muon, we need to repeat the calculation for different nuclear structures until a global minimum in the ground state energy has been found. To do this, we can allow the nuclei to move under the effect of forces defined[7] by

$$F(R) = -\nabla_R E(R) = -\left\langle \psi \left| \nabla_R \hat{H} \right| \psi \right\rangle. \qquad (16.10)$$

We can move the nuclei until the forces become small enough that we judge the system is at equilibrium. This process is called a **geometry optimization**.

16.3 Methods

In carrying out DFT calculations there are several options available, and our final choice of method should take into account the type of system being studied. For example, a crystalline system will require a method that makes use of extended wavefunctions, whereas an isolated molecule or small atomic cluster would often be better described by a localized wavefunction. The first decision in setting up the calculation is therefore the choice of the set of basis functions (or basis set) used to represent the Kohn-Sham single-particle wavefunctions. The two main options are plane waves (PW) or Gaussian type orbitals (GTO). These are discussed in Section 16.4.

When the muon is introduced into the material, a second factor to consider is the resultant charge and spin state. If the muon comes into the system as charge-neutral muonium, then an extra unpaired electron is introduced as well as the muon. The fate of this electron depends on the electronic state of the host and, in general, a spin-polarized calculation[8] would need to be used to account for this. On the other hand, introducing a diamagnetic positive muon will not introduce any extra electron spin, but will change the charge state of the unit cell containing the muon.[9]

The third important decision to be made in setting up a calculation is which exchange-correlation functional to use. Section 16.5 looks at the options available for these.

A further point to note is that when multiple muon sites calculations are being carried out for a particular material, as is usually the case, the use of parallel computing resources allows the set of site calculations jobs to be completed within a reasonable time frame, which can range from minutes to days, depending on the job size.

[7]In classical mechanics these are called virtual forces. The quantum-mechanical version of the argument is known as the Hellmann-Feynman theorem.

[8]In order to include magnetic effects in DFT, the electronic spin must be taken into account. Computations that distinguish between up and down electron spins are called *spin-polarized calculations*.

[9]Note that in periodic systems the periodical boundary conditions cause any charge imbalance to build up to give a divergent total. Therefore extra charge must be balanced by a uniform compensating charge.

Supercells

When placing the muon in a crystalline material for the calculation of a periodic electronic state, care must be taken that a muon does not interact with its replica in the neighbouring cells. It is therefore often necessary to build a **supercell** comprising several copies of the original cell. A single muon is then placed inside this supercell and, if its volume is large enough, the interaction with the replica will become negligible. This is a standard way to deal with impurity problems in DFT, not exclusive to the muon, and represents a compromise between diluting the impurity and retaining a periodic structure. A supercell is not always needed; it may turn out that a structure has a sufficiently large cell that interactions are suppressed.

A supercell may also be used to do a plane wave calculation on a system that is not periodic, such as an individual molecule. In this case the supercell is made to be significantly larger than the extent of the molecule and the extra region around the molecule is kept empty to avoid any interaction between the periodic copies of the molecule. An example of this type of calculation is given in the demonstration of convergence covered in the next section (Fig. 16.1).

Convergence, approximations, and corrections

It is relatively straightforward to establish convergence of the DFT calculation to a stable value in the case of the PW-based methods, since the cutoff energy defining the number of plane waves can be smoothly varied. Care must be also taken in the choice of a sufficient number of k points, although using large supercells reduces this demand. An example showing the convergence of hyperfine parameters for a benzene molecule is shown in Fig. 16.1. In contrast, for calculations with a GTO basis[10], such smooth control of the size of the basis is not possible, as the jumps in basis size and computation time between different basis sets are rather large, e.g. going from double zeta to triple zeta typically boosts the number of primitive Gaussians by 75% and increases the computation time by an order of magnitude. While establishing convergence is essential for confirming the stability of a calculation, we know that it will converge on an incorrect value, due to the exchange-correlation functional being only an approximation to the true functional. Therefore some calibration of final results against experimental data will inevitably be needed. In some circumstances the finite basis set errors can go some way towards cancelling out the errors from having only an inexact functional. This appears to be the case for the combination of the hybrid B3LYP functional[11] with polarized double zeta basis sets, which accounts for the success of this combination in predicting many molecular properties, including muon hyperfine tensors.

Another area where a helpful cancellation may occur is between quantum and thermal corrections. The hyperfine coupling is generally increased by quantum zero-point motion of the muon[12], whereas thermal motion of a molecule usually decreases the coupling. At some tem-

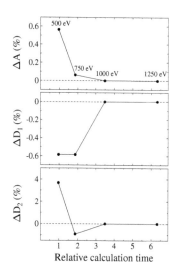

Fig. 16.1 Illustration of the convergence of PW/PBE calculations of the hyperfine parameters for the cyclohexadienyl radical in a benzene molecule versus the cutoff energy. Convergence in all three parameters has been reached at a cutoff energy of 1000 eV.

[10]See Section 16.4.

[11]See Section 16.5.

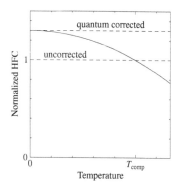

Fig. 16.2 Thermal correction to the hyperfine coupling usually opposes the quantum correction, which can lead to cancelation of the two corrections at a particular temperature T_{comp}.

[12]See Section 16.8.

perature these two corrections can exactly cancel (Fig. 16.2). In the more general case, however, the independent corrections for the quantum effect and the thermal effect both need to be taken into account to correctly describe the coupling and its temperature dependence.

16.4   Basis sets

In order to carry out a DFT calculation, a choice has to be made on the basis set to be used to represent the Kohn-Sham single-particle wavefunctions in the calculation. The choice made will be determined by the type of problem and also by the choice of method, and hence the code used for the calculation. The two main types of basis set used in DFT for muon spectroscopy are plane waves and Gaussian type orbitals.

Plane waves (PW)

For crystalline systems a plane wave basis set is often used. This is based on the idea that valence electrons in solids can be quite similar to free electrons, whose wavefunctions are plane waves. Plane waves form a complete, space-filling set (see below). The plane wave energy E depends quadratically on the wavevector k and the allowed values of k are determined by the periodicity of the cell. Larger E and k values enable the representation of progressively finer details of the wavefunction, so the quality of the basis set is defined by a high energy cutoff E_c for the set of plane waves, which is typically several hundred eV. Localized core electrons are not so well represented by plane waves and accurate coverage would require a very large E_c in the standard approach. To get round this problem, plane waves are usually used in conjunction with **pseudopotentials** to model the effective core potential seen by the valence electrons. These are parameterized for different atoms, with two popular types being norm-conserving pseudopotentials and ultrasoft pseudopotentials. The latter are more complex, but have the advantage of reducing E_c.

Gaussian type orbitals (GTO)

Whereas plane waves are well suited to describing highly delocalized valence electrons, the core electrons are better described by localized atomic orbitals. Sets of Gaussians functions can be used to represent these atomic orbitals and Gaussians provide a particularly efficient basis for evaluating multi-centre integrals in a DFT calculation.[13] Several primitive Gaussian functions are combined in a linear combination to form what is known as a contracted Gaussian function, which can represent the character of each atomic orbital, allowing the relevant pattern of nodes and specific radial dependence to be reproduced. The minimal basis set is known as single-zeta (SZ), but accuracy can be improved by doubling or tripling the number of contracted Gaussian functions to give double-zeta (DZ) or triple-zeta (TZ) basis sets, or the general

[13]However one disadvantage of Gaussians compared to plane waves is that Gaussians generally form an incomplete set, without an obvious route to completeness. Unlike plane waves, the spatial derivatives of the waves are not members of the set, which complicates the implementation of the Hellmann-Feynman theorem. The advantage, however, is that far fewer Gaussians are needed than plane waves to deal with a localized system.

n-zeta (nZ) basis set. Increasing zeta is more effective for valence electrons than for core electrons. This leads to the idea of split valence
basis sets, where only the valence electrons have the specified multi-zeta
representation. The effects of correlation generally require additional
Gaussian functions of higher angular momentum to be included, e.g.
p-functions added to the s-orbitals. These are known as polarization
functions. Where there are loosely bound electrons, the properties of
the outer part of the wavefunction are particularly important and additional functions with low radial exponents can be added. These are
known as diffuse functions. A hierarchy of progressively more accurate
GTOs can be produced by increasing the number of primitive Gaussians
via the zeta level, the addition of polarization functions and the addition
of diffuse functions. Many different GTO basis sets have been developed,
each one typically aimed at making efficient calculations of a particular
physical property with modestly sized basis sets. This abundance of options can be very confusing for someone who is new to the field and it
is then recommended to work within one group of basis sets, at least to
start with. Examples of two such groups of basis sets are given next.

Example 16.3

A commonly used set of split valence basis sets is due to Pople and coworkers. They
have the basic labelling scheme $i - jk$G for DZ and $i - jkl$G for TZ. The first index
i reflects the number of primitive GTOs in the core orbitals. Index j indicates the
number of functions in the split valence representation and k and l give the number
of primitive GTOs used in each valence part. Polarization functions are placed in
brackets after the G, e.g. (d) indicates extra d functions for non-hydrogen atoms
and (d,p) would indicate d functions for non-hydrogen atoms and p functions for
hydrogen. Diffuse functions are indicated in this scheme by placing a + sign just
before the G.

zeta	standard	polarization	diffuse	polarization and diffuse
2	6-31G	6-31G(d,p)	6-31+G	6-31+G(d,p)
3	6-311G	6-311G(d,p)	6-311+G	6-311+G(d,p)

Table 16.1 Examples of Pople's DZ and TZ basis sets.

The polarized split valence basis sets provide a good compromise between accuracy
and speed. A group of these with steadily increasing size due to Dunning is illustrated
in the following table. For calculating spin distributions in muoniated molecular
radicals, the combination of cc-pVDZ with the hybrid B3LYP functional has been
found to work particularly well.

zeta	standard	augmented with diffuse
2	cc-pVDZ	aug-cc-pVDZ
3	cc-pVTZ	aug-cc-pVTZ
4	cc-pVQZ	aug-cc-pVQZ
5	cc-pV5Z	aug-cc-pV5Z

Table 16.2 Some examples of Dunning's correlation-consistent polarized split valence basis sets. Basis sets are defined in this scheme for zeta level up to 7.

16.5 Functionals

The accuracy achievable with DFT is based on having a good approximate representation of the exchange-correlation functional.

Local density approximation (LDA)

The first approach that was applied to approximating the exchange-correlation functional made the assumption that $E_{xc}(\boldsymbol{r})$, the value of the exchange-correlation energy at position $\boldsymbol{r}$, depends only on the local density $n(\boldsymbol{r})$. This energy is split into two: the exchange part E_x uses the exact result for the exchange energy of a free electron gas with uniform density n set to the value of $n(\boldsymbol{r})$. The rest, called the correlation energy E_c, is derived from the local approximation applied to the same free electron gas and the dependence on n has been determined from a combination of exact results and parameterization of quantum Monte Carlo calculations. Some examples of this approach are given the names PW (Perdew and Wang), PZ (Perdew and Zunger), and VWN (Vosk, Wilk, and Nusair).[14] For spin-polarized open-shell systems, the LDA can be generalized to the local spin density approximation (LSDA).

[14] The naming convention used for these functionals is usually based on author initials, often supplemented by a year of publication.

Generalized gradient approximation (GGA)

Improvement can be made on the LDA/LSDA approximation by including not only the local density, but also the local gradient of the density, leading to the GGA methods. An example GGA exchange method is labelled B or B88 (Becke's method, published in 1988) which introduces one additional parameter, whose value is fitted against rare gas data. The correlation part of a GGA functional is generally more complicated than the exchange part, e.g. the correlation functional LYP (Lee, Yang, and Parr) has four parameters that are determined by fitting to the helium atom. The combination of B exchange with LYP correlation is known as the BLYP exchange-correlation functional.[15] A widely used GGA functional in current research is PBE (Perdew, Burke, and Enzerhof). This functional is an example of one that is purely *ab initio*, as it does not rely on any fitting to empirical data.

[15] The names of these functionals are generally built up by combining the abbreviations for the exchange part and the correlation part, with the exchange part placed first.

Hybrid functionals

A further development is provided by the so-called hybrid functionals. These make use of the exact representation of the exchange energy provided by Hartree-Fock (HF) theory[16] to improve on the GGA functionals in DFT by mixing together GGA and HF terms. This can be illustrated by the example of the widely used B3LYP functional.[17]

Example 16.4

The B3LYP hybrid functional has the following form for exchange and correlation terms

$$E_x^{\text{B3LYP}} = (1-a)E_x^{\text{LSDA}} + aE_x^{\text{HF}} + b\Delta E_x^{\text{B88}}, \qquad (16.11)$$

$$E_c^{\text{B3LYP}} = (1-c)E_c^{\text{LSDA}} + cE_c^{\text{LYP}}, \qquad (16.12)$$

where the parameter $a = 0.20$ weights the HF exchange against the LSDA exchange, parameter $b = 0.72$ brings in the GGA exchange correction, and parameter $c = 0.81$ weights the LYP version of GGA correlation against the LSDA value. Note that the 3 in B3LYP refers to the use of three parameters for the hybridization.

---

[16]Relevant details of HF theory can be found in textbooks on quantum chemistry (see Further reading) and also, from a slightly different perspective, in books on many-body physics.

[17]It should be noted that the multi-centre integrals required for evaluating E_x^{HF} are computationally very costly when using a plane wave basis set, whereas they are very efficient with Gaussian basis sets. Hence hybrid functionals are often used in calculations with Gaussian basis sets, but less commonly used in calculations using plane wave basis sets.

16.6 Mixed methods

Using DFT for structural optimization can be very costly in computational time, but some methods are available that can be used to help speed things up considerably. These are known as semi-empirical methods and are generally available for elements in the lower part of the periodic table, e.g. molecular systems. The methods rely on simplifying approximations and parameterizations of the computationally demanding two electron integrals. By combining a structural estimation using a semi-empirical method with a final single point DFT calculation, a useful type of mixed method can be obtained.[18] Two types of semi-empirical methods are currently used in this way for muon site problems.

[18]It could also be called a hybrid method, but we use the term mixed method to avoid confusion with hybrid functionals.

Methods based on HF theory

HF theory was historically a very important method of quantum chemistry in the period before DFT became established and for speeding up HF calculations, a useful series of semi-empirical methods were developed. In these methods a set of empirical parameters is derived for each element covered by the method. This is done by optimizing against a training set of molecular data, which includes properties such as geometry, ionization energy, and electric dipole moment. One such widely used method is PM3 (Parametric Method 3), introduced by J.P. Stewart, which has been refined in stages to the latest version, PM7. These methods are useful for rapidly obtaining reasonable geometries for muoniated molecular radicals, but they are not able to produce accurate spin structures. Thus, in order to produce a reliable spin distribution and corresponding hyperfine parameters, it is necessary to follow on with a single-point DFT calculation after the semi-empirical geometry optimization.

Density functional tight binding

In view of the success of the highly efficient HF semi-empirical approach, it was natural to try and apply similar semi-empirical ideas to the DFT calculation framework. This led to the method known as density functional tight binding (DFTB), which focuses on parameterizing the interactions between pairs of atoms. A significant difference between DFTB and a method such as PM3 is that in the case of DFTB the parameterization is done against full all-electron DFT calculations, rather than against a set of experimental data. The DFTB method is very fast, as each stage in the geometry relaxation only requires a single diagonalization of the energy matrix, rather than the iterative self-consistent-field loop of a usual DFT calculation. Once again, in order to obtain reliable hyperfine parameters, a regular single-point DFT calculation is made using the relaxed geometry from the semi-empirical method.

16.7   Obtaining sites

Crystals

[19]DFT+U is the name for an electronic structure method combining DFT with a site-dependent repulsive potential U (in the spirit of the U parameter in the Hubbard model of interacting electrons). Here we combine DFT with an included muon. (It is also possible to have DFT+$U + \mu$!)

With access to the principles of DFT, how do we address the general problem of computing the muon sites in a crystal? A method has emerged in recent years, known as[19] **DFT+μ**. The idea is to treat the muon as one of the nuclei: essentially a light isotope of hydrogen, and to place the muon at some chosen position in a unit cell, or supercell, of the material. With the muon-containing material defined, we perform a geometry optimization, moving the nuclei under the forces defined by eqn 16.10 and then recalculating the electronic structure, until the forces are minimized. In the general case, this means finding a minimum of the total energy as a function of the position of the muon and all of the nuclei. The function representing the energies as a function of the positions of the particles is called the *potential energy surface* and we seek the minimum of this function, which should provide a candidate muon site.

[20]We often assume that the lowest-energy site has a high probability of being occupied by a muon in an experiment, although this is not necessarily the case.

We have a choice about how to assign the initial muon position in the set up of the calculation. If we place the muon at a random position, then a calculation might well result in us finding a local minimum of the potential energy surface. It is therefore a good idea to carry out multiple calculations, with the muon initialized in different positions, in order to see if the muons end up at one (or a small number) of sites which can then be ordered in energy.[20] We are then faced with a choice of where to place each muon in setting up the series of calculations. To start with, we can reject those starting positions that are closer to an atom than a certain cutoff radius, since the repulsive core potential of the atom produces large repulsive forces that will exclude muons from these regions. The muons can then be placed either randomly, at specific positions of interest, or at a set of sites constrained by some criterion. For example, we might choose to assign initial muon positions quasi-randomly, but reject any starting positions that are symmetry-related to other starting positions, since these are likely to be roughly equivalent and so are unlikely to provide any useful extra information.

Once the initial muon-containing structures have been geometrically optimized, the results need to be interpreted. In general we expect the muons to have converged to a handful of sites, representing the minima of the potential. However, relaxation may still lead to symmetry-equivalent sites among these minima and so some method of identifying symmetry-distinct positions is usually required so that we can separate the muon sites into clusters of distinct sites. An example method is sketched below.

Example 16.5

Using our knowledge of the symmetry of the original crystal without the muon, we reduce the coordinates of the μ^+ or Mu for each structure to the *asymmetric unit*, the minimal volume required to describe (by translation, rotation, reflection, and inversion operations) the entire crystal. We then label each of the relaxed muon-containing structures using its calculated total energy and these reduced coordinates. These four-dimensional labelling vectors define a data space, and we analyse this to find out which structures are close to each other in such a space. We can then use clustering algorithms (such as hierarchical or k-means clustering)[21] to identify a small number of groups of structures that fall closer to each other in the data space, and are thus similar. Each of these clusters is likely to represent a distinct muon stopping site.

The result of an example muon site calculation is shown for the compound FeCrAs in Fig. 16.3. A small number of low-energy muon sites are identified from geometry optimizations. In the case of FeCrAs they broadly coincide with the minima of the unperturbed electrostatic potential. In general the variation of the potential energy in the volume explored by this procedure is in the eV range, like chemical bonds. This is typically rather larger than the energy differences between electron-spin polarized and spin unpolarized states, so that in most cases DFT+μ does not need to include these finer electronic structure aspects for a site search.

[21] Exercise 16.4 explores how clustering algorithms work.

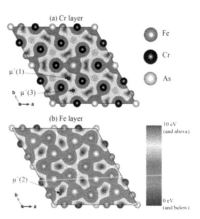

Fig. 16.3 Muon sites in FeCrAs. [Taken from B. M. Huddart *et al.*, J. Phys.: Condens. Matter **31**, 285803 (2019).]

[22]Exercise 16.3 investigates the limits of applicability of the UEP approach.

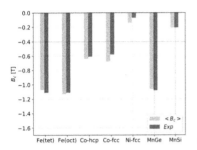

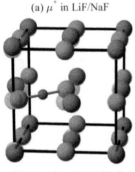

Fig. 16.4 Muon contact hyperfine field B_c for different metallic compounds, experiment vs DFT prediction (including quantum muon effects, discussed in Section 16.8). [Taken from I. J. Onuorah *et al.*, Phys. Rev. Materials **3**, 073804 (2019).]

(a) μ^+ in LiF/NaF

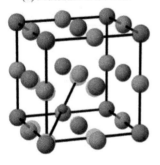

(b) Muonium in LiF/NaF

Fig. 16.5 Calculated equilibrium geometries of dia- and paramagnetic muon states in LiF/NaF. Translucent spheres represent the equilibrium ionic positions before the muon is introduced into the crystal. [Taken from J. S. Möller *et al.*, Phys. Rev. B **87**, 121108(R) (2013).]

Example 16.6

An alternative, and simpler, method for optimizing muon-containing structures is the unperturbed electrostatic potential method (UEP), that was mentioned at the beginning of this chapter.[22] This method can provide an answer *only* in the case of diamagnetic muon sites. The method uses the charge density of the host material obtained from a DFT calculation, but unperturbed by the presence of the muon, to estimate the Coulomb force acting on each of the μ^+ in the initial muon-containing structures. This corresponds to assuming that $\mathbf{R}$ remains constant, and approximating the potential surface as

$$V_{\text{UEP}}(\mathbf{r}_\mu) = \sum_{i=1}^{N} \frac{Z_i}{|\mathbf{r}_\mu - \mathbf{R}_i|} - \int \mathrm{d}^3 r \, \frac{n(\mathbf{r})}{|\mathbf{r}_\mu - \mathbf{r}|}, \quad (16.13)$$

where the sum is carried over all atomic nuclei of the crystal, and the integral over its entire volume, with Z_i charge of each nucleus and $\mathbf{R}_i$ its position. This potential is a direct output of the DFT calculation, being the opposite of the Coulomb potential experienced by electrons, and in the rigid approximation its minimum is the candidate site where the muon feels zero force. Although approximate, the UEP method is extremely fast and works reasonably well in materials in which atomic displacements don't play a major role in stabilising the muon site. High electron density metals are good candidates for this, by virtue of the very short screening distance over which a cloud of conduction electrons can compensate a point-like charge.

Calculation of contact-hyperfine fields

Hyperfine constant calculation, illustrated in Example 16.7, is not limited to molecules. Plane wave codes likewise obtain the contact hyperfine field in periodic magnetic materials, which is defined by means of the spin density of eqn 6.27 and coincides with the spin-resolved density of a polarized-DFT calculation, $n_\uparrow(\mathbf{r}_\mu) - n_\downarrow(\mathbf{r}_\mu)$. It appears to come *for free* from the site search, but unfortunately it relies on the inclusion of realistic spin polarization in the DFT. Therefore, contrary to the site search, the calculation of B_c requires DFT to also self-consistently reproduce the magnetic structure, a task which may be computationally demanding in a supercell that includes the muon.

The contact field is in the direction of the magnetic moment that is producing it. It is of great importance to evaluate it *ab initio*, firstly to determine whether it is negligible or not. The challenge here is the accuracy of the calculation. The early attempts at contact field prediction in simple magnetic metals Fe, Ni, Co, date back to the 1980s. We benefit nowadays from a huge advance in computational performance and the task is feasible in many cases. In particular, good agreement with experiments is obtained for simple (and less simple) metals, as shown in Fig. 16.4.

Example crystal sites

The methods from the last section have been applied to several materials. In alkali fluorides materials we have a way to test the result of the calculations. The experimental observation of F−μ−F stopping states means that the muon must form a linear bond with two fluorine atoms with the form of the signal depending sensitively on the separation of the muon and the surrounding fluorines. The measured $F-\mu-F$ configuration is predicted from the calculations, as shown in Fig. 16.5. For the cases of LiF, NaF, and CaF_2, the calculated bond lengths are within 3% of the experimental values. In fact, these calculations led to a prediction of a similar stopping state in CoF_2, which had been identified in early low temperature experiments (see Note 27) and subsequently confirmed by $F-\mu-F$ high temperature measurements.

DFT techniques have been applied to many more complex materials and plausible muon sites have been identified at positions which have been shown to be entirely consistent with the properties derived from measurements. A particularly complete and convincing case is that of MnSi. Here the muon site computed using DFT methods (shown in Fig. 16.6) is consistent with the symmetry properties derived from Knight-shift measurements and also with the very restrictive details of the complicated magnetic structure of the material.

In many cases, the muon, taken along with its local distortion to the electronic structure, is the object that is sensitive to the magnetic response of a material. (This resembles the notion of a quasiparticle in a metal, which is the particle-like excitation formed from the electron along with its interactions with other electrons.) The presence of the positive muon has the potential to change the local structure and charge distribution, but also, as a result, the local magnetic structure and crystal field levels. The influence of these changes on the magnetism was demonstrated in muon states found in the pyrochlore frustrated magnets $Pr_2B_2O_7$ (B = Sn, Zr, and Hf) where a combination of measurements and muon site computation showed that the muon causes a local distortion to the structure that can indeed alter the crystal field energy levels in a manner that leads to a measurable effect in the muon spectra (Fig. 16.7). In this case, the muon response can be dominated by the muon-induced local distortion, rather than the intrinsic behaviour of the host compound. Although this is a rather extreme case, it is important to bear in mind that it is a possibility and the muon can create a local magnetic environment that leads to the effects that the muon itself subsequently measures.

Another possible example of this is in the molecular spin ladder materials $(Hpip)_2CuBr_4$, where the muon forms states based on $Br-\mu-Br$ bonds (much like the $F-\mu-F$ bonds). This state causes a sizeable local distortion to the atoms neighbouring the muon and consequently to the electronic structure (Fig. 16.8). Although it might be feared that this would prevent the muon from faithfully measuring the properties of the material, this seems not to be the case as the magnetic phase diagram is determined by physics on a length scale that is very long compared to the muon and its distortion. As a result the transitions in the material are observed with μSR at the fields and temperatures found using other techniques.

Fig. 16.6 The muon site in MnSi is found experimentally to be at $(0.532, 0.532, 0.532)$ (indicated by small circles at crystallographically equivalent sites). These positions are enclosed by the regions where the unperturbed electrostatic potential computed by DFT takes a minimum. [Adapted from A. Amato *et al.*, Phys. Rev. B **89**, 184425 (2014).]

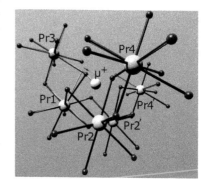

Fig. 16.7 Muon site in $Pr_2Sn_2O_7$. [Taken from F.R. Foronda *et al.*, Phys. Rev. Lett. **114**, 017602 (2015).]

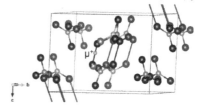

Fig. 16.8 Example muon site in the molecular spin ladder $(Hpip)_2CuBr_4$. [Taken from T. Lancaster *et al.*, New. J. Phys. **20** 103002 (2018).]

At the time of writing strong electron correlation and atoms with large Z, i.e. relativistic effects still constitute something of a challenge for DFT. Therefore DFT+μ also occasionally has difficulties with materials in these two categories, demonstrating that the potential energy surface relevant for muons is not always determined solely by purely chemical effects.

Molecules

In molecules, a muoniated radical state[23] is often the target of the calculation. These states are generally equivalent to adding neutral muonium to an atom with an unsaturated bond. Candidate sites can then be determined from consideration of the chemical structure of the bare molecule. Knowledge of the geometry of chemical bonding allows a reasonable starting position for the added muonium to be set and a geometrical optimization is performed from this starting point using either full DFT or one of the faster semi-empirical methods. A final single-point DFT calculation at the relaxed geometry allows the spin distribution over the molecule to be accurately determined. From this the hyperfine tensors for the muon and the atoms with nuclear spin are obtained. This is done for all of the candidate muon addition sites and the set of hyperfine parameters obtained in this way allows the simulation and interpretation of ALC and TF rotation spectra to be made. In solids both the dipolar part and the contact part of the hyperfine tensor are important, whereas in liquids fast reorientational motion averages the dipolar part to zero and only the isotropic contact part is retained.

The geometry obtained is for the static classical muon site, but significant corrections may be needed for both thermal motion of the molecule and quantum motion of the muon. The quantum motion of the muon is discussed in more detail in Section 16.8. Quantum correction to the hyperfine parameters can be made in several ways. One method is to extend the length of the muon bond within the equilibrium structure by a known amount (of order 5% for a C–Mu bond) to reflect the asymmetry of the quantum motion. The calculated hyperfine coupling then includes the quantum correction for the bond stretching. Another method is to directly apply correction factors to the hyperfine parameters, based on either calibrating against empirical data or else using specific calculations of the effect of the quantum motion. In the case of C–Mu the typical quantum correction factor for the muon contact hyperfine parameter is of order 1.3.

Example 16.7

As an example, we look at the muon sites for the TCNQ molecule. Possible sites for muonium addition to TCNQ are shown in Fig. 16.9. Addition to sites 3 and 4 requires a significant bending of the molecule, which will be hindered in a tightly packed solid-state environment. Of the three remaining sites, site 1, adding to the peripheral nitrogen, has an energy ≈ 1 eV lower than sites 2 and 5. This is the only site found experimentally, revealed as a Δ_1 resonance in solid-state ALC measurements.[24] The resonance line shape indicates a radical with a highly axial hyperfine tensor, characterized by the two significant hyperfine parameters A and D_1. In Table 16.3 we show the experimental values along with calculated values using either PM3 or DFT geometry. As expected, the DFT geometry gives a better overall match to experiment, but the semi-empirical method is more than five times faster and gives a reasonable accuracy that may often be sufficient.

[24]F. L. Pratt *et al.*, Magn. Reson. Chem. **38**, S27 (2000).

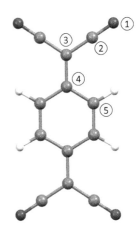

	A (MHz)	D_1 (MHz)	time (s)
Experiment (10 K)	81.9(3)	10.2(4)	
Calculation (DFT geometry)	82.4	9.2	4119
Calculation (PM3 geometry)	76.8	12.4	768
Quantum correction factor	1.10	1.15	

Table 16.3 Comparison of the experimental and calculated hyperfine parameters for radical 1. The calculations were made using the Gaussian09 code with the B3LYP hybrid functional and the cc-pVDZ basis set. The quantum correction factors used here were calculated from the shape of the potential energy surface for the muon. The calculation times are also given for comparison in the final column.

Fig. 16.9 The TCNQ molecule. Carbon atoms are light grey, nitrogens are dark grey, and hydrogens are white. The possible sites for Mu addition are numbered.

For molecular crystals the crystalline environment can lead to additional site displacements, leading to shifts in the hyperfine parameters and splitting of spectroscopic features. Using an alternative method (e.g. PW or DFTB) allows such effects to be modelled.

16.8   Quantum effects

The DFT treatments described have assumed that the muon is a classical particle sitting at its equilibrium position. This is a consequence of the Born-Oppenheimer approximation where we make a clear distinction between the classical nuclei (including the positive muon, which is treated as a light proton) and the quantum electrons. However, this neglects quantum mechanical effects of muons and nuclei, such as zero-point motion, tunnelling, and large zero-point energies (ZPE) ≈ 0.5 eV (Fig. 16.10). These can change the energetic ordering of muon sites, destabilize certain candidate classical muon sites or cause others to merge into a single extended region of enhanced muon density by ZPE

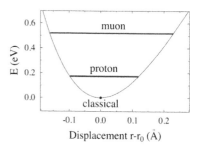

Fig. 16.10 Typical zero-point energies for the proton and muon at a site defined by an interatomic potential. Due to the anharmonic form of the potential, there is an outward displacement of the mean position of the muon compared to that of the proton.

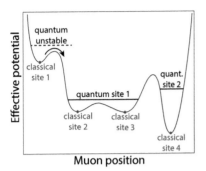

Fig. 16.11 Effects of quantum zero-point energy (ZPE) around classical candidate muon sites: classical muon site 1 is destabilized by its high ZPE (dashed line); classical muon sites 2 and 3 merge into a single quantum site 1 because the ZPE (solid line) overcomes the potential barrier between them; and classical site 4 becomes an unfavourable high-energy quantum muon site 2 due to its high ZPE (solid line). In practice, the muon could also experience quantum tunnelling between quantum sites 1 and 2 due to the modest energy barrier between them.

overcoming potential barriers between them (Fig. 16.11). All of these effects are mass dependent and therefore especially pronounced for light particles like the muon, which is ≈ 9-times lighter than hydrogen. The result is that the muon wavefunction is expected to be spread over an appreciable volume in the material, rather than concentrated at a single point as is tacitly assumed in the Born-Oppenheimer approximation. All observable quantities such as internal magnetic fields and muon coupling constants have, in principle, to be averaged over the whole extended muon wavefunction, not just sampled at a single classical point as in the Born-Oppenheimer approximation. The significant spatial extent of the wavefunction of the implanted muon is known to be relevant in many systems. Well known examples are the quantum diffusion of muons in metallic systems such as copper or of muonium in insulating systems such as solid nitrogen. In these cases band-like coherent transport of the muon or muonium occurs at low temperature (see Section 8.5).

There are several proposals for dealing with this problem, ranging from the simple strategies of approximating the potential around the muon to approximately compute its wavefunction, through to more sophisticated methods based on path-integral molecular dynamics. These methods are the subject of much current research and show promise for allowing us to reach the next level of sophistication in describing the state of the stopped muon in a solid.

16.9   Sites via experiment

Valuable though it is to have computational techniques to *predict* muon sites, confirmation from *experiment* is important to assess the reliability of the computational techniques. Here we briefly review some of these techniques.

All of the results in the following example stem from the basic principle that the magnetic field at the muon site B_μ may not be exactly the same as the applied field B_{ext}. This is due to contributions from the dipolar field, the hyperfine contact interaction, the Lorentz field, and the demagnetization field. In a transverse field experiment, one is often interested in the difference between the two, $B_\mu - B_{\text{ext}}$, but measured along B_{ext}. This is because the component of $B_\mu - B_{\text{ext}}$ measured perpendicular to B_{ext} makes very little difference to the precession frequency.[25] We therefore define the Knight shift K to be

[25]This is because it only enters at second order: $|B_\mu| = (1 + K)|B_{\text{ext}}| + O(B_\perp^2/B_{\text{ext}})$.

$$K = \frac{(B_\mu - B_{\text{ext}}) \cdot B_{\text{ext}}}{B_{\text{ext}}^2}, \qquad (16.14)$$

and hence $B_\mu \approx (1 + K)B_{\text{ext}}$.

Example 16.8

One of the contributions to $B_\mu - B_{\text{ext}}$ is the dipolar field B_{dip}, which is given by

$$B_{\text{dip}}^\alpha = \sum_j \sum_\beta D_j^{\alpha\beta} m_j^\beta, \qquad (16.15)$$

where the first sum is taken over all the magnetic moments m_j within the Lorentz sphere and the dipolar tensor for the jth moment at position r_j is

$$D_j^{\alpha\beta} = \frac{\mu_0}{4\pi R_j^3} \left(\frac{3R_j^\alpha R_j^\beta}{R_j^2} - \delta^{\alpha\beta} \right), \tag{16.16}$$

where $R_j = r_j - r_\mu$ and r_μ is the muon position. These expressions are often used to model ordered magnetic arrangements and this can be a very good test of the validity of a muon site.

However, even in the paramagnetic state these expressions can be useful. In an applied field, the moments of a paramagnet become partially polarized and take the value

$$m_j = \frac{\chi B_{\text{ext}} \mathcal{V}_{\text{c}}}{\mu_0}, \tag{16.17}$$

where $\mathcal{V}_{\text{c}}$ is the volume per magnetic ion and $\underline{\chi}$ is the magnetic susceptibility tensor.[26] In this case,

$$B_{\text{dip}} = \underline{\underline{D}}\underline{\chi} B_{\text{ext}}, \tag{16.18}$$

where $\underline{\underline{D}} = \frac{\mathcal{V}_{\text{c}}}{\mu_0} \sum_j \underline{D}_j$ is the total dipolar tensor.

One can play a similar game with the contact hyperfine interaction B_{hf}, which is given by

$$B_{\text{hf}} = \underline{\underline{A}}\underline{\chi} B_{\text{ext}}, \tag{16.19}$$

where $\underline{\underline{A}}$ is the analogous hyperfine tensor. The contact coupling is usually independent of the field direction and so this can be written as a scalar in the majority of cases.

This provides us with all we need to model the effect of rotating a single crystal in a constant magnetic field and predicting the Knight shift for a particular muon site, the anisotropy of the dipolar coupling resulting in angle dependence, while the contact interaction gives an angle-independent contribution.[27]

Another experimental test of the validity of a candidate muon site can be obtained using the contribution of nearby nuclear dipoles to the decay rate of a transverse-field precession measurement. As we have seen, the shift in muon-precession frequency (compared to what is expected from the applied field) gives us the Knight shift. But the damping of the precession signal arises from contributions from nearby nuclear dipoles, which are not ordered, but sometimes add and sometimes subtract from the applied field. In a transverse field measurement, the broadening is given at short times by a Gaussian relaxation function $\exp(-\sigma^2 t^2/2)$, where the parameter σ^2 (also known as the **second moment** M_2 in NMR) is given by

$$\sigma^2 = \frac{1}{3} \left(\frac{\mu_0}{4\pi} \right)^2 \hbar^2 \gamma_\mu^2 \sum_i \gamma_i^2 I(I+1) \frac{(1 - 3\cos^2\theta_i)^2}{r_i^6}. \tag{16.20}$$

This depends on the spin I of the nearby nuclei,[28] and their position r_i and gyromagnetic ratio γ_i. This depolarization rate (the derivation of which can be found in Appendix G) is angle-dependent (θ_i is the angle between the applied field and the vector between the muon and the nucleus) and so this broadening can be measured experimentally. In the case of a zero-field measurement the relaxation takes on the Kubo-Toyabe form with a value of Δ^2 given by

$$\Delta^2 = \frac{1}{3} \left(\frac{\mu_0}{4\pi} \right)^2 \hbar^2 \gamma_\mu^2 \sum_i \gamma_i^2 I(I+1) \frac{(5 - 3\cos^2\theta_i)}{r_i^6}, \tag{16.21}$$

which again can be checked by experiment (and is also derived in Appendix G). Moreover, in some fortunate cases (particularly fluorides) the zero-field signal from the nuclei contains much more structure than a simple Kubo-Toyabe relaxation (which is derived from an assumption of a Gaussian-distributed random distribution of local field components). In those cases, there is much more information to go on to tie down the muon site and understand the local environment.

[26]The magnetization M is related to the applied field by $\mu_0 M = \chi B$, assuming an (isotropic) susceptibility $\chi \ll 1$. The magnetization $M = m/\mathcal{V}_{\text{c}}$ where $\mathcal{V}_{\text{c}}$ is the volume per magnetic ion. Anisotropic cases require a tensor.

[27]A good example of this technique applied to the material CeB_6 is given in A. Amato *et al.*, Hyp. Int. **85**, 329 (1994). The alternative case of an external field applied to an antiferromagnetic CoF_2 single crystal is illustrated in R. De Renzi *et al.*, Phys. Rev. B **30**, 186 (1984).

[28]It is modified in the case of quadrupolar nuclei (see Appendix G).

Chapter summary

- Knowledge of the muon site enables us to work out which fields are being probed and the influence that the muon has on its local environment.

- Density functional theory allows the computation of muon sites by relaxing a structure containing the muon until the minimum energy is found.

- In molecular systems where muoniated radicals can form, the hyperfine tensors calculated from DFT can be compared directly with those obtained experimentally from TF-μSR and ALC spectroscopy.

- Knight-shift measurements (i.e. rotating a single crystal in a magnetic field and measuring the angle-dependence of the precession frequency) can lead to an experimental verification of a calculated muon site, as can measuring nuclear dipolar fields via TF and ZF relaxation.

Further reading

- A recent review of DFT can be found in A. Pribram-Jones *et al.*, Annual Review of Physical Chemistry **66**, 283 (2015). More details are available in F. Giustino, *Materials Modelling using Density Functional Theory*, OUP (2014) and R. M. Martin *Electronic Structure, 2nd Ed.*, CUP (2020). A very good summary is also available in Walter Kohn's Nobel Lecture, published in W. Kohn, Rev. Mod. Phys. **71**, 1253 (1999).

- The use of DFT in chemistry is described by several texts: W. Koch and M. C. Holthausen, *A Chemist's Guide to Density Functional Theory*, Wiley-VCH (2000); C. J. Cramer, *Essentials of Computational Chemistry*, Wiley (2004); F. Jensen, *Introduction to Computational Chemistry*, Wiley (2017).

- Various approaches to DFT calculations applied to μSR are described and reviewed in the following articles: J. S. Möller *et al.*, Physica Scripta **88**, 068510 (2013); P. Bonfà and R. de Renzi, J. Phys. Soc. Jpn. **85**, 091014 (2016); L. Liborio *et al.*, J. Chem. Phys. **148**, 134114 (2018); S. Sturniolo *et al.*, J. Chem. Phys. **150**, 154301 (2019).

- Links to DFT codes and software tools can be found on the book website: `https://musr.org/muon-spectroscopy-book`.

Exercises

(16.1) Estimate how many gigabytes of memory would be required to store the full quantum wavefunction for a system of $N = 10$ particles, with a grid of side $L = 100$ in double precision. Remember that the wavefunction is a complex number and requires two double precision numbers per point, and that a double precision number occupies 8 bytes of memory.

(16.2) If a muon resides in a harmonic (quadratic) potential, how much larger is its zero-point energy compared with that of a proton in the same well? How does the root mean square deviation of the muon's position from the minimum of the well compare with that for a proton?

(16.3) In order to explore the importance of accounting for atomic displacements when looking for the muon site, consider a simple model. Two atoms are separated by a distance L. A muon is inserted between them, and must relax to a minimum of the potential they generate. Each atom contributes to the potential with a Lennard-Jones term,

$$V_{\mathrm{LJ}}(r) = \Delta E \left[\left(\frac{r_0}{r} \right)^{12} - 2 \left(\frac{r_0}{r} \right)^6 \right], \quad (16.22)$$

where r is the distance from the atom, r_0 the equilibrium distance at which the minimum of the well is located, and ΔE the depth of that well. Consider the two cases in Fig. 16.12:

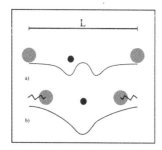

Fig. 16.12 Site scenarios for Exercise 16.3.

(a) $r_0 = 2/5L$ and the two atoms are fixed in place (unperturbed system). The total potential has two distinct energy wells. What is the binding energy of the muon in either of them? Assume the equilibrium distance from the midpoint to be still r_0 for each atom.

(b) With the same value of r_0, consider now the possibility that atoms may be shifted from their equilibrium positions, being attached to them by springs of elastic constant k (i.e. the system is allowed to be perturbed by the muon). How much do they each have to shift in order for the two wells to merge into one? What value does k need to have in order for this configuration to be energetically more stable than the unperturbed one? Consider what this tells us about the regime in which the unperturbed approximation is valid. For a positive muon, will it work better in a crystal that is covalent or ionic?

(16.4) In order to follow the methods used for analysing muon sites in terms of clusters, consider a related problem. The points in Fig. 16.13 are the addresses of various tourist spots in a city with a very regular squared map. In order to design optimal tours among these attractions, build a hierarchical clustering of these points to highlight which ones are grouped closer together. To do this copy the map onto graph paper and apply the following algorithm:

Fig. 16.13 Map for clustering in Exercise 16.4

(1) Calculate the distances between pairs of points (or clusters) and identify the two closest. If there are two pairs with the same distance, choose one at random (use as distance the number of segments you need to walk from one point to the other. For example, the distance between A and B would be 2. This is called the 'Manhattan distance', $d_{\mathrm{M}} = \sum_i |x_i^{(1)} - x_i^{(2)}|$, as opposed to the traditional Euclidean distance, $d_{\mathrm{E}} = \sqrt{\sum_i |x_i^{(1)} - x_i^{(2)}|^2}$);

(2) Remove the two points or clusters and replace them with a cluster that includes them both;

(3) Repeat step 1. To compute the distance between a point and a cluster, or between two clusters, use the distance between their two closest points. Continue until you have only one cluster including all five points.

Numerical modelling

17

The application of computational modelling for estimating muon stopping sites was discussed in Chapter 16. In this chapter we cover some other uses of numerical modelling in muon spectroscopy. First we take an overview of the modelling and tuning of muon beamlines, tasks that are usually carried out by beamline scientists. The second topic is modelling the muon stopping in materials, which is needed in many diverse experimental situations in muon spectroscopy. The final topic is numerical modelling of the evolution of the spin of the implanted muon, which can provide a powerful tool for planning muon experiments and interpreting muon data.

17.1 Beamline optimization

Beamline optimization and modelling can be carried out at two different levels. The basic level treats the beamline as an optical system defined by a series of elements with associated transmission coefficients. The more advanced level uses ray-tracing to take account of fine details in the geometry of the beamline components.

Basic modelling and tuning

The first stage of tuning is to ensure that the beamline will transmit muons as expected, i.e. that the bending magnets deflect muons through the correct angles and the quadrupoles give a focus at or near the desired sample position. There may also need to be an intermediate focus provided at a slit position. As bending magnets, separators, and spin rotators deflect particles of different momentum by different degrees, the overall beamline should ideally be designed to be achromatic, so that muons of different momentum within the pass band are all finally refocused onto the sample. Quadrupole magnets necessarily focus in one plane and defocus in the other (see Fig. 13.5). Bending magnets have different focusing in the bend and non-bend planes, in addition to the bend angle itself. So the two transverse directions have to be considered individually. This is equivalent to having a complex optical system with cylindrical lenses and prisms, rather than just spherical lenses.

The calculation can be done by a straightforward matrix optics method, since the exit trajectory (position, angle, momentum, and time delay) of a particle passing through a beam element is determined by the input trajectory vector multiplied by a matrix (usually 6×6 for a first order calculation). The matrix coefficients depend on the element type and its field value – even a section of plain beam pipe or 'drift space' has a matrix. The response of the whole beamline is just the product of all the element matrices in order. It is then possible to adjust the field strengths to set elements of the whole-beamline matrix to desired values, for example to ensure a point-to-point focus from the muon production target to the sample.

The beam envelope at any point along the beamline can be represented by a 6×6 matrix, encoding the size, divergence, momentum spread, pulse length, and the correlations between them. Again, the output envelope matrix is determined by multiplying the input by the beam element matrices in order. Given an input beam it is possible to constrain the magnet strengths to ensure a required output, for example a small spot or a sufficiently small intermediate beam waist to go through a separator or kicker.

It is also possible to put in steering magnets and check the best location for them, to deflect the beam spot at the sample without too much risk of losing the beam in the apertures upstream. A beamline may have more focusing magnets than constraints on its focus. Adjusting the extra degrees of freedom may allow the beam flux to be maximized. Programs such as TRANSPORT are available to do these calculations and optimization fits. This program also gives a plot of the beam envelope along the beamline, with the location of magnets indicated. This makes it easier to change the layout or identify the best locations of magnets.

Ray tracing and flux optimization

Ray tracing can be used for optimizing the muon flux. The matrix method assumes an elliptical beam envelope. This can be checked against the apertures of the magnets to predict that most of the beam will be transmitted. However the apertures of real magnets and slits have various shapes. A ray tracing program can predict the actual flux, by generating a large number of trial particle vectors within some source envelope, passing them through the beam element matrices in sequence, and at each stage comparing the position against any aperture and stopping that particle if it would hit something.

Outputs include the flux at various positions, including where particles are lost, and histograms giving beam and spot profiles. The effectiveness of a separator can be checked by simulating with positrons of the same source momentum and spot size as the muons. Programs such as TURTLE are available for this; TURTLE uses an input file compatible with TRANSPORT so that both can be used on the same design.

More advanced ray tracing methods can model more effects than simple loss of particles from the beam. Muons may scatter from the beam pipe or the edge of the slits and still travel downstream. Muons may be stopped in the beamline (especially slits or collimators) and their decay positrons can still reach the instrument, causing background counts.

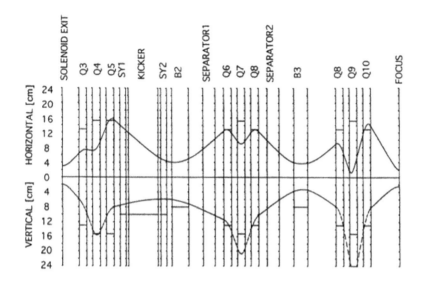

Fig. 17.1 Beam profile modelling results for the RIKEN-RAL beamline, showing the beam characteristics between the decay solenoid exit port and the Port-1 sample position. [Taken from T. Matsuzaki *et al.*, Nucl. Instr. and Meth. A **465**, 365 (2001).]

There may be a beam window which will slow down and scatter the muons slightly. Magnets may have more complex fields than the simple models used in TRANSPORT, so a field map is used. Decay muon beam-lines need to model the transport of the pions, their decay into muons (which is supposed to happen in the solenoid but could be anywhere) and the resulting capture and transport of the muons.

Programs such as G4Beamline and musrSim, both based on the GEANT4 particle physics package, can model detailed beamlines in this way. Both take an input text file describing the beamline, and output histograms of muon spots, flux, polarization, etc.

With any of these modelling programs, magnet values can be scanned and the output parameters plotted. Further optimization of the beam-line tuning is then possible. An example of the beam profile calculated for part of the RIKEN-RAL beamline is shown in Fig. 17.1.

17.2 Muon range profile

Modelling the muon range and stopping profile (see Fig. 15.2) is an important part of ensuring that the muon beam is stopping in the sample, which may require a careful choice of degrader if the sample mass is small. Programs such as musrSim and SRIM/TRIM[1] can be used for this purpose. These use Monte Carlo methods to track the energy loss and eventual stopping of an ensemble of muons passing though specified materials in the beam path. Figure 17.2 compares stopping profiles

[1] SRIM stands for Stopping and Range of Ions in Matter, a group of programs which includes the core program TRIM, standing for TRansport of Ions in Matter.

measured using different degrader materials with the corresponding simulations.

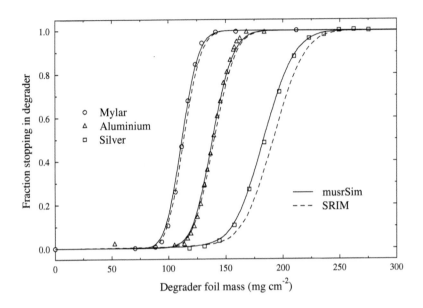

Fig. 17.2 Muon stopping fraction versus degrader mass measured at the sample position in the HiFi instrument at ISIS. Three different materials are compared, Mylar is the polymer polyethylene terephthalate. The lines show simulated stopping profiles made using the musrSim and SRIM programs.

Several features of the stopping process are apparent from these plots. Firstly that the mass efficiency of stopping is much greater for low-Z materials such as aluminium and the organic polymer Mylar, compared to a high-Z material such as silver. The range straggling is given by the width of the transition curve on going from no stopping to full stopping. This range straggling differs even more than the overall range, when comparing low Z and high Z materials. From Fig. 17.2 we can see that an optimum degrader setting, as described in Chapter 15, would be around 75 mg cm^{-2} for Mylar, 110 mg cm^{-2} for aluminium, and 140 mg cm^{-2} for silver. At these optimum degrader settings the amount of sample required to stop the muons is determined by the straggling width. For a sample with stopping characteristics comparable to Mylar, around 25 mg cm^{-2} would be sufficient to stop 80% of the muons, whereas for a sample with stopping characteristics comparable to silver one would need around 50 mg cm^{-2}.

When working with pressure cells, modelling of the range and stopping profile versus muon momentum allows conditions to be optimized for a particular combination of sample and cell (see Chapter 21). For experiments using photoexcitation it is often important to get good overlap of the muons and photons, which requires careful modelling of sample cells (see Chapter 19). Another area where modelling the stopping profile is vital is when using low energy muons (see Chapter 18). Here the muon energy and associated stopping profile is finely tuned to allow the probing of surfaces and interfaces within a sample. TRIM.SP is a special version of the TRIM code that is generally used for modelling low energy muons.

17.3   Muon spin response

Analytical solutions of the muon spin response, such as those described in Chapter 4, can sometimes be impractical because of the numbers of spins involved. Approximations are often made to simplify the solution, but the assumptions made (such as a low or high magnetic field limit) must be noted and the simple result not taken outside its region of validity, which may not be possible if a wide range of field needs to be covered to tie in with experiments. Many real experiments are also carried out on powder or polycrystalline samples, so the modelled response needs to be averaged over orientation. The required integrals can be difficult or impossible to solve analytically, and an alternative method is to solve the muon response numerically without the need for simplifying approximations. Given numerical values of the interactions, magnetic field, etc., one can obtain the muon polarization as a function of time, either as a simple array, or a sum of simple functions such as $\sin(\omega_i t)$ or $\exp(-\lambda_i t)$. It is then straightforward to use this within an orientational averaging loop to generate a powder average.

Computational method

The computations use the density-matrix method, as described in Chapter 4, since this allows a solution in the case that the initial state of some of the spins is unknown. An example of this would be the usual case of having host nuclear spins that are unpolarized just before the muon is implanted. The Hamiltonian is built up from all the specified interactions (Zeeman, hyperfine, dipolar coupling, quadrupolar coupling), and is diagonalized to obtain the eigenvalues (energy levels) and eigenvectors. The initial density matrix is constructed from the specified muon polarization; both this and the spin operator for the detectors are converted into the basis states of the eigenvectors. The evolution of the polarization as a function of time can then be calculated, leading in general to a constant term plus a series of oscillations. To provide an illustration of the computational method, the detailed implementation of the density matrix method by the program QUANTUM is presented in the remainder of this section.

Geometry and powder averaging

As previously noted, calculation of a longitudinal field powder average is often required. In QUANTUM this is done by adding an outer loop to the calculation, where random, or uniformly spaced, orientation vectors are chosen. These set the magnetic field direction, the polarization, and the detector geometry, keeping the anisotropy of the dipolar, hyperfine, and quadrupole interactions fixed in space. For transverse field, a field direction is chosen and then an uncorrelated muon polarization (and detector) direction is chosen, keeping it orthogonal to the field. One special case is in zero field, where the Hamiltonian is independent of direction. The powder average can then simply be calculated as the average of any three mutually orthogonal measurements, for example along x, y, and z.

Time averaging

To reflect actual experimental results, the polarization is averaged over time bins to match the experimental histogram time bin, rather than simply being sampled at discrete points. Additional frequency response corrections may be used to allow for a finite muon pulse width, with a selection of functional forms being available. The limiting case is for integral counting where a single large time bin from 0 to ∞ is used, weighting by the muon lifetime. This integral is much faster to evaluate than generating and summing a time series.

Including dynamics

Dynamics can also be incorporated into the calculations. For multiple muon sites, the system including the muon can be in one of several states, with matching basis states. Transitions between the states can preserve the wavefunction or density matrix through the transition, while transferring population to the new site, although the subsequent evolution will differ, depending on the site. Alternatively, it is possible to depolarize one spin (such as an electron) as part of the transfer. The overall state can be modelled as a 3D 'density matrix' with the third axis being an index over the muon sites. In-place relaxation of one spin will reduce the difference of spin-up and spin-down elements, and reduce the cross terms between them. Such relaxation may also represent muon hopping between equivalent sites with a new set of nearest neighbour atoms, or transport of the muon's neighbours.

Each element of the density matrix evolves based on the values of all the elements and the Hamiltonian. It is therefore possible to first write a set of simultaneous equations to model the simple interaction on each site and then add in the site change and relaxation. The resulting equations form a large matrix that can be solved by diagonalization and the solution is a sum of damped oscillating terms, whose frequencies and relaxation rates are the real and complex parts of the eigenvalues, the amplitudes being calculated from the eigenvectors.

RF resonance

The response of muons to RF fields may be required in some simulations. Rather than using the Hamiltonian directly as the rate of change of the density matrix, it is possible to 'integrate' it over a whole RF cycle. The resulting average Hamiltonian gives the density matrix one RF cycle later, based on the initial value, and disregarding any details of what happens in between. This is a reasonable approximation for high frequencies (not to be observed directly) and relatively weak RF field B_1. Lower frequencies require the details of the spin evolution to be calculated at a series of points throughout the cycle. Taking account of the experimental setup being modelled, it may be necessary to consider the initial phase of the RF waveform as the muon is implanted (or when the RF is turned on) and average over all possible phases. A pulsed stimulus will have a different Hamiltonian for each time regime, for example before, during, and after an RF pulse. These are evaluated in sequence, calculating the density matrix at the end of one time interval and using it as the starting polarization for the next.

Scans

The calculations are often required as a function of some experimental parameter. Any parameter of the model can be varied and the muon response evaluated as a function of that parameter, producing either a single number such as integral asymmetry or else a time series. A typical use would be to scan magnetic field through an avoided level crossing or RF resonance.

Implementation

The QUANTUM program implementing the above calculations works within the MANTID data analysis framework,[2] where it can be used to fit experimental muon data by adjusting model parameters, in addition to predicting the muon polarization from a model. A MANTID table workspace is used to collect the input parameters, and the results of a simulation are put in a workspace which can then be plotted or used for further calculations. In addition, the calculation can be used within a fit function, to extract the optimum values of parameters to reproduce experimental data. This could be either one or more asymmetry time spectra, or a scan of integral asymmetry.

Examples

Three example calculations are now given to demonstrate the use of QUANTUM in different types of problem. In each case the problem is defined by an input table containing two columns: the first column provides keywords and the second column gives their assigned values.

[2]MANTID stands for Manipulation and Analysis Framework for Instrument Data.

Example 17.1

The ALC resonance of a muoniated radical in the solid state can be calculated by QUANTUM with an input table as follows:

Code	Value	
spins	Mu, e, H	*Define the spins to be included in the model*
a(Mu)	500.0, 20.0	*Add the hyperfine constants*
a(H)	100.0, 1.0, 0, 1, 0	*Specify the anisotropy axis, default along z*
lfuniform	100	*Request powder averaging, for LF geometry*
measure	integral	*Specify what to plot*
loop0par	bmag	*Scan the magnetic field*
loop0range	1.5, 2.5, 101	

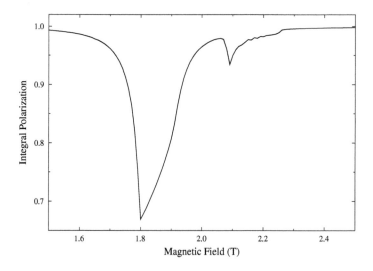

Fig. 17.3 Simulated ALC spectrum for a muoniated radical.

The result is as shown in Fig. 17.3 where the $\Delta m = 1$ line appears at 1.8 tesla and the $\Delta m = 0$ line at 2.2 tesla, both with powder lineshapes.

Example 17.2

An example of nuclear interactions – a muon in the centre of a tetrahedron of four hydrogen atoms which results in a 32×32 Hamiltonian. The result is shown in Fig. 17.4.

Code	Value	
spins	Mu, H1, H2, H3, H4	*Hydrogens numbered for identification*
r(Mu)	0, 0, 0	*Locations of the spins in Å*
r(H1)	1.2, 1.2, 1.2	
r(H2)	−1.2, −1.2, 1.2	
r(H3)	−1.2, 1.2, −1.2	
r(H4)	1.2, −1.2, −1.2	
lf	1, 1, 0	*Specify field and muon polarization direction*
measure	timespectra	*Specify what to plot*
bmag	0.0	*Zero field*

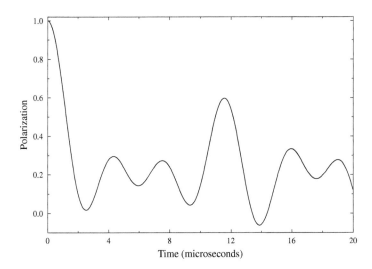

Fig. 17.4 Simulated muon response in the centre of a tetrahedron of hydrogen nuclei.

Example 17.3

The transverse spectrum of a shallow donor in a semiconductor, with a varying ionization rate. This is a powder sample with anisotropy. Time and frequency domain results are shown in Fig. 17.5.

Code	Value	
spins	Mu, e	
dynamic	2	*Enable dynamics, with 2 states*
a(@0,Mu)	0.26, 0.10, 1, 1, 1	*State 0, occupied donor*
a(@1,Mu)	0.0	*State 1, ionized*
pop(0)	1.0	*All muons have an electron to start*
pop(1)	0.0	
convert(0,1)	0.1	*Ionization rate, to be varied*
tfuniform	500	*Powder sample*
bmag	0.01	
measure	timespectra	
loop0par	convert(0,1)	
loop0range	3, 0.03, −5	*Log scale, decreasing*

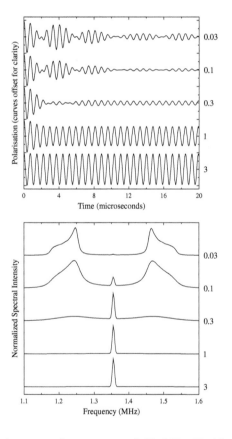

Fig. 17.5 Shallow donor state in a transverse field of 10 mT with varying ionization rates in μs^{-1} given on the right, shown in both time and frequency domains.

Chapter summary

- Modelling and tuning of muon beamlines can be carried out using matrix techniques from optics, or via ray tracing methods.
- Monte Carlo methods for determining the muon stopping profile allow optimal degrader thicknesses to be computed, and pressure-cell and low-energy measurements to be planned.
- Numerical modelling of the muon spin response using the density matrix method allows muon spectra to be predicted. The QUANTUM program enables the polarization function to be computed for a range of interactions between the muon and its environment.

Further reading

- Links to current software and documentation for the methods covered in this chapter are given on the website associated with this book: https://musr.org/muon-spectroscopy-book.

<table>
<tr><td>**18**</td><td># Low energy μSR</td></tr>
</table>

[1]Although significantly smaller areal densities can be measured with the use of degraders, as shown in Chapter 17, as well as using the veto and fly-past techniques discussed in Chapter 15.

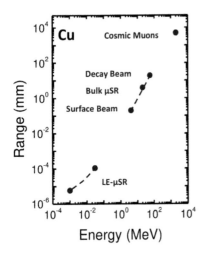

Fig. 18.1 Range of muons in Cu as a function of muon energy.

[2]Note that the name, surface muons, reflects the place of origin of the muons, not the application of muons to near-surface studies, which requires the low-energy muons discussed in this chapter.

[3]The Wikipedia article *Muon Tomography* gives an introduction to the use of cosmic ray muons, including references.

Since muons originate from decaying pions, the muon is born with a kinetic energy of about 4.2 MeV in the rest frame of the pion. This high energy is the reason why muon beams at proton accelerators penetrate deeply into solid matter, reaching depths that range from several hundred micrometres in the case of surface muons to several cm in the case of decay muons, as illustrated in Fig. 18.1. This generally limits the application of μSR to studies of bulk materials with a typical areal density of ≈ 200 mg/cm^2.[1] In order to be able to apply the μSR technique to much thinner nanometer-scale layers (e.g. thin films, heterostructures, interfaces, and near-surface regions), **low-energy muons** with energies in the keV range are required (Fig. 18.1). The production and use of these low-energy muons is the subject of this chapter. We first present the basic principles of low-energy muon (LEM) generation, and give an overview of the facilities at J-PARC and PSI. Next we discuss some special features to be considered in performing low-energy μSR experiments and finally we demonstrate the capabilities of the technique with some science examples.

18.1 Generating slow muons

At the existing μSR facilities, accelerators produce protons at energies between 500 MeV and 3 GeV with currents up to the range of mA. The protons typically impinge on graphite targets, where beamlines close to the targets collect the emitted pions and/or muons. As discussed earlier in Chapter 13, two types of conventional muon beams are available: (i) decay beams, where muons are generated by pions decaying in flight, with kinetic energies of the muon in the range of ≈ 5 to ≈ 60 MeV, and (ii) surface muons, where the muons originate from stopped pions close to the surface of the production target, with energies $\lesssim 4.2$ MeV.[2] Furthermore, surface muons are positively charged, because stopped negative pions will be captured by the carbon nuclei of the target, before decaying into negative muons.

As noted earlier in Chapter 13, besides man-made muons, cosmic ray muons with energies in the GeV range also exist. The large stopping range of these muons, from meters to tens of meters, allows them to be used for radiographic studies of large objects, such as pyramids, volcanos, nuclear reactors, and nuclear fuel rods, and for homeland security,[3] but due to the large range, low intensity at sea level (≈ 100

m^{-2}s^{-1}) and relatively low polarization of $\approx 30\%$, they are not suited for μSR applications.

The developments that led to the generation of low-energy positive muons (LE-μ^+) and enabled the extension of μSR to nanometer-thin objects started in the late 1980s. As of 2021, there is one fully operational user facility, LEM, which has been available since 2006. The LEM facility is located at the Paul Scherrer Institute (PSI) in Switzerland and provides μ^+ with energies in the range of 1 keV–30 keV. At the Japanese Proton Accelerator Research Complex J-PARC, the commissioning of the ultra slow muon (USM) facility started in 2016, delivering μ^+ with energies in the range of 0.1 keV–30 keV.

The generation of low-energy muons with energies in the keV range is a non-trivial endeavour. Starting with an almost fully polarized 4 MeV surface μ^+ beam, one might think of using deceleration by electric fields. However, at a continuous muon source one would need to apply a positive bias of about 4 MV to the experiment, which is unfeasible in practice. As an additional complication, a muon beam has a relatively large position and momentum phase space, as well as an energy distribution with a relative width of several percent. Even if it were possible to apply a bias of 4 MV to the experiment, one would obtain a muon beam with an energy width of at least 40 keV, much larger than the envisaged implantation energy in the range of 0.1 keV–30 keV. Therefore, a technique is required that can compress the phase space of the incoming surface muon beam by many orders of magnitude, while reducing the beam energy down to the keV range. Two phase space compression techniques have been demonstrated to be feasible to slow down a surface μ^+ beam to the 0.2 eV–20 eV range, with conversion efficiencies[4] ε_μ on the order of 10^{-6}–10^{-4}. These thermal or epithermal μ^+ are usually termed **ultra-slow muons** or **slow muons**, respectively. They can be easily accelerated by electrostatic fields to energies up to 30 keV. The muons with energies in the keV range are then called **low-energy muons**.

Two-photon ionization of thermal muonium

At a pulsed muon beam facility, it is possible to use pulsed lasers to generate Lyman-α photons at high power to efficiently excite thermal muonium (Mu) atoms in vacuum to the 2P state, followed by resonant ionization from the 2P state to the continuum with a second laser operated at 355 nm (see Fig. 18.2). This is the scheme pursued at J-PARC, where pulsed muon beams with a 25 Hz repetition rate and a muon pulse width of 100 ns are delivered to the experiments. The availability of a high-power Lyman-α laser is crucial in order to achieve a high ionization probability of $> 50\%$. The required laser energy in a pulse of 10 ns width is 100 μJ, resulting in an instantaneous power of 10 kW. The development of the Lyman-α laser achieved about 8 μJ in 2019 and further improvements are necessary to reach the design parameters of the ultra-slow muon source.

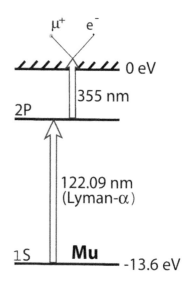

Fig. 18.2 Ionization scheme of thermal muonium. [Taken from P. Bakule and E. Morenzoni, Contemp. Phys. **45**, 203 (2004).]

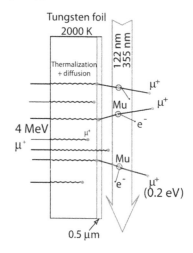

Fig. 18.3 Formation of thermal muonium from a hot tungsten foil, followed by the two-photon ionization, leaving a thermal μ^+ with an energy of 0.2 eV. [Taken from P. Bakule and E. Morenzoni, Contemp. Phys. **45**, 203 (2004).]

[4]The conversion efficiency ε_μ is defined as $\varepsilon_\mu = N_{\mathrm{eV}}/N_{\mathrm{MeV}}$, where N_{eV} is the number of generated μ^+ in the eV range, and N_{MeV} is the number of incoming surface μ^+.

[5]The laser spot size with the achieved pulse energy is of the order of 10 mm², whereas the Mu cloud has an area of order 10 cm², resulting in a small spatial overlap of the Mu atoms with the laser beam.

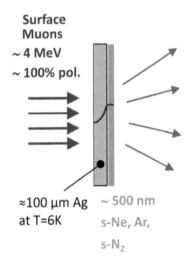

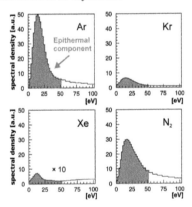

Fig. 18.4 Principle of surface muon moderation in cryosolids.

Fig. 18.5 Energy spectra of emitted epithermal μ^+ from various cryosolids.

[6]The polarization of a surface muon beam is always a bit lower than 100% due to the contamination with so-called 'cloud' muons in the beam with low polarization. These cloud muons originate from low-momentum pions decaying close to the production target. This polarization loss from cloud muon contamination acts alongside the polarization loss due to the range of collection angles discussed in Chapter 13.

At J-PARC a tungsten foil, heated to a temperature of 2000 K, is used for the generation of thermal Mu in vacuum, where $\mu^+ \to$ Mu conversion efficiencies of 3×10^{-3} are reported. The principle of the Mu production and ionization scheme are shown in Fig. 18.3. The overall conversion efficiency ε_μ, taking into account the laser ionization process, can theoretically reach about 10^{-3}. Due to the difficulties in the generation of the Lyman-α light with sufficiently high power, and the spatial and temporal mismatch of the thermal Mu cloud in vacuum with the laser,[5] a conversion efficiency $\varepsilon_\mu \lesssim 10^{-5}$ has been obtained so far. The polarization of μ^+ generated by laser ionization of Mu is $\approx 50\%$, because in Mu half of the μ^+ polarization is lost due to the hyperfine oscillations in the $m = 0$ triplet and singlet states.

Moderation in thin layers of cryosolids

In their search for more efficient positron moderators for the generation of low-energy positron beams, researchers in the 1980s discovered that thin layers of several hundred nanometers of frozen van-der-Waals bound solids, such as rare gas solids (RGS), can yield conversion efficiencies $\approx 10^{-3}$. These efficiencies are superior to the metal moderators that had been used previously. The energy of the emitted moderated positrons is determined by the energy threshold for positronium formation, resulting in an energy spectrum from zero energy to the cut-off threshold energy, which is typically in the eV range for the RGS. This motivated physicists at TRIUMF in Vancouver, Canada, to test whether RGS are also suited for μ^+ moderation. They found conversion efficiencies in the 10^{-5} range for solid argon and less for the heavier rare gases. These results stimulated the development of a low-energy muon beam at PSI at the beginning of the 1990s, where various van-der-Waals bound solids, including the RGS, and solid N$_2$ and O$_2$, were characterized. Unlike positron moderation, there is no sharp cut-off energy in the epithermal μ^+ spectrum, with the energy spectrum having a peak at about 15 eV (see Figs 18.4 and 18.5). In contrast to the Mu ionization method, the polarization of the μ^+ is conserved, resulting in a low-energy μ^+ beam with nearly 100% polarization.[6] The conversion efficiencies of the best moderators are in the 10^{-5}–10^{-4} range, with solid neon having the highest efficiency of $\approx 10^{-4}$, as shown in Fig. 18.6. The active volume of the moderator is determined by the depth from where moderated muons may escape from the layer into vacuum. This escape depth L_{esc} varies between 16 nm for solid N$_2$, 30 nm for solid Ar, and 105 nm for solid Ne. This quantity is determined by fitting the conversion efficiency in Fig. 18.6 to the function

$$\varepsilon_\mu(d) = CL_{\text{esc}} \tanh\left(\frac{d}{2L_{\text{esc}}}\right) + \varepsilon_0, \tag{18.1}$$

where d is the thickness of the film, C is a prefactor proportional to the stop density of the muons, and ε_0 is the efficiency of the blank substrate. The order of magnitude of the conversion efficiency at saturation ($d \gg$

L_{esc}) can be estimated by

$$\varepsilon_\mu \approx \frac{1}{2}(1 - F_{Mu})\frac{L_{esc}}{\Delta R}, \tag{18.2}$$

where the factor of $1/2$ is due to the nearly isotropic angular distribution of the epithermal μ^+ inside the layer, F_{Mu} is the probability that the μ^+ forms Mu before it reaches epithermal energies and ΔR is the width of the stopping distribution of the surface muon beam in the moderator substrate, which is a Ag foil approximately 100 μm thick that supports the solid gas layer (indicated by the solid line in Fig. 18.4). Since ΔR is of the order of 200 μm and L_{esc} about 100 nm at best, the maximum conversion efficiency is estimated to be $\approx 2.5 \times 10^{-4}$ for $F_{Mu} = 0$, in fairly good agreement with the measured efficiency of a neon moderator.

The moderation process in these weakly bound van-der-Waals solids can be described as a slowing down process of muons in a dense gas, where only atomic cross sections for ionization, excitation, charge-exchange, and elastic scattering are relevant. Effective energy loss occurs in the inelastic ionization, excitation, and charge-exchange processes. The best muon moderators are the light RGS, where the cross sections for these inelastic processes become negligibly small once the μ^+ slows down to energies below ≈ 100 eV. The reason for this key characteristic is the mismatch of the velocities of the μ^+ and the valence electrons: the electrons are much faster than the μ^+, which means that the interaction time of the muons with the electrons dramatically decreases in this energy range. This, in turn, leads to a fast drop of all cross sections of the inelastic processes. As a consequence, the energy loss due to inelastic collisions quickly vanishes, and the muon effectively bounces through the layer, losing energy very inefficiently by elastic collisions. This explains that there is no sharp cut-off energy[7] in the energy spectrum of the emitted epithermal μ^+.

In practice, the use of solid neon is difficult due to the very weak bonds between the neon atoms. The impinging surface muon beam may cause sputtering of the layer, generating charged neon atoms which also severely affect the high voltage stability. This (positive) high voltage, on the order of several kV, is applied to the moderator substrate in order to accelerate the epithermal μ^+ to keV energies. The unstable conditions in the layer may then lead to high voltage breakdowns, making a stable facility operation impossible. Therefore, a solid argon moderator is routinely used, even though it has about half of the moderated muon yield of neon. In contrast to the two-photon ionization technique, which works only for pulsed muon beam facilities, the moderation method works for both continuous and pulsed muon beams.

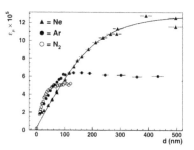

Fig. 18.6 Conversion efficiency ε_μ of solid layers of Ne, Ar, and N$_2$ as a function of layer thickness. The silver substrate foil temperature was 6 K. The solid line is a fit of eqn 18.1.

[7]The peak in the spectrum is a consequence of the fact that at zero energy the μ^+ cannot exit the layer, since their velocity is zero. This means that the spectral density has to drop to zero on approaching zero energy.

18.2 LEM facilities

In this section we present an overview of the two low-energy muon facilities at J-PARC and PSI. Figure 18.7 shows the scheme of the ultra-slow muon (USM) beamline at J-PARC. After generation of the ultra-slow

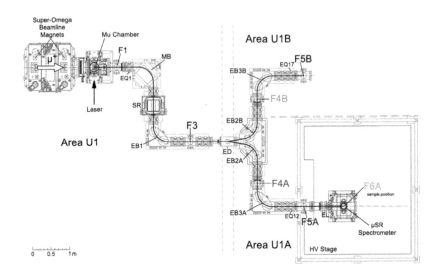

Fig. 18.7 Ultra-slow muon (USM) beamline at J-PARC, delivering μ^+ with energies between 0.1 keV and 30 keV to the μSR spectrometer in area U1A. [Taken from A. D. Pant *et al.*, Nucl. Instr. and Meth. A **929**, 129 (2019).]

muons by Mu laser ionization in the Mu chamber, the μ^+ are accelerated to 30 keV, and transported by electrostatic quadrupoles and bending magnets to the beam deflector ED, sending the beam either in the U1A branch with the μSR spectrometer, or to the U1B branch. The U1B branch will be extended by an accelerator section to post-accelerate the μ^+ up to 1 MeV energy. Due to the expected small phase space of the beam it can be focused to a sub-mm size, forming the basis of the planned ultra-slow muon microscope (USMM). Current developments focus on the optimization of beam transport and the increase of the conversion efficiency by improving the Lyman-α pulse energy, to generate USM at rates $\gg 10^2$ s^{-1}.

The LEM facility at PSI is shown in Fig. 18.8. It routinely uses a solid argon (s-Ar) moderator, producing a moderated muon rate of about 1.1×10^4 s^{-1} for an incoming surface muon rate of 1.9×10^8 s^{-1} (proton beam current of 2 mA). The moderated muons with epithermal energies are electrostatically accelerated up to 20 keV by biasing the moderator substrate up to 20 kV. A set of grids using 100 μm Au-plated tungsten wires on the downstream side is used to generate a nearly homogeneous acceleration field. The low-energy muons are transported by electrostatic **Einzel lenses**, deflected by 90° from the main beam direction by an electrostatic mirror (two planes with 100 μm Au-plated tungsten wires) and finally focused onto the sample by a conical lens. An $\boldsymbol{E} \times \boldsymbol{B}$ field downstream of the electrostatic mirror acts as a spin rotator (between 0° and 180° relative to the beam direction) and as a velocity filter to remove protons and protonated water clusters from the beam.[8] The continuous muon beam at PSI requires a start detector to provide the time stamp when a muon arrives at the μSR spectrometer.

[8]There are actually about four times more protons and water clusters than muons originating from the moderator. The protons and water clusters are from H_2 or water molecules, sticking at the surface of the moderator, where the molecules are dissociated or released by the incoming muon beam. The initial energy of these protons is in the eV range, like the epithermal μ^+.

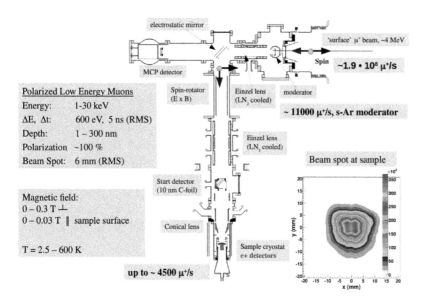

Fig. 18.8 Low-energy muon facility (LEM) at PSI, delivering μ^+ with energies between 1 keV and 30 keV to the sample cryostat region. The distance from the moderator to the electrostatic mirror is 0.5 m, and ≈ 2 m from the moderator to the sample cryostat.

The start detector for keV μ^+ at the LEM facility consists of a very thin 10 nm carbon foil supported on a Cu mesh with 80% transmission. The μ^+ lose about 1 keV in the carbon foil, ejecting on average three secondary electrons on the downstream side of the foil. A grid system generates weak electrostatic fields to deflect these electrons towards a micro-channel plate (MCP) detector, which in turn provides the start signal for the μSR measurement. The detection efficiency of the start detector is about 80%. The interaction of the μ^+ with the foil produces an energy loss distribution with RMS width of ≈ 600 eV in addition to the initial ≈ 5 eV RMS distribution of the epithermal muons. This introduces a time-of-flight distribution of the μ^+ from the start detector to the sample, located 550 mm downstream of the detector. The RMS width of the time-of-flight distribution is about 5 ns, which is the limiting parameter for the overall time resolution of the LEM spectrometer.[9] The transverse phase space of the beam is increased by scattering in the carbon foil. The energy loss distribution and the increase in transverse phase space are both limiting factors for the smallest achievable beam spot size. Muonium formation in the carbon foil leads to a reduction of the beam polarization, reducing the maximum decay asymmetry in the μSR spectrometer. The Mu contamination in the sample region is typically < 10%.

The implantation energy is adjusted by choosing an appropriate extraction energy at the moderator, and a positive or negative electrostatic potential of up to ± 12.5 kV applied to the sample holder plate. Con-

[9]The time resolution at a pulsed muon beam facility is determined by the muon pulse width, which is between 50 ns and 100 ns. At continuous muon beams, the time resolution is determined by the fast scintillators and readout electronics. It is typically between 0.15 ns and 0.5 ns, and can be improved to 0.05 ns for high-magnetic field spectrometers (up to 10 T).

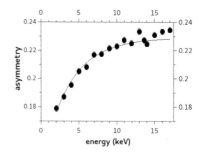

Fig. 18.9 Measured decay asymmetry as a function of implantation energy for a Ag-coated sample plate of the LEM facility. Extraction energy at the moderator is 15 keV.

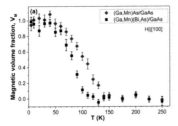

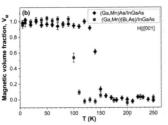

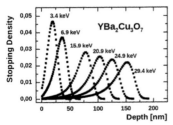

Fig. 18.10 Magnetic volume fraction V_M as a function of temperature for (a) (Ga,Mn)As and (Ga,Mn)(Bi,As) layers grown on GaAs substrates, and (b) for the layers grown on a (In,Ga)As buffer. [Taken from K. Levchenko *et al.*, Scientific Reports, **9** 3394 (2019).]

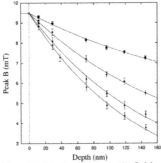

Fig. 18.11 Muon stopping profiles in $YBa_2Cu_3O_{7-\delta}$, calculated using TrimSP.

Fig. 18.12 Peak magnetic field vs implantation depth for $YBa_2Cu_3O_{7-\delta}$ in an applied field 9.5 mT parallel to the sample surface. Curves are for temperatures (from bottom to top) 20 K, 50 K, 70 K, and 80 K. [From T. J. Jackson *et al.* Phys. Rev. Lett. **84**, 4958 (2000).]

trary to conventional μSR spectrometers, one has to take into account in the analysis of the LE-μSR spectra the energy dependence of the observable decay asymmetry, in particular due to backscattered μ^+ from the sample. The backscattered muons capture an electron with high probability to form Mu when leaving the sample. The Mu drifts to the radiation shield of the cryostat, loses its polarization, and eventually stops and decays. The decay positron is detected in the μSR spectrometer, resulting in an energy-dependent reduction of the measured decay asymmetry, as shown in Fig. 18.9. At implantation energies < 5 keV, μ^+ may not even reach the sample and get reflected in the decelerating electric field in front of the sample plate, and stop in the radiation shield, contributing to a background component in the measured decay asymmetry. This contribution has to be taken into account when calculating magnetic volume fractions or diamagnetic fractions in samples at low implantation energies (< 5 keV).

18.3   Science examples

Low energy muons have found important applications in the study of layers and surface states in magnets, superconductors, semiconductors, and elsewhere. In this section we review some examples.

Example 18.1

A straightforward application of low-energy μSR is in the characterization of magnetic films and layers. An example is the class of materials comprising a thin layer of a magnetic material on a non-magnetic substrate. Tuning the muon beam energy allows muons to be implanted into the magnetic layers (or the substrate), so that the intrinsic behaviour of the magnetic component can be isolated. Fig. 18.10 shows the results of measurements of some epitaxially grown dilute magnetic semiconductor layers. These are of interest as they might be useful for spintronics applications (i.e. forming devices where the spin of the electron is exploited in addition to its electric charge). In the results shown, the magnetic volume fraction in the active layers has been extracted as a function of temperature.

Example 18.2

A particular strength of low-energy μSR is the possibility to directly measure magnetic field depth profiles on a nanometer length scale. Figure 18.11 shows calculated muon stopping profiles in the high-temperature superconductor $YBa_2Cu_3O_{7-\delta}$. Note, that with increasing energy, the stopping distributions become broader.

Figure 18.12 shows the first direct measurement of the magnetic field profile in a superconductor with a depth resolution of a few nanometers, allowing for the direct determination of one of the fundamental parameters of a superconductor, the magnetic penetration depth λ. With increasing temperature, the magnetic field penetrates deeper into the superconductor, i.e. λ increases, reflecting a decrease in superfluid density in the superconductor. In this way, the temperature dependence of λ can be determined in a model-free way, which is important for testing microscopic models of superconductivity. (See Chapter 9 for further discussion.)

Example 18.3

A detailed knowledge of defects that exist close to the surface of a semiconductor is important for device applications. Using the capability to implant the low-energy muon ensemble at different depths, the effects of radiation damage to the near-surface region of a semiconductor sample can be investigated. The idea here is that radiation has the potential to change the local electronic properties of a semiconductor and this, in turn, will alter the probability of a particular muon stopping state being realized. Recall that diamagnetic muon stopping states are preferentially found in metals, while paramagnetic muon states occur with greater probability in semiconductors and insulators. The relative populations of these stopping states can be directly detected because the hyperfine coupling of the muon with the electron in muonium causes the muon precession frequencies to differ significantly from the muon Larmor frequency $\omega_L = \gamma_\mu B_0$ in a transverse applied field B_0. This therefore gives access to information about how the local electronic structure is altered by the radiation as a function of depth. Some results are shown in Fig. 18.13. It is found that while the muonium formation is strongly suppressed in the highly defective region of the shallow proton stopping profile, the Mu signal quickly recovers for higher muon energies where the muons reach the untreated semiconductor bulk. This allows estimation of the depth over which radiation damage is significant.

In addition to these examples, low-energy μSR has found uses in a large variety of applications in superconductivity, magnetism, semiconductors, organic materials, near-surface and interface phenomena, and proximity effects. It spans a wide range of topics from fundamental condensed matter physics to the physics of technologically relevant devices.

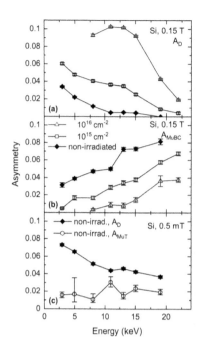

Fig. 18.13 Diamagnetic fraction F_D of a Si sample as a function of implantation energy for different proton-irradiation doses. (b) Change of F_D compared to a non-irradiated sample. [Taken from J. Woerle *et al.*, Phys. Rev. B **100**, 115202 (2019).]

Chapter summary

- In order to probe surfaces and layers, muons with energies in the keV range are required.

- Low-energy muons can be produced through the ionization of thermal muonium or the moderation of the muon beam energy in a cryosolid.

- Applications of low-energy muons have been demonstrated in magnetism, superconductivity, and semiconductor physics.

Further reading

- P. Bakule and E. Morenzoni, Contemp. Phys. **45**, 203 (2004).

- E. Morenzoni, T. Prokscha, A. Suter, H. Luetkens, and R. Khasanov, J. Phys.: Cond. Matt. **16**, S4583 (2004).

- T. Prokscha, E. Morenzoni, K. Deiters, F. Foroughi, D. George, R. Kobler, A. Suter, and V. Vrankovic, Nucl. Instr. and Meth. A **595**, 317 (2008).

- A. D. Pant, T. Adachi, P. Strasser, Y. Ikedo, Y. Oishi, J. Nakamura, W. Higemoto, K. Shimomura, R. Kadono, Y. Miyake, and E. Torikai, Nucl. Instr. and Meth. A **929**, 219 (2019).

Stimulation methods

19

In this chapter we look at how different types of external stimulus can be used in muon experiments. The purpose of stimulation will either be to produce some sort of sample excitation that can then be probed with muons (e.g. using laser spectroscopy), or alternatively, to directly interact with muon spins to enable new types of experiment to be carried out (e.g. using radio-frequency methods). External stimuli may be either transient or steady-state in nature; our focus here will be on transient stimulation techniques and their potential will be illustrated in several case studies. Later sections will discuss laser spectroscopy and radio-frequency methods in greater depth, explaining how these emerging techniques can be applied for new measurements.

19.1 Types of stimulation

There are a number of possible stimuli that might be applied during an experiment. Some will be very familiar, such as temperature, magnetic field, and pressure, and are typically applied in the steady-state as a measurement is carried out. Other stimuli, such as laser flashes, are naturally transient in nature, and can therefore only be applied for brief periods around the time the muons are implanted. Many stimuli, however, such as electric currents or radio-frequency excitation, can be applied either as a transient or steady-state excitation, according to the requirements of the experiment.

Doing more with stimulation

Applying transient stimulations brings some unique possibilities for making muon measurements:

- A transient excitation (known as a pump) can be used to prepare a sample in a specific state, with the muons then used to measure (or probe) the subsequent response. As an example, vibrational or electronic states can be prepared by laser excitation, with muons measuring the properties of these states.

- Transient excitations can be used to manipulate muon spins, with the effect of the perturbation revealed by the subsequent muon signal. Pulsed radio-frequency methods are a good example, where muon spins can be rotated during the short excitation, giving a modified μSR signal after the excitation is turned off.

- The timing of the transient excitation can be adjusted relative to muon arrival, enabling dynamic processes to be studied and kinetics to be determined. Examples can be found in laser spectroscopy, where pump and probe measurements can be used to determine charge carrier recombination times, and in radio-frequency spectroscopy, where slowly formed final states can be investigated.

- There can sometimes be an experimental advantage in applying a normally steady-state stimulation as a transient excitation. This approach might, for example, allow studies to be carried out under conditions that couldn't be continuously sustained, such as using intense magnetic fields, or it might be used to remove a problem with using steady-state conditions, such as charge accumulation at electrodes when applying an electric field.

Stimulation and muon sources

As discussed in Chapter 13, muon sources come in two flavours, characterized by the time structure of the muon beam delivered to the experiments. Of the four established sources, those at ISIS and J-PARC provide pulsed muon beams, where muons arrive in short ($\approx$ 80 ns at ISIS), intense pulses at a frequency of 50 Hz (at ISIS). In contrast, the sources at PSI and TRIUMF provide a continuous beam of muons.

The time structure provided by pulsed sources is particularly suited to stimulation methods requiring intense transient excitations. The excitation can be timed relative to the pulse of muons to provide a very low duty factor (typically less than 0.01%), effectively eliminating heating effects associated with the excitation. It is also straightforward to adjust the timing of the excitation relative to muon implantation, either for 'pump and probe' measurements or to study muon-state kinetics. This is illustrated in Fig. 19.1.

Intrinsic to many pulsed stimulation methods is the need to measure time differential data to long times ($\approx$ 10 muon lifetimes is common) to allow for the timing adjustment and to properly characterize the muon response. In this respect, pulsed sources also have benefits: firstly, all data collection is inherently time differential and therefore no experimental time penalty is involved in analysing data in this manner; secondly, there is a very low measurement background between muon pulses, greatly simplifying data analysis.

For these reasons, pulsed muon sources are often preferred when planning experiments using transient excitations. However, with good engineering and careful experiment design, it is possible to work around many of these problems and allow certain excitation experiments to be run at a continuous facility. This may be useful for two reason: firstly, to take advantage of a measurement method unique to a particular source (such as low-energy muons at PSI); secondly, provided time integral data are sufficient, to exploit the very high data rates typically available at continuous sources for more detailed studies.

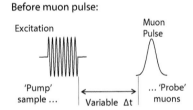

Before muon pulse:

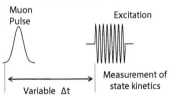

Coincident or after muon pulse:

Fig. 19.1 Timing diagram for an excitation applied at a pulsed muon source, illustrating the synchronization of a transient excitation relative to the muon pulse. The situation where the excitation occurs before muon arrival, a 'pump and probe' experiment (top), and after, to measure the kinetics of slowly formed states, are shown.

19.2   Case studies

Let us examine some specific examples of experiments based on stimulation of the sample using three different types of stimulus, namely electric currents, magnetic fields, and electric fields.

Example 19.1

Pulsed and AC currents. Both pulsed and AC currents have been used to drive the motion of the flux vortices present in the mixed state in type-II superconductors and study their dynamics. Pulsed excitation was used for early experiments at ISIS, matching the pulsed nature of the muon beam and reducing ohmic heating. However, this experimental setup only allowed for a unidirectional driving force. Improved results were obtained from later experiments at PSI, using AC currents to provide an alternating driving force. The geometry of the experiment, defining the relative orientation of the muon beam, the current excitation and the applied magnetic field, is shown in Fig. 19.2(a), with muon spin measurements carried out for PbIn at 4.2 K in a field of 300 G shown in Fig. 19.2(b). Without a driving force, at zero driving current, a static lineshape is measured, the shape as expected for a Type-II superconductor in the mixed state. There is a clear change in lineshape as the driving current is increased towards a maximum of 80 A, as the vortex velocities average out local fields.[1]

[1]Further information about this work can be found in D. Charalambous *et al.*, Phys. Rev. B **66**, 054506 (2002); Phys. Rev. B **73**, 104514 (2006).

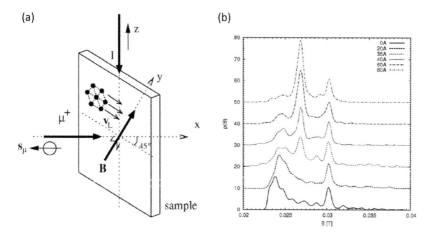

Fig. 19.2 Pulsed currents used to set in motion and study flux vortices in type-II superconductors. (a) The geometry of the experiment, showing the relative orientations of the muon spins, s_μ, the current excitation, I, and the applied magnetic field, B; (b) Lines measured in PbIn at 4.2 K in a field of 0.03 mT for various driving currents. The peak measured at ≈ 0.03 mT corresponds to muons stopped in the cryostat. [Taken from D. Charalambous *et al.*, Phys. Rev. B **66**, 054506 (2002).]

[2]T. Shiroka *et al.*, Physica B **289-290**, 684 (2000).

Example 19.2

Pulsed magnetic fields. Pulsed transverse magnetic fields, synchronized to the muon pulse, have been used to study systems where delayed formation of Mu or other paramagnetic muoniated species occurs. The technique also allows μSR experiments to be carried out at frequencies well above the intrinsic pass band limit of pulsed muon beams, discussed in Chapters 13 and 14, by removing the time uncertainty associated with the finite muon pulse width. A flat loop pulsed field device was developed, shown in Fig. 19.3(a), and large currents could be pulsed (up to 250 A) without issues of ohmic heating. The technique was demonstrated by investigating spin rotation signals for Mu formed in fused quartz, when a pulsed transverse field of ≈ 20 G is applied [Fig. 19.3(b)]. No signal is measured if the field pulse is applied just before the muon pulse (at $t \approx 0$) because the Mu precession frequency (≈ 28 MHz) is well above the muon-pulse-limited ISIS pass band (≈ 6 MHz). However, when the field is delayed to start just after the muon arrival, a damped precession signal is clearly seen [Fig. 19.3(c)]. In the latter case the time structure of the muon pulse is removed, allowing frequencies well above the pass band to be measured.[2]

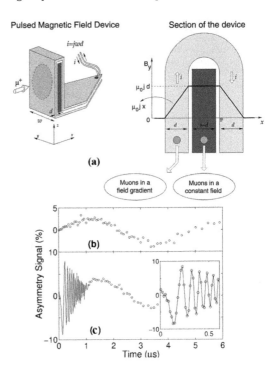

Fig. 19.3 Pulsed transverse magnetic fields allow time-delayed muon species to be investigated and overcome the pass band associated with the finite muon pulse width. (a) The geometry of the field pulsing device. The muon response with the pulsed field applied before (b) and just after (c) the muon pulse, where a damped ≈ 28 MHz signal measured for Mu formed in fused quartz. The slow background oscillation is due to diamagnetic muons.[Taken from T. Shiroka *et al.*, Phys. Rev. Lett. **83**, 4405 (1999).]

Example 19.3

Steady-state and pulsed electric fields. Measurements in applied electric fields (or *E*-fields) have added to the understanding of electron transport in materials such as rare-gas solids and semiconductors. Transport of track electrons created by high-energy muons during thermalization can be modified by applying an external *E*-field, affecting the formation and fractions of final state diamagnetic and paramagnetic species. Early measurements used steady-state *E*-fields. However, it was recognized that accumulation of near-surface charge at the electrodes could screen and reduce the effective *E*-field in the sample. Rapid switching of the polarity of the applied *E*-field was found to resolve the problem, with the diamagnetic fraction measured in GaAs becoming invariant for switching periods less than ≈ 1 s [Fig. 19.4(a)]. The *E*-field dependence of the asymmetries associated with both the diamagnetic and bond-centred Mu fractions are shown in Fig. 19.4(b). The anisotropy in weak *E*-fields arises from the asymmetric spatial distribution of the track electrons relative to the stopped muon. In contrast, strong *E*-fields applied in either direction prevent Mu formation.[3]

[3]D. G. Eshchenko *et al.*, Physics Letters A **264**, 226 (1999).

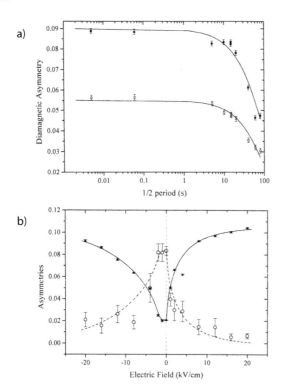

Fig. 19.4 Applied electric fields allow charge transport in materials to be investigated, results for GaAs are shown. (a) The dependence of the diamagnetic asymmetry with switching time and electric field polarity (filled squares $+8$ kV/cm, open squares -8 kV/cm) measured at 50 K; (b) Electric field dependence of both the diamagnetic (filled circles) and bond centred Mu (open circles) measured at 10 K. The half-period of the switched electric field was 5 ms. [Taken from D. G. Eshchenko *et al.*, Physics Letters A **264**, 226 (1999).]

19.3   Photoexcitation

Photoexcited μSR (henceforth referred to as **photo-μSR**) is a light-pump muon-probe experiment in combination with the standard μSR techniques, which can be used to study photo-induced effects on materials or photon-muon interactions. The straightforward concept has been applied to a variety of experimental systems over many years at both continuous and pulsed muon sources. The first successful photo-μSR experiment[4] measured the 1S-2S transition in vacuum Mu using pulsed UV lasers and found that the transition energy was in good agreement with the theoretical prediction of quantum electrodynamics. In material physics research, semiconductors have been a key research field of photo-μSR for reasons that we will discuss later. The first photoexcitation study on silicon and germanium[5] used a pulsed flash lamp (matched to the pulsed beam at KEK). Subsequently, several experiments on semiconductor systems[6] were carried out at TRIUMF and PSI using continuous light sources.[7]

In this section, we will focus on photo-μSR measurements with a pulsed light source, which are evidently best matched by a pulsed muon beams. Compared with other light sources (e.g. flash lamps or LEDs) there are three advantages of using pulsed laser light:

- **High intensity.** Pulsed lasers pack a large number of photons into a single pulse. They can easily achieve a peak intensity of more than a megawatt, even with a compact general-purpose laser system. A high peak intensity (with a low repetition rate) enables a large stimulation to be applied to a material with less local heating. This is key to being able to observe small photo-induced signals.

- **Tuneable and monochromatic light source.** Wavelength tuneability and a very small optical linewidth are essential to enable searching for a resonant level or to excite a specific absorption band. In addition, a narrow linewidth makes the optical design much easier, because every optical component, such as a mirror or waveplate, is optimized for a particular wavelength range. This will allow better accuracy for quantitative estimation of experimental parameters.

- **Light collimation.** As described later, laser sources in photo-μSR experiments are normally located a distance away from a μSR instrument, and hence the generated pump light will need to propagate some distance to reach the sample position. A well-collimated laser beam is therefore essential to transmit light to the sample without too much energy loss. In addition, the spot size of a parallel beam is easy to predict, which then allows an accurate calculation of the photon fluence on the sample.

Most photo-μSR experimental configurations used to date have utilized nanosecond laser systems based on a high energy Nd:YAG laser. Together with the key advantages mentioned above, these technologically mature lasers generate a pulse duration of $\approx$ 10 ns, providing

[4]S. Chu *et al.*, Phys. Rev. Lett. **60**, 101 (1988).

[5]R. Kadono *et al.*, Phys. Rev. Lett. **73**, 2724 (1994).

[6]For more historical details on this subject see K. Yokoyama *et al.*, Phys. Scr. **88**, 068511 (2013).

[7]I. Fan *et al.*, Phys. Rev. B **77**, 035203 (2008) and T. Prokscha *et al.*, Scientific Reports **3**, 2569 (2013).

pulsed photo-excitation within the timescale of pulsed μSR. The laser repetition rate is typically set at a half of the frequency of muon arrival, which enables alternate muon pulses to measure with and without excitation. The low repetition rate allows the investigation of relatively slow effects that can extend out to longer than a millisecond after the laser pulse.[8]

[8]The standard repetition rates of the muon pulse and the laser pulse at ISIS are 50 and 25 Hz respectively. However it is possible to reduce the laser rate even lower to $25/n$ Hz, where $n \geq 2$, i.e. there is essentially no limit on studying laser induced effects on long time scales.

Example 19.4

Photo-μSR setup at HiFi. Fig. 19.5 shows a full view of the photo-μSR setup built on the HiFi spectrometer at ISIS, where a cabin next to the instrument area houses a laser system. The laser generated light is transported through a series of light-tight tubes and mirror boxes, and is eventually incident on the sample. Because of the enclosed sample space within the large superconducting magnet, there are not many geometric options available for light illumination, and the light enters the sample chamber either from the back of the instrument or from the side. Sample geometries for those two cases are illustrated in the inset of Fig. 19.5.

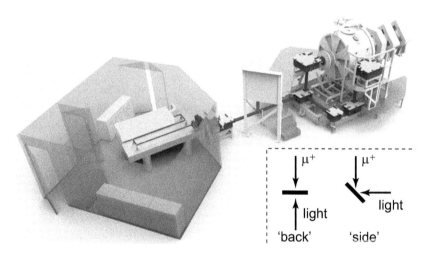

Fig. 19.5 Bird's-eye view of the HiFi Laser and beam transport system at ISIS. The inset shows a sample orientation for the back- and side-pump geometry. [Taken from K. Yokoyama *et al.*, Rev. Sci. Instr. **87**, 125111 (2016).]

As described in the previous chapters, the surface muons are implanted deep in materials; e.g. the implantation depth of 4 MeV muons in crystalline silicon is ≈ 700 μm, with a width distribution (straggling) of 130 μm (FWHM). Therefore, it is important to consider the spatial overlap between muons and a photo-induced effect, since this defines the region of the sample that needs to be optically excited.

As an example, we will consider a case of the back-pump geometry, assuming the muon distribution is positioned at the centre of an optically uniform sample. As shown in Fig. 19.6, the photon flux decays exponentially as the material absorbs light, which can be characterized

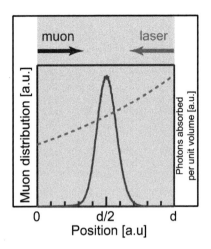

Fig. 19.6 Muon distribution (solid line) and photon flux (broken line) relative to the depth measured from the sample surface on which the muon is incident. The decaying photon flux can be characterized by an equation, $F(z) = F(0)\mathrm{e}^{-\alpha(d-z)}$, where $F(z)$ is a photon fluence, α is an attenuation coefficient, and d is the sample thickness. The muon distribution has been calculated using musrSim. [See K. Sedlak *et al.*, Physics Procedia **30**, 61 (2012). Figure taken from K. Yokoyama *et al.*, Appl. Phys. Lett. **115**, 112101 (2019).]

[9]Relaxation in the μSR time spectra can be observed in Si even with an excess carrier density of 10^{13} cm^{-3}. Note that the minority carrier density in a p-type bulk wafer of a Si solar cell under the sun is in the order of 10^{14} cm^{-3}.

by an attenuation coefficient α. In many cases, it is an advantage to maximize the number of photons absorbed in the region where muons are implanted. There may be a number of ways we can optimize this, but the easiest parameter to change in this case is the pump wavelength. This is because light absorption is associated with vibrational/electronic transitions in material, which strongly depends on the photon energy. A simple trick here is to avoid absorption peaks, and intentionally choose a wavelength for moderate absorption (namely for a lower α). In the case of semiconductors, one can choose a photon energy slightly less than the band gap energy, which excites an electron-hole pair from the band 'tail' states. In practice, an absorption edge has a finite slope even in ultrapure samples because of a small concentration of impurity or defect states distributed in a crystal; these are called the 'tail' states in band structures, which blur the conduction and valence band edges, and extend slightly into the band gap. Therefore, assuming the linewidth of the laser light is sufficiently small, one can tune the attenuation coefficient for a sufficiently low α.

When planning a photo-μSR experiment, it is important to consider the dilute nature of muons ($< 10^5$ cm^{-3}), and how they will interact with the photo-induced phenomena of interest. For instance, as described in Example 19.5, it is known that carrier injection in semiconductors gives a large reduction in muon asymmetry because of the rapid cycling of Mu states upon interaction with the electrons and holes. Mu in semiconductors has a diffusion constant D typically in the order of 10^{-3} cm^2/s at room temperature. However carriers are much more mobile ($D \geq 10$ cm^2/s) and are therefore likely to encounter the localized Mu, which results in the significant reduction of muon asymmetry even with a low-level injection.[9] On the other hand, different types of experiment, e.g. probing excited states of molecules and their reactions with Mu (see Example 19.6), will require pumping with high-power laser light to generate a large number of excited molecules. The concentration of the excited species should be high enough that muons have a chance to encounter them by diffusion.

Needless to say, optical phenomena are of widespread importance across science. Despite the technical difficulties, photo-μSR should provide abundant opportunities for potentially unique experiments in various physical systems, providing a new landscape for μSR applications. In the rest of this section we will illustrate aspects of photo-μSR techniques by looking at a few examples applied to semiconductors and chemistry.

Example 19.5

Carrier kinetics in semiconductors. As discussed above, muon spin depolarization occurs when its parent Mu interacts with excess carriers (electrons and holes), and this depolarization results in a loss of the muon asymmetry signal. This sensitivity can be utilized for studying carrier kinetics in semiconductors. The depolarization mechanism is based on charge, spin, and site exchange interactions, which form a complex network between the charge states of Mu_{BC} and Mu_T defined in Chapter 10.[10] However, the underlying mechanism can be understood as a simple model based on the hyperfine interaction between the muon and bound electron spin. The process is illustrated in Fig. 19.7(a). Starting with Mu^0 in the triplet state, $|\uparrow^\mu\uparrow^e\rangle$, which would maintain its polarization because of a good quantum number, it is ionized by an injected hole and loses its electron. The bare muon subsequently captures an electron, which has its spin either in the $|\uparrow^e\rangle$ or $|\downarrow^e\rangle$ state with equal probability. The regenerated Mu is therefore either in the $|\uparrow^\mu\uparrow^e\rangle$ or $|\uparrow^\mu\downarrow^e\rangle$ state, where the muon spin in the latter state will precess at the hyperfine frequency (in zero field) resulting in depolarization. Mu in the $|\uparrow^\mu\uparrow^e\rangle$ state is the starting point for further cycles. It is obvious that this cycle will go faster with a higher excess carrier concentration and hence gives a faster relaxation of the μSR signal.

[10] K. Yokoyama *et al.*, Phys. Rev. Lett. **119**, 226601 (2017).

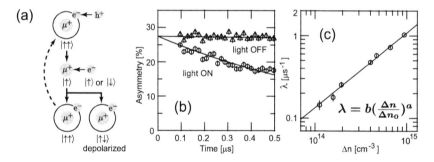

Fig. 19.7 (a) Schematic illustrating a Mu^0 in the triplet state $|\uparrow\uparrow\rangle$, depolarized after the charge exchange interaction with carriers. (b) μSR time spectra taken on a germanium sample for light OFF (triangles) and ON (circles). Solid lines denote a fit to an exponential model. (c) Calibration curve, plotting λ vs. Δn, fit to a power law (solid line). [(b) and (c) taken from K. Yokoyama *et al.*, Appl. Phys. Lett. **115**, 112101 (2019).]

Shown in Fig. 19.7(b) are μSR time spectra from an intrinsic germanium single crystal, where a longitudinal field is applied to the sample. The dark spectrum shows a minimal relaxation, while optical carrier injection induces a relaxation in the spectrum, which can be fitted (empirically) with an exponential function with a relaxation rate λ. The excess carrier density Δn is a known parameter that can be calculated from a photon fluence and attenuation coefficient. Hence, as shown in Fig. 19.7(c), we can attenuate the pump power and obtain a dependence of λ relative to Δn (called a 'calibration curve'), which allows Δn to be calculated for a measured value of λ using the equation shown in Fig. 19.7(c).[11] We can now sweep the delay time Δt between the muon and laser pulse to measure the excess carrier dynamics at a specific depth within a wafer – none of the traditional techniques of lifetime spectroscopy have such an advantage.

[11]For making the calibration curve, Δn is assumed to be constant during the short time window, where λ is obtained. In the case of Fig. 19.7(b) and (c), the window is 0.5 μs long.

To illustrate this unique feature, Fig. 19.8(b) shows a set of depth-dependent carrier lifetime spectra by scanning the muon position through a 1-mm thick silicon wafer. As shown in Fig. 19.8(a), the depth and distribution of implanted muons are simulated based on the amount of material in the muon beam (including muon degraders). It is clear that the excess carriers quickly recombine in the region close to the surface, while Δn decreases with a slower rate at the centre. Generally, a wafer without surface passivation has a very fast surface carrier recombination rate because of a high concentration of impurities and defect states at the surface. In the present case the wafer surfaces are only chemically polished, and are therefore acting as an infinite sink for the injected carriers. With surface passivation, however, it is possible to slow down the decay rate and determine the bulk and surface recombination rate separately – a difficult task with traditional lifetime spectroscopy techniques, which normally measure the superimposed effective lifetime. Surface passivation is a key technology in photovoltaics and power electronics, and there may be a good opportunity for muons to play a role in the development of these applications.

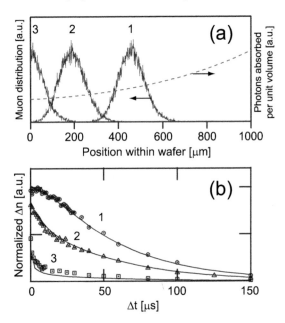

Fig. 19.8 (a) Simulated muon distribution for three implantation depths in a silicon sample using musrSim. Sample geometry and definition of the position are the same as in Fig. 19.6. (b) Carrier lifetime spectra measured at positions in (a). Solid lines: simultaneous fitting to a diffusion equation, assuming an infinite sink on the wafer surfaces. Since this silicon wafer has the bulk carrier lifetime of ≈ 100 μs, we see that the fast surface recombination limits the overall carrier lifetime.

[12]P. Bakule *et al.*, J. Phys. Chem. Lett. **3**, 2755 (2012).

Example 19.6

Gas-phase chemical reactivity. The Mu atom, as a light H-atom isotope, provides a way of investigating quantum mass effects in gas-phase chemical reactivity. There has been much published work investigating reactants in their ground states; however, in one unique study, Bakule *et al.* investigated reaction rates of Mu with vibrationally state-selected reactants.[12] In this case, stimulated Raman pumping is used to pump H_2 gas to the first excited state, $H_2\{1\}$, to allow investigation of the rate of the following reaction:

$$\mathrm{Mu} + H_2\{1\} \xrightarrow{k_{\mathrm{Mu}}\{1\}} \mathrm{MuH} + \mathrm{H}. \qquad (19.1)$$

A reaction cell was designed (Fig. 19.9) with perpendicular ports to admit both muon and laser beams. The laser intensity, the length of the light path and the H_2 gas pressure are all critical parameters for stimulated Raman pumping. The signal amplitude depends on the overlap of the laser and muon beams, with the stopping profile of the muon beam being modelled and adjusted using beam degraders. The laser is pulsed at 25 Hz, allowing alternate ISIS pulses to be collected with and without stimulated Raman pumping, and the results have been compared. A rate constant, $k_{\mathrm{Mu}}\{1\}$, of 9.9×10^{-13} cm^{3}molecule^{-1}s^{-1} at 300 K was determined, a value $\approx 2 \times 10^7$ faster than that measured for H_2 in the ground state and approximately 3 times faster than the $\mathrm{D} + H_2\{1\}$ reaction. Taken together, these results allow models of the potential energy surface for this reaction to be tested.

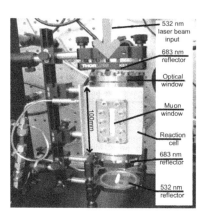

Fig. 19.9 The reaction cell assembly used to investigate the reaction rates of Mu with H_2 gas pumped to its first excited state. The thick arrow indicates the incoming 532 nm laser light, collimated into the shape of an ellipse, as observed on the 532 nm reflector in the figure. The propagation direction of muon and laser light are therefore perpendicular with each other. After passing the cell, the pump beam is reflected back by the reflector to increase the light intensity in the cell. Generated 683 nm light is 'trapped' in an optical resonator (top and bottom mirrors) for efficient Raman pumping. The whole assembly is positioned in a light-tight box, which then slides into the instrument. [Taken from P. Bakule *et al.*, Physica B: Condensed Matter **404**, 1013 (2009).]

[13]Also known as electron paramagnetic resonance (EPR).

[14]T. Coffin *et al.*, Phys. Rev. **106**, 1108 (1957).

[15]i.e. the muon.

19.4 Muon-spin resonance

Muon-spin resonance (μSR) takes its place alongside other more established spectroscopic methods, such as nuclear magnetic resonance (NMR) and electron spin resonance[13] (ESR), for condensed matter studies. In fact, a very early magnetic resonance experiment was carried out with muons in 1957 by Coffin *et al.*[14] This was not for studying materials, but instead designed to accurately determine the magnetic moment of the μ meson.[15] However, as muon spectroscopy has developed into the widely used technique we see today, radio-frequency (RF) techniques have rarely been used. The reason for this, perhaps, relates to the high intrinsic polarization of the implanted muons, which uniquely allows muon spin rotation and relaxation experiments to be carried out without recourse to the complex RF pulse sequences that are commonly used in NMR. However, RF techniques can make new types of muon experiments possible, making them a useful tool to consider alongside other measurements. The purpose of this section is to introduce the concepts of RF μSR and to discuss some of the applications of the technique for muon experiments.

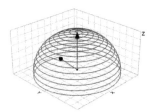

Fig. 19.10 Time evolution of muon moment shown for the case of $\omega_0 = 42.56$ rad/s (corresponding to the resonance frequency for diamagnetic muons in a field of 500 G) and $\omega_1 = 0.85$ rad/s (corresponding to a B_1 field strength of 10 G). The line represents the trajectory of the tip of the moment vector over a time period of 2 μs. The solid arrow indicates the starting orientation of the vector (muons are implanted with their spins collinear with the z-axis), the dotted arrow the orientation after 2 μs.

Example 19.7

Describing the RF resonances. Simple NMR theory can be used to describe the resonance effects expected for muons implanted in a material. For the muon, with a $S = 1/2$ spin in an applied magnetic field B_0, the separation of the two energy levels arising from Zeeman splitting is given by $\Delta E = \hbar \gamma_\mu B_0$, where γ_μ is the gyromagnetic ratio of the muon (851.615 Mrad s^{-1} T^{-1}). The frequency of radiation required to induce a transition between these energy levels is $\omega = \Delta E/\hbar$. Combining these expressions leads us to the equation for the Larmor precession frequency, ω_0, of the muon in an applied field, B_0,

$$\omega_0 = \gamma B_0. \tag{19.2}$$

Therefore, to meet the resonant condition we need to irradiate the system at a frequency ω_0 as defined by eqn 19.2. The classical description of this physics rests on solving the equation of motion for a magnetic moment, $\boldsymbol{\mu}$, placed in a field, B, where it will experience a torque given by:

$$\frac{d\boldsymbol{\mu}}{dt} = \gamma \boldsymbol{\mu} \times \boldsymbol{B}. \tag{19.3}$$

The field vector, $\boldsymbol{B}$, includes both the static field, $\boldsymbol{B}_0 = B_0 \boldsymbol{k}$ and the time varying RF field applied perpendicular to $\boldsymbol{B}_0$, $\boldsymbol{B}_1(t) = B_1 \cos(\omega_0 t)\boldsymbol{i} - B_1 \sin(\omega_0 t)\boldsymbol{j}$ (considering only the component rotating in the same direction as the muon spin). Equation 19.3 can be solved for the muon magnetic moment $\boldsymbol{\mu}$, assuming a starting condition $\boldsymbol{\mu}(0) = \mu_0 \hat{\boldsymbol{z}}$ (muons being implanted with spins collinear with the z-axis), with the components given by

$$\mu_x(t) = \mu_x(0)\sin(\omega_1 t)\sin(\omega_0 t)$$
$$\mu_y(t) = \mu_y(0)\sin(\omega_1 t)\cos(\omega_0 t)$$
$$\mu_z(t) = \mu_z(0)\cos(\omega_1 t). \tag{19.4}$$

These solutions imply that at resonance the muon moment simultaneously precesses about both the static field and RF fields, with the tip of the vector executing a complex nutation motion in time as shown in Fig. 19.10. The motion is described in Example 7.8 invoking the rotating frame and further discussion can be found in Chapter 7, as well as in the magnetic resonance literature.

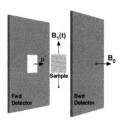

Fig. 19.11 Configuration of a muon spectrometer suitable for RF μSR measurements. Muons are implanted with spins collinear to the static applied field, $\boldsymbol{B}_0$. Detector positions are optimized for longitudinal spectroscopy. The RF field, $\boldsymbol{B}_1$, is applied perpendicular to $\boldsymbol{B}_0$ and generated by a coil wound around the sample.

Muon spin resonance measurements are often carried out using a spectrometer configured for spin relaxation experiments, with muons implanted with the spin polarization collinear to the applied static magnetic field, B_0. Detectors are arranged forward (upstream) and backward relative to the initial muon spin direction to measure the time dependent decay asymmetry of the muon response. The RF coil is wound close to the sample and oriented such that the RF field vector $\boldsymbol{B}_1$, is perpendicular to the static field. The overall layout of the experiment is illustrated in Fig. 19.11.

Either the static field or the RF frequency can be adjusted to meet the resonance condition given by eqn 19.2. However, as the RF coil is generally part of a tuned circuit, it is usually easier to sweep the magnetic field. In practice, data are collected both with and without the RF field, the experiment alternating between the two conditions typically every ten seconds, to clearly identify the effect of the RF field on the system and reduce any artefacts introduced by equipment instability.

Data can be treated in two distinct ways. Firstly, an experiment akin to continuous wave magnetic resonance can be carried out, sweeping the static field and forming a time-integral RF asymmetry at each field point, according to the equation:

$$A_{RF} = \frac{N_F^{on} - N_F^{off}}{N_F^{on} + N_F^{off}} - \frac{N_B^{on} - N_B^{off}}{N_B^{on} + N_B^{off}}, \tag{19.5}$$

where $N_{F,B}^{on}$ are the time-integral positron rates with the RF excitation on and $N_{F,B}^{off}$ with RF turned off, for the forward or backward detectors respectively. In this case, a Lorentzian line shape is typically obtained. This takes the form

$$A_{RF}(B) = A_{RF}(B_0)\frac{\Delta B^2}{(B - B_0)^2 + \Delta B^2}, \tag{19.6}$$

where $A_{RF}(B_0)$ is the on-resonance time-integral asymmetry, B_0 is the on-resonance field, and ΔB is the half-width at half-maximum of the resonance line. Both $A_{RF}(B_0)$ and ΔB depend upon RF power, the inherent muon spin relaxation, and any chemical reaction of the measured species. At a pulsed muon source, the time integration window can readily be adjusted after data collection to focus the final spectrum on prompt or delayed signals. An example time-integral spectrum measured for diamagnetic muons stopped in gaseous N_2 is shown in Fig. 19.12.

Secondly, a time-differential signal can be measured which, at resonance, will correspond to the spin precession around the RF field, $\boldsymbol{B}_1$. In this case, the time evolution of the time-differential asymmetry, $S(t)$, can be defined as follows:

$$S(t) = A_{RF}(t) - A(t) = \frac{1}{2}\left[\frac{N_F^{on}(t)}{N_F^{off}(t)} - \frac{N_B^{on}(t)}{N_B^{off}(t)}\right], \tag{19.7}$$

where $A_{RF}(t)$ and $A(t)$ are the time-differential muon asymmetries with and without RF excitation, respectively. The former being defined during the RF pulse as

$$A_{RF}(t) = A_{RF}^0 e^{-\lambda_1(t-\tau_d)}\cos[\omega_1(t - \tau_d)] \quad (t \geq \tau_d), \tag{19.8}$$

where A_{RF}^0 is the asymmetry of the muon state immediately before the RF pulse is applied, τ_d the delay between muon arrival and RF excitation, and λ_1 a depolarization rate that includes a contribution due to the inhomogeneity associated with the RF field. An example time-differential measurement carried out for diamagnetic muons stopped in boron is shown in Fig. 19.13.

RF techniques for studying slowly formed muon states

A unique area of application for RF techniques is in the study of slowly formed muon states following muon implantation in materials. This section considers two examples, one taken from muonium chemistry where muoniated radicals are measured and a second example taken from semiconductor physics, where H charge states are investigated through the study of the Mu analogue.

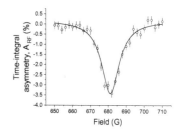

Fig. 19.12 Time-integral RF field scan for diamagnetic muons stopped in gaseous N_2. The RF frequency was 9.2 MHz.

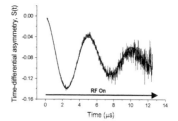

Fig. 19.13 Time-differential measurement carried out at resonance ($B_0 = 1672$ G) for diamagnetic muons stopped in boron. The RF excitation (22 MHz) is applied throughout data acquisition, with the RF field strength, B_1, determined to be ≈ 15 G from data fitting.

High field spin rotation measurements are traditionally used to measure muon hyperfine couplings in organic muoniated radical systems. In this case, the value of the hyperfine coupling constant A_μ can be directly calculated from the splitting of the ν_{12} and ν_{34} lines of the frequency spectrum recorded in the Paschen-Back region of the Breit-Rabi diagram. In its application, however, there are two limitations to this method: firstly, because the frequency spectrum typically extends to hundreds of MHz (depending on the value of A_μ), measurements are only possible at a continuous muon source; secondly, the final radical state needs to be promptly formed. This latter restriction arises from dephasing of the Mu precursor in the transverse field prior to it reacting to form the final radical state (see Chapter 3). In contrast, time-integral RF spectroscopy enables direct measurement of A_μ at a pulsed muon source, while slowly formed states can readily be measured since the initial direction of the muon spin is collinear with the applied field.

[16] I. McKenzie *et al.*, J. Phys. Chem. B **117**, 13614 (2013).

[17] J. S. Lord, Physica B **374-375**, 472 (2006). See also Chapter 17.

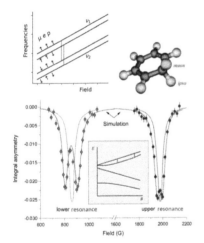

Fig. 19.14 RF spectroscopy of the cyclohexadienyl radical (shown middle). The two resonance absorptions at $\nu_{RF} = 218$ MHz shown at the bottom vs field, correspond to the two transitions in the Breit-Rabi diagram (inset). The blow up at the top shows the splitting, owing to coupling with the ipso proton on the benzene ring. [Data from I. McKenzie *et al.*, J. Phys. Chem. B **117**, 13614 (2013).]

Example 19.8

Determining hyperfine parameters for muoniated radicals. The cyclohexadienyl radical formed by the addition of Mu to benzene is shown in Fig. 19.14 (notice the *ipso* proton). The RF experiment[16] was performed in LF geometry, so that, in the absence of radio frequencies and far from ALC conditions, the muon asymmetry would be maximal and field independent. A working frequency of 218 MHz was selected to match the static ν_{12} transition (see Table 12.3), calculated in the Paschen-Back limit. Because of the non-linear field dependence of the 'muon flip' ν_{12} transition, two resonances (lower and upper field transitions) are anticipated, as shown in the Breit-Rabi diagram, in the inset of Fig. 19.14, bottom. A scan of the LF was performed to cover this region and a further splitting of each of the resonance lines was observed.

Example 4.15 proves useful here, indicating that the nuclear hyperfine coupling associated with the *ipso* proton, the second largest after the muon hyperfine, cannot be neglected. The *ipso* spin is entirely decoupled in the Paschen-Back limit, but the RF placed the experiment in a slightly lower field regime, where the non diagonal elements in eqn 4.71 cannot be completely neglected. Notice that the splitting is more marked in the lower field resonance. Indeed here the departure from the Paschen-Back condition is largest and the two replicas of the radical ν_{12} transitions with oppositely oriented *ipso* spin, shown in Fig. 19.14, top, are not exactly degenerate. Data fitting using the QUANTUM package[17] allows both the muon and proton hyperfine parameters to be determined, with $A_\mu = 514.4(1)$ MHz and $A_p = 128.5(3)$ MHz, respectively.

These data refer to neat benzene, where the final state radical species is promptly formed ($< 10^{-10}$ s). However, in dilute aqueous solutions the formation time of the radical is greatly extended. This makes direct TF detection impossible because of dephasing. The RF methods described here were essential to measure the effect of the solvent on the hyperfine parameters.[16]

While, as discussed above, chemical kinetics can be responsible for a slowly formed product state, there are many other mechanisms that can lead to either the delayed formation of the final muon charge state or to state conversion. One example of the latter can be found in muon spectroscopy of semiconducting materials where delayed charge states can be formed through electron capture or loss.

Example 19.9

Final state measurements and kinetics. For muons stopped in high-purity silicon, at low temperatures the diamagnetic μ^+ fraction is close to zero, with the majority of muons thermalizing as paramagnetic species, either as cage centred Mu (Mu_T) or bond centred Mu (Mu_{BC}, originally labelled Mu*). Transverse field muon spin rotation experiments show no change in the diamagnetic fraction until ≈ 200 K, above which there is a gradual increase in the fraction, reaching ≈ 0.5 of the available polarization at room temperature [shown in Fig. 19.15(a)].

a)
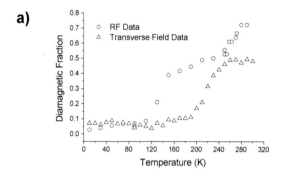

b)
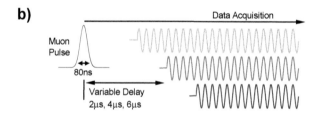

c)
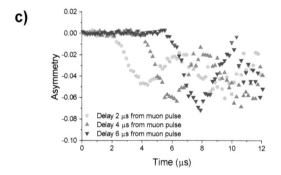

Fig. 19.15 Muon states in high purity silicon: (a) contrasting temperature dependence of diamagnetic fraction for time-integral RF and spin rotation measurements; (b) timing diagram for a delayed RF measurement determining state kinetics; (c) a series of time-differential RF measurements at the diamagnetic resonance condition, with the RF excitation delayed by 2 μs, 4 μs, and 6 μs relative to muon implantation. [Adapted from S. P. Cottrell *et al.* Physica B **326**, 248 (2003).]

[18]B. D. Patterson, Rev. Mod. Phys. **60**, 69 (1988).

Time-integral RF measurements of the diamagnetic species, carried out over a similar temperature range, show a similar step in the diamagnetic fraction; however, in this case the increase comes at lower temperatures, around 130 K. The temperature at which the step in the RF data occurs coincides with a reported rapid increase in the depolarization rate of the Mu_{BC} species,[18] due to a Mu_{BC} to μ^+ state transition. The combined data sets suggest that Mu_{BC} is promptly formed up to at least 200 K; however, above $\approx$ 130 K the state is short lived, and quickly forms the diamagnetic final state seen in the RF measurements. While the spin rotation measurements are only sensitive to promptly formed species, the amplitude of the RF signal is sensitive to both prompt and delayed states owing to the precursor Mu_{BC} species having the muon spin parallel to the applied field.

For favourable state conversion rates (of the order of μs^{-1}), it is possible to directly follow conversion of the muon charge state by delaying RF excitation relative to muon implantation. A timing diagram for the measurement is shown in Fig. 19.15(b). It should be clear that this is an experiment uniquely suited to a pulsed muon facility. Results are shown for silicon at 132.5 K in Fig. 19.15(c), where spectra are recorded with the RF excitation delayed relative to muon arrival by 2 μs, 4 μs, and 6 μs in successive measurements. The series of time-differential RF measurements carried out at the resonance condition clearly show a build-up in the diamagnetic signal amplitude as a function of delay.

[19]Y. Morozumi *et al.*, Physics Letters A **118**, 93 (1986).

Although the technique just described in the example above is excellent at providing a direct measure of state conversion kinetics, it can be difficult to implement for two reasons: firstly, the finite rise time and limited B_1 amplitude of the RF pulse have to be accounted for, introducing uncertainty into the experiment; secondly, the state conversion rate needs to be a good match to the muon time window for the change in final state to be seen. An alternative approach to measuring state conversion kinetics that removes both of these difficulties was demonstrated by Morozumi *et al.*[19] This technique also uses the time-differential RF method, measuring the on-resonance diamagnetic signal under continuous RF excitation conditions, and modelling the total diamagnetic signal as the sum of prompt and delayed components.

Pulsed RF techniques

Many of the RF methods discussed so far in this chapter are directly analogous to the continuous wave (CW) techniques used in the early days of NMR. During the 1970s, however, NMR was revolutionized by the development of pulsed RF Fourier spectroscopy. In this method, short, intense RF pulses are used for broadband spin excitation, with the subsequent free precession signal encoding the system response and allowing the components of the complex spectra to be revealed by Fourier transformation. This new technique provided greater sensitivity, higher resolution, and gave an excellent method for measuring transient properties of spin systems, while removing lineshape distortions that are common with CW measurements. Therefore, an important question for muon spectroscopy is whether similar methods can be exploited for muon measurements.

A key concept in pulsed RF techniques is the use of RF pulses with carefully controlled duration that rotate the muon spins by a specific angle about the B_1 field. In NMR, a 90° RF pulse is fundamental to the detection of resonance phenomena.

[20]S. R. Kreitzman, Hyp. Int. **65**, 1055 (1990).

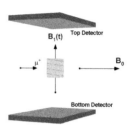

Example 19.10

RF manipulation of muon spins. With reference to Fig. 19.10, it can be seen that the application of a 90° RF pulse will rotate spins that are initially collinear with the z-axis to end up lying in the x-y plane. In this example, with a B_1 field of 10 G and $\omega_1 = 0.85$ rad/s for diamagnetic muons, the pulse length required for a 90° spin rotation will be ≈ 1.85 μs. Comparison of the time required to complete this rotation with that of the muon lifetime highlights the need for high power RF for efficient experiments (this is even more important if multi-pulse sequences are envisaged). An early demonstration of RF muon spin manipulation was carried out at a continuous facility by Kreitzman.[20] However the method is best suited to pulsed muon sources, which can manipulate the many muons in one pulse together, making efficient use of the RF power and minimizing the sample heating.

Considering the simple spectrometer defined in Fig. 19.11, muons are implanted with spins that are collinear with the static field, $\boldsymbol{B}_0$, the 90° RF pulse is then applied and muon spins are rotated leaving them perpendicular to $\boldsymbol{B}_0$. For this type of experiment, however, the detectors illustrated in Fig. 19.11 are not in the right position to detect the spin rotated muons which precess around $\boldsymbol{B}_0$ once the RF pulse has finished. In this case the spectrometer should be reconfigured to have alternative detector positions, using a transverse arrangement as illustrated in Fig. 19.16.

Fig. 19.16 Configuration of a muon spectrometer suitable for pulsed RF μSR measurements. In this case detector positions using a transverse configuration are required for efficient detection of positrons from spin rotated muons (compare this with Fig. 19.11).

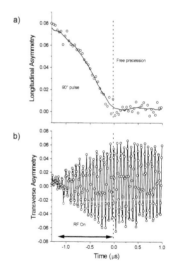

Example 19.11

Spin rotation by RF pulse. An illustration of the signal measured when a 90° pulse applied to diamagnetic muons, μ^+, stopped in InP is shown in Fig. 19.17. The muon response during the ≈ 1 μs RF pulse and also for the first microsecond of the following free precession signal is shown (the RF field is turned off at $t = 0$ μs). Signals measured simultaneously in longitudinal (Fig. 19.11) and transverse (Fig. 19.16) detector configurations are shown for comparison. Immediately following muon implantation, a maximum longitudinal asymmetry is measured; however, this is reduced to zero at the end of the RF pulse. Conversely, the transverse asymmetry increases to reach a maximum at the end of the pulse. The free precession signal is equivalent to a conventional spin rotation measurement, with the signal depolarization providing similar information.

Whereas the longitudinal detector configuration reflects the low frequency precession of muons about the RF field, B_1, the transverse configuration allows the high frequency precession of spin rotated muons about the static field to be directly measured. In the previous example, with $B_0 = 1060$ G, the transverse precession frequency is ≈ 14 MHz, a frequency well beyond the intrinsic ISIS pass band of ≈ 6 MHz, arising from the implanted muon pulse width (≈ 80 ns). This extended frequency capability arises because the time structure of the muon pulse is

Fig. 19.17 Implementation of a 90° RF pulse for diamagnetic muons in InP at 120 K. A 14 MHz, 1 μs RF pulse is applied, finishing at $t = 0$ μs, with a free precession signal (about the 1060 G static B_0 field) from spin rotated muons following at later times. Signals measured in (a) longitudinal (see Fig. 19.11) and (b) transverse (see Fig. 19.16) detector configurations are shown for comparison. [Taken from S. P. Cottrell *et al.*, Physica B **290**, 693 (2000).]

removed as muons are implanted with their spins collinear to B_0 before the RF pulse is applied. Therefore, even the simplest RF pulse sequence, a single 90° pulse, allows a totally new type of experiment to be carried out at a pulsed muon source.

NMR-style pulse sequences

[21]R. R. Ernst *et al.*, *Principles of Magnetic Resonance in One and Two Dimensions*, Oxford University Press (1987).

[22]E. L. Hahn, Phys. Rev. **80**, 518 (1950).

Much of the excitement of using pulsed RF techniques in NMR rests on the possibility of putting together sequences of RF pulses that can modify the system Hamiltonian to extract specific information. Ernst *et al.*[21] refer to this as 'bordering on [spin] sorcery', and continual development in this area has led to the present day explosion of NMR methods. It is therefore pertinent to ask whether any of this magic can help the μSR spectroscopist.

An important limit on how much sorcery is likely to be possible is set by the muon lifetime (≈ 2.2 μs), which implies that the manipulations must be done quickly. Certainly, the use of high power RF is necessary to achieve the required spin rotations within short width RF pulses. Pulsed muon sources clearly have an intrinsic advantage here, minimizing RF heating effects, since the RF power can be off for most of the time (low duty factor), while allowing low background time differential data to be easily measured out to at least 10 muon lifetimes. High data rates are, however, also crucial for completing these measurements without undue expenditure of beamtime. This is perhaps one area where the next generation of intense pulsed muon sources will make a big impact.

Probably the simplest pulse sequence to implement is the **spin echo**, which was first developed by Hahn[22] in 1950. Echo sequences are important in NMR for measuring the true T_2 (spin-spin) relaxation rate of the sample material and, perhaps more importantly for μSR, are central to measurements of spin diffusion, whether due to macroscopic flow or microscopic hopping of the muon. Spin echo sequences therefore allow resonant signals to be used to determine spin diffusion, potentially removing the complication of fitting multiple component relaxation models to data sets to extract hop rates.

Example 19.12

Spin echo. The development of an echo sequence for muons stopped in boron at room temperature is shown in Fig. 19.18. In this material, almost all muons thermalize into a diamagnetic state and are static to very high temperatures. A simple free precession signal following a ≈ 1.2 μs $90°$ RF pulse is shown in Fig. 19.18(a), the data being collected in a resonant field, B_0, of 1672 G (≈ 22.7 MHz), but shown in a 20 MHz rotating reference frame. The signal decays because of the dipolar coupling between the muon and neighbouring boron nuclei, causing a dephasing of the precessing signal. In this sense, the measurement is equivalent to a high field spin rotation experiment. In Fig. 19.18(b) a similar free precession signal is created, but in this case a second $90°$ RF pulse is applied after ≈ 3 μs, forming a Hahn echo at ≈ 5 μs, twice the separation of the two RF pulses. The second RF pulse corrects for the spin dephasing associated with the spread of local fields and refocuses the muon spins to create the echo. Today, more usually a $180°$ RF pulse is used for this type of experiment, since it maximizes the amplitude of the echo. This sequence is shown in Fig. 19.18(c), with the echo now formed at ≈ 6 μs, with an amplitude of approximately double that of the Hahn echo.

More complex techniques, such as double resonance, are now very common in NMR, with many experimental applications, including cross-polarization and RF pulse decoupling. The second of these is demonstrated in the following example, where continuous wave RF decoupling is used to cancel the dipolar interaction between the muon and the neighbouring nuclei. The method works by applying a continuous high power RF decoupling field at the nuclear resonance frequency to 'stir' the nuclear spins, while simultaneously measuring the muon free precession signal following a $90°$ RF pulse.

Example 19.13

RF pulse decoupling. Results obtained from a test experiment studying muons stopped in $Ca(OH)_2$ are shown in Fig. 19.19. Muons implanted into this material thermalize to form a large diamagnetic fraction, and are known to be static at 300 K where this measurement was made. A free precession signal in a 1600 G static B_0 field was created by applying a ≈ 1.2 μs $90°$ RF pulse at an RF frequency of 20.8 MHz [Fig. 19.19(a)]. Dipolar coupling to neighbouring protons in the material gave a depolarization rate of ≈ 0.12 μs^{-1}. Repeating this measurement with the addition of a 40 G RF decoupling field at the proton resonance (≈ 6.5 MHz) resulted in a clear reduction in the damping of the free precession signal (now ≈ 0.02 μs^{-1}) as the proton coupling was removed [Fig. 19.19(b)]. Separate experiments carried out at intermediate RF decoupling fields resulted in depolarization rates that fell between the two values shown here. The technique has also been used to help determine the muon site in a proton conductor[23] and to identify the diffusing species in an ionic conductor.

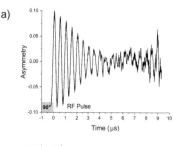

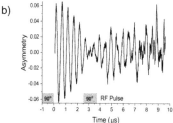

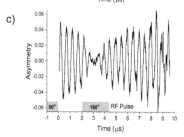

Fig. 19.18 (a) Free precession signal following a ≈ 1.2 μs RF pulse for diamagnetic muons stopped in boron at room temperature, with a static B_0 field of 1672 G and at an RF frequency of ≈ 22.7 MHz; (b) Formation of a Hahn echo following a $90°$–$90°$ RF pulse pair; (c) Formation of a spin echo following at $90°$–$180°$ RF pulse pair. The data are shown in a 20 MHz rotating reference frame. [Adapted from S. P. Cottrell *et al.*, Appl. Mag. Res. **15** 469 (1998).]

[23]N. J. Clayden *et al.*, Phys. Chem. Chem. Phys. **8**, 3094 (2006)

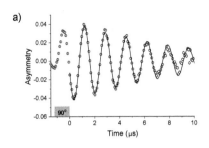

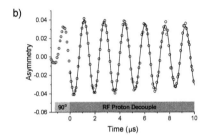

Fig. 19.19 (a) Free precession signals measured for diamagnetic muons stopped in $Ca(OH)_2$ at 300 K following a ≈ 1.2 μs 90° RF pulse, with a static B_0 field of 1600 G, and at an RF frequency of 20.8 MHz; (b) A similar measurement with the addition of a 40 G RF decoupling field at the proton resonance frequency (6.5 MHz). The relaxation rate is reduced from 0.12 μs^{-1} to 0.02 μs^{-1} when the RF decoupling field is applied. The data are shown in a 20 MHz rotating reference frame. [Adapted from S. P. Cottrell *et al.*, Physica B **290**, 693 (2000).]

Chapter summary

- Controlled stimulation of a sample enables various types of modulation spectroscopy to be carried out using muons.
- Transient stimulations with controlled delay times enable pump-probe μSR measurements to be made and allow the manipulation of the spin populations in an experiment.
- Photoexcitation using pulsed lasers provides an extension to muonium spectroscopy, that can be used, for example, in investigations of semiconductors and chemical reactions.
- Radio frequency measurements involve the resonant control of spin dynamics and can be used to isolate the behaviour of a specific spin or relaxation process.

Further reading

- S. P. Cottrell *et al.*, Appl. Magn. Reson. **15**, 469 (1998).
- R. R. Ernst, G. Bodenhausen and A. Wokaun, *Principles of Magnetic Resonance in One and Two Dimensions*, OUP (1987).

- S. R. Kreitzman, Hyp. Int. **65**, 1055 (1990).
- K. Yokoyama *et al.*, Phys. Scr. **88**, 068511 (2013).
- W. Koechner, *Solid-State Laser Engineering 6th ed.*, Springer (2006).
- C. P. Slichter, *Principles of Magnetic Resonance*, Springer (2010).

Exercises

(19.1) The ISIS accelerator produces a pulse of muons every 20 ms. Assuming a 40 μs long 1 kW RF pulse is applied coincident with each muon pulse, what is the duty cycle for this experiment? What is the average RF power delivered to the sample? At continuous muon sources, duty cycles of 50% are common, what would be the average RF power in this case? Compare these powers to the typical cooling power of a ^{4}He cryostat ($\approx$ 20 W at 100 K) and consider how feasible it will be to run this system for RF experiments.

(19.2) A common circuit used to create the RF field, B_1, for many RF μSR experiments is shown in Fig. 19.20.

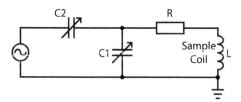

Fig. 19.20 Circuit diagram for Exercise 19.2.

(a) For a simplified case where capacitor $C2$ is omitted, the sample coil (with inductance L and resistance R), together with the tuning capacitor, $C1$, form a parallel resonant circuit. For this configuration, assuming $L = 1$ μH and $R = 0.1$ Ω, what is the value of $C1$ for the circuit to be resonant at a frequency of 20 MHz? What is the circuit impedance at resonance?

(b) To avoid unwanted reflections on the transmission line between the RF amplifier and resonant circuit, the line should be terminated by adjusting the impedance of the RF circuit to 50 Ω resistive. This can be achieved by adding a series capacitor $C2$. Derive an expression for the impedance, Z_0, of the full circuit. Using the condition that Z_0 has a resistance of 50 Ω and a reactance of 0, deduce expressions for $C1$ and $C2$. Assuming identical sample coil parameters ($L = 1$ μH and $R = 0.1$ Ω) and a similar working frequency ($\approx$ 20 MHz), determine values of $C1$ and $C2$ required to tune and match this circuit. (This circuit has been considered by P. D. Murphy and B. C. Gerstein, Report IS-4436, Ames Laboratory, Iowa State University, Iowa, 1978 https://doi.org/10.2172/6895301).

(c) For tuned circuits, Q_L is an important parameter that provides a measure of the energy stored in the system and defines how the circuit responds to pulsed stimuli. For the circuit discussed here, $Q_L = \omega L/R$. At a frequency of 20 MHz, what would the Q_L be for the coil discussed above? How does the Q_L change if R is increased to 10 Ω? The time constant for the rise/fall time of the RF stimulus is given by $2Q_L/\omega$; what is this response time for the two conditions considered above? Which of these conditions is likely to be most suitable for pulsed RF measurements? Why?

(19.3) Assuming the RF coil is operating at a frequency of 20 MHz, what is the static field, B_0, required for resonance of diamagnetic muons? If a pulse width of 1 μs is required for a 90° spin rotation of diamagnetic muons (a 90° pulse), what B_1 field strength is needed? Assuming a two-pulse RF sequence is used to form an echo 5 μs after muon arrival, what fraction of implanted muons would be expected to live to this time?

(19.4) The cyclohexadienyl radical, formed by Mu addition to benzene, has a muon hyperfine coupling,

A_μ, of 514.8 MHz at 293 K. If proton couplings on the ring are ignored, the energy levels describing the four states of this system are given by the Breit–Rabi diagram for a two spin-1/2 system. Assuming an RF spectrometer operating at 120 MHz, calculate the resonance fields for all possible transitions that can be observed for this system.

(19.5) As discussed in Fig. 19.6, the pump wavelength can be optimized for the largest spatial overlap of muons and photons.

(a) Assuming that the muon distribution can be approximated by a Gaussian function, $\exp[-(z - z_0)^2/(2\sigma^2)]$, where z_0 represents the centre of the peak and $2\sqrt{2\ln 2}\,\sigma$ gives the FWHM, show that

the number of photons absorbed in the volume where muons are implanted is proportional to

$$\alpha e^{-\alpha d}\int_0^d e^{-\frac{(z-z_0)^2}{2\sigma^2}+\alpha z}\,\mathrm{d}z. \qquad (19.9)$$

(b) Integrate eqn (19.9) numerically and plot it as a function of α.

(c) Assuming a Si wafer with $d = 1.0$ mm and the muon distribution is centred at $z_0 = 0.5$ mm with its FWHM $= 130\,\mu$m, what α will give the maximum overlap?

(d) Based on an absorption spectrum from literature, e.g. G. G. Macfarlane *et al.*, Phys. Rev. **111**, 1245 (1958), what is the optimum wavelength?

High magnetic fields

20

In this chapter we describe muon experiments performed in high magnetic field. What does high mean? The Earth's magnetic field is approximately 50 μT, and a typical applied magnetic field in a muon experiment is usually in the range[1] $B = 0.1$–500 mT. Such fields can be generated relatively easily using a conventional magnet, typically consisting of a pair of Helmholtz coils of copper wire connected to a power supply. For magnetic fields in excess of these values it is necessary to use a superconducting magnet. We will denote the range 1–10 T as 'high magnetic field'.

The chapter is structured as follows: in Section 20.1 we present a motivation for using high magnetic fields, and in Section 20.2 we explain the experimental challenges for their use with muons, and how they can be overcome. Finally, we present some examples of the use of high fields in Section 20.3.

[1]This translates to 10–5000 G, and at many muon facilities the control software is setup for the cgs unit gauss, rather than tesla. The conversion factor is 1 G = 10^{-4} T.

20.1 Why high fields?

Why should anyone want to perform a μSR experiment at high magnetic field? Below we list several reasons.

(i) *Accessing level crossing resonances:* We have already seen in Chapter 12 that high magnetic fields are necessary for the study of avoided level crossing resonances, the shape and position of which can give important information about muonium states and their dynamics in chemical systems. In such experiments, the magnetic field is applied in the longitudinal direction and so can be performed at both pulsed and continuous sources.

(ii) *Measuring fluctuations:* High-field LF measurements can be used to probe fluctuations of the local environment by studying the field-dependence of the relaxation of the muon spin. This relaxation rate is reduced as the energy required to flip the muon spin increases and turns out to provide a map of the power spectrum of the fluctuations.[2] In the simplest cases, this field dependence is Lorentzian, allowing the average fluctuation rate (or inverse correlation time) to be measured directly from its width and leading to a characteristic inverse-square law dependence at high field. The faster the magnetic fluctuations or motional hop rate, the higher the field required to suppress the relaxation and to define the relevant correlation time.

[2]See Chapter 7.

(iii) *Inducing new phases:* Many interesting materials enter new phases when subjected to large magnetic fields. A relatively simple example is a magnetic system composed of $S = 1$ moments (e.g. Ni^{2+}) for which crystal field splitting results in a singlet ground state separated from the lowest-energy excited state. If the gap between them is closed by a magnetic field, then the system can transition from a non-magnetic state to one in which magnetic order has been induced. This can be interpreted in some cases as a Bose-Einstein condensation transition of the quasiparticles. Phase transitions often lead to dramatic changes in the spectrum measured using TF μSR. In the case that a magnetic phase transition takes place, we also expect the fluctuation spectrum to change considerably, and so measurements of the dynamics with LF μSR can also be used to detect field-induced changes in phase.

(iv) *Studying superconductors:* μSR is particularly useful in studying the vortex lattice in superconductors (see Chapter 9). High magnetic fields are needed for many novel superconductors which have large values of the upper critical field H_{c2}, above which superconductivity is destroyed; these materials often have interesting and unexplored superconducting properties that persist up to high field. Moreover, in highly anisotropic systems the vortex lattice is not a system of rigid rods (as in the isotropic case) but is more like a weakly coupled stack of quasi-two-dimensional 'pancake' vortices, each pancake confined to a superconducting plane. The phase diagram of these anisotropic vortex lattices is very rich and studying them as a function of field, temperature, and field-orientation can be very informative. Another important consideration is that the coherence length ξ can be determined by a careful measurement of the lineshape in transverse field. In high field, the number of vortices per unit area increases, and thus so does the normal state volume, reducing the condensation energy and this can lead to the radius of the vortex cores, and hence ξ, shrinking in high field. This effect can be studied with μSR.

(v) *Determining Knight shifts:* In a magnetic field, the muon Knight shift is dependent on the local magnetic susceptibility at the muon site and hence allows μSR to access the magnetization distribution inside a sample. High magnetic field allows a more accurate determination of the Knight shift which is a small effect.

(vi) *Condon domains:* In metallic samples, a magnetic field can cause **de Haas-van Alphen oscillations** in the magnetization as Landau levels pass through the Fermi energy. These oscillations in M are a function of $B = \mu_0(H + M)$, and there is the possibility of a feedback reaction called the magnetic interaction, in which the oscillating magnetization $M(B)$ caused by the effect also feeds it by contributing to B. If the oscillations in M are sufficiently rapid (corresponding to a large Fermi surface area), then M can be a multivalued function of H, corresponding to an instability. This leads to the formation of what are known as **Condon domains** in which the magnetization takes two different values. These can be observed using μSR in certain pure metallic samples at very low temperature.

20.2 Muons and high magnetic fields

In this section we explore some of the challenges of using muons with high magnetic fields. The first thing to note is that the gyromagnetic ratio of the muon implies that a of field 1 T will give rise to a precession at 135.5 MHz; in 10 T this is 1.355 GHz. This implies that ultra-fast detectors and electronics are needed when measuring precession in this field range. This consideration also shows that measuring precession signals will only be possible at a continuous muon source.[3]

High magnetic fields are deployed routinely for neutron scattering experiments and also (rather less often) for measurements using X-rays. It seems natural that they can be used with μSR, but there is an important complication: *muons and positrons are charged!* Both neutrons and photons are not. This has some very important implications as the application of high fields can lead to a varying fraction of the incoming muon beam reaching the sample, as well as to a significant loss or distortion of positron counts.

In the presence of a magnetic field $\boldsymbol{B}$ the Lorentz force on a particle with charge q is written[4]

$$\boldsymbol{F} = q\boldsymbol{v} \times \boldsymbol{B}, \tag{20.3}$$

where $\boldsymbol{v}$ is the particle's instantaneous velocity. The Lorentz force has the tendency to cause a charge to execute cyclotron motion with an orbit radius r given by[5]

$$r = \frac{Ev_{\mathrm{T}}}{qBc^2}, \tag{20.4}$$

where E is energy and v_{T} is the component of velocity perpendicular to the magnetic field.

Example 20.1

In a constant field we have an equation of motion for the momentum of $\dot{\boldsymbol{p}} = q\boldsymbol{v} \times \boldsymbol{B}$, so for a field B applied along the z-direction the non-relativistic equations of motion are

$$\dot{v}_x = \omega_{\mathrm{c}} v_y \quad \dot{v}_y = -\omega_{\mathrm{c}} v_x \quad \dot{v}_z = 0, \tag{20.5}$$

where the cyclotron frequency ω_{c} is defined as $\omega_{\mathrm{c}} = qB/m$. Integrating the equations of motion, we find

$$\begin{aligned} x &= x_0 + r\sin(\omega_{\mathrm{c}} t + \alpha) \\ y &= y_0 + r\cos(\omega_{\mathrm{c}} t + \alpha) \\ z &= z_0 + v_{0z} t, \end{aligned} \tag{20.6}$$

where x_0 and y_0 are constant positions, v_{0z} is a the constant velocity along z, and α is a constant phase.

The incoming muon beam in a high-field μSR experiment is always directed along the principal direction of the applied magnetic field[6] and any modification of the muons' trajectories arises mainly from the fringing field[7] of the magnet. In this configuration, the transverse compo-

[3]A typical pulsed muon source is restricted to measuring precession signals below $\approx$10 MHz. Pulsed muon sources can still be used to measure high-field μSR in *longitudinal mode*.

[4]The muon and positrons are moving at relativistic speeds in the experiment which must be taken into effect. The relativistic equations of motion can be derived for a four-velocity u^μ which yields

$$m\frac{\mathrm{d}u^\mu}{\mathrm{d}\tau} = qF^{\mu\nu}u_\nu, \tag{20.1}$$

where $F^{\mu\nu}$ is the Faraday tensor and τ is the proper time. Substituting for u^μ and $F^{\mu\nu}$ gives the equation of motion for a particle with three velocity $\boldsymbol{v}$ of

$$\frac{\mathrm{d}}{\mathrm{d}t}(m\gamma\boldsymbol{v}) = q(\boldsymbol{v} \times \boldsymbol{B}). \tag{20.2}$$

In addition to this, the acceleration of the particles leads to energy losses through radiation damping, although this effect is relatively small. The relativistic contributions are described in detail in Landau and Lifshitz, vol II.

[5]The Lorentz force $qv_{\mathrm{T}}B$ must balance the centripetal acceleration $\gamma mv_{\mathrm{T}}^2/r$. Equating these gives $r = \gamma mv_{\mathrm{T}}/qB$ and the relativistic equation for energy $E = \gamma mc^2$ then gives the result in eqn 20.4.

[6]This is the case even in experiments carried out in large transverse fields, as it minimizes the deflection of the beam. Since the muon polarization is created antiparallel to the beam direction, in order to achieve an initial muon-spin polarization perpendicular to the field, the polarization is altered using a spin rotator (see Chapter 13).

[7]Since $\boldsymbol{\nabla} \cdot \boldsymbol{B} = 0$, field lines form loops in space in such a way that there are no sources and sinks of the field.

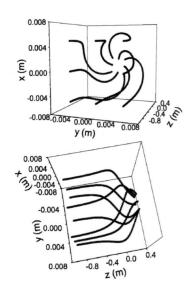

Fig. 20.1 Simulated effect of magnetic field on an incoming muon beam of area 1 cm^2 in a field of 0.45 T generated by Helmholtz coils. [Adapted from P. J. C. King *et al.*, Physica B **326**, 260 (2003) and T. Lancaster, D.Phil thesis (Oxford, 2004).]

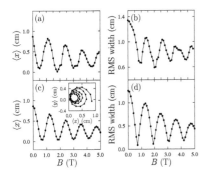

Fig. 20.2 The simulated and measured effect of an applied magnetic field generated by a superconducting solenoid on the beam spot at PSI. [Adapted from T. Lancaster *et al.*, Nucl. Instr. and Meth. A **580**, 1578 (2007).]

nents of the magnetic field at the magnet edges, B_x and B_y, will initially alter the trajectory of the muons causing them to acquire components of velocity transverse to the z-direction. These components will generally cause the muons to execute cyclotron orbits approximately about the large field component B_z near the magnet centre. The net result is that the radius of the muon spot on the sample will oscillate as a function of magnetic field as the muon beam focuses and defocuses as a function of the applied field. This can lead to a large field-dependent background in the experimentally measured data which needs to be understood, modelled, and subtracted.

Example 20.2

This effect of the applied field on the incoming muon beam can be straightforwardly simulated and measured, as shown in Figs 20.1 and 20.2. Even in a relatively modest field of $B_0 = 0.45$ T, Fig. 20.1 shows that the applied field causes a sizeable change to the trajectories of the muons in the incoming beam. Figure 20.2 shows a measurement (upper panels) and numerical simulation (lower panels) of the muon beam spot size in the sample plane as a function of the applied magnetic field, along with the measured (a) and simulated (c) position of the centre of the beam spot. The inset shows the simulated variation of the position of the spot centre with applied field. The RMS widths of the muon intensity distribution [measured (b) and simulated (d)] both display oscillatory behaviour. This wandering of the beam spot and throbbing effect on its size causes field-dependent variations in the illumination of the sample plane in the measurement.

After the muons are implanted in the sample, they decay. The main decay product of interest are the emitted positrons which emerge, at highly relativistic speeds, with a range of possible energies. In low fields one can assume that the positrons travel in straight lines and so it is easy to work out the muon polarization from the angular dependence of the detected positrons. However, in the presence of a large magnetic field one would expect cyclotron motion of the positrons and so they will not travel in straight lines. Moreover, as decay positrons are often emitted with large components of their momentum perpendicular to the principal magnetic field, the effect of the field is greater on the positron ensemble than on the muon beam.

Example 20.3

The results of a simple simulation are shown in Fig. 20.3 for eight radially directed positron paths in fields of 0.45 T and 3 T. The observable cyclotron motion at the higher field is particularly inconvenient since it allows a positron to potentially interact with more than one detector, distorting the value of asymmetry. There is also the possibility of large numbers of positrons being directed onto a small element of solid angle, which is undesirable, particularly at a pulsed source where it can lead to some positron counts being missed owing to detector deadtime limitations unless the detectors are very finely pixelated.

Figure 20.4 which shows the result of a simulation of the effects of a longitudinal field as measured by a (fictional) spherical detector bank, of radius 0.05 m, that surrounds the sample. The figure shows (a) the number of positrons $N(\theta)$ detected as a function of polar angle θ from an ensemble of polarized muons using a spherical detector [shown inset in (b)] in an applied uniform magnetic field of 0 T and 5 T. Also shown as a function of angle is: (b) the asymmetry $A(\theta)$ and (c) a figure of merit for the detectors, $N(\theta)A(\theta)^2$. The idea behind the figure of merit NA^2 is that it is a single quantity that tells us about both the asymmetry measured by a detector (i.e. the size of the signal) along with the number of counts (which should be maximized to reduce uncertainties in the measurement). The design of a detector array involves attempting to maximize this quantity for the available detector pixels.

The quantities plotted in (a), (b), and (c) don't take into account the difference in available solid angle as a function of θ and so, more usefully, the number of events (d), asymmetry (e), and figure of merit (f) are also shown normalized by solid angle.

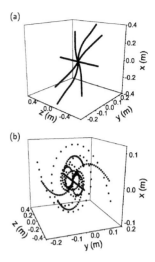

Fig. 20.3 Effect of magnetic field on positron paths measured at (a) 0.45 T and (b) 3 T. [Adapted from P. J. C. King *et al.*, Physica B **326**, 260 (2003) and T. Lancaster, D.Phil thesis (Oxford, 2004).]

Clearly, the influence of the field on the particles needs to be taken into account when designing a high-field spectrometer. It is advantageous to have the detectors close to the sample so that the emitted positrons have less space to wander round the field lines before being detected. However, this places the detectors closer to the field centre and so it is necessary to have detectors that can operate in magnetic field. Photomultiplier tubes, for example, rely on a cascade of electrons between charged plates, and so can be strongly affected by magnetic fields. Avalanche photodetectors are a better solution since they are only very weakly affected by magnetic fields.

When operating a high field spectrometer there are a number of practical considerations that need to be kept in mind when planning the experiment.

- Because of the focusing effects described above, the data rate may be quite field dependent and so, for example, it may be slower to achieve the same statistics at high field compared to low field.

- Superconducting magnets do not ramp up as quickly as Helmholtz magnets; they have to be swept slowly. Moreover, the maximum sweep rate is likely to be significantly lower at higher fields, and so experiments need to be planned to make the best use of the available beamtime. For example, repeatedly jumping back and forth between very high-field measurements and low-field measurements will involve very long waits while the magnet ramps up and down and should be avoided, unless the experiment absolutely demands it.

- High-field superconducting magnets are clearly designed to operate at large magnetic fields, but can struggle at very low field. This is because the magnets can become very slightly magnetized after operating at high magnetic field, even when the current is reduced to zero. This results in a very small remanent field (of order of a few mT) being present, so that one never achieves a perfect zero-field unless the superconducting magnet is quenched or some form of field compensation is employed, involving actively applied additional magnetic fields to give a resultant field of zero at the sample position.

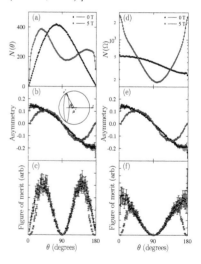

Fig. 20.4 Effect of magnetic field on the positron distribution measured by a spherical detector. [Adapted from T. Lancaster *et al.*, Nucl. Instr. and Meth. A **580**, 1578 (2007).]

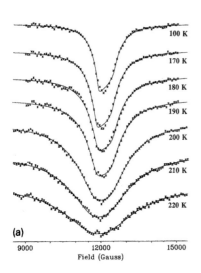

(a)

9000   12000   15000
          Field (Gauss)

Fig. 20.5 ALC spectra obtained with C_{60} at different temperatures. [Taken from E. Roduner *et al.*, Chem. Phys. **192**, 231 (1995).]

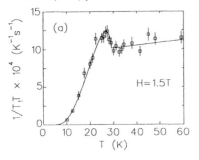

Fig. 20.6 $1/T_1T(\equiv \lambda/T)$ in Rb_3C_{60} at a magnetic field of 1.5 T. The solid curve is a fit to the Hebel–Slichter theory. (b) Arrhenius plot of $T_1^{-1}(\equiv \lambda)$ in a longitudinal magnetic field of 0.3 T. [Taken from R. F. Kiefl *et al.*, Phys. Rev. Lett. **70**, 3987 (1993).]

[8]The quasiparticle excitations of a metal are electrons excited above the Fermi level. These have properties that are *renormalized* through the interactions with the rest of the electrons. In a superconductor, the analogous excitations are called Bogoliubov quasiparticles, whose energies are separated from the ground state by the energy gap Δ.

20.3   Science at high field

Once the technical challenges have been overcome, it is possible to exploit the high field spectrometers in a number of science areas. We outline some examples here.

Example 20.4

The avoided level crossing spectroscopy of crystalline C_{60} provides a good example of the need for high magnetic fields. In C_{60} muonium can add chemically to a double bond to form a free radical on the surface of the fullerene that is very sensitive to the dynamics of the molecules. ALC spectra obtained at different temperatures with C_{60} in the region of the $\Delta M = 1$ transition are shown in Fig. 20.5. The resonance broadens with increasing temperature and at 250 K it is no longer detectable, owing to a phase transition at 260 K which is accompanied by a considerable change in the dynamics. In the high-temperature phase, the C_{60} molecules tumble though all possible orientations fairly randomly; in the low-temperature phase the molecules jump between two orientations following a thermally activated behaviour. The solid line in Fig. 20.5 is a fit to a dynamic stochastic model for spherical rotational diffusion in the low-temperature regime, from which a measure of the activation energy of 176 meV can be obtained.

Example 20.5

The fulleride material Rb_3C_{60} provides an interesting use of high magnetic field to investigate the electronic properties of a superconducting system. From Chapter 7 we have the result that the muon relaxation rate varies as

$$\lambda \propto T \lim_{\omega \to 0} \sum_{\mathbf{q}} |D(\mathbf{q})|^2 \frac{\chi''(\mathbf{q}, \omega)}{\omega}, \tag{20.7}$$

where $D(\mathbf{q})$ encodes the coupling between the spin probe and system, and χ'' is the imaginary part of the susceptibility. This can be used to derive an expression for the muon-spin relaxation which is valid for either an electron fluid or superconductor given by

$$\frac{\lambda}{T} \propto \int dE \left(-\frac{df(E)}{dE}\right) |\mathcal{M}|^2 g(E)^2, \tag{20.8}$$

where E is energy, $|\mathcal{M}|^2$ is the matrix element for the spin-scattering of the quasiparticle excitations[8] in the metal or superconductor, $f(E)$ is the Fermi function, and $g(E)$ is the density of states. In a metal, in the usual regime where temperature is much smaller than the Fermi temperature, eqn 20.8 predicts the constant behaviour $\lambda/T \propto g(0)^2$, which is the Korringa law. The scattering matrix elements for metals and BCS superconductors are the same, but for the BCS superconductor a difference comes from the quasiparticle density of states, which is given by the strongly energy-dependent function $\propto |E|/\sqrt{E^2 - \Delta^2}$, where Δ is the superconducting energy gap. This results in so-called **Hebel–Slichter** behaviour for the ratio between normal (n) and superconducting (s) responses of the relaxation rate, given by

$$\frac{(\lambda/T)_s}{(\lambda/T)_n} = \frac{1}{2} \int_{E=\Delta}^{\infty} dE \left(-\frac{df(E)}{dE}\right) \frac{E^2}{E^2 - \Delta^2}. \tag{20.9}$$

This ratio, with its dependence on the temperature-dependent gap parameter $\Delta(T)$, often leads to the observation of a maximum in λ/T as a function of temperature just below the superconducting T_c, known as the **coherence peak**. The observation of the coherence peak that occurs in a BCS superconductor is due both to the lack or renormalization of the matrix element and to the strongly energy-dependent density of states. It is not observed in some non-BCS superconductors and so it is an important feature in the classification of a superconducting system.

In order to observe this effect, the relevant coupling is between those muons bound in endohedral muonium states (i.e. muonium trapped inside the C_{60} molecular cages) and the quasiparticles of the system. Muon-spin relaxation is then achieved as follows: (i) a quasiparticle from the metal or superconductor scatters the local electron spin bound in the Mu state at a rate ν_s; (ii) the resulting change in the hyperfine field at the muon site causes both the bound electronic spin and the muon spin to flip (i.e. a flip-flop process). The transition giving rise to the muon-spin relaxation is therefore that between the muonium states $|2\rangle$ and $|4\rangle$ which, in the high field limit, are written $|2\rangle = |\downarrow^\mu\uparrow^e\rangle$ and $|4\rangle = |\uparrow^\mu\downarrow^e\rangle$. In the case that $\nu_{24} \gg \nu_s$ (i.e. small ν_s or high field) we obtain a field dependence of the relaxation rate of

$$\lambda(B) \approx \frac{\nu_s(2\pi A)^2}{2\left[(2\pi A)^2 + \gamma_e^2 B^2\right]}, \qquad (20.10)$$

where B is the applied field and A is the muon hyperfine coupling. The field dependence is then very informative in assessing the coupling between the muonium and electronic systems, while the temperature dependence tells us whether there is a coherence peak.

Example 20.6

Condon domains. When the magnetic field B increases inside a metal there are many properties which are periodic in $1/B$, including the magnetization (the de Haas-van Alphen effect) due to Landau levels moving through the Fermi surface, resulting in an oscillatory occupation of the highest occupied Landau level. Inside the sample, $B = \mu_0(H+M)$ but M is the oscillating magnetization of the electrons. The susceptibility $\chi = \partial M/\partial H$ is related to $\chi_B = \mu_0\partial M/\partial B$ by[9]

$$\chi = \frac{\chi_B}{1-\chi_B}, \qquad (20.11)$$

and so an instability occurs if $\chi_B > 1$, something which can occur if there is a strong magnetic interaction (meaning the de Haas-van Alphen effect is so strong that there is a large feedback effect: the oscillatory magnetization, due to B, also makes a substantial contribution to the B-field that causes it). Note also that $\mu_0\partial H/\partial B = 1 - \chi_B$ and this quantity is negative when $\chi_B > 1$, implying a multivaluedness of $B(H)$. This is analogous to the famous discontinuity in the p-V diagram for a van der Waals gas, which also leads to a separation between two thermodynamic phases (liquid and gas), though here there is more than one discontinuity, situated periodically along the B-field axis.

The instability is avoided by having a discontinuous jump in properties inside the sample. Thus for a plate-like sample perpendicular to the applied $\boldsymbol{B} = \mu_0\boldsymbol{H}$ the instability is avoided by the formation of Condon domains, alternating regions of diamagnetic ($B_1 < \mu_0 H$) and paramagnetic ($B_2 < \mu_0 H$) phases, with the proportion of the two domains varying with the applied H so that the average $\bar{B} = \mu_0 H$ to fix the boundary condition. Note that the Condon domains only form at certain magnetic fields where the oscillatory magnetization is large enough to induce this feedback effect.

Condon domains were first observed using NMR[10] and have also been detected with micro-Hall measurements,[11] but μSR has been shown to be a very effective technique for studying them. Measurements for beryllium are shown in Fig. 20.7 and reveal a difference in magnetic field ΔB of ≈ 2.9 mT between two Condon domains.

[9]See Exercise 20.2.

[10]J. H. Condon and R. E. Walstedt, Phys. Rev. Lett. **21**, 612 (1968).

[11]R. B. G. Kramer *et al.* Phys. Rev. Lett. **95**, 267209 (2005).

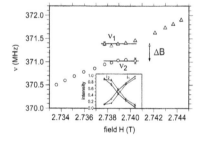

Fig. 20.7 Condon domains in Be observed in the Fourier transform of the transverse-field precession data. The presence of two distinct frequencies close to 2.74 T indicates the existence of separate domains. As shown in the inset, the intensity of one domain grows at the expense of the other as the field increases. [G. Solt *et al.*, Phys. Rev. Lett. **76**, 2575 (1996).]

Chapter summary

- The application of high magnetic fields potentially allows muon measurements to access level crossings, high-frequency fluctuations, and new phases of matter.
- The influence of magnetic field on charged particles means that the application of high fields alters the trajectories of the incoming muon beam and outgoing positron ensemble.
- High fields can cause background contributions to μSR signals owing to changes in the muon spot size and intensity, and to changes in the distribution of detected positrons.

Further reading

- L. D. Landau and E. M. Lifshitz, *Classical Theory of Fields* (volume II of Landau and Lifshitz), Pergamon (1975).
- The physics behind Hebel-Slichter behaviour is discussed in P. Coleman *Introduction to Many-Body Physics*, CUP (2015).
- Condon domains are described in V. S. Egorov, Phys. Usp. **53**, 755 (2010).

Exercises

(20.1) (a) Starting from eqn 20.1, outline the steps that lead to the relativistic analogues of eqns 20.5.

(b) Fill in the missing steps to prove eqns 20.6.

(20.2) By writing $\delta B = \mu_0(\delta H + \delta M)$, show eqn 20.11 and hence that $\mu_0 \partial H / \partial B = 1 - \chi_B$.

Muons under pressure

21

Pressure is an important thermodynamic variable in physics. Under applied pressure the interatomic distances in a material shrink, leading to changes in interatomic interactions. Investigating the response of a material to pressure allows us to probe the interplay of interactions and to determine microscopic parameters. Pressure tuning is therefore widely used to identify critical points, study phase transitions, and establish phase diagrams. In this chapter we shall discuss the requirements and challenges of performing μSR experiments under pressure and, to illustrate the discussion, we describe in detail the high-pressure facility at the Paul Scherrer Institute (PSI) in Switzerland. Gas pressure systems are then briefly discussed, before turning to some science examples and discussing the outlook for pressure studies.

21.1 Requirements

In order to make measurements under applied pressure, the sample must be put into a pressure cell. Implanting muons through the cell walls and into the sample presents us with a first problem. In ambient pressure μSR experiments surface muons are used. These are (close to) monochromatic, with a well-defined μ^+ momentum around 29.8 MeV/c, and can be stopped in thin samples at a depth of $\sim$ 160–200 mg/cm^2 (e.g. for a metallic sample of copper with density 8.96 g/cm^3, the stopping distance is only 0.2–0.3 mm). Consequently, if we want muons to reach a sample loaded inside a pressure cell, the muon energy needs to be much larger. This requires the use of a highly energetic decay muon beam, as introduced in Chapter 13, where the muons are obtained from pions leaving the production target with high energy. By adjusting the beamline to vary the muon momentum, we aim to choose an optimum momentum that maximizes the ratio between the number of muons stopping inside the sample and the number of the muons stopping in the pressure cell walls (this is the signal to background ratio in the measurement). As shown in Fig. 21.1, this ratio depends not only on the muon momentum, but also on the width of the muon beam.

Unfortunately, the size of the muon beam cannot be decreased indefinitely. Moreover, this size also depends slightly on the muon momentum. Obtaining the optimal values of the momentum is therefore a tricky operation that involves considering the dispersion of the beam momentum, convoluted with the scattering of muons along their path, which crosses

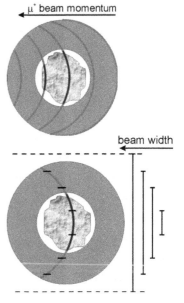

Fig. 21.1 Transverse section of a piston-cylinder pressure cell. The sample is in the centre. Muons stopped in the sample and the pressure cell are shown by darker and lighter lines, respectively. *Top*: Stopping muons inside the sample and the cell for different values of muon momentum. *Bottom*: Stopping muons inside the sample and the pressure cell for different values of the beam width. [Adapted from R. Khasanov *et al.*, High Pressure Research **36**, 140 (2016).]

[1]Work at PSI was dedicated to improving the characteristics of piston-cylinder cells, with the aim of using them in the existing cryogenic environment (where the maximum outer diameter of the pressure cell is limited to about 26 mm).

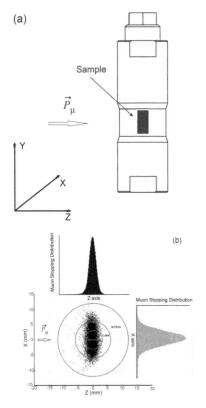

Fig. 21.2 (a) Cylindrical pressure cell with the sample inside. Muons are implanted parallel to vector $\vec{P}_\mu$. (b) Cross section (X-Z plane) of the pressure cell consisting of the inner and the outer cylinders made of CuBe and MP35N alloys. The graphs represent the muon stopping distributions in parallel (*top*) and perpendicular (*right*) directions to the muon beam. The energy of implanted muons is 44 MeV. The simulations are made by using TRIM.SP. [Adapted from Z. Shermadini *et al.*, High Pressure Research **37**, 449 (2017).]

the walls of the cryostat, the walls of the pressure cell and the sample. These latter effects cause a dispersion of order of a few millimetres and limit the minimum possible dimension of the beam spot at the sample (see Fig. 21.2). Consequently, the main considerations for performing muon pressure experiments may be summarized as follows.

- Samples of large dimensions need to be used in order to have a sizeable signal from the sample inside the pressure cell, compared to the background. This is a particular feature of μSR, in contrast to other techniques such as resistivity, a.c. and d.c. magnetization, specific heat, where samples of less than 1 mm^3 can be measured (thus allowing the use of anvil-type pressure cells). At present, the only choice for μSR experiments in the sample environment geometry available at PSI is the piston-cylinder-type pressure cell, which allows a relatively large sample space to be used.[1]

- Test experiments reveal that more than 50% of muons can be stopped in the sample when using a piston-cylinder pressure cell with 7 mm inner diameter, and around 30% for 5 mm cell. However, 7 mm cells sustain less pressure than 5 mm ones, so the choice of the pressure cell is dictated by the maximum pressure needed and the complexity of the μSR signal.

- After choosing the type of the cell, the optimization of the beam optics, in particular the momentum, is the first important step in performing the experiment.

21.2 The PSI setup

In this section an experimental setup to carry out μSR experiments at a high-energy muon beamline is described. Our detailed example from PSI uses the μE1 beamline, the General Purpose Decay (GPD) spectrometer and piston-cylinder-type pressure cells.

Decay beamline

We start with a description of the μE1 decay beamline at PSI. As described in Chapter 13, positive (μ^+) and negative (μ^-) muons are obtained from pion (π^+ and π^-) decay. The high-energy muon beam is obtained from pions that leave the target at high energies. These are collected over a certain solid angle by quadrupole magnets and directed on to an 8 m-long decay section consisting of a superconducting solenoid with a field of $\approx$ 5 T. If the pion momentum is not too high, a large fraction of pions decay before reaching the far end of the solenoid. In this respect the exit of the solenoid may be viewed as a diffuse source of muons. Usually either the decay muons produced along (forward muons) or opposite to (backward muons) the direction of the pion momentum are selected, in order to keep a large degree of polarization. However, the polarization of a high-energy muon beam in the laboratory frame is limited to around 80% in any case.

A schematic view of the high-energy muon beamline μE1 installed at PSI is shown in Fig. 21.3. The beam optics consist of several magnets used to transport the muon beam to the measuring station. Each magnet is tunable, allowing us to control precisely both the beam momentum and spread. The magnets are either quadrupole focusing magnets (Q) or bending magnets (A) that play an additional role in selecting the momenta. Slits (F) are used to limit the lateral beam extension and also the intensity. The superconducting solenoid (the so-called μ−channel) is the source of the muons. Note that in contrast to surface muons which have spins oriented antiparallel to their momentum, the backward muons have spins aligned *parallel* to the momentum. The characteristics of μE1 beamline are summarized in Table 21.1.

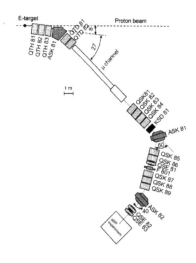

Fig. 21.3 The μE1 beamline. The superconducting solenoid is the source of muons formed from pions decaying in flight inside the solenoid. Quadrupolar magnets (Q) focus the beam; bending magnets (A) play a role of momentum selective elements. Slits (F) limit the lateral beam extension and the intensity of the muon beam.

Momentum acceptance (FWHM)	3%
Pion momentum range [MeV/c]	125–200
Muon momentum range [MeV/c]	60–125
Rate of negative muons [mA^{-1}s^{-1}]	$3 \times 10^7 - 6 \times 10^7$
Spot size (FWHM)	39×28 mm

Table 21.1 Characteristics of μE1 beamline at PSI.

Spectrometer

The measurements under pressure are made using the GPD spectrometer, which is permanently installed on the μE1 beamline. This instrument is designed to perform experiments in ZF, LF, and TF configurations over a wide temperature range, making use of different cryostats. The spectrometer is equipped with a water-cooled Helmholtz-coil magnet providing a maximum field $\simeq 0.6$ T. To select between the longitudinal and the transverse field geometry, the magnet can be rotated by $90°$. Sample rotation is provided for angle-dependent studies of single crystals. The size of the incoming muon beam is defined by passive lead collimators.

Pressure cells

There are many requirements on the sample environment that must be met in order to perform μSR measurements under pressure. The most important, of course, is the availability of a large sample-volume hydrostatic pressure environment. In addition, low temperatures (i.e. sub-Kelvin) are often required. Another important factor is the possibility of performing pressure scans without actually opening the cell. In addition to the obvious time saving (essential when using a costly particle beam) this is important: (i) for calibration, since each opening of the cell can lead to a slight change of the sample position or cause a redistribution of the sample, which would require a new calibration; (ii) the physical properties of the sample might change during each fresh pressurization. These points require the use of piston-cylinder pressure

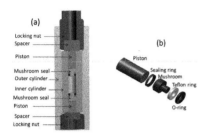

Fig. 21.4 (a) The PSI pressure cell; (b) Detail of the pressure seal. [Adapted from R. Khasanov *et al.*, High Pressure Research **36**, 140 (2016).]

cells for which the mechanical properties can be either calculated (using the long cylinder theory) or simulated (using finite-element analysis).

A fully-assembled double-wall pressure cell is shown in Fig. 21.4. The main components of the cell are: (i) a cylindrical double-wall body, (ii) pistons, (iii) mushroom seals, (iv) locking nuts, and (v) spacers. The latter are needed to support the pistons and prevent their rotation when the nuts are tightened. Diameters and heights of the outer and the inner cylinders are $\varnothing = 24$ mm, $h = 72$ mm, and $\varnothing = 15$ mm, $h = 44$ mm, respectively. Two types of cells with inner diameter 6 mm and 7 mm were produced. With both pistons completely inserted, the maximum sample height is $\simeq 12$ mm. The body of the cell, the top and the bottom locking nuts, and the mushroom pieces are made of MP35N alloy (Ni – 35%, Co – 35%, Cr – 20%, and Mo – 10% in weight percent). All the pieces were heat treated for four hours at 590°C after machining. Pistons and spacers are made of non-magnetic tungsten carbide. The sealing rings were made of fully hardened copper-beryllium alloy.

Cell fabrication

The body of the pressure cell consists of two parts, the inner and the outer cylinders, which are shrink-fitted into each other. Two alternative methods for the fabrication of the pressure cell were tried. In the first approach, the inner cylinder was prepared with a diameter ≈ 0.10 mm larger than the inner diameter of the outer cylinder. By heating the outer cylinder up to 590°C and cooling down the inner one to liquid nitrogen temperature, it was possible to fit them into each other. By returning to ambient temperatures, the inner cylinder remains under radial compression from the outer cylinder. In the other approach, the outer surface of the inner and the inner surface of the outer cylinders were machined conically with the angle of $\approx 1°$. The diameter difference of both fully assembled cylinders is ≈ 0.12 mm. The pressure cell is produced by mechanically inserting the inner cylinder into the outer one by using a hydraulic press. This process is similar to the situation of a cold/warm fitting of the cylinders as described above. The friction between the two parts is high enough to keep the cell fully assembled. The pressure cells produced using both methods were found to behave very similarly in terms of the maximum achievable pressure, as well as with respect to the pressure loss caused by friction between the sealing system and the pressure cell walls.

Stress distribution

Figure 21.5 shows a section of the cell with the corresponding principal stress and the shear stress distributions, computed using finite-element analysis. These calculations were made for extreme conditions, i.e. when the stress at the border between two cylinders reaches the ultimate tensile strength of the MP35N alloy (2.07 GPa). The results reveal that the theoretical limit for a double-wall cell made of MP35N alloy is of the order of ≈ 3.5 GPa. In reality, however, by reaching such pressure values

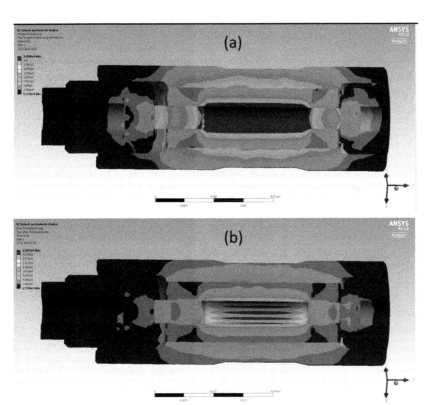

Fig. 21.5 (a) Maximum principal stress and (b) shear stress in a double-wall MP35N pressure cell. The calculations are made using the ANSYS R15.0 software. [Adapted from R. Khasanov *et al.*, High Pressure Research **36**, 140 (2016).]

the pressure cell will simply explode. For safety reasons, the maximum applied pressure is always limited to 3.2 GPa. The applied pressure at closing time is calculated as $p_{\text{closing}} = F_{\text{Load}}/S$, where F_{Load} is the force applied by the hydraulic press and S is the surface area of the piston. Note that p_{closing} is usually much larger than the effective real pressure inside the cell due to friction effects.

Operation

The process of loading the cell is permanently monitored by using a specially-designed computer-controlled system allowing us to measure *in situ* the displacement x of the piston and the radial expansion of the pressure cell as a function of loading force F_{Load}. Special attention is paid to maintain the pressure cell in the elastic regime, i.e. by ensuring that the slope dx/dF_{Load} of the loading curve $x(F_{\text{Load}})$ is almost constant. A sudden change of the slope dx/dF_{Load} would correspond either to a leak in the cell or would signal that we are exceeding the local stress level above the elastic regime. In both cases the loading process must be immediately stopped and the pressure in the cell released to zero.

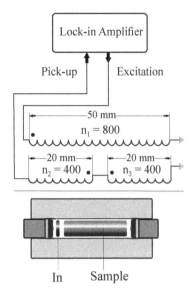

Fig. 21.6 The pressure measurement system. A.c. coils consisting of the excitation and two pickup coils are mounted on a separate cylindrical holder with inner diameter 27 mm. The cell is placed inside the coil such that the indium pressure sensor is located at the center of one pick-up coil. The superconducting transition of the indium probe leads to the abrupt change output a.c. voltage.

The exact pressure inside the cell is determined by monitoring the pressure-dependent shift of the superconducting transition temperature (T_c) of a small piece of indium. The volume of the pressure probe is much lower than that of the sample to avoid an additional background contribution to μSR signal. For the superconducting transition temperature of indium, the following relation holds:

$$T_c(p) = T_c(0) - 0.3812p + 0.0122p^2, \qquad (21.1)$$

where p is the pressure in GPa and $T_c(0) \approx 3.40$ K is the transition temperature of indium at ambient pressure.

The superconducting transition of the pressure probe is determined by a.c. susceptibility measurements. A coil surrounding the pressure cell (the input coil) is fed with an a.c. current. A second coil, inside the first one, made up of two sub-coils (one centered at the position of the pressure probe) wound in opposition, is used as the pick-up coil. The signal of the pick-up coil is compensated above T_c but becomes uncompensated when the superconducting transition is crossed upon lowering the temperature. The lack of compensation is detected using a lock-in amplifier which compares the phase and amplitudes of the input signal with the ones of the pick-up signal (see Fig. 21.6). Note that the pressure is determined at temperatures corresponding to the T_c of indium (i.e. ≈ 3 K) and so, when performing experiments at higher temperatures, we must take into account the temperature-dependent pressure drop of the pressure transmitting medium.

Pressure transmitting medium

Hydrostaticity at low temperature is the key feature for a liquid used as a pressure transmitting medium. The reason is that a hydrostatic pressure is a thermodynamic parameter and the results obtained under such conditions can readily be compared with theory. For our experiments under pressure, either Daphne 7373 oil or a 1:1 mixture of n-pentane/isoamyl alcohol are commonly used. Daphne oil is easier to manipulate, due to its higher viscosity and lower toxicity. When hydrostaticity at higher pressures is needed, alcohol mixtures are recommended, but special care should be given to the preparation of the pressure cells in order to avoid leakage as these mixtures have a low viscosity.

21.3 A gas-pressure setup

Besides the technique of using piston-cylinder clamp cells that was described in detail in the previous section, another technique has been developed for high pressure μSR studies using the ARGUS instrument at the RIKEN-RAL/ISIS facility in the UK. This uses helium gas as the pressure medium and allows continuously variable pressure to be applied to the sample without the need to remove the pressure cell from the cryostat.[2] An external compressor and intensifier rig is connected

[2] I. Watanabe *et al.*, Physica B **404**, 993 (2009).

to the cell via a stainless steel high-pressure capillary and pressures up to 0.64 GPa (6.4 kbar) can be applied to the sample with this system (Fig. 21.7). The cell is made from a 49 mm-diameter cylinder of CuBe. The sample space is a cylindrical chamber with a diameter of 18 mm and a depth of 9 mm, which is designed to match the circular geometry of the muon beam (Fig. 21.8). The maximum pressure here is lower than for a clamp cell design, so the technique is particularly suited to studies of materials whose properties change significantly at low pressures. Typical examples would be organic materials and strongly correlated systems.

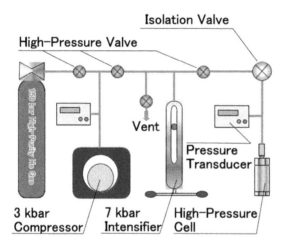

Fig. 21.7 Gas-pressure system used for high-pressure μSR measurements at ARGUS.

Fig. 21.8 The ARGUS gas pressure cell. (a) The rear of the cell showing the sample loading port, (b) the front of the cell showing the flat window facing the muon beam, and (c) the cell with the central gas capillary attached and the thermometry installed.

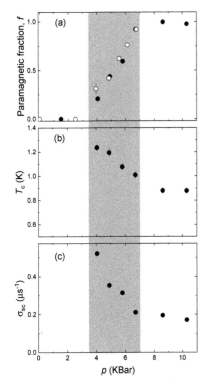

Fig. 21.9 Pressure evolution of (a) the nonmagnetic volume fraction, (b) the superconducting transition temperature T_c, and (c) the superconducting contribution to the Gaussian relaxation rate $\sigma_{sc} \propto \rho_s$ of CrAs. The grey area represents the phase coexistence region. Below $\simeq 0.35$ GPa (3.5 kbar) the sample is fully magnetic. Above $\simeq 0.7$ GPa (7 kbar) the superconductivity develops in 100% of the sample volume. [Taken from R. Khasanov *et al.*, Sci. Rep. **5**, 13788 (2015).]

21.4   Science examples

We now turn to some examples demonstrating what can be done with a high-pressure μSR setup.

Example 21.1

At ambient pressure CrAs is characterized by a relatively high Néel temperature of $T_N \simeq 270$ K. The magnetic moments are ordered in a helical magnetic structure. Under pressures (p), T_N decreases by approximately a factor of three approaching $p \simeq 0.7$ GPa, above which the magnetism completely disappears. Superconductivity sets in for pressures exceeding $\simeq 0.4$ GPa, revealing a range $0.4 \lesssim p \lesssim 0.7$ GPa where superconductivity and magnetism coexist.

Muon measurements under pressure provide an additional microscopic view of the magnetic and the superconducting properties of CrAs. Muon results suggest that the pressure-induced transition in CrAs from the magnetic to the superconducting state is characterized by a phase separation into macroscopic-sized magnetic and superconducting volumes. The less conducting magnetic phase provides additional carriers to the superconducting parts of the material. This scenario naturally explains the substantial increase of both the superconducting transition temperature T_c (from 0.9 K to 1.2 K) and the superfluid density $\rho_s(0)$ (up to $\simeq 150\%$), in the phase coexistence region (see Fig. 21.9). In addition, the superfluid density is found to scale with T_c as $\rho_s \propto T_c^{3.2(2)}$, suggesting a conventional BCS mechanism for the Cooper pairing in CrAs.

Example 21.2

Many elemental or binary materials undergo a superconducting phase transition under applied pressure. An example is the Bi-III phase of bismuth, which is achieved at pressures above 2.7 GPa. Within this phase a superconducting transition occurs below a transition temperature of $T_c \approx 7$ K. The results of muon measurements of the superfluid density $\rho_s(T)$ in the pressure-induced superconducting phase are shown in Fig. 21.10. These suggest that the superconducting phase results from a strong-coupling mechanism.

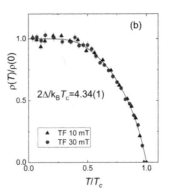

Fig. 21.10 The temperature dependence of the superfluid density of Bi-III. [Taken from R. Khasanov *et al.*, Phys. Rev. B **98**, 140504(R) (2018).].

Example 21.3

The molecule-based magnet $[Ru_2(O_2CMe)_4]_3[Cr(CN)_6]$ shows AF order below 33 K and the magnetic structure is modelled as two inter-penetrating ferrimagnetic lattices, with a weak 10 mK AF exchange coupling between them. The internal field under ambient pressure can be estimated from LF-μSR data taken on the ARGUS instrument (Fig. 21.11a).[3] In these data, the fractional drop of the asymmetry at 1.8 K compared to the asymmetry at 50 K is defined as ΔA. This is equal to to the fractional amplitude of the oscillatory component of the polarization (oscillations are not resolved directly here due to the limited time resolution of the pulsed facility instrument). The internal field is estimated from ΔA using the expression[4]

$$\Delta A(b) = A_{\rm osc}(b) = \frac{1}{4} + \frac{1}{4b^2} - \frac{(b^2-1)^2}{8b^3} \log\left|\frac{b+1}{b-1}\right|, \qquad (21.2)$$

where b is the ratio of the applied field B to the internal field B_0. At ambient pressure a metamagnetic transition takes place around 120 mT; above this field the two lattices are no longer in antiparallel alignment. Fitting the data to eqn 21.2 for the region below the metamagnetic transition gives a B_0 estimate of 39 mT, which is fully consistent with the precession frequency measured directly with ZF-μSR at PSI. On applying pressure, the field dependence of ΔA is strongly suppressed (Fig. 21.11b), indicating a significant increase in B_0. The fits in Fig. 21.11b give B_0 values of 0.17(8) T at 0.138 GPa and 0.4(2) T at 0.6 GPa.

21.5   Outlook

In this chapter we have summarized the status of high-presssure research using muons, concentrating on a case study of the pressure facility at PSI. We have described the requirements for μSR under pressure, the decay beamline, the μSR spectrometer, the construction, and the use of the double-wall pressure cell. Further improvement of μSR pressure cells is likely in the near future. Specific potential improvements include: (i) finding new materials suitable for minimizing background, (ii) development of new pressure cell designs (including the McWhan-type cell, three wall cells, etc.) and (iii) the possible use of anvil-type pressure cells. Preliminary measurements using boron-nitride and aluminium oxide anvils from the Paris-Edinburgh cell reveal that sufficient muons (up to ≈ 25 %) could stop in the sample for viable measurements, therefore potentially opening a rich seam of novel μSR experiments at much higher pressures than currently available. There have also recently been other advances[5] in applying uniaxial stress of around 1 GPa using a piezoelectric device. Such a device is capable of applying 1000 N to a sample and measuring magnetic susceptibility in situ. Altogether, it seems likely that the combination of μSR and pressure will see important advances in the near future.

[3] T. Lancaster *et al.*, Phys. Rev. B **84**, 092405 (2011).

[4] F. L. Pratt, J. Phys.: Condens. Matter **19**, 456207 (2007).

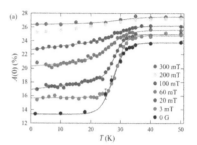

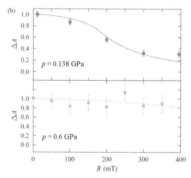

Fig. 21.11 (a) The T dependence of the LF asymmetry in $[Ru_2(O_2CMe)_4]_3[Cr(CN)_6]$ for a range of fields at ambient pressure. (b) The LF dependence of the fractional drop in asymmetry at two different pressures. The lines show fits to eqn 21.2. [Taken from T. Lancaster *et al.*, Phys. Rev. B **84**, 092405 (2011).]

[5] S. Ghosh *et al.*, Rev. Sci. Instrum. **91**, 103902 (2020); V. Grinenko *et al.* Nat. Phys. **17**, 748 (2021).

Chapter summary

- Carrying out μSR under pressure requires thick-walled pressure cells and high energy muons that can penetrate the walls and stop in the sample.
- This involves having sufficient sample mass, high momentum muons from a decay beamline, and carefully designed pressure cells.
- High pressure μSR can provide important insights into magnetism and superconductivity, and offers potential applications in many other areas of condensed matter physics.

Further reading

- M. I. Eremets, *High Pressure Experimental Methods*, OUP (1996).
- S. Klotz, *Techniques in High Pressure Neutron Scattering*, CRC Press (2013).
- W. Wang *et al.*, Rev. Sci. Instrum. **82**, 073903 (2011).
- R. Khasanov *et al.*, High Pressure Research **36**, 140 (2016).
- Z. Shermadini *et al.*, High Pressure Research **37**, 449 (2017).
- I. Watanabe *et al.*, Physica B **404**, 993 (2009).

Negative muon techniques

22

Although the vast majority of muon spectroscopy experiments use positive muons, **negative muons** also have their uses. The field of negative muon techniques is split into two main areas. The first is muon-spin relaxation **spectroscopy** using negative muons (μ^-SR), which typically uses the same spectrometers as for μ^+SR to detect the primary muon decay particles, which this case are electrons rather than positrons. The second area is **elemental analysis**, which is quite different in character, involving the detection of X-rays, rather than charged particles. These X-rays are emitted during the formation of muonic atoms and can provide important information about the atomic composition within different regions of a sample. In this chapter we meet both of these techniques.

22.1 μ^-SR spectroscopy

As we have seen, there have been a huge number of positive muon-spin rotation and relaxation (μ^+SR) studies on assorted materials, including antiferromagnets, ferromagnets, superconductors, semiconductors, battery and fuel cell materials, organic polymers, and biomaterials. Nevertheless, the μ^-SR technique is less common in materials science, mainly due to the fact that we need at least 36 times larger statistics[1] for μ^-SR than for μ^+SR. Indeed, although μ^-SR enjoyed a big community of users over 30 years ago, the number of published works has slowly decreased over the years (Fig. 22.1), where the above disadvantage might be considered the main reason behind this trend. A recent development is a new μ^- beam in J-PARC. This, together with the introduction of a new multi-detector counting system, has drastically increased the counting rate of μ^-SR to over 2×10^4 events/s (72 Mevents/hr). A comparable, but slightly lower counting rate (about 20 Mevents/hr) is also achieved in ISIS. As a result, a complete μ^-SR experiment can be achieved within a reasonable time period, meaning that the main problem with μ^-SR has largely been overcome in J-PARC and ISIS. A good μSR experimentalist should understand the advantages and disadvantages of both μ^+SR and μ^-SR, and use them accordingly. The following sections focus on μ^-SR and, in particular, its features and future prospects.

[1]The reason for this will be discussed shortly.

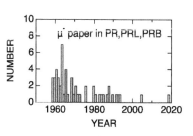

Fig. 22.1 The number of μ^-SR and related papers that have appeared in Physical Review, Physical Review Letters, and Physical Review B since 1960.

Comparing μ^+SR and μ^-SR

Just as for the positive muon, it is the parity violation in the decay of the weakly interacting pion that is responsible for the spin polarization of the negative muon. In this case the spin is parallel to the momentum instead of antiparallel. This allows us to measure internal magnetic fields in solids with $\mu^\pm$SR, even in a zero magnetic field (ZF).

μ^+SR

Let's remind ourselves of the main advantages of μ^+SR over some the other techniques:

- Both μ^+SR and μ^-SR provide information on local magnetic environments in solids caused by nuclear and electronic moments.
- The unique time window of μ^+SR (and μ^-SR) covers a gap of the time range between NMR and neutron scattering.
- A wide range of momentum can be used, enabling measurements on samples from thin films to bulk materials.
- Various sample environments are available, i.e. at temperatures from 20 mK to 1500 K, at pressures from ultra high vacuum to 2.3 GPa, and at magnetic fields from 0 to 9.5 T.
- Samples are not activated by irradiation with positive muons.

Like any experimental technique, there are disadvantages. These include:

- The muon sites are ambiguous, particularly at high temperatures. This is despite progress with the DFT$+\mu$ method (Chapter 16), which predicts the sites at 0 K. Above a certain temperature, a μ^+ starts to diffuse in a material, and we cannot yet reliably predict such temperatures.
- The implanted μ^+ may alter its local structural environment.
- The positive muon behaves as a light isotope of the proton ($m_\mu \approx m_p/9$) in solids. This leads to difficulties in observing H dynamics in solids with μ^+SR, because the muon probe itself is often moving when the H is moving.

Despite the above disadvantages, μ^+SR is widely used in condensed matter physics and solid state chemistry, especially because the first three advantages listed are unique compared to other techniques.

μ^-SR

The particular advantages of μ^-SR can be summarized as:

- The negative muons behave as a heavy electron ($m_\mu \approx 207 m_e$) in solids and are captured by a nucleus, so as to make a muonic atom. Thus, the μ^- site is very clear, i.e. at the lattice site.

- Such a muonic atom is stable, even above the chemical decomposition temperature of the target material for μ^-SR.

- μ^-SR can provide a useful tool for observing dynamics in solids.

However, μ^-SR also has the following disadvantages:

- At least 5/6 of the μ^- spin polarization is lost during the cascade of the μ^- from the outermost shell orbit to the inner orbits of a muonic atom. In order to measure the μ^-SR spectrum with the same reliability to that of μ^+SR, we need ≈ 36 times greater statistics and hence a much longer counting time.

- Since the lifetime of μ^- depends on the nucleus on which μ^- is captured, the μ^-SR spectrum consists of multiple components with different lifetimes (see Fig. 22.2). This means that the spectrum is very complex and, as a result, it is very difficult to analyse the μ^-SR spectrum with conventional μ^+SR methods.

- Even using the most intense negative muon beam, a large amount of sample is still needed to achieve an acceptable counting rate.

- When a nucleus captures a μ^-, it can be considered a nucleus with $Z - 1$.

We will discuss further the current status on these factors in the next section.

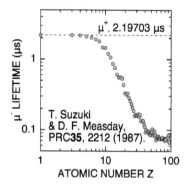

Fig. 22.2 The relationship between the lifetime of μ^- and the atomic number of the nucleus on which the μ^- is captured. [Data from T. Suzuki *et al.*, Phys. Rev. C **35**, 2212 (1987).]

Characteristics of μ^-SR data

Similar to the positive-muon case, μ^-SR deals with histograms. However when n different lifetime components are present in the μ^-SR case, the histogram is given by the sum of these components

$$N(t) = \sum_{i=1}^{n} N^i \exp(-t/\tau_i)[1 + A_i P_i(t)], \qquad (22.1)$$

where N^i is the scaling rate parameter of the ith component, τ_i is the lifetime of the ith component, A_i is the asymmetry of the ith component, and $P_i(t)$ is the muon spin polarization function of the ith component.

Example 22.1

Let's consider the case where $n = 2$. Then, the histograms for forward and backward counters are represented as

$$N_F(t) = N_F^1 \exp(-t/\tau_1)[1 + A_1 P_1(t)] + N_F^2 \exp(-t/\tau_2)[1 + A_2 P_2(t)],$$
$$N_B(t) = N_B^1 \exp(-t/\tau_1)[1 - A_1 P_1(t)] + N_B^2 \exp(-t/\tau_2)[1 - A_2 P_2(t)]. \qquad (22.2)$$

Here, we also define the following parameters

$$\alpha_1 \equiv N_F^1/N_B^1 \qquad \alpha_2 \equiv N_F^2/N_B^2$$
$$\beta^F \equiv N_F^2/N_F^1 \qquad \beta^B \equiv N_B^2/N_B^1. \qquad (22.3)$$

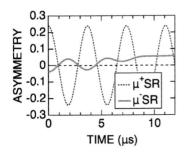

Fig. 22.3 Simulated μ^+SR and μ^-SR spectra in a transverse field with $H = 20$ Oe. For μ^+SR, $n = 1$, $A_1 = 0.24$, and $P_1(t) = 1$. In contrast, for μ^-SR, $n = 3$, $A_1 = 0.24 \times 1/6$, $P_1(t) = 1$, $A_2 = 0$, $A_3 = 0$, $\alpha_1 = 0.9$, $\alpha_2 = 0.85$, $\alpha_3 = 0.8$, $\tau_1 = 2.2$ μs, $\tau_2 = 1.8$ μs, and $\tau_3 = 30$ μs. The initial μ^+ spin direction is antiparallel to its momentum, while the initial μ^- spin direction is parallel to its momentum. As a result, the initial phase of μ^-SR is different from that of μ^+SR by 180°. In this figure, the logical detector definition for μ^+SR is used for defining asymmetry, i.e. forward is upstream and backward is downstream.

[2] J. Sugiyama *et al.*, Phys. Rev. Lett. **121**, 087202 (2018).

Assuming that the first component is predominant, i.e. $N^1 \gg N^2$, the asymmetry $A(t)$ is generally given by

$$A(t) = A_0 P(t) = \frac{N_F(t) - \alpha_1 N_B(t)}{N_F(t) + \alpha_1 N_B(t)}$$

$$= [A_1 P_1(t) + B(t)] \times C(t), \tag{22.4}$$

with

$$B(t) = \left[\frac{\beta^F - \beta^B}{2} + \frac{\beta^F + \beta^B}{2} A_2 P_2(t) \right] \exp\left(-\frac{t}{\tau_2} + \frac{t}{\tau_1} \right), \tag{22.5}$$

and

$$C(t) = \left\{ 1 + \left[\frac{\beta^F + \beta^B}{2} + \frac{\beta^F - \beta^B}{2} A_2 P_2(t) \right] \exp\left(-\frac{t}{\tau_2} + \frac{t}{\tau_1} \right) \right\}^{-1}. \tag{22.6}$$

This means that, even if $P_1(t)$ is time independent, the μ^-SR asymmetry spectrum exhibits a relaxation behaviour due to the lifetime difference between two components (Fig. 22.3). Besides $P_2(t)$, each parameter in eqns (22.5) and (22.6) is a time-independent constant. Therefore, if $P_2(t)$ is known, one can, in principle, fit the μ^-SR spectrum using muon analysis software.

Example 22.2

To illustrate how the internal field can be probed with μ^-SR, we shall discuss the example[2] of the candidate hydrogen storage material MgH_2. By following similar principles, the methods shown here can be applied to other systems. Key to this example is the fact that, since muonic hydrogen is free from electrostatic forces due to its charge neutrality, the muonic hydrogen in this system starts to move in the lattice. The captured μ^- is then transferred onto (or recaptured by) a nearby metal nucleus. This means that the μ^-SR spectra of metal hydrides mainly consists of the μ^- signal captured by the metal. If the atomic number of such a metal is smaller than that of Ca, and does not have a nuclear magnetic moment, the captured μ^- on the metal should provide information on the nuclear magnetic field at the metal site due to the surrounding hydrogen. This upper limit of atomic number is because the lifetime of the μ^- captured by Ca is 0.3327(15) μs ($=\tau_{Ca}$), and we need to measure the μ^-SR spectrum up to at least 5.5 μs to know the behaviour of a nuclear magnetic field. Since the number of the μ^- with τ_{Ca} at $t = 5.5$ μs decreases down to $\exp(-5.5/0.3327) \approx 7 \times 10^{-8}$, it is very difficult to complete the measurements for such μ^- within a reasonable time.

Figures 22.4 and 22.5 show the time histograms of the forward and backward counters [N^F and N^B] measured in TF, demonstrating the difference between the μ^+SR and μ^-SR spectrum for MgH_2. Because only one component exists for μ^+SR, the histogram decreases smoothly with time along with the TF oscillatory signal. In contrast, the histogram for the μ^-SR spectrum consists of five components with different lifetimes, which were assigned as μ^- captured by Mg, C, O, Pb, and an unknown component, probably due to residual contaminating particles in the beam line. The components from the μ^- captured by C and O come from the sample enclosure, which was made of PET, while the Pb component comes from the collimator used to adjust the spot size of the μ^- beam. As a result, the μ^-SR asymmetry spectrum shown in Fig. 22.6 exhibits a nonlinear background signal caused by the different α values for each component. This background is found to be independent of the applied magnetic field.

Hence, the μ^-SR asymmetry spectra were fitted in the time domain between 0.4 and 5.5 μs together with subtracting the nonlinear background (Fig. 22.7). Specifically, the ZF and LF spectra were fitted by combining a dynamic Gaussian Kubo-Toyabe function (G_{KT}^{Dyn}) and an exponential relaxation function

$$A_0 P(t) = A_{KT} G_{KT}^{Dyn}(t, \Delta, \nu, B_{LF}) + A_T e^{-\lambda_T t}, \tag{22.7}$$

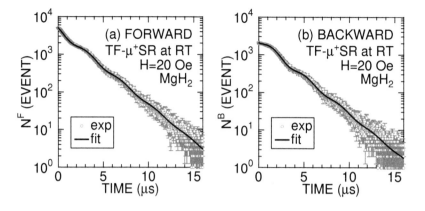

Fig. 22.4 Transverse field (TF) μ^+SR spectra of MgH$_2$ for the (a) forward counter and (b) backward counter. The magnitude of the TF was 2 mT.

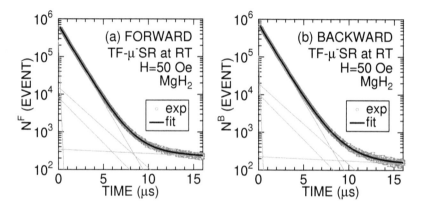

Fig. 22.5 TF-μ^-SR spectra in MgH$_2$ for (a) the forward counter and (b) the backward counter. The magnitude of the TF was 5 mT. Light grey lines indicate the five fitted components. [Taken from J. Sugiyama *et al.*, Phys. Rev. Lett. **121**, 087202 (2018).]

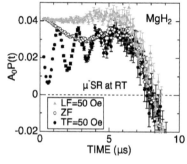

Fig. 22.6 TF, ZF, and LF μ^-SR asymmetry spectra for MgH$_2$ measured at room temperature.

where A_0 is the initial ($t = 0$) asymmetry, A_{KT} and A_T are the asymmetries associated with the signals from the μ^- captured on Mg observing H_{int} caused by nucleus and electron, respectively; Δ is the static width of the local field distribution at the disordered sites; ν is the fluctuation rate of the fields; and λ_T is the exponential relaxation rate. Δ [ν] corresponds to a spin-spin relaxation rate ($1/T_2$) [a spin-lattice relaxation rate ($1/T_1$)]. Using common A_{KT}, A_T, ν, Δ, and λ_T for the ZF and three LF spectra, the fit yielded $A_{KT} = 0.0123(4)$, $A_T = 0.0094(4)$, $\Delta = 0.520(7)$ μs^{-1} [equivalent to a field distribution width of $0.611(8)$ mT], $\nu = 0.10(3)$ μs^{-1}, and $\lambda_T = 0.40(2)$ μs^{-1}. Dipole field calculations predict $\Delta^{calc} = 0.5807$ μs^{-1} [0.6819 mT] at the Mg site, consistent with the fitted Δ from μ^-SR. Thus, we conclude that μ^- can be used to observe the internal field B_{int} at Mg. Because $\nu \approx \Delta/5$, B_{int} fluctuates even at room temperature. This is probably because of the thermal vibration or hopping of H, as proposed by H-NMR.

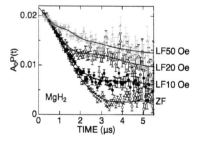

Fig. 22.7 ZF and LF μ^-SR asymmetry spectra for MgH$_2$ shown over the fitting range of 0.4 to 5.5 μs after subtraction of the nonlinear background. The spectra were fitted using eqn 22.7. [Adapted from J. Sugiyama *et al.*, Phys. Rev. Lett. **121**, 087202 (2018).]

The example of MgH$_2$ demonstrates how the negative muon spin rotation and relaxation (μ^-SR) technique can be used to study a fluctuating nuclear magnetic field in a solid. This shows that the dynamic behaviour of light elements (ions) in solids can be investigated with μ^-SR from the fixed viewpoint of the lattice site of the muonic atom. Considering the lifetime and the required time range for such measurements, the μ^- captured on light elements with $I = 0$, such as ^{12}C, ^{16}O, ^{24}Mg, ^{28}Si, etc. would be useful for such purpose. It is possible that combining μ^-SR and μ^+SR could clarify possible ambiguities found with μ^+SR data, thus using μSR itself to determine the limits of what can be done using a μ^+SR measurement.

22.2 Elemental analysis

In this section we shall explore the effects of charge rather than spin on implantation in a material. The muon has a mass ≈ 207 times that of an electron and so we can treat a negative muon as a heavy electron, as far as the energy level structure of the muonic atom is concerned. The formation of muonic atoms leads to the emission of high energy X-rays upon implantation of negative muons into any material. These X-rays are characteristic of the atom which captured the negative muon. Consequently the elements at the muon stopping site can be detected and, by using the ratio of the intensities, the composition can be determined at the depth of muon implantation. This makes compositional analysis using negative muons an extremely useful tool to determine the depth-dependent composition of any material.

Elemental analysis is the process of determining the composition of the material. There are many techniques that can determine this composition, such as X-ray fluorescence, scanning tunnelling microscopy, laser ablation, mass spectrometry, neutron diffraction, and neutron prompt gamma analysis. Some of these techniques are destructive, whilst others are only sensitive to the surface. Moreover, each technique has different sensitivities to composition and/or elements, so developing new techniques with unique and complementary characteristics is vitally important.

The unique characteristics of elemental analysis using negative muons allows for the depth dependence of composition to be determined. In addition and most importantly, this composition profile can be measured completely nondestructively. This technique has been applied to a diverse range of science, e.g. cultural heritage (Bronze Age artefacts, Mary Rose artefacts, swords, cannon balls, sundials, coins), solar cells, piezoelectric devices, Li batteries, and advanced engineering. Depth dependent imaging of the lateral elemental profile is also possible. In this section, we will review the key aspects of the technique, illustrated with some recent examples.

Implantation of the negative muon

When a negative muon is implanted into a material it is captured by an atom in the specimen under examination. In the capture process the final energy loss is transferred to Auger electrons. These are electrons that are ejected from the atom with energies of several kilo-electron volts. The muon is initially captured in a high energy state ($n = 5$) and then cascades down to the 1s state. This process is illustrated for Cu in Fig. 22.8.

The local environment dictates the exact details of the cascade, this may be physical or chemical. The emission energies for most elements have been measured by a series of experiments. These energies can be determined theoretically using a point nucleus approximation for the muon energy levels in the muonic atom, similar to that used for the electrons in a normal atom:

$$E_{n,j} = -\frac{m_\mu c^2}{1 + m_\mu/M}\frac{(Z\alpha)^2}{2n^2}\left[1 + \left(\frac{Z\alpha}{n}\right)^2\left(\frac{n}{j+1/2} - \frac{3}{4}\right)\right], \quad (22.8)$$

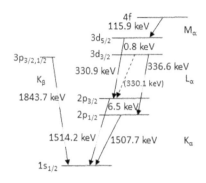

Fig. 22.8 An example cascade for copper.

where n is the principal quantum number, j is the quantum number for total angular momentum, m_μ is the muon mass, M is the mass of the nucleus, Z is the atomic number, and $\alpha = 1/137$ is the fine structure constant. So, for example, the predicted energy of the K_α emission can be derived from eqn 22.8 by taking the difference $E_{2,1/2} - E_{1,1/2}$. From eqn 22.8, it can be seen that the energy of the emitted muonic X-ray is proportional to the particle mass, therefore the emitted X-ray is some 207 times greater in energy than that of the normal X-ray fluorescence transition. This is a key point for measuring deep beneath the surface, as these muonic X-rays can vary from tens of kilo to mega electron volts. Figure 22.9 shows the measured spectrum from silver. This shows that many transitions are observed that can be identified.

The K_α emissions from various elements are shown in Fig. 22.10, illustrating that this technique is easily sensitive to all elements from lithium upwards and, most importantly, to elements next to each other on the periodic table. Finally, the muon will either decay into an electron and two neutrinos or be captured by the nucleus. If the negative muon is captured by the nucleus it can undergo a reaction with a proton, p:

$$\mu^- + p \rightarrow n + \nu, \quad (22.9)$$

where ν is a neutrino and n is a neutron. The neutron is emitted with an energy of 5.2 MeV for protons that are not tightly bound. However, for protons that are tightly bound a rapid conversion to a neutron occurs and the momentum transfer implies that the capturing element is in an excited state. As this excited state decays to a lower energy, one or more neutrons and γ rays are emitted. There should be the possibility of using this information to determine the isotope of the capturing element.

Multi-element capture

For a multi-element target material, the probability of each element capturing the muon depends on the electronic configuration, e.g. for a material with two elements Z_1 and Z_2 with composition $(Z_1)_m(Z_2)_n$ the

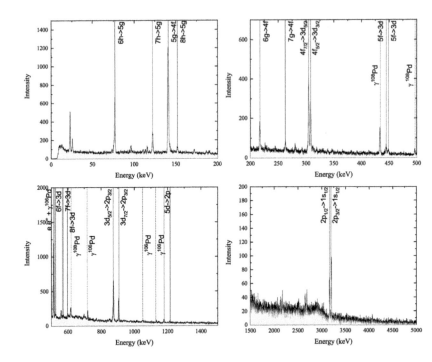

Fig. 22.9 The muonic X-ray spectra from silver. This shows the range of emissions that can be measured.

capture ratio for the elements is defined as

$$A\left(\frac{Z_1}{Z_2}\right) = \frac{nN_\mu(Z_1)}{mN_\mu(Z_2)}, \tag{22.10}$$

where $N_\mu(Z)$ is the proportion of muons captured on Z, which as an initial estimate is assumed to be proportional to the total number of electrons, i.e. proportional to Z. However, some inconsistencies have been discovered, in particular, with oxide materials. Models have been developed to account for these errors by introducing a correction term for the number of loosely bound electrons

$$\sigma = \left(1 - \frac{E_{\rm b}}{E_0}\right)^{1/2}, \tag{22.11}$$

where $E_{\rm b}$ is a binding energy for the electrons and E_0 is a cutoff that is of order 80 eV. This gives a significant improvement, but does not account for the valence electrons. The following equation, with its additional terms including density and valency, takes this into account

$$A\left(\frac{Z_1}{Z_2}\right) = 0.6\rho\left(1+\alpha\rho\right)\left(\frac{Z_2}{Z_1}\right)^{\frac{1}{8}}\left(1+5.53V^{5.45}\times 10^{-5}\right), \tag{22.12}$$

where ρ is the density (g/cm^{-3}), V the valency, and parameter α depends on the material. In the case of oxides $\alpha = -0.164$ when $Z_1 \le 18$

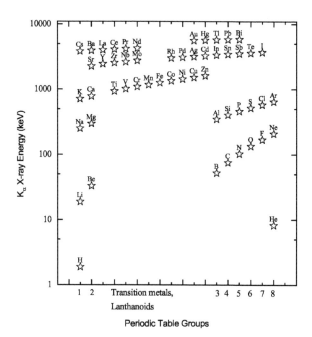

Fig. 22.10 The K_α X-ray emission energy for selected elements in the periodic table.

and $\alpha = 0$ otherwise. For chlorides $\alpha = -0.222$ when $Z_1 \leq 18$. However, in trying to determine quantitatively the exact composition of a material, it is prudent, if not essential, to measure known standards with similar chemical compositions.

As mentioned earlier, the muonic X-rays are emitted over a broad range of energies, from tens to thousands of kilo-electron volts. Therefore, we require a detector with a broad efficiency and sensitivity, along with a high resolution. Currently, high purity germanium detectors fulfil this requirement. Figure 22.11 shows the development setup currently used at the ISIS pulsed neutron and muon source. This shows four detectors with a range of performances and the sample in the centre. The flexibility of the setup allows for any shape/size of sample to be studied.

Another key aspect is utilizing a mono-momentum beamline setup (with a energy spread of usually 4%). By controlling the beamline momentum, the depth dependence of composition can be determined. This is shown in Fig. 22.12 in which a test sample with a known layered structure has been investigated.

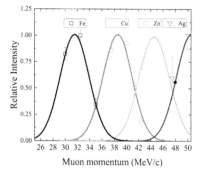

Fig. 22.11 The development instrument at the ISIS neutron and muon source.

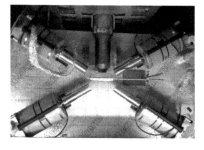

Fig. 22.12 The depth-dependent intensity of the muonic X-rays for a layered sample of iron/copper/zinc/silver.

Science examples

Over the past 50 years, there has been periodic use of negative muons for elemental analysis. In this section, we will review some recent examples illustrating the breadth of science covered by the technique.

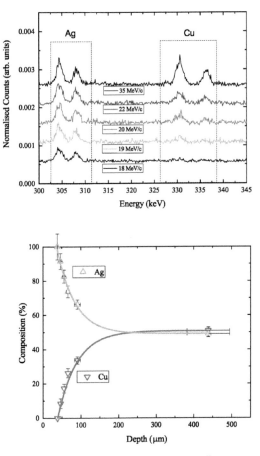

Fig. 22.13 Top: energy spectra of muonic X-rays for different momenta; the counts are normalised to the total number of muons and the spectra are vertically offset for clarity. Bottom: Percentage silver and copper composition as a function of depth. [Taken from B. Hampshire *et al.* (unpublished).]

Example 22.3

Cultural heritage. This is an area that has, in recent times, exploited the unique capabilities of compositional analysis to probe beneath the surface, reflecting the strong desire not to damage the artefact. Recently B. Hampshire *et al.*[3] have measured the depth dependence composition of a Roman coin. The measurement of debasement of coinage from a pure precious metal composition can show the fiscal health of the economy. In particular, Roman 'silver' coins could be heavily debased, containing up to 80% copper. The metal surfaces were deliberately enriched by a heat treatment and oxidization process and after this treatment the coins were struck. In the example of Hampshire *et al.*, a Roman Julia Domna (211-217 CE) coin was examined. The surface was found to be near 100% silver whereas the centre was near 50% silver and 50% copper (see Fig 22.13).

[3]B. Hampshire *et al.* (unpublished).

Example 22.4

Meteorites. Meteorite samples are rare and often need to be kept in an inert environment to ensure that no local chemical environment effects are observed. In a proof-of-principle experiment samples from Allende and Murchison meteorites were investigated.[4] These samples were sealed in quartz tubes, however, a significant signal from carbon suggested that the Murchison meteorite contains organic material (as well as Mg, Si, Fe, Ca, and S), whereas the Allende meteorite contained only Mg, Si, Fe, K, and Ca.

Example 22.5

Biomaterials. The investigation of biomaterials exploits the ability to observe low Z material. Riedy *et al.*[5] used tissue equivalent samples (Shonka plastic and TE liquid) to measure the composition of carbon, nitrogen, oxygen, fluorine, and calcium. The results from these experiments showed that values of 0.15% or 1500 ppm are measurable and that the compositions agree very well by those obtained from other techniques (see Table 22.1). An interesting note is the slight disagreement with the measured value of nitrogen and Reidy *et al.* suggested that this might be used to distinguish normal tissue, malignant and benign tumours.

Example 22.6

Piezoelectric materials. In the ferroelectric relaxor, $Pb(Mg_{1/3}Nb_{2/3})O_3$, the surface shows different structural properties to the bulk. Neutron diffraction shows an increase in strain, and energy-dependent X-ray diffraction shows different lattice parameters as the depth is increased. A Raman scattering study of the soft modes suggests that there may be a gradient in composition near the crystal surface, in particular, a gradient of Mg and Nb near the crystal surface. A negative muon experiment[6] has shown that there is no difference between Mg and Nb from 100–300 μm. However, below 100 μm a change in intensity has been observed.

Example 22.7

Isotope analysis. The energy of the K_α peaks changes, depending upon the isotope. For light Z this shift is too small to be measured with current detectors. However, for heavier Z elements, lead for example, a shift can be observed. Early experiments on lead did not show distinct peaks, but did show a broadening, however, more recent experiments[7] have shown a splitting and have managed to fit each peak associated with a different isotope (see Fig. 22.14). Moreover, on measuring natural lead, an abundance of $50.4 \pm 5.3\%$ for ^{208}Pb was measured, which is consistent with the typical isotopic abundance of 52.4% found in natural lead. Another possible method for isotopic analysis is to make use of the gamma emission from the nuclear capture of the negative muon.

[4]K. Terada *et al.*, Sci. Rep. **4**, 5072 (2014).

[5]J. Reidy *et al.*, IEEE Transactions on Nuclear Science **22**, 1780 (1975).

Element	Expected Value	Measured Value
Shanka Plastic		
H	n.d.	n.d.
C	87.9(3)	87.8(3)
N	3.76(6)	3.7(1)
O	6.65(4)	6.5(3)
F	0.85(2)	0.9(2)
N	0.9(2)	0.5(2)
TE liquid		
H	n.d.	n.d.
C	14.7(2)	14.1(6)
N	5.00(5)	3.9(2)
O	80.3(8)	81.9(6)

Table 22.1 A summary of tissue equivalent samples. [Taken from J. Reidy *et al.*, IEEE Transactions on Nuclear Science **22**, 1780 (1975).]

[6]K. L. Brown *et al.*, J. Phys.: Cond. Matter **30**, 125703 (2018).

[7]K. Ninomiya *et al.*, J. Radioanal. Nucl. Chem. **320**, 801 (2019).

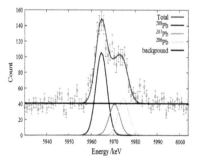

Fig. 22.14 The K_α peaks for a range of lead isotopes. this clearly shows the shift in the peak postions. [Taken from K. Ninomiya *et al.*, J. Radioanal. Nucl. Chem. **320**, 801 (2019).]

Example 22.8

Imaging. The resolution in terms of implantation depth along the beam axis is well defined by the momentum. However, the available resolution in the sample plane perpendicular to the beam is low, as the beam spot is relatively large. In order to measure an image of the sample plane, a high pixel density CdTe detector array can be used. This can either be placed directly behind the material in question or alternatively it can be used further away in conjunction with an aperture to form a pin-hole camera. An example setup and an obtained image[8] is shown in Fig. 22.15. Currently, these CdTe detectors are optimized for low-energy X-rays, so one can only utilize peaks with energies less than ≈ 150 keV.

[8] A. Hillier *et al.*, JPS Conf. Proc. **21**, 011042 (2018).

Chapter summary

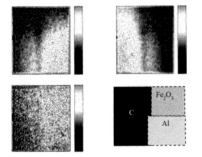

Fig. 22.15 The image reconstruction from a standard sample (bottom right). Top left shows Al. Top right shows C and the bottom left shows Fe and O. [Taken from A. Hillier *et al.*, JPS Conf. Proc. **21**, 011042 (2018).]

- Negative muons can be used to provide a μ^-SR technique that complements regular μ^+SR.

- Although μ^-SR presents many experimental challenges when compared to μ^+SR, it provides some unique benefits, in particular removing ambiguities about the exact muon site and eliminating the possibility of diffusive motion of the muon probe.

- Characteristic X-rays are emitted when muonic atoms are formed by atomic capture of negative muons. This forms the basis of a method for performing elemental analysis of materials using negative muons.

- Muon-based elemental analysis offers a non-destructive depth-tunable technique that that finds broad application across many areas of research in science and cultural heritage.

Further reading

- D. F. Measday, Physics Reports **354**, 243 (2001).
- R. Engfer *et al.*, Atomic Data and Nuclear Tables **14**, 509 (1974).
- E. Borie and G. A. Rinker, Rev. Mod. Phys. **54**, 67 (1982).

Exercises

(22.1) Derive eqns 22.4–22.6 from eqns 22.2–22.3.

(22.2) Calculate the energy of the muonic K_α emission for copper using eqn 22.8 and compare with the measured value shown in Fig. 22.8.

Part V

Complementary techniques

μSR is even more powerful when performed together with other experimental techniques that provide complementary information. This part describes those complementary techniques, both those that can be carried out 'at home' in the university laboratory and those that, like μSR, are performed at large facilities.

- μSR is very closely related to magnetic resonance, both using nuclei (nuclear magnetic resonance, or NMR for short) and electrons (electron spin resonance, or ESR for short). Other relevant bulk techniques can be performed on samples and include magnetization, susceptibility, and heat capacity. We review these techniques in Chapter 23. All of the techniques in this chapter are regularly performed in university laboratories.

- μSR is carried out in large international facilities, not in university laboratories, and this is also true for X-ray scattering (at least, using synchrotron sources) and neutron scattering. In Chapter 24 we review these techniques and show how the information they give can be complementary to that obtained by μSR.

23

μSR versus other resonance and bulk techniques

[1]We defer to Chapter 24 some other bulk experimental techniques, based on neutrons and X-rays, that can only carried out at large facilities; the experimental techniques described in this chapter (with the notable exception of β-NMR, which we will describe in Example 23.2, that is performed at TRI-UMF, a large facility) can be carried out in a university research laboratory.

In this chapter we contrast μSR with some other experimental techniques. First we outline the main advantages and disadvantages of muon spectroscopy compared to other **magnetic-resonance techniques**, which shows the complementarity among them. We shall introduce the fundamental quantities that are probed by the different techniques and then point out the main differences in the methods used to record the signal. We shall argue that μSR is not just another resonance method, but an experimental technique that, in many cases, is the best (or even the only) approach to solve certain problems. We then go on to describe Mössbauer spectroscopy, which is a related form of resonance spectroscopy. Finally μSR will be compared to **magnetization** techniques, again pointing out the complementary nature of the local and bulk methods, all of them necessary to accurately investigate a system.[1]

23.1 Magnetic resonance

We have described some of the ideas behind magnetic resonance in Chapter 7. The general concepts defined there apply to a wide range of related techniques. From an experimental viewpoint these techniques are naturally divided into two classes. There are those like electron spin resonance (ESR) and nuclear magnetic resonance (NMR) that exploit Faraday's law to measure the macroscopic ensemble magnetization of the spin probes (with samples inside cavities or coils); and also those that resort to particle-counting techniques, such as μSR, and β-NMR (described below). The complete polarization of the muon beam in a μSR experiment stands in stark contrast to the very weak polarization of the nuclei or electrons available in an NMR or ESR experiment. However, in all of these methods we study the time evolution of the spin polarization, or magnetization, no matter which probe is being used: muons, nuclei (stable or unstable), or electrons.

NMR is perhaps the most natural comparison with μSR since the muon, once implanted in a material, acts as a light proton before it decays. In the following example we will summarize the essentials of a pulsed NMR experiment to highlight the differences with μSR.

Example 23.1

In a pulsed NMR experiment, the magnetization M of the nuclear spin ensemble nutates (as discussed in Example 7.8) during an RF pulse, with the RF field $B_1 \perp B_0$. The pulse duration τ produces a total nutation angle $\phi = \gamma B_1 \tau$. The simplest example is when $\phi = \pi/2$ and the magnetization ends perpendicular to $\hat{z}$. The same coil that is used to apply the RF pulse, tightly wound around the sample, experiences an induced EMF due to Faraday's law as M undergoes a free precession around B_0 (known as the free induction). Large nutation angles require a resonant condition between the nuclear Larmor frequency and the RF pulse frequency, following the same principle illustrated in Fig. 7.4. Typically, efficient coupling to the coil in a resonant circuit is required both to produce large enough B_1 fields and to maximize the amplitude of the recorded free induction.[2]

This implies that an NMR experiment for a sample with a broad frequency spectrum, as is typically the case in the solid state, requires sequential data acquisitions while scanning either field or frequency. This contrasts with μSR, where the whole spectrum can be acquired in a single data acquisition. The measurement of T_1^{-1} and T_2^{-1} NMR relaxation rates also require distinct preparation conditions in the two cases and repeated data acquisitions in both, whereas in μSR all of the information is typically contained in a single measurement. On the other hand, complex RF pulse sequences allow the separation of interactions and the selective measurement of dynamic and static contributions to the relaxations, a feature that is not directly achievable with muons.

[2] Note also that in NMR, the static magnetic field may well be the hyperfine field at the nuclear site of a magnetically ordered sample in zero applied field.

From the local point of view, the muon's time evolution is determined by the same interaction Hamiltonian that determines the time evolution of proton spins in NMR experiments. We have introduced these notions in Chapter 4 and now generalize them slightly in terms of the following Hamiltonian:

$$\hat{\mathcal{H}} = \hat{\mathcal{H}}_Z + \hat{\mathcal{H}}_{en} + \hat{\mathcal{H}}_{nn}, \qquad (23.1)$$

where $\hat{\mathcal{H}}_Z$ is the Zeeman term describing the interaction with the external magnetic field $B_0 = B_0 \hat{z}$. The Zeeman interaction is necessary in NMR to generate the nuclear magnetization and hence the signal, but not in μSR, where the spin polarization is inherent in the muon production mechanism and it routinely makes zero-field experiments possible. The second term ($\hat{\mathcal{H}}_{en}$) represents the hyperfine interaction between the muon (or nuclear) spins and the electronic spins, which allows us to probe the electronic and magnetic properties of matter. Finally, $\hat{\mathcal{H}}_{nn}$ represents the dipole-dipole interaction between the muon (or nuclear) spin and the surrounding nuclear spins, which depends on the local structure and on the muon site. For nuclei with a spin $I > 1/2$, the interaction between the nuclear electric quadrupole moment and the electric field gradient generated by the surrounding charge distribution must also be included in $\hat{\mathcal{H}}_{en}$. This may be used as a probe of charge distributions and excitations in solids. We recall that although the $S = 1/2$ muon itself does not have an electric quadrupole moment, it may interact with higher spin nuclei that do, which allows the detection of nuclear electric quadrupolar level splittings, e.g. by means of the avoided level crossing μSR experiment (see Chapter 4).

Although the Hamiltonian for the muon and proton spins might be the same, it is notable that the initial state of the system will, in general, not be. Specifically, owing to the nature of their method of production, muon spins come almost completely spin polarized. The magnetization $\boldsymbol{M}$ of an ensemble of nuclei will be determined by Curie's law $\mu_0 M = CB_0/T$, meaning that the nuclear signal intensity is stronger at low temperature and at high fields.

Example 23.2

Another resonance technique related to NMR and μSR is β-**NMR** (short for β-detected NMR). As explained in Appendix E, β decay is the name for a particle- or nuclear-decay process proceeding via the weak interaction where an energetic electron or positron is emitted. Of course, muon β decay make μSR possible. The idea of β-NMR is very similar: unstable, spin-polarized nuclear spins are implanted in a sample where the nuclear spin interacts with the local environment, precessing in a magnetic field at a rate determined by the gyromagnetic ratio. In order to read out the information probed, we use the fact that the nucleus undergoes β decay, which results in an asymmetric distribution of decay products (i.e. emission preferentially along the direction of the spin polarization), whose detection allows us access to the spin polarization at the time of decay.

In contrast to conventional NMR, where the polarization of the nuclear ensemble is limited by Curie's law to very modest values, the large spin polarization that can be achieved in artificially produced beams of unstable nuclei allows very sensitive probing of small interactions. Different nuclei provide a choice of nuclear spin, gyromagnetic ratio, and lifetime, potentially allowing us to probe very much longer timescales to those accessed in μSR. Some common choices are compared with the positive muon in the table.

isotope	spin	lifetime (s)	$\gamma_{\mathrm{I}}/(2\pi)$ (MHz/T)
μ^+	1/2	2.2×10^{-6}	135.5
^{8}Li	2	0.842	6.30
^{11}Be	1/2	13.8	22.0
^{11}Be	1/2	122	10.8

The choice of unstable nuclei available is limited primarily by its lifetime that must not be too short on the timescale of spin dynamics, but short enough to allow a measurement. It further requires a large initial spin polarization and of high intrinsic asymmetry in the nuclear decay. A low mass of the nucleus offers the important advantage of limiting sample radiation damage, favouring the measurement of its pristine properties. One of the most often used nuclei is ^{8}Li, which has an intrinsic asymmetry of $1/3$ (identical to the muon) and, with its $S = 2$ spin, also possesses a quadrupole moment, enabling a broader range of interactions with solids.

Like μSR, the β-NMR experiment can be carried out in zero applied field or in a chosen applied field. Measurements of $1/T_1$ can be straightforwardly carried out by measuring the polarization as a function of time. It is also possible to excite the system with RF radiation in order to achieve resonances where, typically, the applied RF frequency is steadily scanned around the NMR frequency as time-integrated spectra are measured. An example of data from such an experiment is shown in Fig. 23.1.

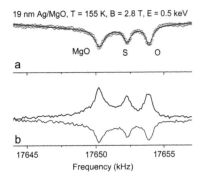

19 nm Ag/MgO, T = 155 K, B = 2.8 T, E = 0.5 keV

MgO   S   O

a

b

17645   17650   17655

Frequency (kHz)

Fig. 23.1 Resonances for ^{8}Li implanted into a 19 nm thick Ag film on MgO. The three resonances are due to Li ions implanted in MgO and in two distinct cubic sites in Ag (labelled S and O). The lower panel shows helicity-resolved data. [Figure taken from W. A. MacFarlane, Sol. State. Nuc. Mag. Res. **68–69**, 1 (2015).]

All of the terms in eqn 23.1 determine the time evolution of the muon-spin polarization, and of the NMR and β-NMR signals. Similarly, the time evolution of an ESR signal in a pulsed experiment is determined by the same terms, with the difference that the last term is replaced by the interaction among the electronic magnetic moments. Accordingly, the quantities to be measured by the different techniques are basically the same: the paramagnetic shift (Knight shift in metals) of the resonance line, the longitudinal relaxation associated with an exchange of energy with the local excitations described by T_1, the spin-lattice relaxation time, or the decay of the polarization in transverse field due to a modulation of the local field, associated for example with the dipolar interaction with the nuclei or with the field modulation induced by the flux lines lattice (FLL) in a superconductor. In spite of these similarities, the different experimental methods used to detect the signal can be more or less suitable to study a certain systems and give rise to a complementarity between the different techniques.

Example 23.3

We can now compare the main advantages and drawbacks of the two particle-detected techniques, μSR and β-NMR, compared to the standard resonance techniques, NMR, ESR, in the context of the topics explored in this book. Many of these points are explained further in the examples in this chapter.

- **μSR and β-NMR**
    - **Drawbacks**
        * Possible perturbation to the system (less important for muons).
        * Expensive.
    - **Advantages**
        * Single particle detection.
        * Large spin polarization allowing ZF experiments.
        * Beam penetration depth can be reduced down to few nm allowing thin-film experiments.
        * Small moment detection.
        * Access to much higher relaxation rates.
- **NMR and ESR**
    - **Drawbacks**
        * Electromagnetic wave detection problematic in metals and superconductors, particularly for ESR.
        * A large external magnetic field is usually required to generate even a low thermal-equilibrium spin polarization (for nuclei, typically 10^{-5} in a few tesla at room temperature).
        * For the most part, only bulk samples can be measured.
    - **Advantages**
        * Negligible perturbation.
        * Relatively low cost.
        * Many developments and applications.

The distinct time scales made accessible by each technique make the techniques complementary to each other. In particular, the finite lifetime of the muon does not allow us to detect slow relaxation processes (anything exceeding a few tens of microseconds). Extremely long times (e.g. tens of seconds) can be characteristic of weakly interacting systems, or of low-temperature phases where the density of excitations is low. This is not an obstacle for NMR. On the other hand, the deadtime of μSR experiments is significantly shorter than that of NMR experiments (from a few to tens of μs) allowing measurement of much shorter relaxation times.[3] Accordingly, while in NMR experiments we can hardly measure a relaxation rate $1/T_1$ larger than a few 0.1 μs^{-1}, in μSR experiments with continuous muon beams one can detect $1/T_1$ of tens of μs^{-1}.

[3]Keep in mind that the relaxation time is not directly the characteristic timescale of the dynamics under investigation, which can be much shorter than any deadtime (see e.g. Fig. 2.4).

23.2　When the muon is a plus

In this section we present some examples of cases where μSR proves to be more suitable than other resonance techniques. We shall refer in most cases to the comparison with NMR.

- **Good nuclei are missing**: There are several systems containing nuclei that are either not NMR active, that is, they have $I = 0$, or they have[4] a very small gyromagnetic ratio γ_n, or again a very small natural abundance N.

[4]The NMR signal amplitude at a given magnetic field is proportional to the so-called receptivity, scaling as $\gamma_n^3 I(I + 1)N$.

Example 23.4

CeRu$_2$ is an important heavy fermion superconductor whose nuclear isotopes are characterized either by a low abundance (Ce) or by a large hyperfine coupling (Ru). The Ru nuclei yield a rather fast relaxation, preventing the detection of the NMR signal by wipeout. Hence, it is not possible to investigate the superconducting state of this material with NMR. In β-NMR one can artificially introduce the nuclear probe, however the intrinsic properties these exotic superconductors are significantly affected by ion induced defects which may act as pinning centers for the FLL, while microwaves used in ESR can hardly penetrate the sample. Thus, μSR remains the best practical approach to investigate the local superconducting properties of this material.

- **Too fast for NMR**: One of the drawbacks of pulsed NMR is the length of the acquisition deadtime after the RF pulses, associated with magneto-acoustic coil vibrations[5] (known as ringing) which last at least a few μs. When the deadtime is on the order of magnitude or longer than the relaxation times the wipeout effect is complete and no measurement is possible. Despite the larger gyromagnetic ratio, and thanks to both weaker interactions and to much shorter deadtimes, μSR is complementary to NMR in many of these cases.

[5]See the book by Fukushima and Roeder listed at the end of the chapter.

Example 23.5

Wipeout typically happens in the paramagnetic phase of materials with a magnetically ordered ground state with the good NMR nuclei of the magnetic ions, such as many transition metals, rare earths, and actinides. The hyperfine interactions are so strong that even the fastest fluctuation limit remains in many cases unobservable. Critical fluctuations on the paramagnetic side of the transition are routinely observed by μSR.

Recall from Section 5.2 that the condition that leads to wipeout depends both on the strength of the coupling, represented by a Larmor frequency ω_0, and on the timescale τ of the dynamics involved. Let us call ω_M the maximum observable frequency of a given instrument. Wipeout roughly amounts to $\omega_0^2 \tau > \omega_M$ when $\omega_0 \ll \tau^{-1}$, and $\tau^{-1} > \omega_M$ when $\omega_0 \gg \tau^{-1}$. Beware that the second regime may also affect the muon polarization detection.

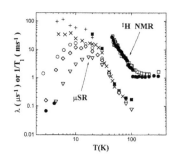

Example 23.6

Wipeout also takes place for nuclei of atoms subject to negligible magnetic coupling. Specifically, in a single-molecule magnet like Mn_{12}-acetate, spin fluctuations slow down on approaching the $S = 10$ ground state, causing a fast relaxation not only of the ^{55}Mn nuclei, but also of the ^1H nuclei. The detection of the proton NMR signal is prevented below 40 K, while μSR measurements allow us to accurately study the spin fluctuations, driven by the spin-phonon coupling, both in zero or in applied magnetic fields (see Fig. 23.2).

Fig. 23.2 Temperature dependence of ^1H NMR $1/T_1$ and of the muon longitudinal relaxation rate (λ) at different magnetic fields in Mn_{12}-acetate. For the NMR data $H = 0.33$ T (open squares) and 0.73 T (closed circles). For μSR the data refer to $H = 0$ (closed squares), 0.025 T (crosses), 0.1 T (X), 0.15 T (open circles), 0.2 T (open diamonds), and 0.37 T (open triangles). [Figure reproduced from A. Lascialfari *et al.*, Phys. Rev. Lett. **81**, 3773 (1998).]

We have seen in Section 7.2 that second order phase transitions involve the slowing down of the fluctuation rates that typically lead to very fast nuclear and muon relaxations on approaching the $\omega_0 \approx \tau^{-1}$ condition.

Example 23.7

In heavy fermion compounds such as $YbCu_{4.4}Au_{0.6}$, close to the quantum critical point (QCP) the fluctuations get particularly enhanced and cause a divergence of the relaxation rate on decreasing temperature. Hence the detection of the ^{63}Cu nuclear quadrupole resonance (NQR) signal is prevented below $T \simeq 1$ K. On the other hand, with μSR it was possible to investigate both the critical behaviour of the fluctuations close to the QCP (see Fig. 23.3) as well as the crossover to a Fermi-liquid phase upon applying a magnetic field.

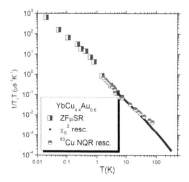

Fig. 23.3 Temperature dependence of the ^{63}Cu NQR $1/T_1$ (half circles), of the square of the macroscopic dc susceptibility (points) and of the muon zero-field relaxation rate (squares) in $YbCu_{4.4}Au_{0.6}$. The ^{63}Cu NQR $1/T_1$ and susceptibility data have been rescaled in order to match the μSR relaxation rate at 4.2 K. μSR allows us to probe the divergence of the relaxation rate at the quantum critical point. [Taken from P. Carretta *et al.*, Phys. Rev. B **79**, 020401(R) (2009).]

- **The advantage of true zero-field experiments** The applied magnetic field is an important tunable parameter in itself, granting access for instance to fluctuation rates in dynamical relaxation (see e.g. Chapter 8). However, often the magnetic field significantly perturbs the system: in ferromagnets it suppresses the second order phase transition; in single-molecule magnets it varies the energy level separation; in quasi-1D molecular conductors new field induced phases may appear; phase transitions, including quantum phase transitions, can be induced by applying a magnetic field, and so on. In these cases it is important to determine the intrinsic properties of the true ground state, which requires experiments in zero magnetic field. This type of experiment is straightforwardly performed with μSR thanks to the fully spin-polarized muon beam. High spin polarization is not usually the case in NMR and ESR where the magnetic field is required to achieve a non-zero initial nuclear or electron spin polarization. An exception is perhaps nuclear quadrupole resonance (NQR), only available when sensitive nuclear species with a non-zero electric quadrupole moment happen to be among the constituents of the sample.

- **Looking deeply inside metals and superconductors**: In metals and superconducting materials the radio-frequency field, used to control the nuclear magnetization in NMR experiments, penetrates only over a finite length. Besides causing a reduction of the intensity of the NMR signal this means that NMR allows us to probe only a finite depth from the surface, corresponding to the skin depth in a metal, or to the Campbell penetration depth in a superconductor, typically of the order of a few hundreds of nanometres. In the case of ESR, the microwave penetration is even lower. This may lead to artefacts in the determination of certain quantities as the second moment of the field distribution in the flux lines lattice of a superconductor. Moreover RF fields may also themselves drive the motion of the flux lines. On the other hand, in the case of μSR these drawbacks are absent since the signal is generated by the counting of positrons emitted throughout the whole sample volume. For this reason μSR is one of the most suitable techniques to study the superconducting phase of materials as described in more details in Chapter 9.

- **Weak order parameters**: In many systems quantum fluctuations are enhanced and lead to a suppression of the order parameter. This is for example the case for magnetic phases close to a quantum critical point. A reduced average magnetic moment is often hardly detectable by several techniques, such as neutron scattering and in some cases also NMR. However, μSR being typically sensitive to local fields of a few gauss can detect local moments with a magnitude even below 0.1 μ_{B}. Moreover, as mentioned above, if the experiments are performed in zero field there is no perturbation of the magnetic ground state and an accurate evaluation of the order parameter is possible (see Fig. 23.4).

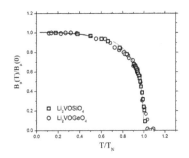

Fig. 23.4 Temperature dependence (normalized by the Néel temperature T_{N}) of the local field at the muon in $\mathrm{Li_2VOSiO_4}$ and in $\mathrm{Li_2VOGeO_4}$, normalized to the extrapolated zero temperature value. These zero-field experiments allow a precise characterization of the order parameter, in particular of the critical behaviour close to T_{N} and of its reduction due to spin waves for $T \to 0$. [Taken from P. Carretta *et al.*, Phys. Rev. Lett. **88**, 047601 (2002).]

[6]The effect was discovered in 1957 by Rudolf Mössbauer (1929–2011, Nobel Prize 1961).

23.3   Mössbauer spectroscopy

The **Mössbauer effect**,[6] consists of the recoilless decay of an unstable, but sufficiently long-lived nuclear isotope, that eventually leads to the emission of a γ-ray photon with extremely sharp energy definition. In a crystal lattice a sizeable fraction of these decays can be recoilless, since the linear momentum of the projectiles from the decay can be transferred to the crystal as a whole. (This is because crystal momentum $\hbar \boldsymbol{k}$ is conserved modulo $\hbar \boldsymbol{G}$, where $\boldsymbol{G}$ is a reciprocal lattice vector). In this way the energy distribution of the photon is not Doppler broadened by phonons (this broadening is typically ~ 100 meV) but only due to Heisenberg uncertainty arising from the relevant excited-state lifetime.

Example 23.8

The best-known Mössbauer nucleus is ^{57}Fe. A source contains ^{57}Co, a gamma-emitting, unstable species that decays into an excited state of the ^{57}Fe nucleus through a K-electron capture (the same process that we describe for negative muons in Chapter 22). This Fe isotope decays into its ground state via the γ cascade emission whose dominant path is shown in Fig. 23.5.

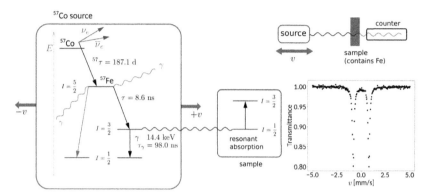

Fig. 23.5 The principle of the Mössbauer technique. Left: source; center: sample (absorber); top right: detection; bottom right: example of a spectrum.

The last photon in the cascade is the $h\nu = 14.431$ keV γ used in the detection and its lifetime corresponds to the remarkably small uncertainty $\Delta E/E \approx 2 \times 10^{-13}$ in the rest frame of the source lattice. This is the observed linewidth and it sets the timescale of the experiment. For comparison, it corresponds to the Doppler shift produced by a source moving at $v = 0.06$ mm/s (see Exercise 23.1). Therefore it is sufficient to place the source on the membrane of a vibrating loudspeaker to modulate the γ energy, with amplitude of order $10^2 \times \Delta E$. A Fe-containing sample, with the 0.02 natural abundance of ^{57}Fe, only absorbs a recoilless fraction of these γ resonantly if its energy exactly matches the nuclear excitation in the sample. A simple γ counter will record the transmittance versus the velocity v of the source, showing negative peaks when the absorption tales place.

With such a high intrinsic resolution one can easily observe local changes of the nuclear environment with respect to that of the source. They are due to the same interactions that lead to the NMR couplings and, to a large extent, to the μSR couplings. Typical examples are shown in Fig. 23.6. The transition on the left produces the photon at the source, those on the right absorb it at the sample. The first case on the right represents the change in the ground and excited nuclear states induced by the electron density at the nucleus, influenced by its chemical surroundings. This is called the isomer or chemical shift and is the analogue of NMR chemical shifts (also present for muons, but far too small to be measured). The second is due to the fact that the excited state has four spin $I = \frac{3}{2}$ levels, and is subject to a non-zero electric quadrupole. This produces the same splitting as observed in NMR, and only indirectly affects muons, as we saw in Example 4.20. In the Fe example the electric field gradient of the crystal splits the four levels into two pairs,

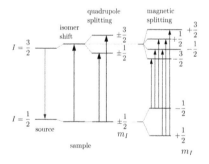

Fig. 23.6 Three types of change in the neighbourhood of the nucleus, producing the isomer shift, and the quadrupole and magnetic splittings.

according to the absolute value of the magnetic quantum number m_I, as shown in the figure, producing two distinct absorptions. The spectrum shown in Fig. 23.5 represents this case. Lastly, the rightmost case in Fig. 23.6 shows what happens if a hyperfine field is present at the nucleus. The Zeeman interaction splits the ground state into a doublet and the excited state into a quadruplet of levels. The optical transition selection rule $\Delta m_I = 0, \pm 1$ produces six lines in this case. The splitting is proportional to the hyperfine field at the nucleus, conceptually the same situation that we encounter for the muon.

We recall that the hyperfine coupling always exists for non-zero spin nuclei. However, in the paramagnetic state of a magnet at high temperature, the field at the nucleus fluctuates at a frequency J/h given by the exchange energy J, typically a few tens of meV. What kind of spectrum does Mössbauer then detect in a magnet above T_c? This is dictated by the lifetime τ_γ, which is roughly 100 ns in the example in Fig. 23.5. Therefore $\tau = h/J \ll \tau_\gamma$ and the absorbing state experiences the vanishing time average of the fast fluctuating hyperfine field. In the language of Chapter 8, if we assume a simple correlation function of the hyperfine field B_{hf} of $\langle B_{hf}(0)B_{hf}(t)\rangle = B_{hf}^2 e^{-t/\tau}$, then when $\tau \ll \tau_\gamma$ (which occurs at temperatures above the phase transition[7]) the ^{57}Fe spectrum changes from a sextet of lines to a doublet.

Besides ^{57}Fe, there are dozens of gamma ray source nuclei in the periodic table, with roughly three times as many Mössbauer-active nuclei (absorbers). Other commonly used ones are ^{119}Sn and ^{151}Eu. However, not many gamma sources have conveniently long lifetimes. Very unstable Mössbauer isotopes (for instance ^{61}Ni, ^{155}Gd, ^{161}Dy) can be produced at special isotope-separation facilities, such as ISOLDE at CERN, RIBF at RIKEN, or ISAC at TRIUMF. Mössbauer was a rather common laboratory technique in the past, partly because of its simple, low cost apparatus. Nowadays few laboratory spectrometers have survived, but the unstable isotope sources can be replaced by highly monochromatised synchrotron radiation. At present, Mössbauer facilities are operated at, for example, ESRF (Grenoble, France), APS (Argonne, USA), and SPRING-8 (Sayo, Japan). The disappearance of many lab-based spectrometers has been compensated for by the availability of dedicated beamlines with much shorter acquisition times.

23.4 Bulk techniques

Many condensed matter experiments are typically carried out in stationary, homogeneous systems, whose response to an external perturbation neither depends on when, nor on where it is applied inside the sample. Accordingly, if one applies a perturbing field $\boldsymbol{H}(\boldsymbol{q}, \omega)$, the system will respond with a response function[8] $\chi(\boldsymbol{q}, \omega)$.

In case of bulk techniques this means that, since the perturbation (e.g. the magnetic field in a magnetometry experiment) is constant over the sample volume, the response function is probed at $\boldsymbol{q} = 0$, namely

[7]This implies that below T_c the spectrum measures splittings proportional to the average of the hyperfine field over fast excitations (spin-waves or critical fluctuations), i.e. they reproduce the temperature dependence of the order parameter, as in μSR.

[8]See Chapter 7 and the book by R. White.

it is the uniform susceptibility $\chi(0, \omega)$. The value of the frequency ω is set by the experimental condition; for example, it is zero in a d.c. magnetization measurement and it may be in the kHz or MHz range in an ac susceptibility measurement.[9] On the other hand, local probes like the muons are sensitive to the local field fluctuations. Since this local field may be decomposed in its Fourier components

$$\boldsymbol{h}(t) = \sum_i A_i \boldsymbol{S}_i(t) = \sum_{i,\boldsymbol{q}} A_i e^{i\boldsymbol{q}\cdot\boldsymbol{r}_i} S_{\boldsymbol{q}}, \qquad (23.2)$$

with A_i the hyperfine coupling with the spin S_i at the ith site. One immediately realizes that with μSR we are probing the q-integrated dynamic spin susceptibility. In fact, when the longitudinal muon spin relaxation λ is associated with spin-lattice relaxation processes one can write

$$\lambda = \frac{\gamma^2 k_{\mathrm{B}} T}{2\hbar} \frac{A^2}{N} \sum_{\boldsymbol{q}} \left[\frac{\chi''(\boldsymbol{q}, \omega)}{\omega} \right], \qquad (23.3)$$

with $\chi''(\boldsymbol{q}, \omega)$ the imaginary part of the dynamic spin susceptibility at the muon Larmor frequency ω and at the wavevector $\boldsymbol{q}$ and N is a constant. Here A is the hyperfine coupling which, for simplicity is taken as q independent. This expression is an extension of what we derived in Chapter 7. Hence, local probe and bulk techniques allow the determination of complementary information when spin correlations (described by a q-dependent susceptibility) are present.

[9] Although these considerations apply to any response function, in the following we will concentrate on the spin susceptibility. We further note that in many glassy magnetic systems one can measure a difference between the static ($\omega = 0$) susceptibility of a sample that has been cooled in zero-field and one that has been cooled in a field, often below some characteristic temperature where this glassy (history-dependent) behaviour becomes more pronounced. Such an effect can then be followed up with a.c. susceptibility ($\omega \neq 0$) to see how the system responds as a function of temperature to a faster rate of change the applied field. μSR can then be useful for exploring such effects at higher frequency.

Example 23.9

A paradigmatic example is provided by single molecule magnets, such as $[\mathrm{TbPc_2}]^0$. Figure 23.7 shows the characteristic correlation times as determined by a.c. susceptibility measurements at different magnetic fields, and by μSR longitudinal relaxation measurements at $\mu_0 H = 1$ kG. We assume that the complex dynamics of the system are described by a dispersion of correlation times for the spectrum of the q modes of the system. Susceptibility probes the much longer time components of the spectrum, since only $q = 0$ modes are probed, while λ probes the q-integrated susceptibility.

At high temperatures a thermally activated trend is observed, followed at low temperatures by a quantum tunnelling regime, producing a plateau in τ_c. It is quite evident that the values of τ_c derived from the μSR relaxation rate are shorter than the ones derived from a.c. susceptibility at the same magnetic field. This is due to the fact that more q modes contribute to the former quantity.

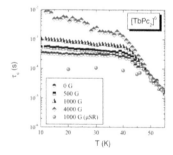

Fig. 23.7 Comparison of the temperature dependence of the correlation times by a.c. susceptibility and μSR. [Taken from F. Branzoli *et al.*, Phys. Rev. B **82**, 134401 (2010).]

The static uniform susceptibility $\chi(\boldsymbol{q} = 0, \omega = 0)$ can be measured directly by means of bulk magnetization measurements and also by μSR Knight shift measurements So, what is the advantage of the more expensive μSR experiment? First of all, in Knight shift measurements one measures the *local* static uniform spin susceptibility. Here *local* refers to the fact that one is probing only the susceptibility associated with those electron spins that are coupled to the muon. This means that the Knight shift is selective and is not affected by extrinsic contributions (e.g. those due to small amounts of impurity phases) as magnetization

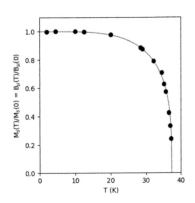

Fig. 23.8 Reduction of the sublattice magnetization M_s with temperature in CoF_2, as deduced by zero-field μSR. The very small initial reduction up to $T/T_N = 0.4$ is typical of the large exchange anisotropy in this antiferromagnet. Taken from R. De Renzi *et al.*, Phys. Rev. B **30**, 197 (1984).

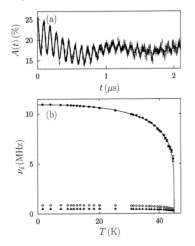

Fig. 23.9 (a) Muon spectra and (b) the fitted precession frequencies for NaFeAs. In this compound there are three muon sites, but although there is a different coupling between each muon site and the order parameter, all three frequencies follow the same temperature dependence. After D. R. Parker *et al.* Phys. Rev. Lett. **104**, 057007 (2010).

measurements very often are. Moreover, in sufficiently pure samples, the plot of the Knight shift versus the macroscopic d.c. susceptibility (the Clogston-Jaccarino plot with temperature as an implicit parameter) directly provides the hyperfine coupling tensor value as the proportionality constant between the two quantities.

In the magnetically ordered phase, the complementarity between local and bulk techniques emerges whenever the order parameter is characterized by a wavevector $Q \neq 0$ as is the case, for example, in an antiferromagnet (AF). While bulk magnetization measurements cannot access the sublattice magnetization, corresponding to the order parameter of the AF, the muon in zero field will experience a local field proportional to the sublattice magnetization (see Fig. 23.8 and Fig. 23.9 for examples with one and several muon sites respectively).

μSR has a particular advantage over bulk techniques when studying low-dimensional magnetic systems, as the following example shows.

Example 23.10

Consider, for example, a chain of antiferromagnetically coupled spins. This is a one-dimensional system and so will not order at any non-zero temperature, but its magnetic correlation length ξ will grow on cooling, and this is accompanied by a broad peak in the heat capacity centred on a temperature $\approx J/k_B$, where J is the exchange constant, reflecting the fact that the entropy of the spins decreases as they start to correlate. Note however that even as they start to correlate on cooling, their dynamics will be too fast to pick up using μSR. In a crystal containing such antiferromagnetically coupled chains there will likely be a small interchain exchange interaction J' and this will result in three-dimensional long-range ordering at T_N, but since $J' \ll J$ we will have $T_N \ll J/k_B$. The three-dimensional ordering transition is however associated only with a rather small change in entropy and thus gives rise to a tiny peak in the heat capacity, the size of which decreases as $|J'/J|$ decreases, and it is completely dwarfed by the broad maximum that occurs around J/k_B that is associated with the correlations. The argument works for quasi-two-dimensional magnetic systems as well as quasi-one-dimensional magnetic systems. This effect can be demonstrated in Monte Carlo simulations for both one- and two-dimensional magnetic systems and implies that identifying three-dimensional ordering in very anisotropic magnets using heat capacity (or indeed any thermodynamic probe) can be challenging. In contrast, the transition from a non-long-range ordered state, even one with dynamic correlations of large spatial extent, to a 3D long-range ordered state is rather straightforward using μSR, since the oscillatory signal that results from magnetic order is incredibly easy to spot.

An example of this is shown in Fig. 23.10, which plots heat capacity and μSR data for the organic coordination polymer $[Cu(HF_2)(pyz)_2]BF_4$ (where pyz=pyrazine). The broad peak in the heat capacity [Fig. 23.10(a)] is associated with the antiferromagnetic correlation of spins in planes of magnetic Cu ions with an exchange constant of $J/k_B = -2.85$ K. However the long-range ordering of the spins which occurs at 1.54 K (indicated by the vertical dashed line), and is driven by the weaker coupling between the planes, is barely visible in the heat capacity data since it is associated with very little thermodynamic weight (the spins have already begun to correlate so there is little entropy left to expel). However, the muons still see rapidly fluctuating spins until long-range order sets in, at which point the precession signal (here comprising two frequencies) is unmistakeable and allows the order parameter to be followed with ease [Fig. 23.10(b)].

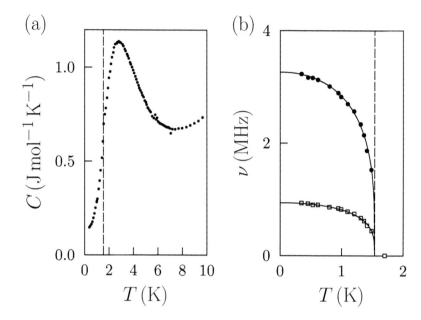

Fig. 23.10 (a) Heat capacity and (b) μSR precession frequencies measured in [Cu(HF$_2$)(pyz)$_2$]BF$_4$ (where pyz=pyrazine). [Adapted from J. L. Manson *et al.*, Chem. Commun. 4894 (2006).]

Sometimes, a commonly used experimental technique, aiming at extracting a bulk property, can be strongly influenced by surface effects. A paradigmatic example is the determination of the lifetime of photo-induced carriers in silicon. Results from techniques, such as photoconductance and luminescence, often used to measure the lifetime, depend significantly on the surface quality and the lifetime of the carriers at the surface can be significantly different from the one in the bulk. On the other hand, since the muons penetrate deeply inside the sample they are sensitive to the bulk lifetime only. In fact, Yokoyama *et al.* (Section 19.3) have shown that the longitudinal relaxation in Si significantly increases upon light irradiation with a power-law dependence on the photoelectron carrier density. Hence, by following the behaviour of the muon relaxation once the light is switched off, it is possible to determine the bulk lifetime of the carriers.

The study of dynamics with μSR or with bulk techniques may also lead to quite different results. For example, the ionic diffusion can be studied with d.c. conductivity measurements which allow one to measure the drift of the ions generating a current between two electrodes. This means that if local ionic hopping motion between different lattice sites is taking place without causing a global drift of the ions between the electrodes, d.c. conductivity measurements will not detect it. However, since the local hopping motions will cause fluctuations in the magnetic field at the muon sites, they will affect the relaxation rates and allow the characteristic correlation time for the random hopping of the ions to be derived.

A similar situation occurs when comparing the flux-line dynamics in a superconductor detected by means of resistivity measurements and by means of μSR. Here the dissipative motion detected by transport measurements is associated with the drift of flux-line bundles throughout the superconductor volume, driven by the Magnus force generated by the Lorentz force acting on the electron pairs. On the other hand, from the narrowing of the μSR spectra we detect the local field fluctuations induced by the flux-line dynamics, which are present even without applying an external current.

Finally, we note that while bulk techniques and μSR often give complementary information, in many cases the bulk techniques (a class which should also include structural characterization, electrical, and thermal conductivity, and many other methods in addition to the ones described here) provide the necessary and unavoidable preliminary characterization of the system to be investigated, before a more complex and powerful technique such as μSR is applied.

Chapter summary

- μSR is similar in many respects to other spin resonance techniques, each of which has access to a particular range of time scales. μSR has the advantage of a high intrinsic sensitivity owing to full spin polarization, but is relatively expensive.

- Magnetometry gives the average response to a field applied across the entire sample. It is far cheaper than μSR and is generally a necessary step in the characterization of a magnetic material.

Further reading

- J. A. Weil and J. R. Bolton, *Electron Paramagnetic Resonance. Elementary Theory and Practical Applications*, John Wiley and Sons (2007).

- P. Carretta in *NMR-MRI, μSR and Mössbauer Spectroscopies in Molecular Magnets*, eds. P. Carretta and A. Lascialfari, Springer-Verlag (2007).

- E. Fukushima and S. B. W. Roeder, *Experimental Pulsed NMR: A nuts and bolts approach*, Addison-Wesley (1981).

- R. R. Ernst, G. Bodenhausen, and A. Wokaun, *Principles of Nuclear Magnetic Resonance in One and Two Dimensions*, OUP (1987).

- W. A. MacFarlane, Solid State Nucl. Mag. Reson. **68–69**, 1 (2015) [a review of β-NMR].

- U. Gonser (ed.), *Mössbauer Spectroscopy*, Springer (1975).

- Y. Yoshida and G. Langouche, *Modern Mössbauer Spectroscopy*, Springer (2021).
- C. V. Topping and S. J. Blundell, J. Phys.: Condens. Matter **31**, 013001 (2019) [a review of a.c. susceptibility].
- R. M. White, *Quantum Theory of Magnetism*, Springer-Verlag (1983).

Exercises

(23.1) Calculate the velocity required to Doppler-shift a Mössbauer source of an amount equal to the ^{57}Fe intrinsic linewidth, given in Example 23.8 ($\Delta E/E \approx 2 \times 10^{-13}$).

<table>
<tr><td>

24

</td><td>

X-rays, neutrons, and μSR

</td></tr>
</table>

[1]The energy of a photon in keV is given by $E = 12.3984/\lambda$ where λ is the wavelength in Å.

(a)

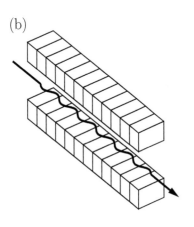

(b)

Fig. 24.1 (a) Synchrotron emission of X-ray radiation is confined to a cone of semiangle $1/\gamma$ for an electron in a circular orbit with velocity v [and $\gamma = (1 - v^2/c^2)^{-1/2}$]. (b) Schematic of a wiggler in which an electron is subjected to an alternating magnetic field.

Muon sources are frequently located close to sources of neutrons and X-rays. For example, the SμS muon facility at the Paul Scherrer Institute (PSI) is adjacent to SINQ (neutrons) and the Swiss Light Source (SLS, X-rays), while at the Rutherford Appleton Laboratory, the ISIS facility is a source of both pulsed muons *and* neutrons, and the Diamond Light Source (X-rays) is on the same site. The previous chapter has examined how muons can be combined with experimental techniques that are likely to be available in a university laboratory. This chapter focuses on the connections that can be made between μSR, and both **neutron** and **X-ray scattering** techniques, particularly important given that these techniques tend to be available at the same large international facilities.

24.1 X-rays

X-rays are photons with an energy in the range of about 1–100 keV. This means that they have a wavelength of roughly[1] 12–0.12 Å. Thus the wavelength of X-rays matches quite well with an interatomic spacing. While laboratory sources produce X-rays by accelerating electrons onto a metal target, much brighter beams of X-rays are available at synchrotron sources. At such sources, electrons are accelerated by a linear accelerator (LINAC) and, after further acceleration by a synchrotron, injected into a storage ring. On their path around the ring, the electrons will emit synchrotron radiation when accelerated [Fig. 24.1(a)] and to optimize the emission they pass through **wigglers**, regions of alternating magnetic field in which the electrons are shaken from side to side [Fig. 24.1(b)], producing a brilliant beam of X-ray synchrotron radiation which can be fed into an experimental area.

When an X-ray interacts with a sample, the dominant effect is that the X-ray is scattered by the electrons (Thomson scattering). The interference between the scattered waves from different atoms produces a diffraction pattern whose form is the Fourier transform of the electron density distribution. Thus if the X-ray is scattered from momentum $\hbar \mathbf{k}_i$ to $\hbar \mathbf{k}_f$, resulting in a wave vector change $\mathbf{Q} = \mathbf{k}_f - \mathbf{k}_i$, then the elastic scattering cross section can be written

$$\frac{d\sigma}{d\Omega} \propto \sum_{\mathbf{G}} \delta(\mathbf{Q} - \mathbf{G})|F(\mathbf{Q})|^2, \tag{24.1}$$

where

$$F(\boldsymbol{Q}) = \sum_i f_i(\boldsymbol{Q}) e^{i\boldsymbol{Q}\cdot\boldsymbol{r}_i}, \qquad\qquad (24.2)$$

and the sum is over atoms in the unit cell, f_i is the form factor of atom i (which is proportional to the atomic number Z), and $\boldsymbol{G}$ is a reciprocal lattice vector. Thus the scattering only occurs when $\boldsymbol{Q}$ matches a reciprocal lattice vector. This effect gives rise to the field of X-ray crystallography, originating in the work of von Laue and the Braggs in the early twentieth century. By the mid-twentieth century this technique had matured to the point where the structure of large molecules, such as haemoglobin and DNA, could be inferred from this technique. When a new material is synthesized in a chemistry lab, the solving of the structure using X-ray crystallography is often the first step in its characterization. The very bright synchrotron sources now available allow the structure of very complex structures to be determined within a fraction of a second, so that many users of these facilities are now biologists rather than physical scientists.

Recently, people have started to exploit the weaker processes in the scattering which are sensitive to the spin and orbital components of the moments in the sample. These processes are 10^5–10^7 times weaker than Thomson scattering, and so require a synchrotron source. The great advantage of looking at these weaker scattering processes is that it is possible to separate orbital and spin magnetization. Moreover, the highly focused beams allow very small samples to be studied and the high wave vector resolution can be useful for studying rather subtle changes in spin structure. A major challenge though is in interpreting these experiments since understanding the cross-section and structure factors of all the complicated light-matter interactions involved in these experiments is highly non-trivial.

Magnetic scattering is a very weak process and so X-rays were not used in studies of magnetism until rather recently. The game-changer was the discovery of resonant magnetic scattering[2] which showed that an enhancement of several orders of magnitude of the magnetic scattering intensity occurs when the photon energy is tuned close to the absorption edge of one of the elements inside the sample. This gives the possibility of element specificity, i.e. the ability to see which atom in your sample you are probing. In addition, the resonant process is sensitive to electronic states with specific orbital character, and this has led to experiments which probe charge, spin, and orbital degrees of freedom. An important recent development in this context is Resonant Inelastic X-ray Scattering (RIXS), a technique in which both the energy and the momentum change of the scattered X-ray photon are measured. A key advantage of the technique is that it is element specific, because the incident X-ray energy is tuned to the binding energy of a core level of a particular element in the sample. The inelastic scattering of X-rays is a very weak process, and so the effect would not show up with a laboratory X-ray setup; RIXS experiments require a high-brilliance X-ray source and so can only be performed at synchrotron radiation sources.

[2]D. Gibbs *et al.*, Phys. Rev. Lett. **61**, 1241 (1988).

[3]The mass of a neutron is $m_n = 1.674927 \times 10^{-27}$ kg, a little larger than the proton mass $m_p = 1.672622 \times 10^{-27}$ kg.

[4]An example of a reactor source is the Institut Laue Langevin (ILL) in Grenoble; an example of a spallation source is ISIS in Oxfordshire.

24.2   Neutrons

The neutron is an uncharged particle with a mass[3] very similar to that of a proton, and so about nine times larger than the mass of a muon. Beams of neutrons are available at reactor sources (as a consequence of nuclear fission) or at spallation sources (as a consequence of bombarding a heavy metal target with a high energy proton beam).[4] Free neutrons have a lifetime of 880 s, which is much longer than any typical experimental timescale for a single neutron, and so in contrast to µSR we do not need to worry about the neutron lifetime. The neutron beam is moderated by passing through a medium such as water, liquid methane, or graphite, the temperature of which fixes the distribution of the neutron flux (see Fig. 24.2). A wavelength of just under 2 Å is obtained for a moderator at room temperature (and corresponds to neutrons travelling at a little more than $2 \, \mathrm{km \, s^{-1}}$).

Because the neutron is uncharged, it interacts only weakly with matter and penetrates a few centimetres into most materials. The main scattering mechanism is from the nucleus of the atoms in the sample. Because the nucleus is very small compared to the electron clouds, it is possible to measure diffraction with much larger wave vector than with X-rays. As for X-rays, the scattering cross section is given by $d\sigma/d\Omega \propto \sum_G \delta(\boldsymbol{Q} - \boldsymbol{G})|F(\boldsymbol{Q})|^2$, where this time

$$F(\boldsymbol{Q}) = \sum_i b_i e^{i\boldsymbol{Q} \cdot \boldsymbol{r}_i}, \tag{24.3}$$

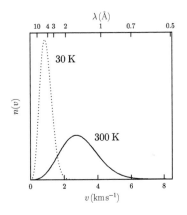

Fig. 24.2 The distribution of neutrons from moderators fixed at two different temperatures as a function of the neutron velocity v (or wavelength $\lambda = h/m_n v$ on the top axis).

where b_i is the scattering length of atom i. The scattering length depends on the interaction between the neutron and the nucleus of an atom, and has a much more irregular dependence on the atomic number Z than for X-rays. Neutrons are very sensitive to light atoms (which are harder to see with X-rays), particularly hydrogen and are also sensitive to different isotopes (which have the same Z).

The neutron may not have charge, but it does have spin. It is a spin-$\frac{1}{2}$ particle (like the muon) and its small magnetic moment makes it sensitive to ordered magnetic moments in the sample.[5] The magnetic scattering structure factor is given by

$$\boldsymbol{F}_{\mathrm{M}}(\boldsymbol{Q}) \propto \sum_i \boldsymbol{m}_i f_i(\boldsymbol{Q}) e^{i\boldsymbol{Q} \cdot \boldsymbol{r}_i}, \tag{24.4}$$

[5]Magnetic scattering from neutrons is from the spin-polarized electrons in the sample, and so this does not persist as far out in wave vector as the nuclear scattering. This effect of the magnetic form factor (essentially the Fourier transform of the magnetic electron density around a magnetic atom) cutting off the high-Q scattering can be useful for distinguishing between magnetic and phonon inelastic signals).

[6]$\boldsymbol{F}_{\mathrm{M}}^\perp(\boldsymbol{Q}) = \hat{\boldsymbol{Q}} \times (\boldsymbol{F}_{\mathrm{M}}(\boldsymbol{Q}) \times \hat{\boldsymbol{Q}})$ which leads to

$|\boldsymbol{F}_{\mathrm{M}}^\perp|^2 = \sum_{\alpha\beta}(\delta_{\alpha\beta} - \hat{Q}_\alpha \hat{Q}_\beta)F_{\mathrm{M},\alpha} F_{\mathrm{M},\beta}^*.$

where $\boldsymbol{m}_i$ is the magnetic moment of atom i, and the scattered intensity turns out to be proportional to $|\boldsymbol{F}_{\mathrm{M}}^\perp(\boldsymbol{Q})|^2$ where $\boldsymbol{F}_{\mathrm{M}}^\perp(\boldsymbol{Q})$ is the component of the structure factor perpendicular to the scattering wave vector.[6] Thus neutron magnetic diffraction experiments can be enormously valuable in determining magnetic structures.

Most experiments are carried out with unpolarized neutrons, but it is possible to select out neutrons with a particular polarization whose polarization can then be manipulated with a spin-flipper. The polarization of the scattered beam can then be controlled, and then both the spin-flip and non-spin-flip scattering cross sections can be measured, allowing the nuclear and magnetic scattering to be separated.

Small-angle scattering or reflectometry studies can be performed to study films and multilayers. When the scattering angle is small, the component of momentum transferred to the sample is small and thus the corresponding wavelength is large, resulting in sensitivity to structures ranging from a nanometre to several hundred nanometres.[7] Such measurements are highly complementary with low-energy muon experiments and it is not uncommon to see studies which combine X-ray and neutron reflectometry with low-energy muons. Small-angle neutron scattering (SANS) has been extensively used to study the structure of superconducting vortex lattices and this technique is also complementary to the muon experiments described in Chapter 9.

Neutrons are particularly effective for studying excitations in materials, particularly because the thermal neutron energies (typically 1-100 meV)[8] are well matched to typical energies of phonons and magnons. Inelastic neutron scattering is widely used to study phonon and magnon dispersions, and has provided a lot of important information about unusual excitations in quantum magnets. Because the neutrons are highly penetrating, they are able to probe the bulk of a sample but because the inelastic scattering is rather weak it is often necessary to have a very large single crystal sample to do a careful experiment. This is an area where RIXS has the potential to be extremely effective since X-ray beams can be focused down on to much smaller samples (and RIXS also has orbital sensitivity) but neutrons currently win on instrumental resolution.

[7]X-ray reflectometry is also a very valuable technique, although this is only able to detect the structure. A technique such as polarized neutron reflectometry (PNR) can measure the magnetic structure of films and multilayers.

[8]With certain spectrometers it is possible to study excitations up to 1 eV and, with techniques such as neutron spin-echo, down to around $0.1\,\mu$eV.

24.3   Where do muons fit in?

There are many more X-ray synchrotrons and neutron sources around the world than there are muon sources. That is as it should be, as both *scattering* techniques have a wider utility, primarily because of their important role in structure determination. Although muons can be used to give information about magnetic structure, it is more indirect and relies on various assumptions about the muon site and the nature of the coupling. Diffraction techniques are more direct and give direct k-space information, so magnetic neutron diffraction would normally be the first choice to investigate a magnetic structure.

It is worth noting though that neutron techniques have their own problems. Neutrons have difficulty with samples containing hydrogen due to the incoherent scattering from this element; it is common to have to deuterate samples for neutron scattering experiments, which is both expensive and inconvenient. Moreover, there are other elements which are strongly neutron-absorbing (such as cadmium, iridium, and boron, to take just three). This can either prevent the experiment from taking place, or result in the sample becoming radioactive and unable to be returned to the experimenter following an experiment, sometimes for many years. Muons have none of these problems.

Inelastic neutron experiments are very time-consuming and require an enormous single crystal sample (or a mosaic of co-aligned single crystals) of many grams, while the sample requirements for a muon experiment are much more modest. In studying excitations, neutrons and muons are very complementary: neutrons probe dynamics usually in the 1–100 meV range, with characteristic frequencies in the THz. Muons are much more sensitive to dynamics in the MHz and GHz (see Fig. 2.4 in Chapter 2).

Thus, as always in science, which technique is best depends on the problem you are trying to solve. But very often, research problems are solved by using a *combination* of complementary techniques, where each probe can play to its strengths to deliver the best science outcome. Different samples of nominally the same chemical composition can sometimes show subtle variations; hence, measurements using complementary techniques *on the same physical sample* can be particularly effective. Moreover, by co-locating the science techniques at particular facilities, one also co-locates the scientists who are experts in these techniques. This can help foster collaborations, utilizing the most appropriate techniques and moving scientific progress forward.

Chapter summary

- X-rays are used to study structures of samples using diffraction. Though the dominant interaction is Thomson scattering from the electron density, magnetic scattering being a much weaker effect, very high brilliance sources of X-rays are available at synchrotrons and these can be used to perform studies of magnetism.
- Neutrons scatter from the nuclei resulting in the ability to probe diffraction out to high scattered wave vector. The neutron has a small magnetic moment allowing magnetic structure determination. Inelastic scattering of neutrons is used to measure excitations.
- Muons are complementary to neutrons and X-rays, and many studies use a combination of techniques.

Further reading

- J. Als-Nielsen and D. F. McMorrow, *Elements of Modern X-ray Physics*, Wiley (2011).
- A. T. Boothroyd, *Principles of Neutron Scattering from Condensed Matter*, OUP (2020).
- J. M. Carpenter and C. -K. Long, *Elements of Slow-neutron Scattering*, CUP (2015).
- G. L. Squires, *Introduction to the Theory of Thermal Neutron Scattering*, Dover (1978).

- There are a series of books published by Springer called *Magnetism and Synchrotron Radiation* that are proceedings of the Mittelwihr schools on this subject and which contain many helpful articles.

Exercises

(24.1) Calculate the wavelength of an X-ray with an energy of 12.4 keV and a neutron with an energy of 82 meV. Comment on the difference.

(24.2) In a time of flight spectrometer, the energy of a scattered neutron is deduced by the time it takes to arrive at a detector. Estimate the time delay in arrival of a 49 meV neutron compared with a 50 meV neutron over the distance in a typical spectrometer (let's say 10 m).

Fundamental constants

A

Bohr radius	a_0	5.29177×10^{-11} m
speed of light in free space	c	2.99792458×10^8 m s^{-1}
electronic charge	e	1.60218×10^{-19} C
muon $g - 2$	$g - 2$	2.33184×10^{-3}
Planck constant	h	6.62607×10^{-34} J s
$h/2\pi =$	$\hbar$	1.05457×10^{-34} J s
Boltzmann constant	k_B	1.38065×10^{-23} J K^{-1}
muon rest mass	m_μ	1.8835×10^{-28} kg
	$m_\mu c^2$	105.658 MeV
electron rest mass	m_e	9.109×10^{-31} kg
	$m_\mathrm{e} c^2$	0.5110 MeV
proton rest mass	m_p	1.6726×10^{-27} kg
	$m_\mathrm{p} c^2$	938.27 MeV
muon-electron mass ratio	m_μ/m_e	206.768
muon-proton mass ratio	m_μ/m_p	0.112609
Avogadro number	N_A	6.022×10^{23} mol^{-1}
fine structure constant $\dfrac{e^2}{4\pi\epsilon_0 \hbar c} =$	α	$(137.036)^{-1}$
permittivity of free space	ϵ_0	8.85419×10^{-12}F m^{-1}
magnetic permeability of free space	μ_0	$4\pi \times 10^{-7}$H m^{-1}
muon gyromagnetic ratio	γ_μ	$2\pi \times 135.53$ MHz T^{-1}
Bohr magneton	μ_B	9.27401×10^{-24} J T^{-1}
nuclear magneton	μ_N	5.05078×10^{-27} J T^{-1}
muon magnetic moment	μ_μ	4.49045×10^{-26} J T^{-1}
		$= 8.89060\mu_\mathrm{N}$
		$= 0.00484197\mu_\mathrm{B}$
neutron magnetic moment	μ_n	$-1.9130\mu_\mathrm{N}$
proton magnetic moment	μ_p	$2.7928\mu_\mathrm{N}$
flux quantum	Φ_0	2.067884×10^{-15} T m^2

Note on units: the SI unit of magnetic flux density (B) is the tesla (T). Regrettably, cgs units are still often used and in these units B is measured in gauss (G). $1\,\mathrm{T} \equiv 10^4$ G so that $1\,\mathrm{mT} \equiv 10$ G. Therefore $\gamma_\mu/2\pi = 135.53\,\mathrm{MHz\,T}^{-1} \equiv 13.553\,\mathrm{kHz\,G}^{-1}$. In SI units, $B = \mu_0(H + M)$, but in cgs units $B = H + 4\pi M$ (and hence in free space $B = H$). In cgs, H is measured in oersteds (Oe) and so in free space $H = 1$ Oe is equivalent to $B = 1$ G. In SI units, H is measured in A m^{-1} and $1\,\mathrm{A\,m}^{-1} \equiv 4\pi \times 10^{-3}$ Oe.

B Nuclear moments

Table B.1 lists the values of the nuclear spin I, magnetic moment μ in nuclear magnetons, and the quadrupole moment Q in 10^{-28} m^2 for isotopes with non-zero nuclear spins and natural abundance greater than 0.1%. Where there are more than one isotope, they are listed together and the abundances are also listed as a percentage. Note that any nucleus with $I = 0$ or $I = \frac{1}{2}$ will have a zero quadrupole moment.

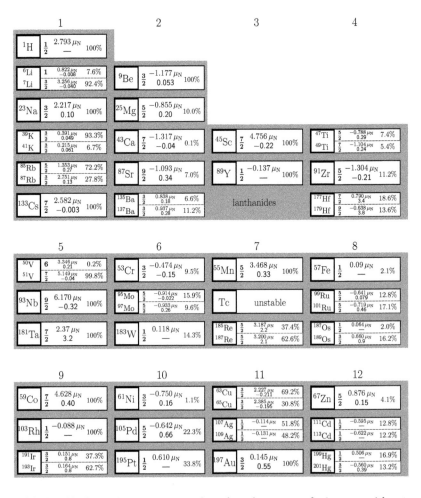

Table B.1 Nuclear spins, moments, and quadrupole moments for isotopes with natural abundances > 0.1%. The numbered columns represent the periodic table groups.

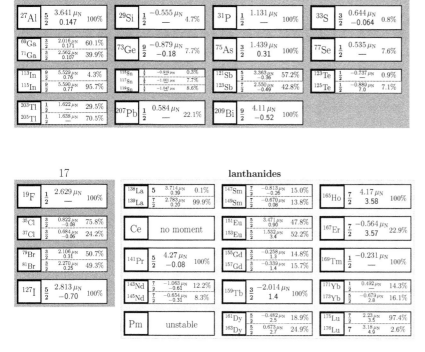

Groups 13–16

Isotope	Spin	μ (μ_N)	Q	Abundance
^{10}B	3	1.801 μ_N	0.085	19.9%
^{11}B	$\frac{3}{2}$	2.689 μ_N	0.041	80.1%
^{27}Al	$\frac{5}{2}$	3.641 μ_N	0.147	100%
^{69}Ga	$\frac{3}{2}$	2.016 μ_N	0.171	60.1%
^{71}Ga	$\frac{3}{2}$	2.562 μ_N	0.107	39.9%
^{113}In	$\frac{9}{2}$	5.529 μ_N	0.76	4.3%
^{115}In	$\frac{9}{2}$	5.590 μ_N	0.77	95.7%
^{203}Tl	$\frac{1}{2}$	1.622 μ_N	—	29.5%
^{205}Tl	$\frac{1}{2}$	1.638 μ_N	—	70.5%
^{13}C	$\frac{1}{2}$	0.702 μ_N	—	1.1%
^{29}Si	$\frac{1}{2}$	−0.555 μ_N	—	4.7%
^{73}Ge	$\frac{9}{2}$	−0.879 μ_N	−0.18	7.7%
^{115}Sn	$\frac{1}{2}$	−0.919 μ_N	—	0.3%
^{117}Sn	$\frac{1}{2}$	−1.001 μ_N	—	7.7%
^{119}Sn	$\frac{1}{2}$	−1.047 μ_N	—	8.6%
^{207}Pb	$\frac{1}{2}$	0.584 μ_N	—	22.1%
^{14}N	1	0.404 μ_N	0.020	99.6%
^{15}N	$\frac{1}{2}$	−0.2831 μ_N	—	0.4%
^{31}P	$\frac{1}{2}$	1.131 μ_N	—	100%
^{75}As	$\frac{3}{2}$	1.439 μ_N	0.31	100%
^{121}Sb	$\frac{5}{2}$	3.363 μ_N	−0.36	57.2%
^{123}Sb	$\frac{7}{2}$	2.550 μ_N	−0.49	42.8%
^{209}Bi	$\frac{9}{2}$	4.11 μ_N	−0.52	100%
^{16}O	0	0 μ_N	—	100%
^{33}S	$\frac{3}{2}$	0.644 μ_N	−0.064	0.8%
^{77}Se	$\frac{1}{2}$	0.535 μ_N	—	7.6%
^{123}Te	$\frac{1}{2}$	−0.737 μ_N	—	0.9%
^{125}Te	$\frac{1}{2}$	−0.889 μ_N	7.0	7.1%

Group 17

Isotope	Spin	μ (μ_N)	Q	Abundance
^{19}F	$\frac{1}{2}$	2.629 μ_N	—	100%
^{35}Cl	$\frac{3}{2}$	0.822 μ_N	−0.08	75.8%
^{37}Cl	$\frac{3}{2}$	0.684 μ_N	−0.06	24.2%
^{79}Br	$\frac{3}{2}$	2.106 μ_N	0.31	50.7%
^{81}Br	$\frac{3}{2}$	2.270 μ_N	0.25	49.3%
^{127}I	$\frac{5}{2}$	2.813 μ_N	−0.70	100%

lanthanides

Isotope	Spin	μ (μ_N)	Q	Abundance
^{138}La	5	3.714 μ_N	0.39	0.1%
^{139}La	$\frac{7}{2}$	2.783 μ_N	0.20	99.9%
Ce		no moment		
^{141}Pr	$\frac{5}{2}$	4.27 μ_N	−0.08	100%
^{143}Nd	$\frac{7}{2}$	−1.063 μ_N	−0.61	12.2%
^{145}Nd	$\frac{7}{2}$	−0.654 μ_N	−0.31	8.3%
Pm		unstable		
^{147}Sm	$\frac{7}{2}$	−0.813 μ_N	−0.26	15.0%
^{149}Sm	$\frac{7}{2}$	−0.670 μ_N	0.08	13.8%
^{151}Eu	$\frac{5}{2}$	3.471 μ_N	0.90	47.8%
^{153}Eu	$\frac{5}{2}$	1.532 μ_N	3.4	52.2%
^{155}Gd	$\frac{3}{2}$	−0.258 μ_N	1.3	14.8%
^{157}Gd	$\frac{3}{2}$	−0.339 μ_N	1.4	15.7%
^{159}Tb	$\frac{3}{2}$	−2.014 μ_N	1.4	100%
^{161}Dy	$\frac{5}{2}$	−0.482 μ_N	2.5	18.9%
^{163}Dy	$\frac{5}{2}$	0.673 μ_N	2.7	24.9%
^{165}Ho	$\frac{7}{2}$	4.17 μ_N	3.58	100%
^{167}Er	$\frac{7}{2}$	−0.564 μ_N	3.57	22.9%
^{169}Tm	$\frac{1}{2}$	−0.231 μ_N	—	100%
^{171}Yb	$\frac{1}{2}$	0.492 μ_N	—	14.3%
^{173}Yb	$\frac{5}{2}$	−0.679 μ_N	2.8	16.1%
^{175}Lu	$\frac{7}{2}$	2.23 μ_N	3.5	97.4%
^{176}Lu	7	3.18 μ_N	4.9	2.6%

Table B.1 (continued from previous page).

C   Negative muon lifetimes

Table C.1 The lifetime of negative muons (in nanoseconds) implanted in various elements. For comparison, the lifetime of the free muon is 2196.981(2) ns [from the Review of Particle Physics, Chin. Phys. C **40**, 199991 (2016)]. A collection of results of measured negative muon lifetimes is given in T. Suzuki, D. F. Measday, and J. P. Roalsvig, Phys. Rev. C **35**, 2212 (1987) and the numbers in this table have been obtained by combining the measurements listed in that paper. Where data from multiple isotopes were presented, only the most naturally abundant isotope has been used.

1	2	3	4	5	6	7	8	9	10	11	12	13	14	15	16	17	18
1 H 2194.80(0.06)																	2 He
3 Li 2187.47(0.80)	4 Be											5 B 2097(3)	6 C 2030(1)	7 N 1920(2)	8 O 1795(2)	9 F 1461(5)	10 Ne 1461(9)
11 Na 1204(2)	12 Mg 1069(2)											13 Al 864(2)	14 Si 759(1)	15 P 616(1)	16 S 555(1)	17 Cl 561(2)	18 Ar 537(32)
19 K 435(1)	20 Ca 336(1)	21 Sc 317(3)	22 Ti 329(1)	23 V 280(2)	24 Cr 259(2)	25 Mn 231(1)	26 Fe 206(1)	27 Co 186(1)	28 Ni 157(1)	29 Cu 164(1)	30 Zn 161(1)	31 Ga 163(2)	32 Ge 167(1)	33 As 153(1)	34 Se 163(1)	35 Br 133(1)	36 Kr
37 Rb 137(3)	38 Sr 132(2)	39 Y 120(1)	40 Zr 110(1)	41 Nb 92(1)	42 Mo 104(1)	43 Tc	44 Ru	45 Rh 95.8(6)	46 Pd 96.0(6)	47 Ag 88.5(6)	48 Cd 90.6(6)	49 In	50 Sn 90.7(8)	51 Sb 92.4(9)	52 Te 104(1)	53 I 85.5(6)	54 Xe
55 Cs 88(2)	56 Ba 94.9(6)	*	72 Hf 75(1)	73 Ta 75.5(6)	74 W 76.5(8)	75 Re	76 Os	77 Ir	78 Pt	79 Au 72.8(5)	80 Hg 76(1)	81 Tl 76(1)	82 Pb 74.7(4)	83 Bi 74.7(4)	84 Po 73.5(4)	85 At	86 Rn
87 Fr	88 Ra	†	104 Rf	105 Db	106 Sg	107 Bh	108 Hs	109 Mt									

* Lanthanides:

57 La 89.9(7)	58 Ce 84.0(6)	59 Pr 72.1(6)	60 Nd 78.4(7)	61 Pm	62 Sm 79(1)	63 Eu	64 Gd 80.6(8)	65 Tb 76.2(7)	66 Dy 79(1)	67 Ho 74.9(6)	68 Er 74(2)	69 Tm	70 Yb	71 Lu

† Actinides:

89 Ac	90 Th 78.0(3)	91 Pa	92 U 77.5(2)	93 Np 72.0(7)	94 Pu	95 Am	96 Cm	97 Bk	98 Cf	99 Es	100 Fm	101 Md	102 No	103 Lr

Answers to selected problems

(2.1) The average path length is the product of the muon velocity βc and the dilated muon lifetime $\gamma\tau_\mu$. This gives a value for the muon path length of 4.9 km. Without time dilation the path length would be reduced to 0.6 km.

(2.2) For measuring fast fluctuations, the relaxation rate λ is proportional to A^2/ν, with the constant of proportionality being 1 for TF mode and 2 for LF mode (see Chapter 5). The upper limit for the fluctuation rate is determined by the smallest measurable relaxation rate in LF mode, giving $\nu = 2A^2/\lambda_{\min} = 8{\times}10^9 \mathrm{s}^{-1}$.

(2.3) The relative amplitude factor is given by the average phase factor over the detector segment, i.e. for N detectors this is $N/2\pi \int_{-\pi/N}^{\pi/N} \cos\theta\,\mathrm{d}\theta = \mathrm{sinc}(\pi/N)$. This leads to amplitude factors of 90.0 % for $N = 4$, 97.4 % for $N = 8$, and 99.4 % for $N = 16$. Thus both $N = 8$ and $N = 16$ will keep the amplitude loss from dephasing to below 3 %.

(3.1) Writing $E = Ar^{-2} - Br$ we have $\mathrm{d}E/\mathrm{d}r = -2Ar^{-3} + Br^{-2} = 0$ yielding $r = 2A/B = a_0 n^2/Z$. The minimum value of $E = r^{-2}(A - Br) = -Ar^{-2} = -(\hbar^2/2ma_0^2)(Z^2/n^2)$ which equates to the expression given.

(3.2) Use $A_\mathrm{D} = A_\mathrm{H}\frac{3}{2}\frac{g_\mathrm{D}}{g_\mathrm{H}}\left(\frac{a_{0,\mathrm{H}}}{a_{0,\mathrm{D}}}\right)^3$, $A_\mathrm{T} = A_\mathrm{H}\frac{g_\mathrm{T}}{g_\mathrm{H}}\left(\frac{a_{0,\mathrm{H}}}{a_{0,\mathrm{T}}}\right)^3$, and $A_\mathrm{Mu} = A_\mathrm{H}\frac{g_\mathrm{Mu}}{g_\mathrm{H}}\left(\frac{a_{0,\mathrm{H}}}{a_{0,\mathrm{Mu}}}\right)^3$.

(4.1) Differentiating $|\boldsymbol{\mu}|^2 = \boldsymbol{\mu}\cdot\boldsymbol{\mu}$ as product gives $2\boldsymbol{\mu}\cdot\mathrm{d}\boldsymbol{\mu}/\mathrm{d}t = \gamma\boldsymbol{\mu}\cdot\boldsymbol{\mu}\times\boldsymbol{B} = 0$. Hence $|\boldsymbol{\mu}|$ is time-independent. A similar argument can be used to show $\boldsymbol{\mu}\cdot\boldsymbol{B} = 0$.

(4.2) The results are obtained using straightforward substitution. For the last part, use $\langle\sigma^z(t)\rangle = \langle\psi(t)|\sigma^z|\psi(t)\rangle = \cos^2\theta(\cos^2\frac{\omega t}{2} + \sin^2\frac{\omega t}{2})\sin^2\theta(\cos^2\frac{\omega t}{2} - \sin^2\frac{\omega t}{2})$ and hence the final result.

(4.3) Since $(\sigma^x)^2 = (\sigma^y)^2 = (\sigma^z)^2 = 1$, then $(\sigma^\alpha)^n$ is equal to σ^α if n is odd and equal to the identity matrix if n is even. The final result is obtained by expanding $\exp(-(\mathrm{i}/\hbar)\hat{\mathcal{H}}t)$ as a Taylor series.

(4.4) The algebra involved in this problem [and also the next four problems (4.5–8)] is straightforward and just involves using the expressions given in the main text.

(4.9) (a) Adding $I = \frac{1}{2}$ and $S = \frac{1}{2}$ one gets a singlet $J = 0$ and a triplet $J = 1$. Adding again $I = \frac{1}{2}$ to $J = 1$

one gets $F = \frac{3}{2}, \frac{1}{2}$, adding $I = \frac{1}{2}$ to $J = 0$ one gets a second $F = \frac{1}{2}, \frac{1}{2}$. (b) The central block is obtained setting $\nu_{\pm\pm} = 0$ in eqn 4.70, swapping back rows and columns 3,4. The outermost blocks are trivial. The eigenvalue equation for the two identical 3x3 blocks is now easy and eigenvalues are given in Example 4.15. (c) The result is given in Example 4.15.

(4.11) This expression (which is derived by simple substitution of the density matrix into the expression for $P_z(t)$) could be relevant to a transverse-field experiment if it were carried out with the muon implanted with its spin initially aligned along the x-direction and its polarization subsequently monitored along the z-direction. What would make this a transverse-field experiment however would be that the magnetic field should be transverse to the initial muon spin (so aligned along the z-direction for example) and that is something that would need to be included in the Hamiltonian actually used.

(4.12)
$$P_z(t) = \frac{\mathrm{Tr}(\sigma_\mu^z\sigma_\mu^z(t))}{\mathrm{Tr}(\mathbb{1})} = \frac{\mathrm{Tr}(\sigma_\mu^z e^{\mathrm{i}\mathcal{H}t/\hbar}\sigma_\mu^z e^{-\mathrm{i}\mathcal{H}t/\hbar})}{\mathrm{Tr}(\mathbb{1})}, \tag{D.1}$$

(4.13) One way to make this problem a lot easier is to start with a basis in which the muon is either up or down but the fluorines (which are identical) are in the states $|\uparrow\uparrow\rangle$, $|\downarrow\downarrow\rangle$, $(|\uparrow\downarrow\rangle + |\downarrow\uparrow\rangle)/\sqrt{2}$, and $(|\uparrow\downarrow\rangle - |\downarrow\uparrow\rangle)/\sqrt{2}$. The resulting 8×8 matrix then neatly separates into four 1×1 blocks and two 2×2 blocks which are straightforward to diagonalize.

(5.4) (a) The probability to survive from time t_1 to t' without a collision is $R_{z,0}(t' - t_1)$. The probability of a collision in the next interval $\mathrm{d}t'$ is $\nu\mathrm{d}t'$. So we can construct the probability for $\ell + 1$ collisions between times t_2 and t_1 as being that for: (i) zero collisions from t_1 to t', (ii) a collision in the period between t' and $t' + \mathrm{d}t'$, and then (iii) ℓ subsequent collisions, which (reading right to left) looks like

$$R_{z,\ell}(t_2 - t')\nu\mathrm{d}t'R_{z,0}(t' - t_1). \tag{D.2}$$

We integrate this with respect to t' to tell us the probability of the initial collision occuring at any point between t_1 and t_2. so we write

$$P_z(t_2 - t_1) = \int \mathrm{d}t'\, R_{z,\ell}(t_2 - t')\nu R_{z,0}(t' - t_1), \tag{D.3}$$

which yields the given recursion relation on substituting $t_2 = t$ and $t_1 = 0$.

(b) For the second expression, substitution yields

$$P_z(t_2 - t_1) = R_{z,0}(t_2 - t_1)$$
$$+ \int_{t_1}^{t_2} dt' R_{z,0}(t_2 - t')\nu R_{z,0}(t' - t_1)$$
$$+ \int_{t_1}^{t_2} dt' R_{z,1}(t_2 - t')\nu R_{z,0}(t' - t_1)$$
$$+ \int_{t_1}^{t_2} dt' R_{z,2}(t_2 - t')\nu R_{z,0}(t' - t_1) + \dots \qquad (D.4)$$

Using the linearity of the integration operation, we can rewrite this as

$$P_z(t_2 - t_1) = R_{z,0}(t_2 - t_1)$$
$$+ \int_{t_1}^{t_2} dt' \nu R_{z,0}(t' - t_1) \left[R_{z,0}(t_2 - t') \right.$$
$$\left. R_{z,1}(t_2 - t') + R_{z,2}(t_2 - t') + \dots \right]. \qquad (D.5)$$

The term in the square bracket is $P_z(t_2 - t')$. Substituting $t_2 = t$ and $t_1 = 0$ yields eqn 5.30.

(c) Using the usual notation for a Laplace transforms $\mathcal{L}[f(t)] = \tilde{f}(s)$, the identities needed are (i) the shifting theorem $\mathcal{L}\left[e^{at}f(t)\right] = \tilde{f}(s-a)$ and (ii) the convolution theorem $\mathcal{L}[f(t) * g(t)] = \tilde{f}(s) \cdot \tilde{g}(s)$. Applying these to eqn 5.30 we obtain

$$\tilde{P}_z(s) = \tilde{P}^s_z(s+\nu) + \nu\tilde{P}_z(s)\tilde{P}^s(s+\nu), \qquad (D.6)$$

which can be rearranged to yield eqn 5.36.

(6.1) Minimizing the free energy we have

$$\frac{\partial F}{\partial M} = 2a_0(T - T_c)M + 4bM^3 = 0. \qquad (D.7)$$

Solutions are $M = 0$ (valid for $T > T_c$) and $M = \pm(a_0/2b)^{\frac{1}{2}}(T_c - T)^{\frac{1}{2}}$ (valid for $T < T_c$), so $\beta = 1/2$.

(6.2) A Gaussian sphere of radius r encloses magnetic charge equal to $\rho_m \times \frac{4}{3}\pi r^3$ and equating this to $\int \boldsymbol{H} \cdot d\boldsymbol{S} = H \times 4\pi r^2$ gives $H = \rho_m/3$. $\boldsymbol{H}$ points everywhere out of the surface and so $\boldsymbol{H} = \rho_m r/3$. Now with two oppositely charged spheres (displaced by $\boldsymbol{d}$) superimposed, then eqn 6.49 gives $\boldsymbol{H} = -\frac{\rho_m \boldsymbol{d}}{3}$. This situation is equivalent to a dipole moment per unit volume (or magnetization) equal to $\rho_m \boldsymbol{d}$ and hence $\boldsymbol{H} = -\boldsymbol{M}/3$. This calculation gives the result for a magnetized sphere, but we actually want the effect of a spherical cavity in an infinite magnetized medium. By the principle of superposition (and using the fact that an infinite uniformly magnetized medium would give $\boldsymbol{H} = 0$ since $\nabla \times \boldsymbol{M} = 0$ everywhere) we deduce that for the spherical cavity $\boldsymbol{H} = +\boldsymbol{M}/3$, and inside the cavity $\boldsymbol{B}_L = \mu_0 \boldsymbol{H}$ and the result follows.

(7.1) Using the definition of the Fourier transform we have

$$\int \frac{d\omega}{2\pi} \delta\tilde{M}(\omega)e^{-i\omega t}$$
$$= \int dt' \frac{d\omega}{2\pi} \frac{d\nu}{2\pi} \tilde{\chi}(\nu)e^{-i\nu(t-t')}\delta\tilde{H}(\omega)e^{-i\omega t'}$$
$$= \int dt' \frac{d\omega}{2\pi} \frac{d\nu}{2\pi} \tilde{\chi}(\nu)e^{-i\nu t}\delta\tilde{H}(\omega)e^{-i(\omega-\nu)t'}$$
$$= \int \frac{d\omega}{2\pi} \frac{d\nu}{2\pi} \tilde{\chi}(\nu)e^{-i\nu t}\delta\tilde{H}(\omega)2\pi\delta(\omega - \nu)$$
$$= \int \frac{d\omega}{2\pi} \tilde{\chi}(\omega)e^{-i\omega t}\delta\tilde{H}(\omega). \qquad (D.8)$$

Matching Fourier components, we conclude $\delta\tilde{M}(\omega) = \tilde{\chi}(\omega)\delta\tilde{H}(\omega)$.

(7.2) (c) Starting with $x(t) = f_0|\tilde{\chi}(\omega)|\cos(\omega t - \phi)$ we have

$$\dot{x} = -\omega f_0|\tilde{\chi}(\omega)|(\sin\omega t \cos\phi - \cos\omega t \sin\phi), \qquad (D.9)$$

and so

$$P = -\frac{1}{T}\int_0^T \omega f_0^2|\tilde{\chi}(\omega)|(\sin\omega t \cos\phi - \cos\omega t \sin\phi)\cos\omega t$$
$$= \frac{1}{2}\omega f_0^2|\tilde{\chi}(\omega)|\sin\phi. \qquad (D.10)$$

Since $\tan\phi = \chi'/\chi''$, we have $\sin\phi = \chi''/|\tilde{\chi}|$ and so $P = \frac{1}{2}\omega f_0^2\chi''$.

(7.5) The integral we need to do is

$$\mathcal{I} = \int_0^\infty d\Delta_j \frac{\gamma_\mu}{\sqrt{2\pi}\Delta_j} e^{-\frac{\gamma_\mu^2 B_i^2}{2\Delta_j^2}} \sqrt{\frac{2}{\pi}}\left(\frac{a}{\Delta_j^2}\right) e^{-\frac{a^2}{2\Delta_j^2}}$$
$$= \frac{\gamma_\mu a}{\pi}\int_0^\infty \frac{d\Delta_j}{\Delta_j^3} e^{-\frac{(\gamma_\mu^2 B_i^2 + a^2)}{2\Delta_j^2}}. \qquad (D.11)$$

Make the substitution $x = \Delta_j^{-1}$ to find

$$\mathcal{I} = \frac{\gamma_\mu a}{\pi}\int_{x=0}^\infty dx\, x e^{-\frac{(\gamma_\mu^2 B_i^2 + a^2)x^2}{2}}. \qquad (D.12)$$

This is a standard Gaussian integral of the form

$$\int_0^\infty dx\, x e^{-Ax^2} = \frac{1}{2A}, \qquad (D.13)$$

so we find

$$\mathcal{I} = \frac{\gamma_\mu}{\pi}\frac{a}{(\gamma_\mu^2 B_i^2 + a^2)}, \qquad (D.14)$$

which is $p(B_i)$, as claimed.

(8.1) Subtract the baseline λ_0 from the data and play with $T_{max} \approx 145$ K, $\tau_\infty \approx 10^{-12}$ s, $D \approx 5.7$ MHz. BPP fitting without constraints (such as here $\omega = 2\pi\nu_{12}$) is hopeless.

(8.2) Inverting eqn 8.25 with $a = 2(\pi A_p^2)/(\lambda - \lambda_0)$ one obtains $\tau^{-1} = a/\omega^2(1 + \sqrt{1 - (\omega/a)^2})$ for $T < T_{max}$ and $\tau^{-1} = a/\omega^2(1 - \sqrt{1 - (\omega/a)^2}$ for $T < T_{max}$). Beware that the argument of the square root must be positive, i.e. A_p must be large enough.

(9.1) The curl of Ampère's law is $\nabla \times \nabla \times \boldsymbol{B} = \mu_0 \nabla \times \boldsymbol{J_s}$. Using eqn 9.4 and $\nabla \cdot \boldsymbol{B} = 0$ and also the London eqn 9.1, we obtain $-\nabla^2\boldsymbol{B} = -\mu_0\frac{ne^2}{m}\nabla \times \boldsymbol{A}$, i.e. $\nabla^2\boldsymbol{B} = \lambda^2\boldsymbol{B}$.

(9.2) To check that in the reference frame of Fig. 9.4 the vector potential may be written $\boldsymbol{A} = B_0 y \hat{\boldsymbol{x}}$ it suffices to calculate its curl using

$$\det \begin{pmatrix} \hat{\boldsymbol{x}} & \hat{\boldsymbol{y}} & \hat{\boldsymbol{z}} \\ \frac{\partial}{\partial x} & \frac{\partial}{\partial y} & \frac{\partial}{\partial z} \\ A_x & A_y & A_z \end{pmatrix}. \tag{D.15}$$

(9.3) If the order parameter has a homogeneous phase $\psi = \psi_0 e^{i\theta}$, with $\boldsymbol{\nabla}\theta = 0$ (without loss of generality we may consider it real), the term in parenthesis in eqn 9.12 vanishes and the second term, assuming $n_s = |\psi_0|^2$, can be rewritten as $\boldsymbol{J}_s = -\boldsymbol{A}/(\mu_0 \lambda^2)$.

(9.4) One square unit cell of parameter $a_\square$ contains one fluxon, i.e. $\Phi_0 = B a_\square^2$, hence $a_\square = (\Phi_0/B)^{\frac{1}{2}}$. In an equilateral triangular lattice of parameter $a_\triangle$ the unit cell is a parallelogram of area $a_\triangle^2 \cos \pi/3$, formed by two adjacent triangles. Therefore $a_\triangle = (\Phi_0/B \cos \pi/3)^{\frac{1}{2}}$.

(9.6) The second moment is $\langle B^2 \rangle - \langle B \rangle^2$. The average is obtained by integrating over one cell, which yields $\langle B \rangle = \Phi_0/|\boldsymbol{a} \times \boldsymbol{b}| = B_0$, using

$$\frac{1}{v_c} \int_{\text{cell}} \mathrm{d}\boldsymbol{r}\, e^{i\boldsymbol{Q}\cdot\boldsymbol{r}} = \delta_{\boldsymbol{Q},0}, \tag{D.16}$$

where $v_c = |\boldsymbol{a} \times \boldsymbol{b}|$. Equation 9.24 is obtained by substituting the sums in the cell integral twice, using the suggested identity, and noting that the $Q = 0$ term reproduces B_0^2.

(9.7) The Fourier transform of eqn 9.21 is a sum of Gaussian functions $\tilde{g}(\omega) = \sum_i (f_i/\sqrt{2\pi}\sigma_i) e^{-(\omega - \gamma_\mu B_i)^2/2\sigma_i^2}$. The first and second moments are calculated as $\langle \omega \rangle = \int_{-\infty}^{\infty} \omega \tilde{g}(\omega) \mathrm{d}\omega = \sum_i f_i \omega_i$ where $\omega_i = \gamma_\mu B_i$, and $\langle \omega^2 \rangle - \langle \omega \rangle^2$, respectively. Finally, $\langle \omega^2 \rangle$ is calculated substituting $\tilde{g}(\omega)$ twice and using $\int_{-\infty}^{\infty} \mathrm{d}\omega\, \omega^2 e^{-\alpha\omega^2} = -\frac{\mathrm{d}}{\mathrm{d}\alpha} \int \mathrm{d}\omega\, e^{-\alpha\omega^2} = \pi^{\frac{1}{2}}/(2\alpha^{\frac{3}{2}})$. Recall that B is real $(B^2 = \boldsymbol{B} \cdot \boldsymbol{B}^*)$ and $\int_{\text{cell}} \mathrm{d}\boldsymbol{r}\, e^{i(\boldsymbol{Q}-\boldsymbol{Q}')\cdot\boldsymbol{r}} = v_c \delta_{\boldsymbol{Q},\boldsymbol{Q}'}$.

(9.9) The exercise asks us to demonstrate that $\int_{\Delta_k}^{\infty} \frac{\partial f}{\partial E_k} \frac{E_k}{\sqrt{E_k^2 - \Delta_k^2}} \mathrm{d}E_k = \int_0^{\infty} \frac{\partial f}{\partial E_k} \mathrm{d}E_k = f(\infty) - f(0) = -\frac{1}{2}$, remembering that in the BCS theory, $E = 0$ coincides with the Fermi surface, i.e. $f(0) = \frac{1}{2}$.

(11.1) The value of Δ from the six octahedral neighbours at $r = a/2$ is 0.426 MHz, increasing to 0.436 MHz when the eight cube corner neighbours are also included. When the quadrupolar factors are included, these two Δ values become 0.381 MHz and 0.389 MHz, the latter result being in excellent agreement with experiment.

(11.2) Using the numbers given leads to the diffusion coefficient $D_{\text{Li}} = 9 \times 10^{-10}$ cm^{2}s^{-1}, which is in close agreement with the published values.

(12.1) The reduced mass (μ) of a one-electron atom is given by $1/\mu = 1/m_X + 1/m_e$ where m_e is the mass of the electron and m_X is the mass of the other particle. This can also be written as $\mu = m_X m_e/(m_e + m_X)$. In positronium $m_X = m_e$ and so $\mu = m_e/2$. The muon, proton, deuteron, and triton are much heavier than the electron $(m_X \gg m_e)$, so $\mu \approx m_e$ for them. The actual values are given in the following table.

Atom	m_X (m_e)	μ (m_e)	Difference from H (%)
Ps	1	0.5	-49.95
Mu	206.7682830	0.995187	-0.43
H	1836.15267343	0.999456	0.00
D	3670.48296788	0.999728	0.03
T	5496.92153573	0.999818	0.04

The ionization energy is equal to the negative of the energy of the 1s orbital and is $\mu/m_e hcR_\infty$ where R_∞ is the Rydberg constant. The Bohr radius a is given by $a = (m_e/\mu)a_0$ where a_0 has a value of 52.918 pm. The ionization energies and Bohr radii of Ps, Mu, H, D, and T are listed in the following table.

Atom	I.E. (eV)	Bohr radius (pm)
PS	6.803	105.835
Mu	13.540	53.174
H	13.598	52.947
D	13.602	52.932
T	13.603	52.927

Chemical properties of an atom are governed by the ionization energy and Bohr radius. The ionization energy and Bohr radius of Mu are very similar to the corresponding values of H (i.e. within 0.5%), while the ionization energy and Bohr radius of Ps differ by almost a factor of 2. It is for this reason that Mu behaves like a light isotope of hydrogen while Ps is not.

(12.2) O$_2$ must be removed from sample as it reacts rapidly with Mu via Heisenberg spin exchange. When studying Mu kinetics this leads to rapid damping of the Mu precession signal.

For the case considered in the problem, $\lambda = k_{\text{Mu}}[\text{O}_2]$ has a value of 4.68 μs^{-1} for the reaction of Mu with O$_2$ in water. This is very rapid and will make it very difficult to observe damping due to other reactions. This reaction also interferes with observing muoniated free radicals. It causes depolarization of the Mu precursor, which results in negligible spin polarization being transferred to the muoniated radical. The muoniated radical will also react with O$_2$ via Heisenberg spin exchange, which causes rapid electron spin flips and broadens μSR signals beyond detection.

(12.3) The first step is to determine the diamagnetic muon frequency, which is $\nu_D = 0.01355$ MHz G^{-1} × 5000 G = 67.75 MHz. The radical frequencies are equally spaced about the diamagnetic. The peak at 32.93 is actually at -32.93 MHz. You can verify this by looking at the difference between the radical frequency and the diamagnetic:

$$67.75\,\text{MHz} - (-32.93\,\text{MHz}) = 100.68\,\text{MHz}$$
$$168.46\,\text{MHz} - 67.75\,\text{MHz} = 100.71\,\text{MHz} \tag{D.17}$$

The muon hyperfine coupling constant is given by the difference between the two radical frequencies, and so

$$|A_\mu| = 168.46\,\text{MHz} - (-32.93\,\text{MHz}) = 201.39\,\text{MHz}.$$
$$(\text{D.}18)$$

The muon hyperfine coupling constant of the muoniated radical is negative as it is an α muon that resides in the nodal plane of the singly occupied molecular orbital. At 1 kG the radical frequencies will be split by hyperfine coupling to the two protons. One proton will split each frequency into a doublet and the second proton will result in a 1:2:1 triplet.

(12.4) Use $A_\text{p} = -Q\rho^\pi$ to calculate the spin density in the π system where Q is a proportionality constant that is ≈ 70 MHz. We consider only the α protons, which are H(1) – H(5).

Nucleus	A_p (MHz)	
H(1)	−25.14	0.36
H(2)	7.47	−0.11
H(3)	−36.19	0.52
H(4)	7.47	−0.11
H(5)	−25.15	0.36

The sum of the spin density around the ring is ≈ 1. This is because there is one unpaired electron in the π system.

(13.1) If you assume for simplicity that the number of muons passing through each unit area at 15 km is approximately equal to those that reach each unit of area on the surface then $N_\mu \approx 10^{14}/\text{s}$. More complicated assumptions give a slightly higher number but of the same magnitude.

(13.2) Counting time: Fraction of muons still to decay after 30 μs is $N/N_0 = \text{e}^{-t/\tau_\mu} = 1.2 \times 10^{-6}$. If 10^8 muons per hour are implanted then roughly 30 seconds counting time is needed to record a muon decaying after 30 μs.
Rate of cosmic ray muons hitting the instrument: Assuming 1 muon/cm^2/min then for a detector array of 600cm^2, 10 muons/second will hit the detectors.
How long after muon implantation do these counts match those from implanted muons: The incoming rate of 10^8/hour is around 28×10^3/second. If we identify $N = 10$ and $N_0 = 10^8/\text{hour} \approx 28 \times 10^3$ then solve the above equation to find $t = 17.4$ μs.
In the case of a realistic detector duty factor: Multiply the cosmic ray muon rate by the duty factor 10/s $\times 30$ μs/20 ms to get a new value of $N = 0.015$ and repeat the above working to find $t = 31.7$ μs. This slightly overestimates the real value as individual cosmic rays can trigger multiple detectors.

(13.3) What is the smallest field that can be measured: Assume that an eighth period of the oscillation needs to coincide with 25 μs for a clear signal $\gamma_\mu B t = \pi/4$. $B = 1/8 \times 25$ μs $\times 135.5$ MHz/T = 37 μT. Smaller fields or field changes can be measured but this shows why instruments carefully compensate the Earth's magnetic field of ≈ 50 μT.

What is the largest field that can be measured at a pulsed source: Consider muons arriving at the front and back of the incoming pulse. The first muons to arrive will precess before the last muons arrive. If we assume that to reduce the signal to zero we need the first muons to precess through 2π before the last muons arrive, using the previous expression $B = 1/100$ ns $\times 135.5$ MHz/T = 73.8 mT.
Detector resolution limited field: Using the previous expression with $t = 200$ ps gives $B = 36.9$ T. Since the positron spiralling radius in such a field would be around 1 mm this time resolution is currently only useful for muonium studies.

(13.4) Equate the centripetal force $F = mv^2/R$ to the Lorentz force $F = qvB$ and simplify.

(13.5) Speed of surface muons: For a kinetic energy $E_\mu = 4$ MeV the total energy of the muon is $4 + 105.7$MeV. The Lorentz factor $\gamma = 109.7/105.7 = 1.038$, which leads to $v/c = 0.27$.
Speed of a positron with the same momentum: From the main text, $p = 29$ MeV/c. This gives a total energy $\sqrt{p^2c^2 + m^2c^4} = 29$ MeV and $\gamma = 29/0.511 = 57$, which leads to $v/c = 0.9998$.
Electric field needed in separator: The electric and magnetic forces on the muons will cancel for $E = vB = 3.24$ MV/m. Since ≈ 100 mm plate separation is required to allow the beam through this corresponds to $\pm \approx 200$ kV DC power supplies.
Spin rotation from separator: We can estimate the spin rotation by considering the muon transit time through the separator and the magnetic field, $\phi = \gamma_\mu BL/v = 48°$.

(15.1) Straightforward integration gives $\int_0^\infty f\text{e}^{-ft}\,\text{d}t = 1$ and $\langle t \rangle = \int_0^\infty tf\text{e}^{-ft}\,\text{d}t = f^{-1}$. Including the dead time the probability $P(t)$ of detection becomes zero if $0 \leq t \leq \tau$ and $f\text{e}^{-ft}/[\int_\tau^\infty f\text{e}^{-ft}\,\text{d}t] = f\text{e}^{-f(t-\tau)}$ if $t > \tau$. Then $\langle t \rangle = \int_\tau^\infty tf\text{e}^{-f(t-\tau)}\,\text{d}t = f^{-1} + \tau$ as required. If our measurement is $N_\text{exp}/T = \langle t \rangle^{-1}$ then rearranging this result with $N/T = f$ gives the desired result. This problem assumes a non-paralysable deadtime in which a count arriving during the deadtime is simply lost (rather than it resetting the detector so that it remains disabled for a further τ, irrespective of how much it had already recovered – which is the paralysable case).

(16.1) $\approx 1.6 \times 10^{52}$ GB.

(16.2) $m_\mu \approx m_\text{p}/9$ and so the zero-point energy $\propto \omega \propto m^{-1/2}$ will be three times bigger. This is $\propto \langle x^2 \rangle$ and hence $x_\text{rms} = \sqrt{\langle x^2 \rangle}$ is $\sqrt{3}$ times bigger.

(16.3) (a) $\approx 1.17\Delta E$; (b) $0.1L$; $k \leq 83\Delta E/L^2$; Covalent, because a ionic crystal has both weaker bonds, meaning lower k, and higher attraction between anions and a positively charged muon, meaning higher ΔE.

(19.5) (c) $\alpha = 20$ cm^{-1}.

(20.1) (b) For a relativistic case we have the equations of motion in terms of components the momentum p_i

$$\frac{dp_x}{d\tau} = qB\frac{dy}{d\tau}$$
$$\frac{dp_y}{d\tau} = -qB\frac{dx}{d\tau}, \qquad (D.19)$$

where τ is the proper time. In order to express these in terms of coordinate time t we can use

$$\frac{d}{d\tau} = \frac{dt}{d\tau}\frac{d}{dt} = \gamma(v)\frac{d}{dt}, \qquad (D.20)$$

where $\gamma(v) = (1 - |v|^2/c^2)^{-\frac{1}{2}}$ for a three-velocity with magnitude $|v|$. This allows us to write relativistic equations of motion

$$\frac{d}{dt}\left(\gamma m\frac{dx}{dt}\right) = qB\frac{dy}{dt}$$
$$\frac{d}{dt}\left(\gamma m\frac{dy}{dt}\right) = -qB\frac{dx}{dt}. \qquad (D.21)$$

These therefore differ from the non-relativistic versions owing to the factor γ. This factor is large for the emitted positrons in muon decay.

(20.2) Start with $\delta B = \mu_0(\delta H + \delta M)$ and divide both sides by $\mu_0 \delta M$. This gives $\mu_0^{-1}(\delta B/\delta M) = (\delta H/\delta M) - 1$ and hence $\chi_B^{-1} = \chi^{-1} + 1$ from which the first result follows. The last part is obtained by starting with $\delta B = \mu_0(\delta H + \delta M)$ and divide both sides by δH, giving $\mu_0(\delta H/\delta B) = 1 - \mu_0(\delta M/\delta B) = 1 - \chi_B$.

(22.2) For Cu $Z = 29$ and so $(Z\alpha)^2 = 0.0448$. The term $m_\mu c^2/(1 + m_\mu/M) = 105.47$ MeV, and so $\frac{1}{2}(Z\alpha)^2 m_\mu c^2/(1 + m_\mu/M) = 2.36$ MeV, and the K_α transition (from $n = 2$ to $n = 1$ with j fixed at $\frac{1}{2}$) is about three quarters of this value, giving about 1.77 MeV, of the right order of magnitude but somewhat larger than the value in the figure.

(23.1) The observed photon energy is $E_o = hc\lambda_o$ where λ_o, λ_s are the observed and source wavelengths. The relativistic Doppler shift is

$$\lambda_o = \lambda_s \frac{1 + \beta}{\sqrt{1 - \beta^2}}, \qquad (D.22)$$

where $\beta = \frac{v}{c}$. For laboratory velocities the observed energy reduces to

$$E_o \approx E_s\left(1 - \frac{v}{c}\right). \qquad (D.23)$$

Therefore $(E_s - E_o)/E_s = v/c$ and $v = c\Delta E/E \approx 0.06$ mm/s.

(24.1) For the X-ray $E = hc/\lambda$ and so $\lambda = 0.1$ nm. For the neutron, $\lambda = h/\sqrt{2m_n E}$ and so $\lambda = 0.1$ nm. To give the same wavelength, the photon is associated with much more energy than the neutron.

(24.2) The velocity of a neutron is $\sqrt{2E/m_n} \approx 3$ km s^{-1}. Over $L = 10$ m, the difference in arrival time is $L\Delta E\sqrt{m_n/8E^3} \approx 0.03$ ms.

Muon particle physics

The spin $S = 1/2$ muon is a massive Fermi particle (or fermion) and is unstable, with average lifetime of 2.2 μs. Positive muons are typically produced through the decay of positive pions π^+, which are a bound state $\mathrm{u\bar{d}}$ comprising an up quark u and down antiquark $\bar{\mathrm{d}}$. These particle-physics processes lie at the heart of muon spectroscopy. Here we briefly describe some of the background to understanding the production, properties, and decay of muons.

E.1 Parity violation

First, we note that both muon decay and pion decay involve the **weak interaction**. Other famous processes involving the weak interaction include β^+ decay of atomic nuclei, in which a proton in a nucleus transforms into a neutron via $\mathrm{p^+ \rightarrow n + e^+ + \nu_e}$, and β^- decay[1] ($\mathrm{n \rightarrow p^+ + e^- + \bar{\nu}_e}$). Weak decay processes tend to occur on a relatively long timescale, whereas processes which occur via the strong or electromagnetic interaction tend to be much more rapid.[2] This is because the lifetimes are approximately inversely proportional to the coupling strength of the interaction (so *weak* interactions give rise to relatively *long* lifetimes, fortunate for μSR!). Weak decays have the unusual feature of **violating parity conservation**. This phenomenon leads to a propensity for the emitted positron to emerge predominantly along the direction of the muon-spin when it decayed.

[1] Because the neutron is slightly more massive than the proton, free neutron decay can occur outside the nucleus (with a mean lifetime of 880 s).

[2] For example, neutral pion decay, $\pi^0 \rightarrow \gamma\gamma$, has a mean lifetime of 8×10^{-17} s, since it occurs via the electromagnetic interaction.

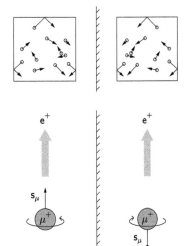

Fig. E.1 (Top) Molecules in a box collide with the walls. (Bottom) Muon decay. Only in the top figure does the process in the mirror obey the laws of physics in our universe.

Example E.1

Most processes in physics conserve parity. For example, in the kinetic theory of gases, molecules in a box collide with the walls and with each other. These collisions do not violate parity so that both the process shown on the left of Fig. E.1(top), and its mirror image on the right, could be observed in nature. The physics for the process in the mirror look exactly the same in our universe, so we can say that the mirror universe is just like our universe.

The same is not true for the process of muon decay, shown in Fig. E.1(bottom). The direction of the muon-spin is reversed in the mirror so that, in the mirror universe, the positrons are emitted predominantly in a direction opposite to that of the muon-spin. The violation of parity means that in our universe only the process on the left-hand side of the diagram is ever observed.

E.2 Standard Model and weak interactions

The **Standard Model** of particle physics describes the properties and interactions of particles like the muon. The Standard Model is described within a framework of quantum field theory in which all particles are excitations of matter fields that are defined over all space and time. From this point of view, the (negatively charged) muon and the (positively charged) antimuon are excitations in the muon field. The muon wavefunction represents a solution to the equation of motion obeyed by all $S = 1/2$ fermion fields, which is known as the **Dirac equation**. Wavefunction solutions of the Dirac equation have four components, and are known as spinors since they encode information about the spin of the particle. However, the four components in a spinor can be grouped into two pairs of two-component parts, reflecting a property of particles known as their **chirality**: that is, those parts of the wavefunction we call left-handed and right-handed.

Example E.2

The Dirac equation for a muon of mass m can be written[3]

$$\left(\sum_\alpha i\gamma^\alpha \partial_\alpha - m \right) \psi(x) = 0, \tag{E.1}$$

where the index $\alpha = 0, 1, 2,$ or 3 labels the components (t, x, y, z). Here γ^α are the set of four 4×4 Dirac matrices and $\partial_\alpha = \partial/\partial x^\alpha$ are derivatives. This equation describes the four components of the wavefunction, but we can rewrite it in two-component form as

$$\left[\begin{pmatrix} 0 & \hat{p}^0 \\ \hat{p}^0 & 0 \end{pmatrix} - \begin{pmatrix} 0 & \boldsymbol{\sigma} \cdot \hat{\boldsymbol{p}} \\ -\boldsymbol{\sigma} \cdot \hat{\boldsymbol{p}} & 0 \end{pmatrix} - \begin{pmatrix} m & 0 \\ 0 & m \end{pmatrix} \right] \begin{pmatrix} \psi_{\mathrm{L}} \\ \psi_{\mathrm{R}} \end{pmatrix} = 0, \tag{E.2}$$

where $\hat{p}^0 = i\partial/\partial t$ is the energy operator, $\hat{\boldsymbol{p}}$ is the momentum operator, and $\boldsymbol{\sigma} = (\sigma_x, \sigma_y, \sigma_z)$ are the Pauli matrices. The components of the wavefunction ψ_{L} and ψ_{R} are the left- and right-handed parts respectively. Massive particles, like the muon, can be thought of as oscillating between being left- and right-handed as a function of time, at a rate proportional to their mass. On the other hand, massless particles are constrained to be in fixed states of chirality, so will be described by non-zero ψ_{L} or non-zero ψ_{R}.

Massless particles in chirality eigenstates are also eigenstates of the **helicity** operator[4]

$$\hat{h} = \frac{\boldsymbol{\sigma} \cdot \hat{\boldsymbol{p}}}{|\boldsymbol{p}|}, \tag{E.3}$$

which has eigenvalues ± 1. Neutrinos are very close to being massless[5] and do not, therefore, change their chirality nor their helicity. We have only ever observed left-handed chirality neutrinos in nature, with no right-handed neutrinos ever having been observed. Similarly we have only observed right-handed antineutrinos and no left-handed ones. For this reason, eqn E.2 implies that neutrinos must always have negative helicity only, while antineutrinos have positive helicity.

---

[3]In this appendix we use units where $\hbar = c = 1$.

[4]Note that helicity depends on the frame in which an observation of a particle's momentum is made. This is not the case for chirality.

[5]The small but finite mass of the neutrino has important implications for cosmology and significant efforts are being devoted towards the challenging task of measuring this mass.

In quantum field theory, particles interact with each other through the emission and absorption of other, force-carrying particles. The Standard Model combines three interactions, each with its own force-carrying particles: (i) the electromagnetic interaction, mediated by the photon; (ii) the weak interaction, mediated by W and Z particles; and (iii) strong interactions mediated by gluons. The production of pions takes place via the strong interaction; the decay of pions and also of muons takes place via the weak interaction. In fact, within the Standard Model, muons (along with electrons, tauons, and their associated neutrinos) are **leptons**, which are a type of particle that interacts via the weak interaction. One of the most important properties of weak interaction is that the W boson (or, equivalently, the W_μ field) only interacts with left-handed particles (or right-handed antiparticles), so it interacts with neutrinos and also with muon particles, but only with the negatively charged muon particle's left-handed parts (or with the positively charged antimuon's right-handed part).[6] The emission or absorption of a W particle by a muon turns it into a muon neutrino as shown in the **Feynman diagrams** shown in Fig. E.2. Feynman diagrams are both a way to understand particle physics interactions and also a shorthand for computing scattering amplitudes. Generally they can be interpreted in terms of time running up the page. Particles are drawn as directed arrows travelling in the positive time direction, antiparticles (such as the positive muon) are down as arrows running in the negative time direction. Some important examples of weak interaction processes are shown in the Feynman diagrams in Fig. E.3.

[6]The parity operator swaps the chirality of particles (turning a left-handed particle into a right-handed one, and vice versa). This is the technical sense in which the weak interaction is said to 'violate parity', as discussed above.

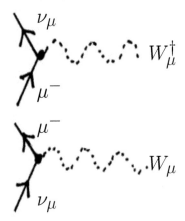

Fig. E.2 Interaction vertices for the weak interaction. (Top) the emission of a W$^-$ boson (equivalent to an interaction with the $W_\mu^\dagger$ field) changes a left-handed, negatively charged muon into a (left-handed) neutrino. (Bottom) The emission of a W$^+$ boson (or interaction with the W_μ field) turns a left-handed neutrino into a left-handed, negatively charged muon.

[7]These are equations for spinor fields (or spinor wavefunctions if you prefer). So, for example μ_L is short for a function $\mu_L(x)$. The bar in this case on a field ψ is short for $\bar\psi = \psi^\dagger \gamma^0$.

Example E.3

The weak interactions shown in Fig. E.2 make a contribution to the amplitude $i\mathcal{A}$ for scattering of $\bar\nu\gamma^\alpha\mu_L$, where the symbols represent the particles' respective Dirac spinors. The left-handed part of a spinor can be projected out using the operator equation $\mu_L = \frac{1}{2}(1 - \gamma^5)\mu$, where $\gamma^5 = \gamma^0\gamma^1\gamma^2\gamma^3$. This allows the interaction contribution to the scattering amplitude to be written in terms of the field operators involved as[7]

$$\bar\nu\gamma^\alpha\mu_L = \frac{1}{2}\left(\bar\nu\gamma^\alpha\mu - \bar\nu\gamma^\alpha\gamma^5\mu\right). \tag{E.4}$$

The first term transforms like a vector V; the second transforms like an axial vector A. For this reason, this part of the weak interaction is often called the $V - A$ interaction.

E.3   Muon production

While there are many elementary particle processes that result in muon production, μSR requires intense beams of muons with sufficiently low energies that they will stop in a typical sample of volume < 1 cm^{-3}. The only decay processes that meet these requirements are the two-body decays of charged pions $\pi^\pm$. These pions are produced in muon facilities by colliding a beam of accelerated protons p$^+$ (of typical energy 600 MeV) into a production target. This produces pions via the reactions

$$p^+ + p^+ \rightarrow p^+ + n + \pi^+,$$

$$p^+ + n \rightarrow p^+ + p^+ + \pi^-, \tag{E.5}$$

where n is the neutron. Pions have spin quantum number $S = 0$ and decay with a time constant of 26.03 ns, via the two-body reactions

$$\pi^+ \rightarrow \mu^+ + \nu_\mu, \tag{E.6}$$

and

$$\pi^- \rightarrow \mu^- + \bar{\nu}_\mu. \tag{E.7}$$

A Feynman diagram for positive pion decay is shown in Fig. E.3.

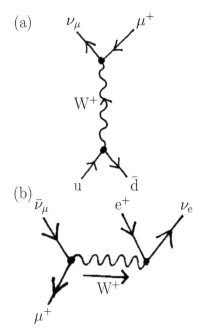

Fig. E.3 Some example electroweak processes. (a) positive pion decay involving the pion constituent quarks u and $\bar{d}$ decaying into $\mu^+ + \nu_\mu$; (b) positive muon decay $\mu^+ \rightarrow \bar{\nu}_\mu + e^+ + \nu_e$.

Example E.4

The conservation of energy and momentum in a two-body decay (from rest) implies that

$$E_\pi = E_\mu + E_\nu. \tag{E.8}$$

Since the π decays at rest and, for the neutrino $E_\nu = |\boldsymbol{p}_\nu|$ we have, from conservation of energy,

$$m_\pi = E_\mu + |\boldsymbol{p}_\nu|. \tag{E.9}$$

Conservation of momentum implies

$$\boldsymbol{p}_\mu = -\boldsymbol{p}_\nu, \tag{E.10}$$

and so on substitution we find

$$m_\pi - E_\mu = |\boldsymbol{p}_\mu|. \tag{E.11}$$

Using the dispersion relation $E_\mu^2 = |\boldsymbol{p}_\mu|^2 + m_\mu^2$, we then obtain

$$E_\mu = \frac{m_\pi^2 + m_\mu^2}{2m_\pi}. \tag{E.12}$$

The kinetic energy of the muon is $E_k = E_\mu - m_\mu$ and we find (restoring factors of c)

$$E_k = \frac{(m_\pi - m_\mu)^2 c^2}{2m_\pi}. \tag{E.13}$$

The muon therefore emerges with a kinetic energy of 4.119 MeV and a momentum of 29.79 MeV/c.

Since neutrinos necessarily possess negative helicity (i.e. their spin is directed antiparallel to their momentum), in order that both angular and linear momentum are conserved in the process in eqn E.6, the muon is also forced to have negative helicity in the pion rest frame. This means that the positive muons produced via pion decay at rest are 100% spin polarized, with their spin directed antiparallel to their momentum.[8]

[8]The practical need to collect muons with different momenta into a beam will lead to an average beam polarization that is slightly less than 100%.

E.4 Muon decay

The feature of muon decay that allows us to read out the polarization of the muon ensemble is that the decay positron is emitted preferentially along the muon spin direction at the time of decay. The dominant muon decay channel is the reaction

$$\mu^+ \rightarrow e^+ + \nu_e + \bar{\nu}_\mu. \tag{E.14}$$

A Feynman diagram for muon decay is shown in Fig. E.3.

Enrico Fermi (1901–1954).

[10]Louis Michel (1923–1999).

Example E.5

In order to compute the decay probability that governs muon decay, we must translate the Feynman diagram for the electroweak interaction to read off the amplitude.[9] Applying the standard set of rules to Fig. E.3(b) we obtain an amplitude

$$i\mathcal{A} \propto \int \frac{\mathrm{d}^4 p}{(2\pi)^4} \left(-\frac{g^2}{8}\right) \left[\bar{\mu}\gamma^\alpha (1-\gamma^5)\nu_\mu\right] \frac{i\eta_{\alpha\beta}}{p^2 - M_W^2} \left[\bar{\nu}_e \gamma^\beta (1-\gamma^5)e\right], \qquad (\mathrm{E}.15)$$

where M_W is the mass of the W particle, g is a coupling constant, and $\eta_{\alpha\beta} = \mathrm{diag}(1,-1,-1,-1)$ is the Minkowski tensor. If we expand out the spinors and compute the density of final states, we can use this expression to construct a decay cross section, whose properties are discussed below.

In the low-energy limit, where momentum is small, this process can be recast in terms of a Hamiltonian

$$\hat{\mathcal{H}}_\mathrm{I} = \frac{g^2}{8} \left[\bar{\mu}\gamma^\alpha (1-\gamma^5)\nu_\mu\right] \left(\frac{\eta_{\alpha\beta}}{M_W^2}\right) \left[\bar{\nu}_e \gamma^\beta (1-\gamma^5)e\right]. \qquad (\mathrm{E}.16)$$

This has the form of Enrico Fermi's celebrated phenomenological theory of β decay, whose Hamiltonian is written

$$\hat{\mathcal{H}}_\mathrm{I} = \frac{G}{\sqrt{2}} J_\mu^\dagger \cdot J_e, \qquad (\mathrm{E}.17)$$

where $(J_\mu^\alpha)^\dagger = \bar{\mu}\gamma^\alpha (1-\gamma^5)\nu_\mu$ and $J_e^\beta = \bar{\nu}_e \gamma^\beta (1-\gamma^5)e$ are particle currents and $G = g^2/4\sqrt{2}M_W^2$ is known as the Fermi interaction parameter.

Ultimately, we want to know the cross section $\mathrm{d}^2\Gamma/\mathrm{d}E\mathrm{d}\Omega$ for positrons to be emitted at a certain energy E and into a certain solid angle $\mathrm{d}\Omega = \mathrm{d}(\cos\theta)\mathrm{d}\phi$. The decay probability Γ of positron emission at a polar angle θ to the parent μ^+ spin direction can be computed from the Feynman diagram in the last example and is found to be a function of E, parameterized by $x = E/E_\mathrm{max}$, where $E_\mathrm{max} = 52.83$ MeV is the maximum possible positron energy consistent with the muon-decay process given in eqn E.14.

Example E.6

Results of this type in the Standard Model are parameterized using the **Michel parameters**[10] which reflect the available phase space for scattering and decay processes. The parameters are called ρ, η, ξ, and δ and, in terms of these the positron decay distribution Γ resulting from muon decay, can be written

$$\frac{1}{x^2} \frac{\mathrm{d}^2\Gamma}{\mathrm{d}x\,\mathrm{d}(\cos\theta)} \propto (3-3x) + \frac{2}{3}\rho(4x-3) + QP_\mu\xi\cos\theta \left[(1-x) + \frac{2}{3}\delta(4x-3)\right], \qquad (\mathrm{E}.18)$$

where P_μ is the muon-spin polarization and Q is its charge. The Standard Model computation predicts the Michel parameters should be

$$\rho = \tfrac{3}{4} \quad \eta = 0 \quad \xi = 1 \quad \xi\delta = \tfrac{3}{4}, \qquad (\mathrm{E}.19)$$

and so we have the prediction for muon decay that

$$\frac{\mathrm{d}^2\Gamma}{\mathrm{d}x\,\mathrm{d}(\cos\theta)} \propto \frac{x^2}{2} \left[(3-2x) - QP_\mu(1-2x)\cos\theta\right]. \qquad (\mathrm{E}.20)$$

This expression contains two quantities of interest: the distribution of the positron energies and of the positron directions.

For a case of most interest to us, where positive muons are spin-polarized along the z-direction with polarization $P_\mu = 1$, we conclude from the previous example that the decay probability Γ of positron emission at an angle θ to the parent μ^+ spin direction is given by

$$\mathrm{d}\Gamma(x,\theta) = E(x)[1 + a(x)\cos\theta]\,\mathrm{d}x\,\mathrm{d}(\cos\theta), \qquad (\text{E}.21)$$

where the asymmetry factor $a(x)$ is

$$a(x) = \pm\frac{(2x-1)}{(3-2x)}, \qquad (\text{E}.22)$$

and the normalized energy spectrum is

$$E(x) = 2x^2(3-2x). \qquad (\text{E}.23)$$

These two distributions are shown in Fig. E.4. The logic here is that a positron is emitted with a particular energy, picking a value of $x = x_0$ from the distribution $E(x)$. This then gives a particular angular distribution $\Gamma(\theta) = 1 + a(x_0)\cos\theta$. The corresponding angular distributions of the emitted positron $\Gamma(x,\theta)$ are shown in Fig. E.5.

We see, therefore, that the known positron distribution allows us access to the polarization of the stopped muons in the sample at the time of their decay. In order to extract the evolution in time of the polarization of the muon ensemble, we record positron emission as a function of both angle and time.

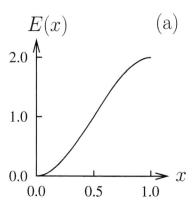

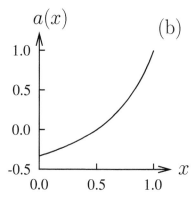

Example E.7

In practice, a positron detector will be fixed in angle and so if the muon ensemble spin-polarization along the symmetry axis d of the detector is $P_d(t)$, then the expression for the number of positrons detected in that detector will contain a factor

$$1 + P_d(t)\int_0^1 a(x)E(x)\mathrm{d}x. \qquad (\text{E}.24)$$

We obtain a time spectrum in each detector of the form

$$N(t) = \eta + N_0\exp(-t/\tau_\mu)\left[1 + A(t)\right], \qquad (\text{E}.25)$$

where N_0 is a normalization constant, η is a time independent background, and $A(t)$ is the asymmetry spectrum, which may be extracted as

$$A(t) = \left(\frac{N(t)-\eta}{N_0}\right)\exp(t/\tau_\mu) - 1. \qquad (\text{E}.26)$$

From the previous example we might expect, considering eqn (E.24), that

$$A(t) = \langle a\rangle P_d(t). \qquad (\text{E}.27)$$

The link between the parameters $A(t)$ and $a(x)$ is not, however, quite as simple as might be imagined. The theoretical average for an on-axis detector with $\theta = 0$ is given by

$$\langle a\rangle = \int_0^1 a(x)E(x)\mathrm{d}x = \frac{1}{3}. \qquad (\text{E}.28)$$

Fig. E.4 Distributions for $E(x)$ and $a(x)$.

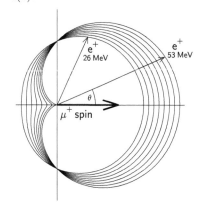

Fig. E.5 The angular distribution of emitted positrons $\Gamma(x,\theta)$ for positron energies ranging from $\xi = 26$ MeV ($x \approx 1/2$) to $\xi = 53$ MeV ($x \approx 1$).

[11]Exclusion of low energy positrons increases the asymmetry, which can compensate for the reduction due to finite solid angle. Positron degraders can be deliberately introduced and optimized to take advantage of this property.

This is not quite found in practice. This is primarily due to the geometrical arrangement of detectors around the sample which, therefore, each cover a non-zero angular range, resulting in some averaging of the asymmetry signal. However, this angular range is necessary so that a reasonable fraction of the emitted positrons can be detected. Also very significant is the exclusion from the detected signal of low energy positrons which stop in thin layers of solid matter such as the wrappings of positron counters.[11] There are also the complicated effects of radiative corrections to the emission spectra and the effects of magnetic field on the positron trajectories that must be considered. This explains why the maximum observed asymmetry in most detectors lies in the approximate range 24-30%.

Further reading

- J. H. Brewer in *Encyclopedia of Applied Physics, Vol. 11: Mössbauer Effect to Nuclear Structure*, VCH (1994).

- F. Halzen and A. D. Martin, *Quarks and Leptons: An Introductory Course in Modern Particle Physics*, Wiley (1984).

- T. Lancaster and S. J. Blundell, *Quantum Field Theory for the Gifted Amateur*, OUP (2014).

Quantum-mechanical polarization functions

The calculation of the polarization function using quantum mechanics is a little lengthy, but simply relies on a time-dependent perturbation theory expansion. This is most easily carried out using the interaction representation, which involves us identifying a free and interacting part of the Hamiltonian. In this appendix we sketch out the main features of this derivation.[1]

[1]In this appendix we use units where $\hbar = 1$. The full treatment is given, using the approach sketched here, in the paper by McMullen and Zaremba, referenced at the end of this appendix.

F.1 Time-dependent perturbations

A time-dependent magnetic field exists at the muon site. Our aim is to compute the time-dependence of the expectation value of the muon spin operator

$$\langle \boldsymbol{S} \rangle = \frac{1}{2} \langle \hat{\boldsymbol{\sigma}} \rangle, \tag{F.1}$$

where $\hat{\boldsymbol{\sigma}} = (\hat{\sigma}_x, \hat{\sigma}_y, \hat{\sigma}_z)$ are the Pauli spin operators. For a field $\boldsymbol{B}_{\text{tot}}(t)$ that fluctuates dynamically about a mean $\boldsymbol{B}_0$, we can write

$$\boldsymbol{B}_{\text{tot}}(t) = \boldsymbol{B}_0 + \delta\boldsymbol{B}(t), \tag{F.2}$$

where $\delta\boldsymbol{B}(t)$ is the time-dependent, fluctuating part. If we assume this fluctuating part is relatively small, we can use the eigenstates of the static part of the interaction as the basis of a perturbation expansion. We express the Hamiltonian $\hat{\mathcal{H}}$ for the muon spin $\boldsymbol{S}$ in this B-field in terms of a free ($\hat{\mathcal{H}}_0$) and interacting ($\hat{\mathcal{H}}'$) part:

$$\hat{\mathcal{H}} = \hat{\mathcal{H}}_0 + \hat{\mathcal{H}}', \tag{F.3}$$

where

$$\hat{\mathcal{H}}_0 = -\gamma_\mu \hat{\boldsymbol{S}} \cdot \boldsymbol{B}_0 \tag{F.4}$$

is the secular term, and we isolate an interaction term

$$\hat{\mathcal{H}}' = -\gamma_\mu \hat{\boldsymbol{S}} \cdot \delta\boldsymbol{B}(t). \tag{F.5}$$

Our coordinate system is defined such that $\boldsymbol{B}_0$ is directed parallel to the z-axis and the initial polarization of the muon is[2] $\boldsymbol{P} = (P_x, P_y, P_z)$.

The free time-evolution operator is given by

$$\hat{U}(t) = \mathrm{e}^{-\mathrm{i}\hat{\mathcal{H}}_0 t}, \tag{F.6}$$

[2]This, of course allows us to work in the TF geometry by setting the polarization to have components $\boldsymbol{P} = (1, 0, 0)$, or the LF geometry where we have $\boldsymbol{P} = (0, 0, 1)$.

and so, in the absence of the perturbation, all operators freely time-evolve according to the Heisenberg prescription $\hat{O}(t) = \hat{U}^\dagger(t)\hat{O}\hat{U}(t)$ or

$$\hat{O}(t) = \mathrm{e}^{\mathrm{i}\hat{\mathcal{H}}_0 t}\hat{O}\mathrm{e}^{-\mathrm{i}\hat{\mathcal{H}}_0 t}. \tag{F.7}$$

A particularly important operator is the time-dependent interaction Hamiltonian, written

$$\hat{\mathcal{H}}_\mathrm{I}(t) = \mathrm{e}^{\mathrm{i}\hat{\mathcal{H}}_0 t}\hat{\mathcal{H}}'\mathrm{e}^{-\mathrm{i}\hat{\mathcal{H}}_0 t} = -\frac{\gamma_\mu}{2}\mathrm{e}^{-\frac{1}{2}\gamma_\mu\hat{\sigma}_z B_0 t}\hat{\boldsymbol{\sigma}}\cdot\delta\boldsymbol{B}(t)\mathrm{e}^{\frac{1}{2}\gamma_\mu\hat{\sigma}_z B_0 t}. \tag{F.8}$$

Example F.1

Heisenberg time evolution means that we have to deal with time-dependent spin operators like $\hat{\sigma}_i(t) = \mathrm{e}^{\mathrm{i}\hat{\mathcal{H}}_0 t}\hat{\sigma}_i\mathrm{e}^{-\mathrm{i}\hat{\mathcal{H}}_0 t}$, or

$$\hat{\sigma}_i(t) = \mathrm{e}^{-\frac{1}{2}\gamma_\mu\hat{\sigma}_z B_0 t}\hat{\sigma}_i\mathrm{e}^{\frac{1}{2}\gamma_\mu\hat{\sigma}_z B_0 t}. \tag{F.9}$$

If, for example, we evaluate[3] $\langle\uparrow_x|\hat{\sigma}_x(t)|\uparrow_x\rangle$ we obtain

$$\begin{aligned}
\langle\uparrow_x|\hat{\sigma}_x(t)|\uparrow_x\rangle &= \langle\uparrow_x|\mathrm{e}^{-\frac{1}{2}\gamma_\mu\hat{\sigma}_z B_0 t}\hat{\sigma}_x\mathrm{e}^{\frac{1}{2}\gamma_\mu\hat{\sigma}_z B_0 t}|\uparrow_x\rangle \\
&= \frac{1}{\sqrt{2}}\langle\uparrow_x|\mathrm{e}^{-\frac{1}{2}\gamma_\mu\hat{\sigma}_z B_0 t}\hat{\sigma}_x\left(|\uparrow_z\rangle\mathrm{e}^{\frac{1}{2}\gamma_\mu B_0 t}+|\downarrow_z\rangle\mathrm{e}^{-\frac{1}{2}\gamma_\mu B_0 t}\right) \\
&= \frac{1}{\sqrt{2}}\langle\uparrow_x|\mathrm{e}^{-\frac{1}{2}\gamma_\mu\hat{\sigma}_z B_0 t}\left[|\downarrow_z\rangle\mathrm{e}^{\frac{1}{2}\gamma_\mu B_0 t}+|\uparrow_z\rangle\mathrm{e}^{-\frac{1}{2}\gamma_\mu B_0 t}\right] \\
&= \frac{1}{\sqrt{2}}\langle\uparrow_x|\left[|\downarrow_z\rangle\mathrm{e}^{\mathrm{i}\gamma_\mu B_0 t}+|\uparrow_z\rangle\mathrm{e}^{-\mathrm{i}\gamma_\mu B_0 t}\right] \\
&= \frac{1}{2}\left(\mathrm{e}^{\mathrm{i}\gamma_\mu B_0 t}+\mathrm{e}^{-\mathrm{i}\gamma_\mu B_0 t}\right) \\
&= \cos\gamma_\mu B_0 t. \tag{F.10}
\end{aligned}$$

Since we also have $\langle\uparrow_x|\hat{\sigma}_y(t)|\uparrow_x\rangle = -\sin\gamma_\mu B_0 t$, we can combine the x and y components and write[4]

$$\langle\uparrow_x|\hat{\sigma}_x(t)|\uparrow_x\rangle + \mathrm{i}\langle\uparrow_x|\hat{\sigma}_y(t)|\uparrow_x\rangle = \mathrm{e}^{-\mathrm{i}\omega_0 t}, \tag{F.12}$$

with $\omega_0 = \gamma_\mu B_0$. If, alternatively, we evaluate $\langle\uparrow_z|\hat{\sigma}_z(t)|\uparrow_z\rangle$ we obtain

$$\langle\uparrow_z|\hat{\sigma}_z(t)|\uparrow_z\rangle = \langle\uparrow_z|\mathrm{e}^{-\frac{1}{2}\gamma_\mu\hat{\sigma}_z B_0 t}\hat{\sigma}_z\mathrm{e}^{\frac{1}{2}\gamma_\mu\hat{\sigma}_z B_0 t}|\uparrow_z\rangle = 1. \tag{F.13}$$

---

[3]We use the notation $|\uparrow_i\rangle$ to denote a spin up along the i-direction, and so on.

[4]This leads us to define

$$\hat{\sigma}_+ = \hat{\sigma}_x + \mathrm{i}\hat{\sigma}_y, \tag{F.11}$$

which is useful below.

The perturbation approach relies on the use of the interaction representation, an intermediate picture between Schrödinger and Heisenberg representations. In this description operators and states both retain some time dependence. As we have seen, the operators evolve, not according to the full Hamiltonian, but rather only to its free, static part, via $\hat{O}(t) = \hat{U}^\dagger(t)\hat{O}\hat{U}(t)$ by means of the evolution operator $\hat{U}(t)$, given in eqn F.6. The states $|\psi(t)\rangle_\mathrm{I}$ evolve according to the interaction part of the Hamiltonian $\mathcal{H}_\mathrm{I}$, defined in eqn F.8, via the equation of motion

$$\mathrm{i}\frac{\partial}{\partial t}|\psi\rangle_\mathrm{I} = \hat{H}_\mathrm{I}(t)|\psi(t)\rangle_\mathrm{I}. \tag{F.14}$$

For simple spin dynamics the interaction representation is the quantum analogue of the classical spin dynamics in the *rotating reference frame*, that we described in Chapter 7, particularly in Example 7.8.

The key to calculations of this sort is to work out how the interacting part of the Hamiltonian affects the time evolution of the operators. The most important operator in this context is, therefore, the interaction representation time-evolution operator $\hat{U}_\mathrm{I}(t)$

$$\hat{U}_\mathrm{I}(t) = T \exp\left[-\mathrm{i}\int_0^t \mathrm{d}\tau\,\hat{\mathcal{H}}_\mathrm{I}(\tau)\right], \tag{F.15}$$

where we use T to denote time ordering. This can be used to compute the time evolution of our spin, since we have

$$\langle \boldsymbol{\sigma}(t)\rangle = \mathrm{Tr}\left[\rho\,\hat{U}_\mathrm{I}^\dagger(t)\hat{\boldsymbol{\sigma}}(t)\hat{U}_\mathrm{I}(t)\right], \tag{F.16}$$

where ρ is the initial density operator of the system.

To evaluate the effect of the perturbation, we expand the operators $\hat{U}_\mathrm{I}$ in eqn F.16 to obtain a perturbation series

$$\begin{aligned}
\langle \boldsymbol{\sigma}(t)\rangle =& \langle \boldsymbol{\sigma}(t)\rangle_0 + \mathrm{i}\int_0^t \mathrm{d}t'\langle[\hat{\mathcal{H}}_\mathrm{I}(t'),\hat{\boldsymbol{\sigma}}(t)]\rangle_0 \\
&+ \int_0^t \mathrm{d}t' \int_0^{t'}\mathrm{d}t''\left\langle\left[\hat{\mathcal{H}}_\mathrm{I}(t''),[\hat{\boldsymbol{\sigma}}(t),\hat{\mathcal{H}}_\mathrm{I}(t')]\right]\right\rangle_0 + \cdots \\
=& \langle \boldsymbol{\sigma}(t)\rangle_{(0)} + \langle \boldsymbol{\sigma}(t)\rangle_{(1)} + \langle \boldsymbol{\sigma}(t)\rangle_{(2)} + \cdots,
\end{aligned} \tag{F.17}$$

where $\langle\,\rangle_0$ denotes the trace with respect to the initial density matrix ρ, and $[\,,\,]$ is the commutator. We make some observations about this expansion below.

Example F.2

The expansion has the basic form shared by many time-dependent perturbation theory computations: the zeroth-order contribution results from muons not interacting with the fluctuating field; the first order contribution comes from muons that interact once; the second order contribution comes from muons interacting twice and so on.

The building blocks of this expansion are terms like

$$\langle 0|H_\mathrm{I}(t')\hat{\sigma}_j(t)|0\rangle = -\frac{\gamma_\mu}{2}\delta\boldsymbol{B}(t')\cdot\left\langle 0\left|\mathrm{e}^{-\frac{\mathrm{i}}{2}\gamma_\mu\hat{\sigma}_z B_0 t'}\hat{\boldsymbol{\sigma}}\mathrm{e}^{\frac{\mathrm{i}}{2}\gamma_\mu\hat{\sigma}_z B_0(t'-t)}\hat{\sigma}_j\mathrm{e}^{\frac{\mathrm{i}}{2}\gamma_\mu\hat{\sigma}_z B_0 t}\right|0\right\rangle, \tag{F.18}$$

where $|0\rangle$ is the initial state of the muon. These amplitudes are built from components like

$$F(i,t',j,t) = \left\langle 0\left|\mathrm{e}^{-\frac{\mathrm{i}}{2}\gamma_\mu\hat{\sigma}_z B_0 t'}\hat{\sigma}_i\mathrm{e}^{\frac{\mathrm{i}}{2}\gamma_\mu\hat{\sigma}_z B_0(t'-t)}\hat{\sigma}_j\mathrm{e}^{\frac{\mathrm{i}}{2}\gamma_\mu\hat{\sigma}_z B_0 t}\right|0\right\rangle, \tag{F.19}$$

which is an example of the muon's propagator, which can be written

$$F(i,t',j,t) = \langle\hat{\sigma}_i(t')\hat{\sigma}_j(t)\rangle. \tag{F.20}$$

In words, the interpretation of the propagator is
- The system starts in a state $|0\rangle$ at time $t = 0$,
- We time evolve it through t in a field B_0,
- We measure the jth component of spin,
- We time evolve it on to time t' in a field B_0,
- We measure the ith component of spin,
- We time evolve it back through $-t'$ in a field B_0,
- We measure the probability amplitude that it is still in state $|0\rangle$. The choice of i comes from the direction of the field fluctuation: a fluctuation in the z-direction gives $i = z$ and one in the x-direction $i = x$. The idea then, is to evaluate the effect of fluctuation in the i-direction on the ability of the muon to propagate between given states.

For example, if we consider the longitudinal polarization, so $|0\rangle = |\uparrow_z\rangle$ and $j = z$, we have

$$F(i, t', z, t) = \langle \sigma_i \rangle. \tag{F.21}$$

We then find $F(z, t', z, t) = 1$ and $F(x, t', z, t) = 0$. Alternatively, if $|0\rangle = |\uparrow_x\rangle$ and $j = x$, we find $F(z, t', x, t) = -i \sin \omega_0 t$ and $F(x, t', x, t) = \cos \omega_0 (t' - t)$.

Higher order contributions follow the same pattern: we must keep evaluating more and more interactions, and evaluating the amplitude that the muon is in the same state.

F.2   Evaluating terms

We split the quantity $\langle \boldsymbol{\sigma} \rangle$ into longitudinal ($\langle \sigma_z \rangle$) and transverse ($\langle \hat{\sigma}_+ \rangle$) parts, where $\hat{\sigma}_+ = \hat{\sigma}_x + i\hat{\sigma}_y$. This enables the results of both geometries to be extracted. Evaluation of the lowest-order term in the series leads to[5]

$$\langle \sigma_+(t) \rangle_{(0)} = P_+ e^{-i\omega_0 t} \tag{F.22}$$

and

$$\langle \sigma_z(t) \rangle_{(0)} = P_z, \tag{F.23}$$

where $P_+ = P_x + iP_y$. In words, a spin polarized transverse to the field B_0 oscillates at frequency $\omega_0 = \gamma_\mu B_0$, and a spin polarized longitudinal to the field does not change.

The first-order correction term yields zero.[6] The second-order term is the important one for our purposes.

[5] These follow from eqns F.12 and F.13 respectively, noting that in the transverse case, the muon has initial spin along x and we project along x, giving

$$\langle \sigma_+(t) \rangle_{(0)} = \langle \uparrow_x | \hat{\sigma}_+(t) | \uparrow_x \rangle,$$

and for the longitudinal case, the muon has initial spin along z and we project along z, so

$$\langle \sigma_z(t) \rangle_{(0)} = \langle \uparrow_z | \hat{\sigma}_z(t) | \uparrow_z \rangle.$$

[6] It will give a frequency shift in the transverse case if the fluctuations are about a nonzero average field.

Example F.3

The combination of commutators in the second-order correction yields up six terms. Separating longitudinal and transverse contributions, the transverse part can be written in terms of the magnetic fields as

$$
\begin{aligned}
\langle \sigma_+(t) \rangle_{(2)} = {} & -\frac{\gamma_\mu^2}{2} e^{-i\omega_0 t} \int_0^t dt' \int_0^{t'} dt'' \left(P_+ \left\langle \{ \delta B_z(t'), \delta B_z(t'') \} \right\rangle_0 \right. \\
& + \frac{P_+}{2} \left\langle \{ \delta \tilde{B}_+(t'), \delta \tilde{B}_-(t'') \} \right\rangle_0 - \frac{P_-}{2} \left\langle \{ \delta \tilde{B}_+(t'), \delta \tilde{B}_+(t'') \} \right\rangle_0 \\
& - P_z \left\langle \{ \delta B_z(t'), \delta \tilde{B}_+(t'') \} \right\rangle_0 + \left\langle [\delta \tilde{B}_+(t'), \delta B_z(t'')] \right\rangle_0 \\
& \left. - \left\langle [\delta B_z(t'), \delta \tilde{B}_+(t'')] \right\rangle_0 \right),
\end{aligned}
\tag{F.24}
$$

where $\delta \tilde{B}_\pm(t) = e^{\pm i\omega_0 t} \delta B_\pm(t)$, $\delta B_\pm(t) = \delta B_x(t) \pm i\delta B_y(t)$, and $\{\ ,\ \}$ denotes the anticommutator.

The longitudinal part is

$$
\begin{aligned}
\langle \sigma_z(t) \rangle_{(2)} = {} & -\frac{\gamma_\mu^2}{4} \int_0^t dt' \int_0^{t'} dt'' \left(P_z \left\langle \{ \delta \tilde{B}_+(t'), \delta \tilde{B}_-(t'') \} \right\rangle_0 \right. \\
& + P_z \left\langle \{ \delta \tilde{B}_-(t'), \delta \tilde{B}_+(t'') \} \right\rangle_0 - P_- \left\langle \{ \delta \tilde{B}_+(t'), \delta B_z(t'') \} \right\rangle_0 \\
& - P_+ \left\langle \{ \delta \tilde{B}_-(t'), \delta B_z(t'') \} \right\rangle_0 + \left\langle [\delta \tilde{B}_-(t'), \delta \tilde{B}_+(t'')] \right\rangle_0 \\
& \left. - \left\langle [\delta \tilde{B}_+(t'), \delta \tilde{B}_-(t'')] \right\rangle_0 \right).
\end{aligned}
\tag{F.25}
$$

Since we are taking equilibrium averages, all of the physics must only depend on the time difference $\tau = t' - t''$. This means we can do one of the two integrals in these expressions.

Doing one of the integrals and some tidying up, we find that the second-order correction term leads to the general results for the contributions to $\langle \boldsymbol{\sigma}(t) \rangle$:

$$\langle \sigma_+(t) \rangle_{(2)} = - e^{-i\omega_0 t} \int_0^t d\tau \left\{ P_+(t-\tau)\Phi_{zz}(\tau) \right.$$

$$+ \frac{1}{2} P_+(t-\tau)e^{i\omega_0 \tau}\Phi_{+-}(\tau)$$

$$+ \frac{iP_-}{4\omega_0}[e^{i\omega_0(2t-\tau)} - e^{i\omega_0 \tau}]\Phi_{++}(\tau)$$

$$\left. + \frac{iP_z}{\omega_0}[e^{i\omega_0(t-\tau)} - 1]\Phi_{z+}(\tau) \right\}, \qquad (F.26)$$

and

$$\langle \sigma_z(t) \rangle_{(2)} = - \frac{1}{2} \int_0^t d\tau \left\{ P_z(t-\tau)[e^{i\omega_0 \tau}\Phi_{+-}(\tau) + e^{-i\omega_0 \tau}\Phi_{-+}(\tau)] \right.$$

$$- \frac{iP_+}{\omega_0}[e^{-i\omega_0 t} - e^{-i\omega_0 \tau}]\Phi_{-z}(\tau)$$

$$\left. + \frac{iP_-}{\omega_0}[e^{i\omega_0 t} - e^{i\omega_0 \tau}]\Phi_{+z}(\tau) \right\}. \qquad (F.27)$$

In equations (F.26) and (F.27) we have defined the correlation function $\Phi_{ij}(t' - t'')$ as

$$\Phi_{ij}(t' - t'') = \frac{\gamma_\mu^2}{2} \langle \delta B_i(t')\delta B_j(t'') + \delta B_j(t'')\delta B_i(t') \rangle_0. \qquad (F.28)$$

This quantity therefore forms the engine-room for the relaxation.

Example F.4

For the longitudinal geometry, $P_x = P_y = 0$ and $P_z = 1$, and, for the case of equivalent x- and y-directions and no cross correlations, this leads to

$$P_z(t) = e^{-\lambda(t)t}, \qquad (F.29)$$

with

$$\lambda(t)t = \int_0^t d\tau(t-\tau)\cos\omega_0\tau \left[\Phi_{xx}(\tau) + \Phi_{yy}(\tau)\right]. \qquad (F.30)$$

When time t is longer than any correlation time,[7] $\lambda(t)$ becomes independent of t and takes the value

$$\lambda = \int_0^\infty d\tau \cos\omega_0\tau \left[\Phi_{xx}(\tau) + \Phi_{yy}(\tau)\right], \qquad (F.31)$$

which is the equation we used in Chapter 7. This expression coincides with what we called T_1^{-1} in Example 7.7. Remember though that we have implied the presence of a large longitudinal field, that provides the *quantization axis* (via the secular part of the Hamiltonian). Care must then be taken in extrapolating this result to low, or zero static magnetic field, where the notion of longitudinal and transverse loses its meaning.

[7] In other words, at times when the correlation functions $\Phi_{xx}(t)$ and $\Phi_{yy}(t)$ have largely disappeared.

For the transverse geometry, $P_x = 1$ and $P_y = P_z = 0$, things are slightly more complicated. For the case of no cross correlations we obtain

$$
\begin{aligned}
\langle \sigma_x(t) \rangle_{(2)} =& \mathcal{R}e \langle \sigma_+(t) \rangle_{(2)} \\
=& -\int_0^t d\tau \left\{ (t-\tau) \Phi_{zz}(\tau) \cos \omega_0 t \right. \\
& + \frac{1}{2}(t-\tau) [\Phi_{xx}(\tau) + \Phi_{yy}(\tau)] \cos \omega_0 (t-\tau) \\
& \left. - \frac{1}{2\omega_0} [\Phi_{xx}(\tau) - \Phi_{yy}(\tau)] \sin \omega_0 (t-\tau) \right\}.
\end{aligned}
\tag{F.32}
$$

We expect the perturbation series to give a polarization function of the form

$$
P_x(t) = e^{-\lambda(t)t} \cos(\omega t + \phi).
\tag{F.33}
$$

At time t longer than any correlation time, and assuming $t \gg \omega_0^{-1}$, and ignoring any out of phase component, eqn F.32 predicts

$$
\lambda = \int_0^\infty d\tau \left\{ \Phi_{zz}(\tau) + \frac{1}{2} \cos \omega_0 \tau [\Phi_{xx}(\tau) + \Phi_{yy}(\tau)] \right\}.
\tag{F.34}
$$

In Example 7.7 we called this term T_2^{-1}. Similar to the longitudinal case, it is important to remember that this transverse case also implies a large static field in the secular part of the Hamiltonian.

Equations F.30 and F.32 are the results used in Chapters 5 and 7.

Further reading

- T. McMullen and E. Zaremba, Phys. Rev. B **9**, 3026 (1978).
- T. Lancaster and S. J. Blundell, *Quantum Field Theory for the Gifted Amateur*, OUP (2014).

The second moment of a spin distribution

When an external magnetic field B_0 is applied to a muon inside a sample, the muon precesses at an angular frequency $\omega_{\text{loc}} = \gamma_\mu B_{\text{loc}}$, where B_{loc} is the local field at the muon site. Even in the absence of strong magnetic interactions, like those of magnetic or superconducting materials, B_{loc} could be a little bit higher or lower than B_0 because of the presence of other nuclei in the sample. This deviation depends on the orientation of the nearby nuclear spins which can cause a net dipolar field to add to or substract from the applied field, resulting in a distribution of local fields. Understanding this is an old problem, dating back to the early days of NMR, since the broadening of NMR absorption lines is caused by the same distribution. In this appendix, we will calculate the broadening effect of nuclear dipolar fields via the so-called *second moment* of the distribution, to be defined below.

G.1 The dipolar interaction

The dipolar Hamiltonian between two different spins, $\boldsymbol{S}$ and $\boldsymbol{I}$ (angular momentum operators, written in units of $\hbar$), can be written

$$\hat{\mathcal{H}}_{\text{dip}} = \frac{\mu_0}{4\pi} \frac{\gamma_S \gamma_I \hbar^2}{r^3} \left[\boldsymbol{S} \cdot \boldsymbol{I} - \frac{3(\boldsymbol{S} \cdot \boldsymbol{r})(\boldsymbol{I} \cdot \boldsymbol{r})}{r^2} \right], \qquad \text{(G.1)}$$

where $\boldsymbol{r}$, with polar coordinates (r, θ, ϕ), is the vector joining the two spins. Van Vleck showed that, by expanding the scalar products of spin operators, this can be written as a sum of six terms

$$\hat{\mathcal{H}}_{\text{dip}} = \frac{\mu_0}{4\pi} \frac{\gamma_S \gamma_I \hbar^2}{r^3} [A + B + C + D + E + F], \qquad \text{(G.2)}$$

the so-called dipolar alphabet, where[1]

$$A = S^z I^z (1 - 3\cos^2\theta),$$

$$B = -\frac{1}{4}(S^+ I^- + S^- I^+)(1 - 3\cos^2\theta),$$

$$C = -\frac{3}{2}(S^+ I^z + S^z I^+)\sin\theta\cos\theta e^{-i\phi}, \quad D = C^\dagger,$$

$$E = -\frac{3}{4}S^+ I^+ \sin^2\theta e^{-2i\phi}, \quad F = E^\dagger. \qquad \text{(G.3)}$$

[1] This can be shown by patiently substituting $\boldsymbol{r} = r\sin\theta\cos\phi\,\hat{\boldsymbol{x}} + r\sin\theta\sin\phi\,\hat{\boldsymbol{y}} + r\cos\theta\,\hat{\boldsymbol{z}}$ and the definition of all spin components in terms of their raising and lowering spin operators in eqn G.1. The dipolar alphabet notation is due to van Vleck.

Notice that the first term of the alphabet, A, does not change the spin of any state it operates on, whereas all the others do. We will come back to this in the high transverse field condition.

Example G.1

Another way to write eqn G.1 is

$$\hat{\mathcal{H}}_{\text{dip}} = -\frac{1}{2}\gamma_\mu \hbar \boldsymbol{\sigma} \cdot \boldsymbol{B}_{\text{dip}} = -\gamma_\mu \hbar \left(\frac{\sigma^z}{2} B_{\text{dip}}^z + \frac{\sigma^+}{4} B_{\text{dip}}^- + \frac{\sigma^-}{4} B_{\text{dip}}^+ \right), \tag{G.4}$$

where $\boldsymbol{S} = \frac{\hbar}{2}\boldsymbol{\sigma}$ is the muon spin[2] (and so $\gamma_S \equiv \gamma_\mu$), with $\sigma_\pm$ as in Appendix F.2. Assuming an identical spin I at each lattice site i we can write out

$$\boldsymbol{B}_{\text{dip}} = \frac{1}{\gamma_\mu} \sum_i \omega_{\text{d}i} \boldsymbol{b}_i, \quad \text{with} \quad \omega_{\text{d}i} = \frac{\mu_0}{4\pi} \frac{\gamma_\mu \gamma_I \hbar}{r_i^3}, \tag{G.5}$$

and

$$b_i^z = I_i^z (3\cos^2\theta_i - 1) + \frac{3}{2} I_i^- \sin\theta_i \cos\theta_i e^{i\phi_i} + \frac{3}{2} I_i^+ \sin\theta_i \cos\theta_i e^{-i\phi_i}$$

$$b_i^+ = 3 I_i^z \sin\theta_i \cos\theta_i e^{i\phi_i} + \frac{3}{2} I_i^- \sin^2\theta_i e^{2i\phi_i} - \frac{1}{2} I_i^+ (3\cos^2\theta_i - 1)$$

$$b_i^- = 3 I_i^z \sin\theta_i \cos\theta_i e^{-i\phi_i} - \frac{1}{2} I_i^- (3\cos^2\theta_i - 1) + \frac{3}{2} I_i^+ \sin^2\theta_i e^{-2i\phi_i}. \tag{G.6}$$

This form can be checked by direct comparison with eqn G.3 and it will be useful later.

---

[2] Note that in this appendix we are writing the muon spin as $\boldsymbol{\sigma}$. We could have used $\boldsymbol{\sigma}_\mu$ to avoid any ambiguity, but we have chosen not to do so in order to avoid 'subscript overload'.

G.2   High transverse field

We start considering this interaction as a small perturbation in a sufficiently large static magnetic field $\boldsymbol{B}_0 = B_0 \hat{z}$, corresponding to the muon and nuclear Larmor frequencies $\omega = \gamma_\mu B_0 \gg \omega_{\text{d}i}$ and $\omega_I = \gamma_I B_0$, respectively.[3] The Hamiltonian $\hat{\mathcal{H}}$, in angular frequency units, is

$$\frac{\hat{\mathcal{H}}}{\hbar} = -\frac{1}{2}\omega\sigma^z - \omega_I \sum_i I_i^z - \frac{1}{2}\boldsymbol{\sigma} \cdot \sum_i \omega_{\text{d}i} \boldsymbol{b}_i, \tag{G.7}$$

[3] Do not confuse the alphabet term B with B_0! Furthermore, notice that our muon Larmor frequency is ω (and not ω_μ) in order to remove a redundant subscript and, also, because it's the muon we want to keep in our main focus.

a sum of three terms, corresponding to: (1) the Zeeman interaction between the magnetic field and the muon spin; (2) the Zeeman interaction between the magnetic field and the nuclear spins; (3) the dipolar interaction $\mathcal{H}_{\text{dip}}$ between the muon and the nuclear spins, rewritten by means of eqn G.5.

Within this Hamiltonian, recalling eqn G.3, the operator A in the dipolar interaction commutes with both Zeeman interactions. It produces a pure shift of the Larmor frequencies, and it is known as the **secular part** of $\mathcal{H}_{\text{dip}}$. The other operators of the dipolar alphabet are known as **non-secular** and in a perturbative approach they produce only second order shifts, which we may neglect if B_0 is sufficiently large.[4] Notice that if the objects corresponding to $\boldsymbol{S}$ and $\boldsymbol{I}$ were the same, as can happen in NMR, the term B in eqn G.3 would also involve no net change in angular momentum since it only contains flip-flop terms ($\boldsymbol{S}$

[4] In the transverse case, it is only the operator A that contributes to the second moment of the muon precession signal. Because it leaves the angular momentum of a state unchanged it will produce a contribution in the transverse field data which is at approximately the Larmor frequency ω; the other terms in eqn G.3 will produce contributions at very different frequencies.

raised and $\boldsymbol{I}$ lowered, or vice versa), in which case B could also be a secular term. However, we will be choosing $\boldsymbol{S}$ to be the muon, and $\boldsymbol{I}$ to be some nuclear spin, which is therefore different and so this situation will not arise.

With our present assumptions ($\omega \gg \omega_{\mathrm{d}i}$) the dipolar interaction reduces to its secular approximation

$$\frac{\hat{\mathcal{H}}_\mathrm{d}}{\hbar} = \frac{1}{2}\sigma^z \sum_i \alpha_i I_i^z, \tag{G.8}$$

where[5] $\alpha_i = \omega_{\mathrm{d}i}(1 - 3\cos^2\theta_i)$, and where $\hat{\mathcal{H}}_\mathrm{d}$ contains only the A term of $\hat{\mathcal{H}}_{\mathrm{dip}}$.

The muon polarization $P_x(t)$ in a transverse field experiment (see Exercise 4.11) is

$$P_x(t) = \frac{\mathrm{Tr}(\sigma^x \sigma^x(t))}{\mathrm{Tr}(\mathbb{1})} = \frac{\mathrm{Tr}(\sigma^x \mathrm{e}^{\mathrm{i}\hat{\mathcal{H}}t/\hbar}\sigma^x \mathrm{e}^{-\mathrm{i}\hat{\mathcal{H}}t/\hbar})}{\mathrm{Tr}(\mathbb{1})}. \tag{G.9}$$

In the secular approximation all the operators in the Hamiltonian G.7 commute with each other, being combinations of σ^z and I_i^z, so that each time evolution operator in eqn G.9 is simply a product of three exponential terms

$$\mathrm{e}^{\mathrm{i}\hat{\mathcal{H}}t/\hbar} = \mathrm{e}^{-\frac{1}{2}\mathrm{i}\omega\sigma^z t}\mathrm{e}^{-\mathrm{i}\sum_i \omega_I I_i^z t}\mathrm{e}^{\mathrm{i}\hat{\mathcal{H}}_\mathrm{d}t/\hbar}, \tag{G.10}$$

that commute as well. We want to work out $\mathrm{e}^{\mathrm{i}\hat{\mathcal{H}}t/\hbar}\sigma^x \mathrm{e}^{-\mathrm{i}\hat{\mathcal{H}}t/\hbar}$: the easy bit to do first is to switch around the nuclear Zeeman and the dipolar exponential terms. This brings the nuclear Zeeman exponentials on both sides of σ^x and since any pair of nuclear and muon operator commute, we can write

$$\mathrm{e}^{-\mathrm{i}\sum_i \omega_I I_i^z t}\sigma^x \mathrm{e}^{\mathrm{i}\sum_i \omega_I I_i^z t} = \sigma^x. \tag{G.11}$$

Next, let's tackle the muon Zeeman interaction.

[5] Our definition of α_i differs by a factor of two from the choice made in the book by Yaouanc and Dalmas de Réotier.

Example G.2

We need to work out

$$f(\phi) = \mathrm{e}^{-\mathrm{i}\sigma^z \phi/2}\sigma^x \mathrm{e}^{\mathrm{i}\sigma^z \phi/2}, \tag{G.12}$$

and a good way to do this is to differentiate $f(\phi)$, so that

$$\frac{\mathrm{d}f}{\mathrm{d}\phi} = \mathrm{e}^{-\mathrm{i}\sigma^z \phi/2}\left(-\mathrm{i}\frac{\sigma^z}{2}\sigma^x + \mathrm{i}\sigma^x \frac{\sigma^z}{2}\right)\mathrm{e}^{\mathrm{i}\sigma^z \phi/2} = \mathrm{e}^{-\mathrm{i}\sigma^z \phi/2}\sigma^y \mathrm{e}^{\mathrm{i}\sigma^z \phi/2}, \tag{G.13}$$

where we have used differentiation of a product and also the commutation relation $[\sigma^z, \sigma^x] = 2\mathrm{i}\sigma^y$. Repeating the process gives

$$\frac{\mathrm{d}^2f}{\mathrm{d}\phi^2} = \mathrm{e}^{-\mathrm{i}\sigma^z \phi/2}\left(-\mathrm{i}\frac{\sigma^z}{2}\sigma^y + \mathrm{i}\sigma^y \frac{\sigma^z}{2}\right)\mathrm{e}^{\mathrm{i}\sigma^z \phi/2} = -\mathrm{e}^{-\mathrm{i}\sigma^z \phi/2}\sigma^x \mathrm{e}^{\mathrm{i}\sigma^z \phi/2} = -f. \tag{G.14}$$

Using the condition that $f(0) = \sigma^x$ (from eqn G.12) and $f'(0) = \sigma^y$ (from eqn G.13) a solution to $\mathrm{d}^2f/\mathrm{d}\phi^2$ is then immediately given by $f(\phi) = \sigma^x \cos\phi + \sigma^y \sin\phi$.

A direct application of this example then gives

$$e^{-\frac{1}{2}i\omega\sigma^z t}\sigma^x e^{\frac{1}{2}i\omega\sigma^z t} = \sigma^x \cos\omega t + \sigma^y \sin\omega t. \tag{G.15}$$

We still have to work out the effect of the dipolar Hamiltonian evolution operators, and we can replicate the trick of the previous example.

Example G.3

We now define the function $g(t)$ by

$$g(t) = e^{i\hat{\mathcal{H}}_d t/\hbar}\sigma^x e^{-i\hat{\mathcal{H}}_d t/\hbar}, \tag{G.16}$$

and differentiate to obtain

$$\frac{dg}{dt} = e^{i\hat{\mathcal{H}}_d t/\hbar}\left(\frac{i}{\hbar}\hat{\mathcal{H}}_d\sigma^x - \frac{i}{\hbar}\sigma^x\hat{\mathcal{H}}_d\right)e^{-i\hat{\mathcal{H}}_d t/\hbar}$$

$$= \frac{i}{\hbar}e^{i\hat{\mathcal{H}}_d t/\hbar}[\hat{\mathcal{H}}_d, \sigma^x]e^{-i\hat{\mathcal{H}}_d t/\hbar}, \tag{G.17}$$

and differentiating twice gives

$$\frac{d^2 g}{dt^2} = -\frac{1}{\hbar^2}e^{i\hat{\mathcal{H}}_d t/\hbar}[\hat{\mathcal{H}}_d, [\hat{\mathcal{H}}_d, \sigma^x]]e^{-i\hat{\mathcal{H}}_d t/\hbar}, \tag{G.18}$$

so that

$$\left.\frac{d^2 g}{dt^2}\right|_{t=0} = -\frac{1}{\hbar^2}[\hat{\mathcal{H}}_d, [\hat{\mathcal{H}}_d, \sigma^x]]. \tag{G.19}$$

A similar result applies to σ^y.

Let's take stock. We have so far shown that

$$P_x(t) = Q(t)\cos\omega t + R(t)\sin\omega t, \tag{G.20}$$

where

$$Q(t) = \frac{\mathrm{Tr}(\sigma^x e^{i\hat{\mathcal{H}}_d t/\hbar}\sigma^x e^{-i\hat{\mathcal{H}}_d t/\hbar})}{\mathrm{Tr}(\mathbb{1})},$$

$$R(t) = \frac{\mathrm{Tr}(\sigma^x e^{i\hat{\mathcal{H}}_d t/\hbar}\sigma^y e^{-i\hat{\mathcal{H}}_d t/\hbar})}{\mathrm{Tr}(\mathbb{1})}. \tag{G.21}$$

[6] At $t = 0$, $\sigma^x\sigma^x = \mathbb{1}$, and $\sigma^x\sigma^y = i\sigma^z$ is traceless, therefore $Q(0) = 1$ and $R(0) = 0$. As time increases, $Q(t)$ decreases slightly from unity and $R(t)$ increases slightly from zero, but does not contribute very much. We may further assume that Q is an even function of time (due to time reversal invariance).

It turns out[6] that $R(t)$ can be ignored, so let's focus on the first term in eqn G.20. With a magnetic field along z, we would expect an oscillatory behaviour, like $P_x(t) = \cos\omega t$ and so the function $Q(t)$ must correspond to the envelope of the oscillation. This is a slowly decaying function since nuclear spins are tiny, so we can approximate it as a time-reversal invariant power series

$$Q(t) = 1 - \frac{1}{2!}M_2 t^2 + \frac{1}{4!}M_4 t^4 - \cdots, \tag{G.22}$$

which only includes the even terms. In particular, if we were to assume a Gaussian envelope function $Q(t) = e^{-\Delta^2 t^2/2}$, we would get that $M_2 = \Delta^2$, $M_4 = 3\Delta^4 = 3(M_2)^2$ and so on.

More generally the moments M_{2n} are defined by

$$M_{2n} = (-1)^n \frac{d^{2n}Q(t)}{dt^{2n}}\bigg|_{t=0}, \tag{G.23}$$

and the **second moment** is

$$M_2 = (-1) \frac{d^2Q(t)}{dt^2}\bigg|_{t=0}, \tag{G.24}$$

a quantity that we conveniently obtained in eqn G.19. Since the trace of a product of operators is invariant under their cyclic permutations,[7] a direct expansion shows that $\text{Tr}(\sigma^x[\hat{\mathcal{H}}_\text{d}, [\hat{\mathcal{H}}_\text{d}, \sigma^x]]) = -\text{Tr}([\hat{\mathcal{H}}_\text{d}, \sigma^x]^2)$. This allows us to deduce

[7] This means $\text{Tr}(A_1 A_2 \cdots A_{n-1} A_n) = \text{Tr}(A_n A_1 A_2 \cdots A_{n-1})$

$$M_2 = -\frac{1}{\hbar^2} \frac{\text{Tr}([\hat{\mathcal{H}}_\text{d}, \sigma^x]^2)}{\text{Tr}(\mathbb{1})}. \tag{G.25}$$

Example G.4

We can now evaluate the commutator

$$\frac{1}{\hbar}[\hat{\mathcal{H}}_\text{d}, \sigma^x] = [\sigma^z, \sigma^x] \sum_i \frac{\alpha_i}{2} I_i^z = i\sigma^y \sum_i \alpha_i I_i^z, \tag{G.26}$$

so that

$$\frac{1}{\hbar^2}[\hat{\mathcal{H}}_\text{d}, \sigma^x]^2 = -\sum_{i,j} \alpha_i \alpha_j I_i^z I_j^z, \tag{G.27}$$

using the fact that $(\sigma^y)^2 = \mathbb{1}$. Hence

$$M_2 = -\frac{1}{\hbar^2} \frac{\text{Tr}([\hat{\mathcal{H}}_\text{d}, \sigma^x]^2)}{\text{Tr}(\mathbb{1})} = \sum_{i,j} \alpha_i \alpha_j \frac{\text{Tr}(I_i^z I_j^z)}{\text{Tr}(\mathbb{1})} = \frac{1}{3} I(I+1) \sum_i \alpha_i^2. \tag{G.28}$$

The final equality comes from noting that the trace of a product of nuclear spin operators $I_i^z I_j^z$ is zero unless $i = j$, in which case we can use $\text{Tr}(I^z)^2 = \frac{1}{3}\text{Tr}\mathbf{I}^2 = \frac{1}{3}I(I+1)\text{Tr}(\mathbb{1})$.

We note in passing that the fourth moment M_4, calculated from eqn G.23 in analogy with the present example, will in general be different from $3(M_2)^2$, in contrast to the case of the Gaussian envelope function. This implies that the relaxation in a single crystal will deviate from a simple Gaussian form (which remains however a very good approximation).

Equation G.28 can be written out in full as

$$M_2 = \frac{1}{3} \left(\frac{\mu_0}{4\pi}\right)^2 \hbar^2 \gamma_\mu^2 \sum_i \gamma_I^2 I(I+1) \frac{(1 - 3\cos^2\theta_i)^2}{r_i^6}. \tag{G.29}$$

In the case of a polycrystalline sample, we can perform a polycrystalline average and use $\langle (1 - 3\cos^2\theta_i)^2 \rangle = \frac{4}{5}$ to obtain

$$M_2 = \frac{4}{15} \left(\frac{\mu_0}{4\pi}\right)^2 \hbar^2 \gamma_\mu^2 \sum_i \frac{\gamma_I^2 I(I+1)}{r_i^6}. \tag{G.30}$$

G.3 Zero field

We can use a similar approach to calculate the second moment in zero field, but in this case the expression for the second moment should be

$$M_2 = -\frac{1}{\hbar^2} \frac{\text{Tr}([\hat{\mathcal{H}}_{\text{dip}}, \sigma^z]^2)}{\text{Tr}(\mathbb{1})}, \tag{G.31}$$

so the relevant Pauli operator is σ^z and we need to deal with the full dipolar Hamiltonian $\hat{\mathcal{H}}_{\text{dip}}$, not just the secular part. In fact, the secular part (which is proportional to σ^z) is the one part that commutes with σ^z and so doesn't contribute to M_2 in this case.

Example G.5

The commutators can be evaluated using the form of $\hat{\mathcal{H}}_{\text{dip}}$ given in eqn G.4. In this case

$$[\hat{\mathcal{H}}_{\text{dip}}, \sigma^z] = -\frac{\gamma_\mu \hbar}{4}[\sigma^+ B_{\text{dip}}^- + \sigma^- B_{\text{dip}}^+, \sigma^z] = \frac{\gamma_\mu \hbar}{2}(B_{\text{dip}}^- \sigma^+ - B_{\text{dip}}^+ \sigma^-). \tag{G.32}$$

Hence

$$[\hat{\mathcal{H}}_{\text{dip}}, \sigma^z]^2 = \frac{\gamma_\mu^2 \hbar^2}{4}\left[(\sigma^+)^2(B_{\text{dip}}^-)^2 + (\sigma^-)^2(B_{\text{dip}}^+)^2 \right.$$
$$\left. - B_{\text{dip}}^- B_{\text{dip}}^+ \sigma^+ \sigma^- - B_{\text{dip}}^+ B_{\text{dip}}^- \sigma^- \sigma^+ \right]. \tag{G.33}$$

Now, using the three identities $(\sigma^+)^2 = (\sigma^-)^2 = 0$, $\sigma^+ \sigma^- = 2(1 + \sigma^z)$, $\sigma^- \sigma^+ = 2(1 - \sigma^z)$, and noting that σ^z is traceless, we have

$$-\frac{1}{\hbar^2}\text{Tr}([\hat{\mathcal{H}}_{\text{dip}}, \sigma^z]^2) = \frac{1}{2}\gamma_\mu^2 \text{Tr}(B_{\text{dip}}^- B_{\text{dip}}^+ + B_{\text{dip}}^+ B_{\text{dip}}^-). \tag{G.34}$$

We recall from eqn G.6 that we can write

$$B_{\text{dip}}^- = \frac{1}{\gamma_\mu}\sum_i \omega_{\text{d}i}(\lambda_i I_i^z + \eta_i I_i^+ + \xi_i I_i^-)$$

$$B_{\text{dip}}^+ = \frac{1}{\gamma_\mu}\sum_i \omega_{\text{d}i}(\lambda_i^* I_i^z + \eta_i^* I_i^- + \xi_i I_i^+), \tag{G.35}$$

where

$$\lambda_i = 3\sin\theta_i \cos\theta_i e^{-i\phi_i}, \quad \eta_i = \frac{3}{2}\sin^2\theta_i e^{-2i\phi_i}, \quad \xi_i = \frac{1}{2}(3\cos^2\theta_i - 1). \tag{G.36}$$

Using $\text{Tr}(I_i^\pm I_i^z) = \text{Tr}(I_i^\pm I_i^\pm) = 0$ and

$$\frac{\text{Tr}(I_i^z I_j^z)}{\text{Tr}\mathbb{1}} = \frac{1}{3}I(I+1)\delta_{ij}, \quad \frac{\text{Tr}(I_i^+ I_j^-)}{\text{Tr}\mathbb{1}} = \frac{\text{Tr}(I_i^- I_j^+)}{\text{Tr}\mathbb{1}} = \frac{2}{3}I(I+1)\delta_{ij}, \tag{G.37}$$

we can compute the second moment for the zero field case

$$M_2 = -\frac{1}{\hbar^2}\frac{\text{Tr}([\hat{\mathcal{H}}_{\text{dip}}, \sigma^z]^2)}{\text{Tr}(\mathbb{1})} = \frac{I(I+1)}{3}\sum_i \omega_{\text{d}i}^2\left(|\lambda_i|^2 + 2|\eta_i|^2 + 2|\xi_i|^2\right)$$

$$= \frac{I(I+1)}{3}\sum_i \omega_{\text{d}i}^2(5 - 3\cos^2\theta_i). \tag{G.38}$$

Hence the zero-field second moment can be written as

$$M_2 = \frac{1}{3}\left(\frac{\mu_0}{4\pi}\right)^2 \hbar^2 \gamma_\mu^2 \sum_i \gamma_I^2 I(I+1)\frac{(5 - 3\cos^2\theta_i)}{r_i^6}. \qquad (G.39)$$

In the case of a polycrystalline sample, we can perform a polycrystalline average and use $\langle(5 - 3\cos^2\theta_i)^2\rangle\rangle = 4$ to obtain

$$M_2 = \frac{4}{3}\left(\frac{\mu_0}{4\pi}\right)^2 \hbar^2 \gamma_\mu^2 \sum_i \frac{\gamma_I^2 I(I+1)}{r_i^6}. \qquad (G.40)$$

However, note that the quantity of interest in zero-field measurements is the width Δ and the start of a Kubo Toyabe relaxation is $\approx 1 - \Delta^2 t^2$, whereas our envelope function is $\approx 1 - M_2 t^2/2$. This means that $\Delta^2 = \frac{M_2}{2}$ and hence

$$\Delta^2 = \frac{2}{3}\left(\frac{\mu_0}{4\pi}\right)^2 \hbar^2 \gamma_\mu^2 \sum_i \frac{\gamma_I^2 I(I+1)}{r_i^6}. \qquad (G.41)$$

Example G.6

If not all the nuclei in the sample are the same, the results derived above are very easy to generalize. The transverse-field (TF) second moment from eqn G.30 becomes

$$M_2^{\mathrm{TF}} = \frac{4}{15}\left(\frac{\mu_0}{4\pi}\right)^2 \hbar^2 \gamma_\mu^2 \sum_i \frac{\gamma_i^2 I_i(I_i+1)}{r_i^6}, \qquad (G.42)$$

where the ith nucleus has spin quantum number I_i and gyromagnetic ratio γ_i. In the same spirit, eqn G.41 for zero-field (ZF) becomes

$$(\Delta^2)^{\mathrm{ZF}} = \frac{2}{3}\left(\frac{\mu_0}{4\pi}\right)^2 \hbar^2 \gamma_\mu^2 \sum_i \frac{\gamma_i^2 I_i(I_i+1)}{r_i^6}. \qquad (G.43)$$

G.4 Quadrupolar coupling

Any nucleus with $I > \frac{1}{2}$ can have a quadrupolar moment which can interact with an electric field gradient (EFG), and the muon will produce an EFG on nearby nuclei (and any nucleus at a site with point symmetry lower than cubic can experience an EFG, even without the presence of the muon). The quadrupolar coupling is an additional term to include in the Hamiltonian,[8] but here we will state some useful results. First, the quadrupolar interaction will be quenched in a transverse field experiment performed at very high field. However, at lower transverse field it needs to be considered. Second, in zero field it is possible to obtain analytical results for the case when the quadrupolar coupling is dominant, and these results depend on whether the nuclear spin is integer

[8] For more details, see O. Hartmann, Phys. Rev. Lett. **39**, 832 (1977) and the treatment in A. Yaouanc and P. Dalmas de Réotier, *Muon Spin Rotation, Relaxation and Resonance*, OUP (2011).

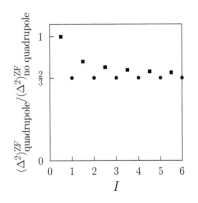

Fig. G.1 The factor in eqn G.45 plotted for integer (circles) and half-integer spins (squares).

[9]The half-integer ratio can also be expressed as $\frac{2}{3}\left(1 + \frac{3(2I+1)}{16I(I+1)}\right)$.

or half-integer. The results are (reverting back to the case of a single nucleus)

$$
(\Delta^2)^{\text{ZF}} = \begin{cases} \frac{4}{9}\left(\frac{\mu_0}{4\pi}\right)^2 \hbar^2 \gamma_\mu^2 \sum_i \frac{\gamma_I^2 I(I+1)}{r_i^6} & \text{(integer } I) \\ \frac{4}{9}\left[1 + \frac{3}{8}\frac{I+\frac{1}{2}}{I(I+1)}\right]\left(\frac{\mu_0}{4\pi}\right)^2 \hbar^2 \gamma_\mu^2 \sum_i \frac{\gamma_I^2 I(I+1)}{r_i^6} & \text{(half-integer } I), \end{cases}
$$
$$(G.44)$$

and the correction needed for the inclusion of the quadrupolar coupling is more easily seen by evaluating the ratio[9]

$$
\frac{(\Delta^2)^{\text{ZF}}_{\text{quadrupole}}}{(\Delta^2)^{\text{ZF}}_{\text{no quadrupole}}} = \begin{cases} \frac{2}{3} & \text{(integer } I) \\ \frac{16I(I+1)+3(2I+1)}{24I(I+1)} & \text{(half-integer } I), \end{cases} \quad (G.45)
$$

which is plotted in Fig. G.1. This shows that, as expected, for $I = \frac{1}{2}$, the quadrupole interaction has no effect (the ratio is one, consistent with the fact that an $I = \frac{1}{2}$ nucleus has no quadrupole moment anyway). As I increases, the ratio tends towards $\frac{2}{3}$ for half-integer spins, which is the value it has for all integer spins. The presence of the quadrupolar interaction can thus lead to a reduction of no greater than one-third in the zero-field value of Δ^2 compared to what would be expected if the quadrupolar interaction were ignored.

A short history of μSR

S. F. J. Cox[1]

Our story starts around the beginning of the 20th century with the concept of the nuclear atom. Just two particles then provided the first Standard Model of atomic and sub-atomic physics, namely the electron, identified in cathode rays by Thomson senior at the very end of the 19th century, together with the proton, discovered by Rutherford in 2019. It was a model that lasted a surprisingly long time, even surviving identification of the neutron in 1932, Chadwick at first believing his particle to be a tightly bound electron-proton pair. The model could accommodate Dirac's strange notion of antimatter, mooted in 1928, and positrons were obligingly discovered by Anderson in cosmic rays, also in 1932.[2]

But cosmic rays provided the first challenge to this simple model. At the beginning of the century, the portable instruments of choice for radiation studies were electrometers. At sea level, their failure to hold an electric charge could be blamed on Earth's natural radioactivity—c.f. Bequerel, Curie &, Co.[3] But the discharge rate seemed to increase with height above ground. Look on the internet for pictures of the splendid characters involved! Our hero must surely be the Austrian Viktor Hess who, in 1912, took his electroscope to an altitude of over 5 km, no less, in a *hydrogen* balloon, where the discharge rate increased three-fold. Hess was one of the first to argue that the offending radiation came from above, i.e. from space, or the upper atmosphere.

By the 1930s, electrometers had been superseded by the Wilson cloud chamber, in which the cosmic rays could be seen from their ionization tracks to comprise high-energy particles. The most abundant seemingly had a mass greater than that of the electron, but less than that of the proton. To measure their momentum, the town electricity supply of Rostock, northern Germany, was commandeered to power a huge coil, by Kunze in 1933, but the cautious fellow did not go as far as claiming a new particle. That claim was made more and more explicitly in papers between 1936 and 1938 by two rival groups on the east and west coasts of the USA—Street and Stevenson of Harvard, and Anderson and Neddermeyer in Pasadena.[4]

The new particle was named for its intermediate mass as the *mesotron*. Its existence was clearly unexpected by Isidor Rabi, who famously asked 'who ordered that?'[5] But the muon's mass made it a tempting candidate for the particle Yukawa had just predicted (1935) as mediator of the strong force, binding protons and neutrons together in atomic nuclei. It would decay to an electron (thereby accounting for nuclear beta decay)

[1] Steve Cox worked at the ISIS Muon Facility, Rutherford Appleton Laboratory, Oxfordshire, UK until his retirement in 2005. This appendix is adapted from his after-dinner talk on the history of μSR. It's written *in memoriam* of Anatole Abragam, Marshall Stoneham, Martyn Symons, Tom Estle, Peter Meier, Brian Webster, and Masa Senba; and with his special personal thanks to Roberto De Renzi and Ola Hartmann at CERN, Bruce Patterson and Dierk Herlach at SIN/PSI, Jess Brewer, Syd Kreitzman, Paul Percival, and David Walker at TRIUMF, Steve Cottrell and James Lord at ISIS, Rod Macrae and Tom Claxton for chemical insights and, for the semiconductor work, Roger Lichti of Texas Tech, Ted Davis of Leicester University, and Rui Vilão and all his colleagues at the Coimbra group.

[2] The positron was thus the first elementary particle to be predicted theoretically, before its observation.

[3] Already in the 1780s, Coulomb had noted a similar problem with his charged spheres!

[4] Anderson used the same cloud chamber as for finding the positron, and seems to get most credit.

[5] Rabi's atomic and molecular beam experiments would be the precursors to condensed matter NMR—nuclear magnetic resonance—as we know it now.

and have about 200 times the mass. In 1940, the mesotron decay was caught on camera, so to speak: a cloud chamber photograph taken by the Aberystwyth group (University of Wales) showed contiguous tracks of both particles. The mesotron mass had by then been refined to around $220\,m_e$, close enough to Yukawa's estimate to qualify as his meson, but by the mid 1940s it was established that it did not have the necessary strong interactions with nuclei.

The conundrum was resolved by a series of photographs taken at high altitude in 1947. The cloud chamber had by now been superseded for these purposes by 'nuclear emulsion'—extra thick photographic plates in which particle tracks are recorded directly. Not content with promising results from the Pic du Midi (French Alps), the Bristol group of Cecil Powell despatched a crazy young Italian named Lattes to 5000 metres up a mountain in Bolivia—almost as high as Viktor Hess in his daring balloon flight—and his plates showed mesotron tracks preceded by those of a heavier parent particle. Named the pi-meson (pi standing either for Powell or primary particle—we'll never know which for sure), it is indeed Yukawa's elusive meson, and its discovery earned Powell the Nobel prize in 1950.[6]

For a while, the mesotron was renamed the mu-meson. It is still called this in Abragam's classic (1961) book on nuclear magnetism (where the origins of µSR and measurement of the muon magnetic moment are succinctly described in two sentences!). Nowadays the names are abreviated to just *pion* (π) and *muon* (μ). Their mass ratio is 1.3. The name meson has now been given the more specific meaning in particle physics of a quark-antiquark pair. So the strongly interacting pion still qualifies, but not the weakly interacting muon which, like the electron, is now classed as a lepton. The 'events' that Powell's group recorded in 1947 were $\pi \to \mu$ decays, but they were cautious in the attribution, so much so that another Bristol colleague—the eccentric Charles Franck—wondered if they might be fusion events in the hydrogenous emulsion gel, triggered by the capture of *negative* muons (μ^-). The notion of muon-induced fusion was conceived! But that's another business which, even though it has much been studied on the RIKEN-RAL beams at ISIS, and elsewhere, is beyond the remit of this article. The rest of our story concerns *positive* muons (μ^+).[7] ISIS, in this context, is the name adopted for the neutron and muon sources at the Rutherford-Appleton Lab, coming on line in 1984 and 1987, respectively. It's not an acronym or abbreviation, but is the affectionate local name for the River Thames at Oxford, in turn named for the ancient Egyptian goddess!   RIKEN is the Japanese Institute for Physical and Chemical Research, which sponsored the second muon-beam complex at ISIS (1995).

It's remarkable that so much was established using just naturally occurring (i.e. cosmic-ray) muons: their position in the decay sequence $\pi \to \mu \to e$, the existence of both charge states, and good values for mass and lifetime. Knowing what we know now, muon spin rotation could also have been developed (up to a point) using cosmic-ray muons. Indeed, some teaching laboratories offer demonstration experiments, stopping

[6]A kindly man, much concerned with the social responsibility of scientists, Powell was still at Bristol while I was there, but dropped dead on retirement. Not a good advertisement for retirement.

[7]One could correctly and pedantically call them anti-muons, but we never do.

cosmic muons in a kilogram or so of copper, that can show their radioactive decay curve after a counting time of a few hours. These get a reasonable value of muon lifetime, $\tau_\mu = 2.2$ μs.[8] And a few days of counting will show muon spin precession in a magnetic field applied to the copper 'sample'. But, for quicker experiments on small samples, beams of artificially generated muons are needed.

It was particle physics that drove the development of particle accelerators, of course. Our energetic chap Lattes pops up again in California, associated with pion production at the Berkeley cyclotron; this appears to be the first of such machines to be built for American university campuses: Berkeley 1948, Columbia (New York) 1950, Chicago 1951.

And then, in 1956, theoreticians Lee and Yang of Columbia University made the heretical suggestion[9] that parity might not be conserved by the Weak Interaction, so that beta-decay processes would have no mirror image. Another Chinese-American colleague—the experimentalist nicknamed 'Madame' Wu, or 'The Chinese Madame Curie'—suggested that this could be tested by looking for asymmetry in the beta decay of polarized ^{60}Co nuclei. Pauli was not impressed. He wrote 'I don't believe God is left-handed; I am willing to bet that the experiment will give a symmetrical result' (or words to that effect).

Pauli lost his bet. In the three-day period January 15–17, 1957, three papers[10] arrived at the editorial office of The Physical Review. (No Phys. Rev. Letters in those days.) The first was by Wu et al., confirming the ^{60}Co result.[11]

More importantly for us, the other two papers reported equally compelling evidence for anisotropy in muon-decay. I'll take the last to arrive first. Friedman and Telegdi had used photographic emulsion to record some thousand $\mu \to$ e decays at the Chicago synchrotron and found more positrons thrown forward (with respect to the incoming beam) than backward. By implication, the muons must have been spin-polarized, so that parity is violated in pion decay too. Remarkably, they thought the positron asymmetry was lower than expected, and blamed this on some of the muons being depolarized in atomic-like states formed by capturing electrons! Muonium was born! It's a notable insight, but a rather unfortunate name (probably inspired by Landau's *mesonium*), which we're stuck with.[12] In an accompanying paper, Breit and Hughes set out the basic properties of muonium, and the role of the hyperfine interaction in the depolarization.[13] So now we must call hydrogen *protium*, to distinguish it from muonium and the heavier isotopes deuterium and tritium. Hughes would go on to champion the precise measurement of muonium properties as a test of quantum electrodynamical (QED) calculations, both particles in muonium behaving as point-like.[14] The comparison for protium is still spoiled by uncertainty in the proton size!

And then there was Garwin, Lederman, and Weinrich, who used pions from the Columbia cyclotron, stopping them one by one in a 'target' of graphite. The muon-decay positrons were detected, not in photographic emulsion but in scintillator 'telescopes', much as we still use today. Theirs is the paper most often cited as the origin of muon spin

[8] This corresponds to a radioactive *half*-life of 1.5 μs, but the μSR community habitually cites τ_μ, the *mean* lifetime.

[9] T. D. Lee and C. N. Yang, Phys. Rev. **104**, 254 (1956).

[10] C. S. Wu et al., Phys. Rev. **105**, 1413 (1957); R. L. Garwin et al., Phys. Rev. **105**, 1415 (1957); J. I. Friedman and V. L. Telegdi, Phys. Rev. **105**, 1681 (1957).

[11] Lee and Yang later shared the Nobel Prize. But no prize for Madame Wu! One is reminded of the similar stories of Lisa Meitner and controlled nuclear fission, and of Rosalind Franklin and The Golden Helix.

[12] The difficulties of naming exotic and anti-matter atoms, and the legitimacy of muonium as a hydrogen isotope, are discussed by D. C. Walker [*Muon and Muonium Chemistry*, Cambridge University Press (1983) Chapter 1].

[13] G. Breit and V. W. Hughes, Phys. Rev. **106**, 1293 (1957). That's Breit of the 1931 Breit-Rabi expressions for the hyperfine-coupled electron-proton spin states in atomic hydrogen, so he should know. But see R. Ferrell and F. Chaos, Phys. Rev. **107**, 1322 (1957) for the familiar B-R diagram of muonium energy levels, and the predicted recovery of muon polarization, plotted against applied field.

[14] The statement is not that they have no spatial extent, but that they obey QED without form-factors to describe their interactions.

rotation, because they did apply a magnetic field to set the muon spins precessing. They didn't plot an oscillating function of time but, instead, a slightly oscillating function of the applied magnetic field, obtained by summing the positron count over a range of elapsed times following trigger signals from the incoming particles. We'd call this now a form of (gated) integral counting, but it was sufficient to infer anisotropy in the muon decay. By all accounts, whereas Wu's ^{60}Co experiment required a lengthy set-up (at the National Bureau of Standards), that of Garwin *et al.* was a sudden brainwave, cobbled together, and completed in a long weekend! 19 data points; 20 counts per minute; 20 minutes per point.

[15]R. A. Swanson, Phys. Rev. **112**, 580 (1958).

The first muon spin rotation and relaxation signals appear to be those due to Swanson, published barely a year later (1958).[15] Working at the Chicago synchrotron, now also equipped with single-particle detectors, he surveys muon-decay asymmetry in a useful list of materials. He notes: 'substances that are good electrical conductors depolarize little or not at all; insulators and ionic conductors give partial depolarization'. His precession signal in graphite has maximal amplitude and is essentially undamped, whereas that in boron nitride shows the same initial amplitude, but a subsequent decrease consistent with dephasing in the spread of local fields at the muon stopping sites, originating from the B and N nuclear moments. This is spin-spin relaxation (analogous to T_2 relaxation in NMR). Presumably Swanson's chief purpose at the time was to identify the best target materials for studying the basic properties of muons at rest, but clearly this was the starting point for muon and muonium studies in condensed matter too. It would be a while before longitudinal-field or spin-lattice (T_1) relaxation was studied, e.g. due to molecular dynamics in muonium-substituted organic radicals, or muonium depolarization in semiconductors and ionic conductors.

So we already have muon spin rotation and relaxation when, back at the rival Columbia machine, muon spin *resonance* was equally quickly demonstrated (also 1958). Coffin, Garwin *et al.*[16] used a fixed rf frequency, pulsed with some delay to each muon's arrival, a swept field monitored by proton NMR and, again, integral counting. The resonance curve is loud and clear and displayed as a function of the relative proton NMR frequency. At resonance, the ratio f_p/f_μ is 0.314, giving a direct (presumably the first) determination of muon magnetic moment: $m_\mu \simeq 3.2\, m_p$, within 0.2% of the modern value.

[16]T. Coffin *et al.*, Phys. Rev. **109**, 973 (1958).

Their corresponding value for muon *g*-factor, relating its spin and magnetic moment, is 2.005, so, like that of the electron, close to the integer value 2 for Dirac particles.[17] This is yet another of the fortunate properties of the muon for our purposes: it means that its Larmor and cyclotron frequencies are almost identical: as our muon beams are transported in magnetic bending and focusing magnets, their polarization remains (almost) locked to the beam direction and so is conserved to the point of implantation in the sample. Then the processes of slowing down and thermalization are essentially electrostatic (ionization, primary, and secondary electron ejection—all spin-independent and rapid) and the low muon mass means there is relatively little radiation damage,

[17]The modern value is 2.00233184.

and that usually left behind in the muon's terminal track. So polarization is again conserved until data-taking begins. Well, except when muonium is involved.

Measured originally at CERN[18] in Geneva in the 1960s and again at the Brookhaven lab (Long Island, New York) around the turn of the millenium, the quantity $g - 2$ is is still giving the theoreticians headaches, showing a slight discrepancy with calculations of the radiative corrections. It is most precisely determined by the deviation of muon spin and momentum after multiple circuits of a storage ring. Relativistic time dilation gives these muons an effective lifetime of over 60 μs, allowing their muon spin rotation signal to seen out to almost a millisecond! The same phenomenon allows cosmic-ray muons to reach us from the upper atmosphere. They are zapping through you at about one per second as you read this.

Garwin *et al.* well knew what they had started, ending their 1957 paper with the perceptive comment that muons '...will become a powerful tool for exploring magnetic fields in ... interatomic regions'. Let's now set this in the context of conventional magnetic resonance, and make some comparisons. We have the Berkeley group to thank for introducing the catchy name 'μSR', standing for muon spin rotation, relaxation, and resonance (and championed tirelessly ever since by the inimitable Jess Brewer). Unveiled in their first μSR Newsletter in 1972, it was clearly coined to resemble 'ESR' (electron spin resonance) and to underline basic analogies in the methods.

Both ESR and NMR were successfully demonstrated at the end of World War II, in 1944–45. This was no coincidence, but owed much to wartime electronic developments. A good deal was known about the behaviour of spins in condensed matter well before that, of course: the thermodynamics of paramagnetism from 1905 onwards (Langevin, Brillouin, Curie—something of a French monopoly here!) and the more atomistic picture from the 1920s onwards—electron spin, nuclear moments, atomic spectra, hyperfine interactions. There was even a good understanding of spin-lattice relaxation already by the 1940s, thanks largely to experiments in Leiden (the Netherlands) on adiabatic demagnetization.

The first ESR signal is attributed to Zavoisky, working in 1944 at Kazan, one of Russia's oldest university cities. It's a broad weak signal, recorded at the rather low frequency of 100 MHz. It's hard enough finding a new magnetic resonance signal by sweeping the field even now, when you know it should be there, so this was quite an achievement. ESR sensitivity increases with frequency and so, at Oxford University's Clarendon Lab, *microwave* spectrometers were quickly built, using the so-called X-band (9 GHz, 3 cm wavelength) equipment developed for radar. It is remarkable that the recently top-secret components were so quickly made available for academic research, together with detailed handbooks on microwave design.[19] The Oxford group of Bleaney & Co. concentrated on magnetic ions in inorganic solids, for a while making the subject their own.[20] The first ESR spectra of organic radicals are

[18]CERN stands for Conseil Européen pour la Recherche Nucléaire but is officially translated as the European Organization for Nuclear Research.

[19]I believe the iconic solid-copper 3 cm waveguides in the ESR spectrometer I used at Bristol in the 1960s were largely government surplus!

[20]Some would prefer to call this *EPR*—electron *paramagnetic* resonance—reserving *ESR* for paramagnetic centres where orbital magnetism is less important, and the *g*-factor closer to the free-electron value of 2. This is generally the case for organic radicals.

[21]The almost simultaneous publications are E. Roduner, P. W. Percival, D. G. Fleming, *et al.*, Chem. Phys. Lett. **57**, 37 (1978), and C. Bucci, G. Guidi, *et al.*, Chem. Phys. Lett. **57**, 41 (1978). (The Parma group submitted first but the Zurich group had the clearer spectra!)

[22]V. W. Hughes *et al.*, Phys. Rev. Lett. **5**, 63 (1960); K. Ziock *et al.*, Phys. Rev. Lett. **8**, 103 (1962).

[23]Not sure why that took so long, but removal of dissolved oxygen is crucial to avoid spin-exchange depolarization.

[24]Don Fleming, Paul Percival, and David Walker.

[25]Chapters 12 and 19.

[26]This role and nomenclature was belatedly accepted and endorsed in 2001 by the International Union for Pure and Applied Chemistry—I believe thanks largely to David Walker and myself browbeating the IUPAC secretary Koppenol at a Radiation Chemistry conference! See W. H. Koppenol, Pure Appl. Chem. **73**, 377 (2001) for the recommended usage, including definitions of such strange terms as *muono* versus *muonido*, *muonated* versus *muoniated*, etc.

[27]A. Schenck *et al.*, Chem. Phys. Lett. **12** 544 (1972).

reported from the USA: those of stable radicals by Pake in California from 1952 and of transient free radicals created by *in situ* radiolyisis by Fessenden and Schuler in Pittsburgh from 1960. It's these latter, the ethyl radical in liquid ethene and suchlike, that are mimicked in the first μSR studies of organic radicals, for which we have to wait until 1978 for the work of the Zurich group at SIN (the then Swiss Institute for Nuclear Research, now PSI, the Paul Scherrer Institute) and of the Parma group at CERN,[21] quickly followed by programmes at all available muon facilities. Emil Roduner would later (1988) be awarded the prestigious Werner prize and medal for his rôle in implementing this hugely important activity in chemistry.

But we are getting too far ahead. Although muonium had been blamed for loss of initial polarization from the outset (1957), and this hypothesis supported by recovery of the polarization in high fields— the Ferrell and Chaos (1957) method variously known as decoupling, quenching, or repolarization—the expected triplet-state precession signal could not at first be seen in solids or liquids. It was first found in gases—N_2O and Ar.[22] Don Fleming measured muonium reaction rates in Br_2/Ar mixtures at Berkeley as early as 1976, went on to perform the classic Mu + H_2 studies at TRIUMF, Vancouver, and is still coming for beamtime to ISIS! (TRIUMF was named for its original sponsors as the Three Universities Meson Facility, but is now a national, not to say international, Canadian lab). Also in 1976, the Zurich group detected muonium in degassed water.[23] All this following on from studies of Radiation Chemistry, pursued into the 1980s, notably by the Vancouver groups.[24] Interest turned from the nature of the radiation track to focus on the thermalized end-of-track products, which μSR was superbly able to do: *Muonium Chemistry* was born. Short-lived though it may be, muonium is to be used as a light isotope of hydrogen, as deuterium is used as a heavy isotope (Mu:H:D = $\frac{1}{9}$:1:2). This enhances and clarifies isotope effects in molecular structure and dynamics, as well as in reaction kinetics, emphasizing the importance of nuclear zero-point energy throughout.[25]

A word on nomenclature is in order here. In condensed matter, there are no such animals as free muons, just as there are no free protons. Those that do not manage to form atomic muonium lower their energy by seeking out sites of high electron density. So it is useful to reserve Greek μ for the incoming energetic particle, and switch to Roman Mu in any chemical context, emphasizing its role as a hydrogen isotope.[26] So, in oxides, the diamagnetic muon site is invariably adjacent to oxygen: one could write $\mu^+ + O^{2-} \rightarrow (OMu)^-$, a species analogous to the hydroxyl ion. Likewise in μSR spectra for ice or water, a muonium signal coexists with that of a so-called diamagnetic fraction, identified already in 1971[27] by the proton couplings as the muonium-substituted molecule, HMuO. The thermalized muons must initially stick to lone electron pairs on oxygen to form an ionic species, quickly transformed to the neutral species by proton exchange: $H_2O:Mu^+ + H_2O \rightarrow HMuO + H_3O^+$. The fractions have been much studied in ice and water by the Vancouver group of Paul Percival, and lately even in the supercritical phase, leading to a new branch of muonium chemistry able to characterize transient free radicals under hydrothermal conditions.

As another example, the non-magnetic muonium centres in fluorides, commonly written FμF, should properly be written [FMuF]$^-$, analogous to the bifluoride ion [FHF]$^-$.[28] Noticed by wild exchanges of polarization with the close F nuclei in signals from the alkali and alkaline earth fluorides in 1986.[29] These states provide a good testing ground for calculations of structure and the effects of muon zero-point energy, involving what Steve Blundell calls 'the quantum muon', lately pursued in his Oxford group. Other ionic states resembling the interstitial proton or hydride ion can be written simply as Mu$^+$ and Mu$^-$.[30] But, when speaking rather than writing, be careful that the audience does not confuse Mu$^-$ (i.e. [μ^+ee]) with μ^- (i.e. the *negative* muon).[31]

Back to 1945–1946, NMR was achieved almost simultaneously by two groups, again on opposite sides of the USA, but apparently unaware of each other's work. Both detected proton resonance: protons in water at Stanford and protons in a hydrocarbon at Harvard. The west-coast team explain that the rf field drives precession of the nuclear moments, and that this in turn induces a voltage in their rf circuit; the east-coast guys describe it as absorption of rf photons, inducing transitions between the spin states. This is presumably the first example of the duality of semi-classical and quantum mechanical descriptions that continues to be useful throughout magnetic resonance, including μSR.[32] The leaders of the two groups, Bloch and Purcell, shared the Nobel prize in 1952, and NMR techniques now pervade physics, chemistry, and medicine.

A muon spin rotation signal resembles the free induction decays of pulsed NMR, but the signal is spontaneous–it does not require 90° pulse preparation or tuned-circuit detection; rf-driven muon spin resonance is more akin to the methods of broad-band NMR, although some of the pulsed techniques of spin-locking or spin echoes can be imitated–albeit with severe limitations from the muon lifetime.[33] In NMR, studies of Knight shifts in metals date from 1949, and chemical shifts from 1950. Muon Knight shifts (K_μ) reflect the local electronic response to the presence of the muon's positive charge, and have been widely surveyed since the late 1970s, at first exploiting a stroboscopic μSR method using the fine time structure of the SIN beam.[34] K_μ can be particularly large in semimetals, notably Sb, perhaps representing incipient muonium formation,[35] and an associated Korringa relaxation can sometimes be seen. The muon lifetime precludes measurements of the much smaller chemical shifts in muonium-substituted organic molecules, however. And there is no direct counterpart of NQR (nuclear quadrupole resonance), since the muon has spin-$\frac{1}{2}$. But, in 1984, Abragam suggested how quadrupole splittings on adjacent nuclei might be measured by a form of level-crossing resonance (LCR). Having a greater magnetic moment than all other nuclei, the muon Zeeman energy can always be tuned to match the combined Zeeman and quadrupole energy of a neighbour, allowing resonant cross-relaxation between them. The resonance

[28] No chemist would write FpF! But FμF seems to have stuck, and is even used in this book (Chapter 4). Bifluoride is no ordinary hydrogen bond. It's much stronger, the proton (or muon) charge pulling the F$^-$ ions closer together. Coulson realized this could not be just electrostatics, and that a degree of covalency was implied. (FHF)$^-$ can be described as a donor (F$^-$)—acceptor (HF) pair, or as 3-centre 4-electron molecular orbital. In contrast, bond-centred muonium in silicon and certain other semiconductors is a 3-centre 3-electron bond. The extra unpaired electron weakens the original Si–Si bond, pushing the Si atoms further apart. And of course it makes this defect centre paramagnetic, rather than diamagnetic.

[29] J.H. Brewer et al., Phys. Rev. B **33**, 7813(R) 1986.

[30] See Chapters 3 and 10.

[31] See Chapter 22.

[32] See Chapter 4.

[33] See Chapters 19 and 23.

[34] See e.g. M. Camani, Phys. Rev. Lett. **42**, 679 (1979).

[35] O. Hartmann et al., Hyp. Int. **6**, 47 (1979); A. Schenck, Hyp. Int. **8**, 445 (1981).

[36]S. R. Kreitzman *et al.*, Phys. Rev. Lett. **56**, 181 (1986).

[37]This work was taken up by, amongst others, Roduner, Kiefl, Kreitzman, Macrae, Percival, Walker: see Chapter 12.

[38]J. W. Schneider *et al.*, Phys. Rev. B **41**, 7254(R) 1990.

[39]These latter known as 'Arizona beams' after the originating group, but now more commonly as 'surface beams'. They are not to be confused with ultra-slow muon beams now available for surface-science studies (Chapter 18).

[40]R. S. Hayano *et al.*, Phys. Rev. B **20**, 850 (1979).

[41]See Chapters 6 and 7.

in copper metal provided proof of principle in 1986;[36] resonances with quadrupolar nuclei such as B, N, Ga, As, or with ^{17}O in suitably enriched samples (e.g. of ice or cuprate superconductors) have subsequently allowed muon site determination in a variety of diamagnetic environments. In terms of the total energy of all spins concerned, these resonances are only seen where degeneracies are lifted–hence the somewhat pedantic term 'avoided level crossing' (ALC). The notion was quickly extended to allow multiple nuclear hyperfine splittings in organic radicals to be resolved and measured.[37] And likewise in the vicinity of paramagnetic muonium defect centres in solids–for some particularly beautiful examples, look for the cuprous halide spectra by Schneider and colleagues.[38]

Concerning the relative sensitivities, ESR and NMR rely on Curie's Law to establish a measurable starting polarization. At room temperature, and in a field, say, of 3 T, the thermal-equilibrium polarizations are of order 10^{-2} for unpaired electron spins, and 10^{-5} for nuclear spins. Factoring in their very different magnetic moments for microwave or rf detection, a sample must contain at least 10^{10} paramagnetic centres to get a decent ESR spectrum, or 10^{17} nuclear spins for NMR. Lowering the temperature may not help if spin-lattice relaxation is slow, as it invariably is for nuclear spins. In μSR, on the other hand, the initial polarization is intrinsically high, by the nature of the pion decay, *irrespective of the field or temperature of the sample*. It is 60-80% from pions decaying in flight in the beamline, and close to 100% from those decaying at rest on the surface of the production target.[39] The subsequent evolution of polarization within the sample is monitored, not by electromagnetic induction at the tiny energy of the relevant muon-spin transition (e.g. 10^{-4} eV for Larmor precession at 1 MHz), but at the enormously higher energy of the muon decay (50 MeV), via single-particle detection of the emitted positrons. Thanks to this amplification, μSR signals may be accumulated with just a million or so implanted muons for simple Larmor precession signals, or 10^7–10^8 muons for more complex radical spectra. This is a huge sensitivity *per spin*. And the equal sensitivity of μSR to both diamagnetic and paramagnetic species combines aspects of NMR and ESR which–as for the Mu and HMuO signals in water–are often visible in the same spectra. Accumulation rates depend on beam intensity, of course, and we shall see how these have increased dramatically, and continue to do so.

A happy consequence of all this is that zero-field (ZF) measurements (i.e. those in null external field) are possible with μSR, in a direct manner that is essentially impossible for ESR and NMR. The iconic example is Kubo-Toyabe relaxation–the hypothetical function derived by these authors in 1966, but not observed or finding any application until rediscovered[40] by the Tokyo μSR group of Hayano, Yamazaki, *et al.*, working at the TRIUMF lab in 1979 while facilities in Tokyo were under construction. ZF measurements have become an important speciality, not least for weak or fragile magnetism (heavy fermions, spin-glasses).[41]

Accelerators in the 1960s, both in America and by then also in Soviet Russia (Dubna near Moscow, Gatchina near Leningrad), produced poor beams, but a flood of new particles. This prompted the building of more powerful and expensive machines in the 1970s, again in North America (LAMPF, the Los Alamos Meson Physics Facility in New Mexico, and TRIUMF in Vancouver) but now also in western Europe (SIN near Zurich and new accelerators at CERN in Geneva). This was the era of the so-called *meson-factories*. They were grand and expensive facilities, built by particle physicists for their own purposes. But by the time they were ready, particle physics had moved on, and needed higher energies still. Happily, the machines optimized for pion production became useful for medium-energy nuclear physics, and beamtime was also allocated–somewhat grudgingly–to a growing community, mostly with a background in nuclear methods but wanting now to explore the use of muons in solid state physics, and in chemistry.

A gradually increasing publication rate during the 1960s and '70s began to define the subject areas that are familiar today. They group broadly into the use of the muon as a probe of internal fields in magnetic materials–a microscopic gaussmeter, so to speak–and its use as a lightweight proton analogue, or of muonium as a light isotope of hydrogen. Muonium had by now been identified in a number of insulating solids, including quartz and ice, and–an example of the new Russian activity–frozen CO_2.[42] Russian theoretical work in the 1960s and '70s on muonium spin dynamics was far ahead of its time, deriving expressions for spin-exchange and charge-exchange depolarization, for instance, that were only much later observed.[43] Unfortunately, the English translations are hard-going. Some three decades later, the topic would be reworked and greatly clarified by the late Masa Senba (Canada).[44]

In 1972, Gurevich at Dubna observes muon diffusion in copper. In 1973, Brewer *et al.*, at Berkeley discover 'anomalous muonium' in silicon. Also in 1973 (a good year!) Foy *et al.* at SREL–the Space and Radiation Effects Lab, Virginia, USA–measure the interstitial fields in ferromagnetic nickel and iron.[45] In 1975, spin-glass studies (CuMn, AuFe) begin, also by the Virginia group. Studies of ferromagnetic Co, Gd, antiferromagnetic Cr, and numerous rare earth elements quickly follow in Europe. Perhaps more important than the internal field is the remarkable divergence in relaxation rate seen towards the critical temperatures, e.g. in MnSi, by Hayano *et al.* working at TRIUMF in 1978. The relaxation peaks are extreme examples of the T_1-minima familiar in NMR: classic NMR studies of antiferromagnetic CoF_2 and MnF_2 were later repeated and broadened using μSR at CERN by the Parma group (De Renzi, Bucci *et al.*). The comparison amply validates μSR as a probe of critical fluctuations, approaching T_N more closely (as well as extending *below* T_N) and bridging the time-scale gap with neutron scattering. Striking anisotropy in the critical spin correlations in Gd, anticipating the nature of the ordered state, would be revealed at ISIS.[46]

Also in 1978, a real sense of community was achieved at the first international μSR conference, held in Switzerland. The triennial conference series continues to bring together scientists of apparently disparate disciplines, and its proceedings are a succinct summary of the developments

[42] G. G. Myasischeva *et al.*, Zh. Eksp. Teor. Fiz. **53**, 451 (1967) [JETP **26**, 298 (1968)].

[43] V. G. Nosov and I. V. Yakovleva, Nucl. Phys. **68**, 609 (1965); I. G. Ivanter and V. P. Smilga, JETP **27**, 301 (1968); **28**, 796 (1969); **33**, 1070 (1971); **34**, 1167 (1972).

[44] See for example M. Senba, J. Phys. B **24**, 3531 (1991).

[45] Already in 1944, in Quebec, Rasetti had studied the deflection of cosmic ray muons through magnetized iron!

[46] Through work by O. Hartmann, G. M. Kalvius, P. Dalmas de Réotier, A. Yaouanc, and others.

[47]The International μSR conferences have been held in Rohrschach, Switzerland (1978), Vancouver, Canada (1980), Shimoda, Japan (1983), Uppsala, Sweden (1986), Oxford, UK (1990), Maui, USA (1993), Nikko, Japan (1996), Les Diablerets, Switzerland (1999), Williamsburg, USA (2002), Oxford, UK (2005), Tsukuba, Japan (2008), Cancun, Mexico (2011), Grindelwald, Switzerland (2014), Sapporo, Japan (2017), with the 2020 conference in Parma, Italy postponed until 2022 due to the Covid pandemic. The proceedings of these conferences have been published in various journals and provide a useful set of snapshots charting the development of the field.

[48]This latter, surprisingly, from Abragam. But he became supportive later, approving–on one of my begging missions–the French contribution to the first muon beam at ISIS, and shortly afterwards, in 1984, publishing his idea for muon level-crossing resonance.

[49]The muon being a lepton and the proton a hadron, these analogies would horrify particle physicists, but they are not an issue at thermal energies, where electrostatics dominate the behaviour.

[50]A distinctive glitch in the field dependence of this linewidth as the Cu quadrupole coupling is quenched would later be predicted, by Hartmann of Uppsala (Sweden) in 1977, and used by Schenck *et al.*, at SIN, to define the muon site.

and achievements.[47] Some of the early experiments were speculative, of course. 'These people would stop muons in their sandwiches' a TRIUMF director was heard to say! And there were the all the usual objections to a new technique: 'It'll never work'; 'OK, so it does, but what does it tell us that we don't already know?' Or 'A probe should be passive. Doesn't yours perturb what it's studying?' (well, of course it does, being a charged particle.) And even 'These muons are posing more questions than they're answering!'[48]

To answer these questions or criticisms, and also because they exemplify the topics I have most experience of, in what follows I'll concentrate on solid-state studies where the muons themselves, as prototype light interstitials, or muonium mimicking hydrogen defect centres, are the focus of attention.[49] All such experiments have revealed more than was looked for, including some surprises that profoundly upset received wisdom.

The hydrogen-in-metals community was one of the first to take the new opportunity seriously. Their μSR proposals were not speculative, but had a very specific focus. Hydrogen solubilities and diffusion rates vary enormously from metal to metal, but there were no atomistic pictures of the interstitial sites or hopping mechanisms involved. NMR is of very limited use, frustrated by the rf skin-depth or low concentrations. Muons may be implanted in virtually any material, irrespective of hydrogen solubility and (up to now at least) so few are present in the sample at any moment, that pairwise interations are unimportant. The isolated interstitial is studied. And somewhat counter-intuitively, atomic muonium is not formed in metals. The conduction electrons instead form a screening charge that precludes formation of the bound state. As Jacques Chappert of the Grenoble group put it: 'the muon flirts with all the electrons, but does not marry one'! The same is true of the screened-proton state.

The Gurevich (1972) muon spin rotation signal in copper is damped at liquid-nitrogen temperatures, consistent with a spread of local dipolar fields from the surrounding Cu nuclei.[50] At higher temperatures the damping rate decreases: the line-width is motionally narrowed by the onset of muon diffusion. An Arrhenius temperature dependence shows this to be incoherent hopping, rather than any coherent or band-like motion, but the parameters are wrong for the expected potential barrier between adjacent sites. The activation energy and prefactor are both too low. The muon should be 'self-trapped' by the local lattice distortion and electronic perturbation it has created. Now the theoreticians get interested. They envisage lattice vibrations momentarily equalizing the potential at adjacent sites, allowing the muon to tunnel through the barrier. Stoneham called it 'phonon-assisted tunnelling'. Of course, with its lighter mass, $m_\mu = \frac{1}{9}m_p$, the muon has a greater propensity for tunnelling than the interstitial proton, and so provides a stringent test of theories on the borderline between classical and quantum behaviour, where the low mass of H is already a challenge. Over many years (and despite his favourite quip 'Never mind the data, what is it in theory?') encouragement and constructive criticism from the late Marshall Stoneham–an independent and well-respected theoretician–have been immensely valuable to the μSR community worldwide. Already in 1975 he had drawn attention to The Theory of Defects in Solids in his substantial book of that title, making the topic respectable at a time when undergraduate physics courses were still only teaching the structure and excitations of perfect crystals.

The Uppsala, Jülich and Grenoble groups, working together in the CERN μSR Collaboration in the late 1970s, were fortunate to have access to a very powerful dilution refrigerator, designed for beamline use. Pursuing the copper measurements to lower temperatures, they were perhaps surprised to find the muon mobility exhibiting a minimum around 50 K, and increasing again to lower temperatures–still increasing below 40 mK. Later zero-field experiments, notably at TRIUMF from 1982,[51] confirmed that this was so, zero-field Kubo-Toyabe relaxation being able to distinguish static and dynamic local fields. Initially assumed to be the half-expected cross-over to coherent tunnelling, i.e. band-like motion of these lightweight particles, the effect was later explained by Kondo, no less, as relating to how the electron screening charge struggles to follow the muon at the lowest temperatures. It has long been questioned whether the muon-electron mass ratio is sufficiently large for the usual Born-Oppenheimer (1927!) or adiabatic approximations to apply. This may be the one example in μSR where they do break down–this being most likely near electronic degeneracies. They work fine for all the molecular stuff.[52]

It's worth mentioning here that a similar minimum of mobility is seen for atomic muonium diffusing in certain insulators and semiconductors and notably in the so-called cryocrystals: frozen hydrogen, nitrogen, and all the rare-gas solids.[53] But the hopping rates for this neutral interstitial are very much higher than for muons in metals, and the question of whether coherent band-like propagation is exhibited at the lowest temperatures still divides opinion.

Muon diffusion in metals, and also some metal hydrides, continued to be studied at all available muon sources throughout the 1980s and '90s. The zero-field studies especially benefitted from the arrival of pulsed sources: KEK-BOOM, Tokyo, 1980; ISIS at RAL, 1987.[54] An initial scepticism concerning the suitability of pulsed sources was quickly overcome, as their capabilities were seen to ideally complement those of the continuous beams of PSI and TRIUMF. The Uppsala group had to wait for ISIS muons to come on line to complete their survey with the 'difficult' metals: Pt, Pb, W, Mo, Au, Ag, and Pd, i.e. those with weak nuclear moments, so slow spin relaxation.[55] At ISIS, the relaxation function can be followed to 10 muon lifetimes.[56] Mysteriously, at the time of the 'cold fusion' episode, the palladium samples were recalled to Sweden.

The semiconductor community, on the other hand, was slower to respond to the early muon results. I have a semiconductor textbook from the 1960s on my shelves that tells me there is no hydrogen impurity in Si. Evidence for muonium formation in Si and Ge, by Feher (1960) in America and Andrianov (1970) in Russia, attracted little attention. By

[51]Through the work of C. W. Clawson, R. Kadono, J. H. Brewer, G. M. Luke; other labs quickly acquired DRs, of course

[52]See the work of B. C. Webster.

[53]See the work of V. G. Storchak, and also R. Kadono.

[54]KEK is the Japanese National Lab for High Energy Physics, and BOOM its Booster Meson source.

[55]See work by O. Hartmann, E. Karlsson, S. Harris, and coworkers.

[56]The limit set by cosmic radiation has been optimistically estimated by David Bugg as $50\tau_\mu$!

[57]S. F. J. Cox, M. C. R. Symons, Chem. Phys. Lett. **126**, 516 (1986).

[58]A. Mainwood and A. M. Stoneham, J. Phys. C: Solid State Phys. **17** 2513-24 (1984).

[59]Work by T. A. Claxton, T. L. Estle, S. Estreicher, P. F. Meier, R. Jones, C. G. Van de Walle, and others.

[60]R. F. Kiefl *et al.*, Phys. Rev. Lett. **60**, 224 (1988).

[61]J. W. Schneider *et al.*, Phys. Rev. Lett. **71**, 557 (1993).

[62]See Chapter 16.

[63]E. A. Davis, J. Non-Cryst. Sol. **198**, 1 (1996).

the 1970s, the statement might have been that, if hydrogen is present, it is electrically inactive, and so unimportant. Only the small community of muon specialists got excited by the amazing discovery, by Brewer *et al.* of Berkeley, that muonium in silicon is *metastable*: an unpredicted and highly anisotropic state coexists in the spectra with the anticipated quasi-atomic form. They called it 'anomalous muonium', or Mu*. Zurich and Konstanz groups working at SIN had refined the parameters by 1978, and likewise those of anomalous muonium in Ge and diamond, but it was some while before their implications for local structure were recognized. The muon must be located at a node of the singly occupied orbital, at the centre of a stretched Si-Si bond.[57] No stable site had been found in early simulations,[58] because the available computing power did not allow for distortion of the rigid diamond-type lattice. Knowing what to look for, later simulations quickly confirmed the result,[59] as did measurement of spin density on the adjacent nuclei by ^{29}Si level-crossing resonance,[60] and likewise on ^{13}C in diamond.[61] This was a real scoop for μSR. Nowadays reliable calculations would generally predict the site and electronic structure *before* valuable beamtime is used for confirmation.[62]

By the 1980s, hydrogen impurity in semiconductors was recognized as unavoidable, being introduced during vapour deposition or etching, but useful for passivating unwanted defects. In amorphous Si, this is in fact essential for electronics-grade material[63]. So the electronics community finally had to pay attention to the existence of these esoteric muonium centres, with defect levels firmly in the energy gap. Identification of bond-centred muonium prompted ESR searches for its ordinary hydrogen counterpart, which was duly found by Russian and Danish groups in proton-implanted Si. The hf parameters scale exactly with the ratio of muon and proton magnetic moments, totally validating muonium as a model for interstitial hydrogen.

By the mid 1980s, the installation of spin-rotators in beamlines (derived from particle separators) allowed precession signals to be recorded in high magnetic fields. The high fields quenched unresolved nuclear hyperfine interactions, so that bond-centred muonium could be seen also in a few compound semiconductors with large nuclear moments, namely GaAs and GaP. And for 'normal' muonium, i.e. the isotropic or quasi-atomic state, detectors with sub-nanosecond resolution developed by an exceedingly bright young post-doc, Holzschuh, were able to detect the zero-field hyperfine oscillation in Si, Ge, and even in diamond. Patterson calls it the 'heartbeat' signal; it represents a direct measure of hf constant, which in diamond is close to the vacuum-state muonium value of 4.5 GHz. Sadly, Holzschuh died some years later. Patterson is author of the comprehensive and influential review of μSR studies up to 1986, written as a dissertation to establish his position at Zurich University; his young student, perhaps postdoc, Rob Kiefl, builder of the high-field spectrometer at SIN, would go on to animate the μSR scene at TRIUMF, where he has also now set up β-NMR facilities.

Trapping and detrapping of thermally activated carriers revealed a fascinating interplay of site and charge state of the muonium centres. In the 1990s the corresponding deep acceptor and donor levels were determined, in Si chiefly by rf resonance at TRIUMF and in Ge by spin-lattice relaxation at ISIS,[64] and matched to those being determined for hydrogen. This electrical activity was also strikingly revealed by the effects of illumination, whether by flash-lamp[65] or laser.[66] Then another μSR scoop came with the discovery of a *deep-to-shallow instability* of muonium centres, and by inference monatomic hydrogen centres too, both in semiconductors and insulators: from 1999 onwards shallow-donor levels were discovered in certain sulphides, nitrides, and especially oxides with high electron affinity.[67] Some of the oxides were those being mooted at the time as possible replacements for SiO_2 as a gate dielectric, so that conductivity *caused* by hydrogen impurity would be a serious matter indeed! I like to give the name *muonics* to these studies where muonium is used as a model for the structure and electrical activity of hydrogen centres in electronic materials, but it has not caught on.

A parallel study of spin and charge transport in organic polymers began in 1984, following a pioneering study of solitons in polyacetylene.[68] A survey of numerous polymers and related donor and acceptor molecules was taken up by the Oxford group.[69] Not content with organic metals, even superconductors, this prolific team, with its long succession of DPhil students, went on to an equally wide survey of the newly discovered purely organic ferromagnets, μSR being ideally suited to detecting fragile magnetic ordering at dilution-fridge temperatures.

Now a word on working methods. In the Meson Factory era, μSR rigs were typically the property of individual university groups, assembled in the beamline at the beginning of each beamtime allocation. At CERN, the first day would typically be spent mounting photomultipliers and wrapping light-guides! In the best cases (at SIN), the spectrometers would be craned in as a unit, and craned out at the end of the experiment. Admittedly beamtime allocations were measured in weeks, or even months, rather than days as now, but still it was an inefficient business ('time is data!', remarked an impatient Alex Schenck) and a great obstacle to newcomers. And there was little or no help from accelerator staff (dare I mention especially at SIN). Few groups had developed automatic control of field or temperature scans and (at TRIUMF and KEK especially) it was considered good for the graduate students to sit up all night starting one run at a time! Well, data rates were lower, so that did allow time for analysis, and deciding what best to do next. We had to work like that for a few years at ISIS, too. I remember it well, snatching a few hours sleep now and then in my camper-van.

Of all the technological advances that have enabled new science, the evolution of computing from this time is perhaps the most extraordinary. At CERN in 1980, the μSR group's rack-mounted computer had to be booted with a precious strip of perforated paper 'ticker tape'. Data accumulated on reels of magnetic tape, but had to be transferred to punched cards for analysis on the particle physicists' main-frame ma-

[64] See papers by T. L. Estle, S. R. Kreitzman, R. Lichti, K. H. Chow, and many others.

[65] R. Kadono *et al.*, Phys. Rev. Lett. **73**, 2724 (1994).

[66] See Chapter 19.

[67] See Chapter 10.

[68] K. Nagamine *et al.*, Phys. Rev. Lett. **53**, 1763 (1984), at the Tokyo BOOM facility.

[69] Bill Hayes, Francis Pratt, soon to be joined by a young post-doc named Steve Blundell.

chine. The results came out on miles of Z-fold paper. Manuscript (yes, hand-written) articles were taken the Typing Pool. At Zurich University, Peter Meier surprised us by splitting from the physics department to form his 'Computer Assisted Physics' group. But he was ahead of the game, computer and simulation now being referred to as the third pillar of science, alongside (or perhaps between) theory and experiment. And now, anybody's laptop can be used to control the experiment, collect and analyse the data, and then to write the paper.

By the 1980s, the meson factories had begun to invest in beamlines and equipment specifically for μSR, in response to demand.[70] But I believe it is true to say that the concept of a *User Facility* originates at ISIS–that is, standard instruments permanently installed and ready for new as well as experienced Users, with Instrument Scientists on hand to help and advise.[71] Training courses for new Users are offered occasionally (not to mention the more theoretical Advanced School that is the origin of this monograph). Remarkably powerful programs have been made available for simulating the muon response and fitting the data in complex situations (see Chapter 17).

This vision was perhaps made easier by setting up the new facility from scratch in the mid 1980s.[72] The awkward legacy of Big Science versus Little Science was still with us to some extent, however, even though materials and molecular science have long since entered the big league, with neutron and synchrotron sources.[73] The original plan for a combined muon and medical pion source at ISIS was dropped during construction of the neutron facilities. The request to add muons later was at first refused: it was clear that the Research Council would prefer the money spent on bench-top research in individual universities. Without contributions of equipment, cash, and personnel from a consortium of European and UK groups to kick-start the project, there might be no ISIS muons. Perhaps in an attempt to placate the Research Council, there was a worrying period when it was thought appropriate to charge users the full economic cost of beamtime![74] I understand that only the occasional industrial Users pay these full rates at ISIS now. Admittedly we also had some tense moments when our principle of accepting proposals on merit alone meant that our sponsors did not always get the access they thought they had paid for! But it's a good principle and it has created a wider European and British μSR community. Good proposals are also accepted from further afield (Russia, North America). The model won us several European grants to extend the facility, and access to it. On the very first day of beam in 1987, interested visitors from Tokyo, anticipating the high instantaneous data rates, were waiting at the door to try analogue detection methods! RIKEN would later sponsor the RIKEN-RAL beamline complex (1995, overseen by the indefatigable frequent flyer, Ken Nagamine) as a testing ground for methods to be used at their projected J-PARC facility (the Japan Proton Accelerator Research Complex at Tokai, 2008). Of the contributing authors to this monograph from those early days of ISIS muons, Roberto De Renzi is

[70]See Chapter 13.

[71]This together with a more flexible development area–ours is currently filled with lasers, see Chapter 19.

[72]G. H. Eaton, C. A. Scott, and colleagues.

[73]See Chapter 24.

[74]At tens of thousands of pounds per day, what group is going to apply for a week's time rather than hire another postdoc? Nowadays it's understood that the expensive national labs should be made available to as many scientific areas as possible. At least I hope so.

from the Parma Group, Pierre Dalmas de Réotier from Grenoble, and Rui Vilão from the Coimbra Group in Portugal.

The other μSR labs around the world now also have suites of standard spectrometers, and facility staff to help, of course. SIN, renamed the Paul Scherrer Institute in 1988, has called its centre 'SμS', the Swiss Muon Source, and TRIUMF has its 'Centre for Molecular and Materials Science' (a name I claim some credit for!). J-PARC has survived the 2011 earthquake, resuming User operations early in 2012 at its high-intensity MUSE (Muon Source Extension) as part of a Materials and Life-Sciences facility. Let's call the current period the User-Facility Era! And after a word on beams and data rates I can hurry through just a few more scientific highlights.

Unlike the steady exponential attenuation of beams of light or X-rays, i.e. photons, where intensity is greatest at the surface of a sample, charged particles implant to a certain depth, in a relatively narrow bell-shaped 'range curve'. It is this that makes pion or proton therapy particularly suited for some cancer treatment. The early muon beams, generated by the decay of pions in flight, had relatively high momentum, and spread of momentum, requiring tens or hundreds of grams of sample material. This was ideal for the glass bulbs beloved of chemists for their liquid samples, and is indeed necessary for doing μSR inside thick-walled pressure cells, as for studies of the effects on hyperfine fields in Fe and Ni by the μSR Collaboration at CERN in 1980[75] and now taken to extremes at PSI.[76] It's not so good for gases, and useless for expensive new materials available in only gram or milligram quantities. Here the Arizona or surface beams, used for μSR since the mid 1970s, were a crucial innovation. These have a range centred around 100 mg/cm^2 so, in water for instance, implanting to about 1 ± 0.2 mm, this range scaling roughly inversely with density in other materials–liquids, solids, and gases. The range curve allows the use of moderating material to trim implantation depth to a degree, both for decay and Arizona beams. Collimation or veto detectors can allow measurements on samples a few mm across, or tens of mg in weight, although most are much larger. Position-sensitive detection, imaging the sample, was demonstrated at CERN in the early 1980s[77] and is now being revived. 'Surface beams' proves to be a slightly unfortunate and ambiguous misnomer now because more sophisticated, if inefficient, moderation techniques can produce 'ultra-slow' muons with implantation energies between 0.5 and 30 keV, tuning the implantation depth electrostatically between 5 and 200 nm. These are beginning to be used for proper surface, subsurface and depth-profiling studies, albeit at low data rates.[78]

As for data rate, it was about 1000 per hour for Garwin *et al.* in 1957. At CERN in 1980 it was 100,000–so that a radical spectrum could take 24 hours to accumulate–but it was already a million per hour or more at SIN and TRIUMF. Currently it is in excess of 100 million at ISIS, and all facilities are looking towards 1000 million! This is all very well, but it will be hard to use it efficiently. Good for automated scans of magnetic field or rf frequency, maybe, but somewhat wasted while waiting for

[75]T. Butz *et al.*, J. Magn. Magn. Mater. **15-18**, 671 (1980).

[76]See Chapter 21.

[77]This was done by Cesare Bucci and colleagues.

[78]See Chapter 18.

[79]See Chapter 19.

[80]Just the year before, the editor of Physical Review Letters had said at a conference that 'μSR publications were still below the critical mass, and that μSR was not yet accepted as a generally useful tool, since so few facilities existed'!

[81]This thanks to a formula due to Delrieu of Saclay, derived in the early 1970s already for hypothetical NMR spectra of Type-II superconductors, but presumably not much used. Diffraction methods, in contrast, are sensitive to the square of the field contrast.

[82]See, for example, S. L. Lee *et al.*, Phys. Rev. Lett. **71**, 3862 (1993).

[83]See Chapter 9.

[84]J. I. Budnick *et al.*, Phys. Rev. Lett. **124** 106 (1987).

[85]Although I'm told that battery materials are fast catching up, at least at ISIS!

temperature to change or equilibrate, or thinking what to do next. But these high rates will certainly lead to more thorough and systematic, if not shorter, studies and surveys. At continuous sources especially, such high rates are really only suitable for integral, rather than time-resolved, counting. At pulsed sources, they will benefit parametric studies, e.g. with pulsed illumination, currents, magnetic, electric or rf fields, etc.[79] De Renzi points out that, for most μSR groups, what counts now is muons-per-year, rather than muons-per-hour, i.e. their overall beamtime allocation.

The arrival of high-T_c cuprate superconductors in 1986, perhaps more than anything else, established the reputation of μSR spectroscopy, and brought it to the attention of a wider audience.[80] In the new materials, the magnetic penetration depth could be obtained directly from the μSR linewidth, i.e. its second moment, even for the rather dirty early samples.[81] Coming on line in 1987, this was opportune timing for the new ISIS muon facility. We got off to an uncomfortable start, when our neutron colleagues told us we had got the penetration depth wrong by a factor ten! But the neutron-reflection result was for some reason in error. Our muons had to be right, of course: the μSR frequency spectrum is a simple one-to-one map of internal field distribution in the vortex state. As sample quality improved, the distinctive shape of the distribution was revealed, allowing coherence length–roughly the width of the vortex cores–and the spacing of the flux lines to be determined as well. Studies extended to pinning, wobbling and melting of the flux-line lattices.[82]

μSR undoubtedly provides the most reliable measurement of these length scales, and is well suited to examining small samples of each new candidiate material as it comes along. London penetration depth is intimately related to the pairing mechanism of the superconducting carriers, via their effective mass and their concentration, for which μSR provides a virtually unique calibration.[83] The survey of such a wide variety of materials, from the original cuprates to heavy fermion systems, chevrel phases, and more exotic newcomers, even including some organic superconductors, as well as the discovery of a systematic variation with transition temperature and nature of the condensation which embraces all of these (Uemura plots) would be unthinkable without μSR.

Antiferromagnetic order in La_2CuO_4, the undoped parent compound of the first high-T_c superconductor, was also quickly confirmed by μSR.[84] And likewise in most other families of these materials discovered since then. It's probably true to say that the interplay of magnetism and superconductivity, and the elucidation of their phase diagrams as a function of composition or doping, accounts for the major allocation of beamtime now at all μSR facilities.[85]

I'm not in favour of jumping on bandwagons or following fashions but occasionally it's necessary to respond quickly to new ideas or discoveries, so that schemes for fast-track beamtime allocations would be valuable. With all the present facilities over-subscribed and fully scheduled for months in advance, that's a tricky matter. Another difficult situation arose at TRIUMF, when some applicants for high-T_c beamtime wanted to keep their findings secret. Jess Brewer was adamant that there should be no proprietary data, and that all results should be published in the open literature.[86]

One last example of μSR making unique contributions to the study of a new material: C_{60} 'Buckminster-Fullerene' or 'buckyballs' (named for the quirky architect of geodesic structures!) found its way into muon beams as soon as it became available in sufficient quantities in 1990. Ansaldo and colleagues of Saskatchewan, working at TRIUMF, surprised hot-atom chemists by announcing coexistent atomic and radical states of muonium. These are the organic counterparts of Mu and Mu* in diamond, i.e. also in elemental carbon–just a rather different allotrope! The atomic state is trapped *inside* a C_{60} cage, and is chemically stable. It's dubbed *endohedral* muonium, or Mu@C_{60}. The molecular radical, on the other hand, must be formed by reaction from *outside* the sphere, i.e. with muonium that has thermalized *between* the buckyballs. Whereas most muoniated radicals have been identified by correspondence with species already known to ESR, C_{60}Mu was discovered ahead of its conventional counterpart, and knowledge of its hyperfine constant assisted ESR chemists in the later identification of the elusive monohydride, C_{60}H.

Details of the very different reorientational dynamics in the plastic-crystal phases of C_{60} and C_{70} were revealed by weird shapes of the 'pure muon' level-crossing resonances (measured at PSI), where the applied field tunes out the dominant hyperfine field.[87] Here the radical is used just as a spin label for the dynamical studies. As for its electronic structure, partial double-bond character in C_{60}Mu means that the singly occupied electronic orbital extends all around the sphere. A definitive mapping of the spin-density distribution on each carbon atom was achieved by Paul Percival and colleagues (also in 1995, working at TRIUMF)[88] by detecting the muon-nuclear level-crossing resonances in a liquid solution of almost isotropically pure $^{13}C_{60}$! These fullerene spectra are surely amongst the most remarkable in the μSR literature.

On that note, I'll refer you to the appropriate sections of this book for highlights and developments in these and other topics since the New Millenium! They show that μSR is not just there to check theories, but equally to set puzzles for theoreticians! μSR has always bridged the uncomfortable divides between condensed matter physics, materials science, and chemistry. It has evolved into an indispensable tool in studies embracing metallic, magnetic, and superconducting systems, through hydrogen storage media, electronic, and ionic conductors to novel battery electrolytes, and through dielectrics and semiconductors to solar cells. I'll finish by quoting the outgoing president of the Institute of Physics, Julia Higgins:

> As physicists, we know that our work has the potential
> to help resolve many of the world's most acute problems–
> whether they relate to the environment, climate change,
> healthcare, or supplies of food, water, or energy.

[86]At ISIS, all original data are placed in the public domain three years after the respective measurements.

[87]E. Roduner *et al.*, Chem. Phys. **192**, 231 (1995).

[88]P. W. Percival *et al.*, Chem. Phys. Lett. **245**, 90 (1995).

Participants at the International Advanced School in Muon Spectroscopy held in August 2019 at the Rutherford Appleton Laboratory, UK. This book grew out of the lectures at the School. This photograph is a reminder that, for all of the sophisticated instrumentation that is necessary to perform this technique, muon spectroscopy is ultimately due to people.

Figure credits

The editors are grateful to the following copyright-holders for kindly granting permission to reproduce various figures in this book: The American Chemical Society (Figs 9.31, 11.7, 12.23, 19.14); The American Institute of Physics (Figs 19.6, 19.7); The American Physical Society (Figs 4.23, 6.6, 6.8, 6.9, 6.16, 7.10, 7.11, 8.1, 8.3, 8.4, 8.5, 8.7, 8.11, 8.16, 8.18, 8.19, 9.5, 9.6 9.15, 9.16, 9.17, 9.23, 9.25, 9.26, 9.27, 9.28, 9.29, 9.30, 9.31, 9.32, 10.2, 10.3, 10.4, 10.5, 16.4, 16.5, 16.6, 16.7, 18.12, 18.13, 19.2, 19.3, 20.6, 20.7, 21.10, 21.11, 22.5, 22.6, 22.7, 23.2, 23.3, 23.4, 23.7, 23.8, 23.9); Elsevier (Figs 8.15, 12.16, 12.19, 17.1, 18.7, 19.4, 19.9, 19.15, 19.17, 19.19, 20.1, 20.2, 20.3, 20.4, 20.5, 23.1); Institute of Physics Publishing (Figs 11.4, 12.18, 16.3, 16.8); John-Wiley (Figs 8.6, 12.21, 22.15); Royal Society of Chemistry (Figs 11.5, 11.6, 12.4, 12.15, 12.24, 23.10); Springer-Nature (Figs 9.14, 9.20, 9.33, 9.34, 12.11, 12.14, 18.10, 19.18, 21.9, 22.14); Taylor and Francis (Figs 9.12, 9.13, 12.7, 12.8, 18.2, 18.3, 21.1, 21.2, 21.3, 21.4, 21.5).

Index